The Search for Charm, Beauty, and Truth at High Energies

ETTORE MAJORANA INTERNATIONAL SCIENCE SERIES
Series Editor:
Antonino Zichichi
European Physical Society
Geneva, Switzerland

(PHYSICAL SCIENCES)

Recent volumes in the series:

A Continuation Order Plan is available for this series. A continuation order will bring delivery of each new volume immediately upon publication. Volumes are billed only upon actual shipment. For further information please contact the publisher.

The Search for Charm, Beauty, and Truth at High Energies

Edited by

G. Bellini

Istituto di Fisica dell' Università and
Sezione INFN
Milan, Italy

and

S. C. C. Ting

Massachusetts Institute of Technology
Cambridge, Massachusetts

Plenum Press • New York and London

Library of Congress Cataloging in Publication Data

Europhysics Study Conference on High-Energy Physics (1981: Erice, Italy)
   The search for charm, beauty, and truth at high energies.

   (Ettore Majorana international science series. Physical sciences; v. 16)
   "Proceedings of a Europhysics Study Conference on High-Energy Physics, held
November 15–22, 1981, in Erice, Sicily, Italy"—T.p. verso.
   Includes bibliographical references and index.
   1. Particles (Nuclear physics)—Charm—Congresses. 2. Quantum flavor dynamics—
Congresses. I. Bellini, Gianpaolo. II. Ting, S. C. C. (Samuel C. C.), 1936–
III. Title. IV. Series.
   QC793.3.C53E88 1981   539.7   83-8088

   ISBN-13: 978-1-4612-9657-7   e-ISBN-13: 978-1-4613-2659-5
   DOI: 10.1007/978-1-4613-2659-5

Proceedings of a Europhysics Study Conference on High-Energy Physics held
November 15–22, 1981, in Erice, Sicily, Italy

© 1984 Plenum Press, New York

Softcover reprint of the hardcover 1st edition 1984

A Division of Plenum Publishing Corporation
233 Spring Street, New York, N.Y. 10013

PREFACE

The search for flavored particles is one of the most interesting topics in high energy physics. Many experimental groups are working on this subject, but the solution to many of the problems are still open.

Therefore it seemed very useful that people interested in these problems can probe then in a discussion. This is the aim of this Europhysics Study Conference, which has been organized both as a conference and a workshop.

The present experimental knowledge on branching ratios, lifetimes, cross sections and production mechanisms of flavored particles has been presented in general talks and discussed in the morning sessions, as well as the bases of the theoretical ideas and predictions. The experimental methods: visual detectors, live targets, high resolution vertex detectors, special triggers of search on flavored particles, have been treated in the afternoon panels.

These proceedings contain the talks and panel discussions with the exception of a few small contributions to the panels and talks by C. Baltay ("Search for charm and new flavors with bubble chambers"), G. Alteralle ("Lifetime of charm and new flavors"), P. Monacelli ("Results on charm production from a CERN Beam Dump experiment"), A. Capone ("Experimental study of same-sign dimunon events produced in neutrino and anti-neutrino beams").

We are very grateful to all participants, session chairmen and speakers who contributed to the success of the meeting as well as to A. Zichichi, director of the Center "Ettore Majorana" for their warm hospitality. Special thanks is due to Mrs. Luciana Brogiato, secretary to the Conference, for her compentent and enthusiastic cooperation, also to the staff of "Ettore Majorana" Center for their assistance.

G. Bellini

S.C.C. Ting

CONTENTS

SESSION III

SEARCH FOR CHARM WITH BUBBLE CHAMBERS

SESSION IV

PANEL ON USE OF VISUAL DETECTORS IN THE SEARCH FOR CHARM AND OTHER
FLAVOURS

Chairman: A. Bettini

SESSION V

LIFETIME MEASUREMENTS

SESSION VIII

PANEL ON HIGH RESOLUTION VERTEX DETECTORS TO SEARCH FOR CHARM AND
OTHER FLAVOURS

Chairman: L. Dick

SESSION IX

SEARCH FOR HEAVY FLAVOURS

SESSION X

PANEL ON SPECIAL TRIGGERS TO SEARCH FOR CHARM AND HEAVY FLAVOURS

Chairman: L. Foà

SESSION XI

HEAVY FLAVOR PRODUCTION IN e^+e^- ANNIHILATION

Hinrich Meyer

University of Wuppertal
Gaussstr. 20
5600 Wuppertal 1

ABSTRACT

A brief review is given of the production characteristics of the
particles containing heavy quarks in e^+e^- annihilation.

Experiments at e^+e^- storage rings have been a very rich part
of the field, an exclusive source of information on new quark
flavors. Two systems of heavy quark antiquark bound states are
observed, the ψ family with two resonances $\psi(3095)$, $\psi'(3685)$ containing the charmed quark c, and the Y family with three resonances
Y (9.46), Y' (10.02), Y" (10.36) formed by the b quark. The third
heavy quark (t) system expected to show up at high energies remains to be discovered above 38 GeV.

In this talk the resonances will not be discussed only the two enhancements ψ (3.772) and Y (10.54) just above the threshold for
open charm and beauty production.

A look at table 1 shows that there are five e^+e^- storage rings in
operation that continue to provide collisions in the charm and
beauty energy range and fifteen mostly general purpose detectors
that can be used for ongoing detailed studies of particles with
heavy flavors.

The cross-section for production of heavy new quarks is known once
the threshold region has been passed, it is given by

$$\Delta R = 3 \cdot e_q^2$$

Table 1

Storage Ring	E_{max} (GeV)	Particles resonances	open	Detectors old	future
SPEAR	7.4	ψ	charm	MARK I, MARK II, Chrystal Ball, DELCO	MARK III
DORIS I	10.02	ψ, Y	charm	PLUTO, DASP, LENA	
DORIS II	11.2	Y	charm, B		Chrystal Ball, ARGUS
CESR	16	Y	charm, B		CUSB, CLEO
PEP	30		charm, B	MARK II, MAC, HRS,TPS,S DELCO	
PETRA	38, 42 (1983)	(Top)?	charm, B, (New parti-cles)?	PLUTO,JADE, MARK J,TASSO, CELLO	

where e_q is the charge of the new quark (q), the factor 3 counts the color degrees of freedom and ΔR is related to the cross-section for quark pair production $\sigma_{q\bar{q}}$ through

$$\sigma_{q\bar{q}} = \Delta R \cdot \sigma_{\mu\mu}$$

with

$$\sigma_{\mu\mu} = \frac{4\pi}{3} \cdot \frac{\alpha^2}{S} = 86.7/E^2 \text{ (nb)}$$

and E in GeV is the total center of mass energy. This gives for the charm quark (c) $\sigma_{c\bar{c}} = 4/3 \cdot \sigma_{\mu\mu}$ and for the beauty quark (b) $\sigma_{b\bar{b}} = 1/3 \cdot \sigma_{\mu\mu}$. The experiments on the total cross-section ratio (R) although still rather imprecise are in good agreement with this expectation on cross-sections (see Fig.1).

The cross-section in the energy region above charm threshold is very complicated up to about 4.5 GeV from whereon the cross-section is rather flat (see Fig. 2). The threshold behaviour for pointlike spin 1/2 particles is given by the factor $f \equiv \beta \cdot (3-\beta^2) / 2$ where β is the velocity of the quark in question. At 4.5 GeV for a charm quark of mass $m_c = 1.5$ GeV we have $\beta = 0.745$. for the b-quark at the same velocity we have E= 13.5 GeV. Structures similar to the energy region from 3.9 - 4.5 GeV, reduced however by the ratio of quark charges squared (1/4) are then to be expected between 10.6 and 13.5 GeV. From Fig.1 we see that cross-section measurements of the necessary density in energy points and statistical precision are not yet available apart from a small energy range around the Y-resonances.

The structures in R are certainly related to quasielastic production of specific charm states, like $e^+e^- \to D^*\bar{D}^*$ or $D \, D^*$. Theoretical estimates along this line that seem to be in qualitative accord with the data is shown in Fig. 3.

The threshold region for charm particles is therefore dominated by $\bar{c}u + c\bar{u}$ and $c\bar{d} + d\bar{c}$ channels, while the pickup of the next heaviour quark s that leads to F production is most likely suppressed. In fact copious production of F particles, especially at a particular energy point has not been observed. This renders the possibility of observing particles containing both the b and c quark even more unlikely. Nevertheless special searches near the threshold expected around 12 GeV should be performed.

Searches for structures in R have continued after the observation of the three Y-states both above the Y-resonances and below. No new features have been seen the corresponding limits on Γ_{ee} are given in Fig. 4a and Fig. 4b.

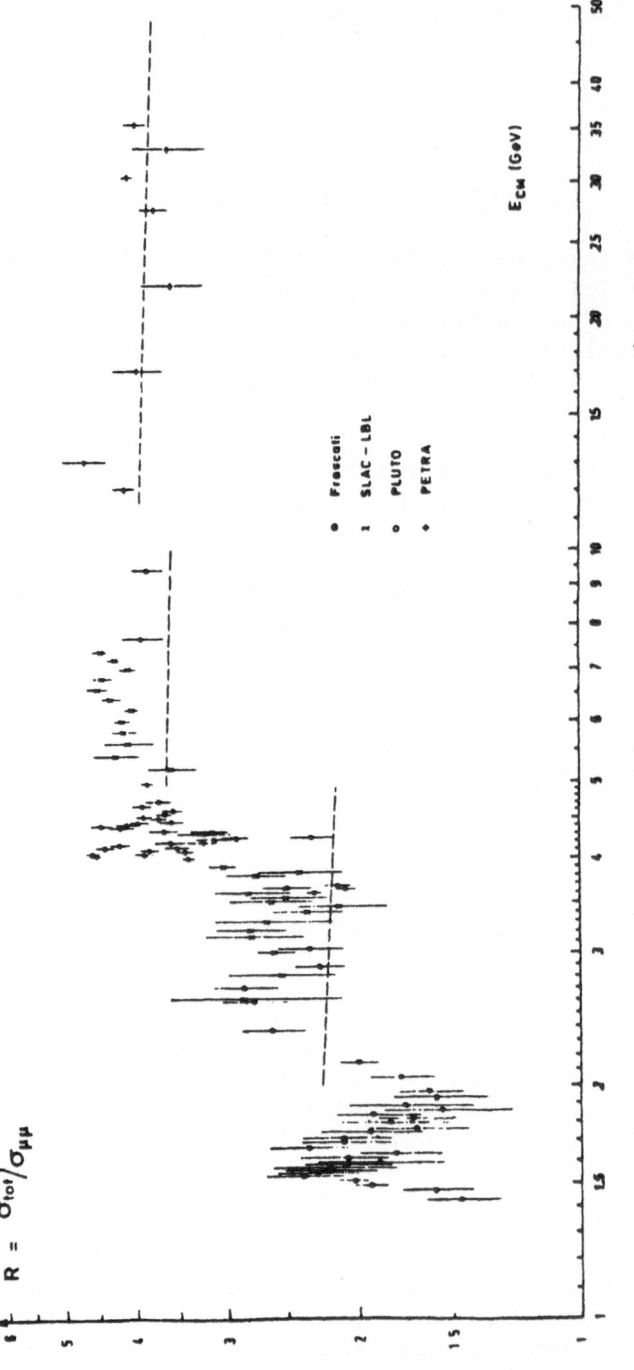

Fig. 1: Total cross-section measurements for hadron production in $e^+ e^-$ annihilation (Ref. 1)

a)

(Ref. 2)

b)

(Ref.3)

c)

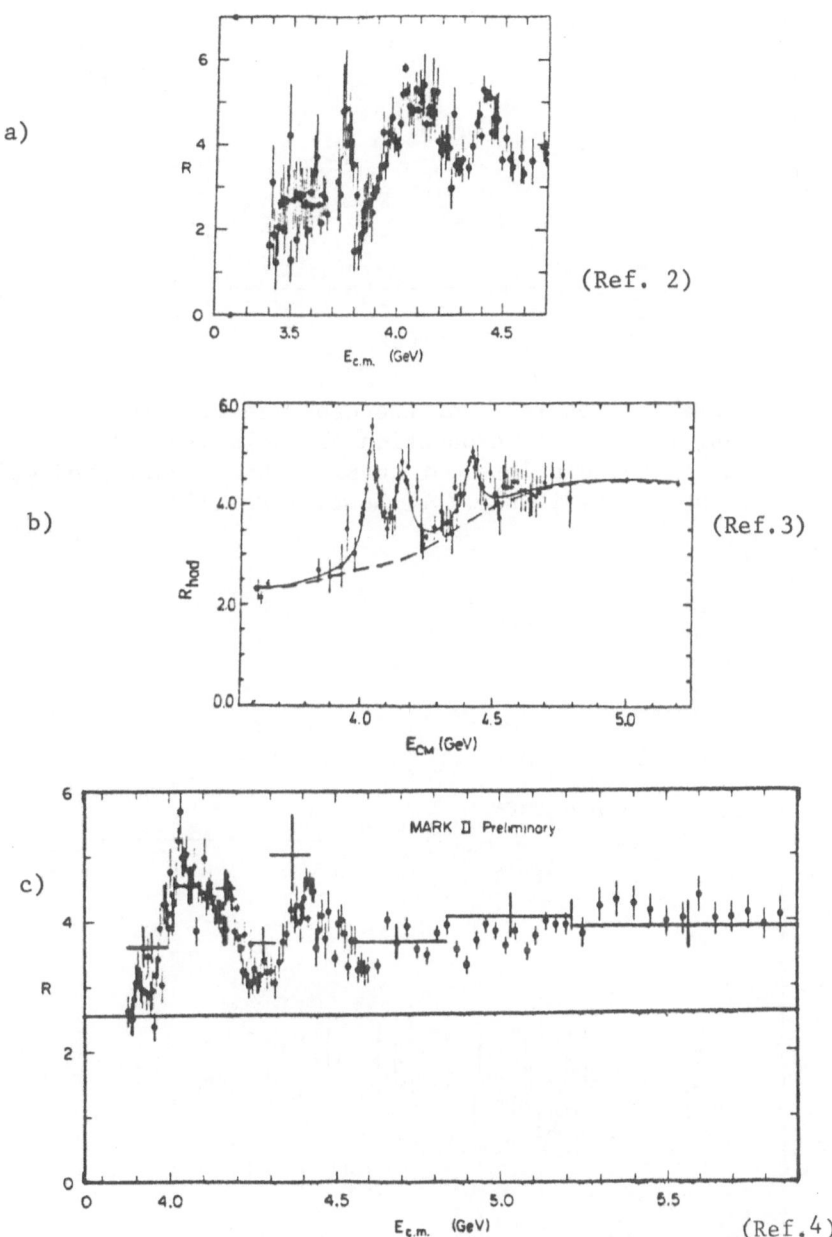

(Ref.4)

Fig. 2: The cross-section in the charm threshold region, data
are from the MARK I Detector (a) the DASP Detector(b)
and the MARK II Detector (c).

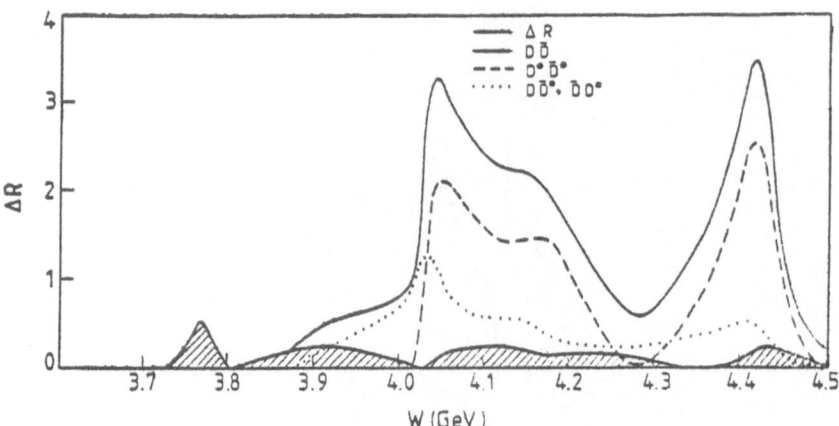

Fig. 3: Theoretical estimate of the cross-section for
        quasielastic charmproduction channels.(Ref.5)
        Similar calculations for quasielastic beauty production
        is not yet available.(see however Ref.11)

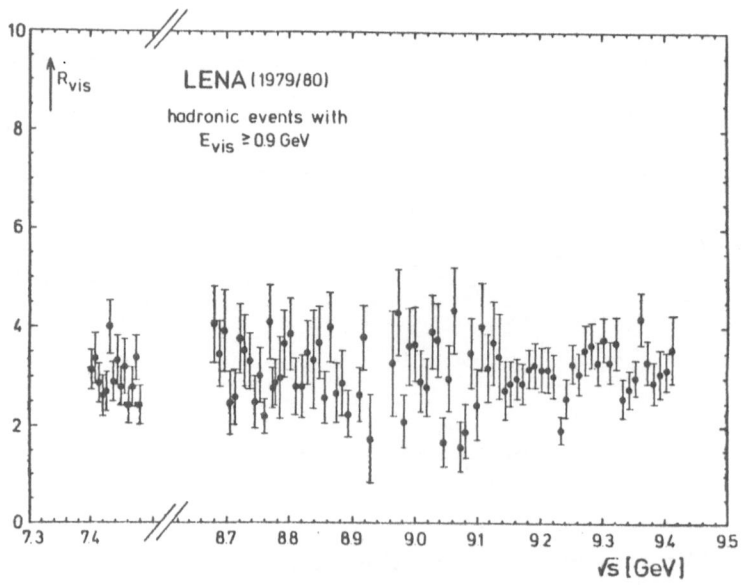

Fig. 4a from Ref. 6

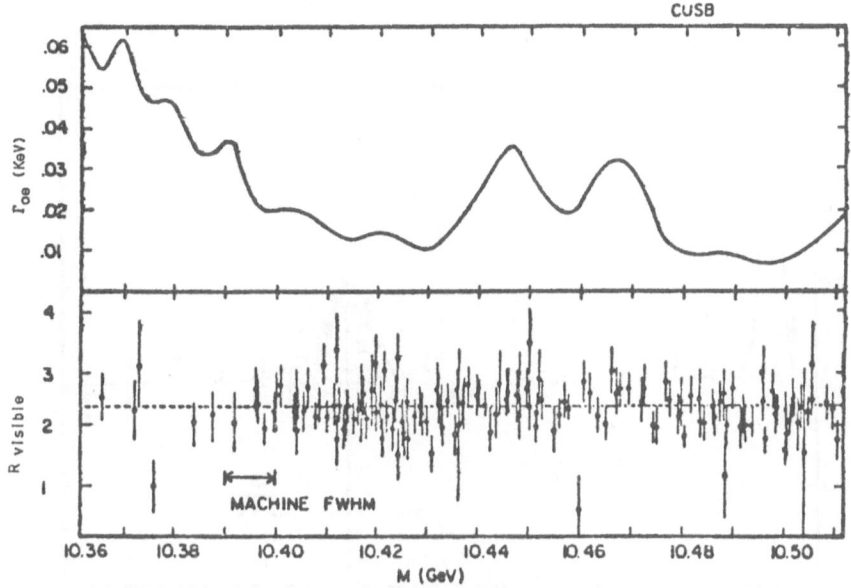

Fig. 4b from Ref. 7

Two very interesting structures one in the charm region and a further state in the beauty region have been observed at 3.772 GeV and 10.54 GeV as seen in the R plots of Fig. 5a and 5b. The enhancements are about $\Delta R = 1.4 - 1.7$ with similar width of ~ 50 MeV. For the ψ (3.772) it is known to be entirely due to $D^0 D^{\bar{0}} + D^+ D^-$ production.

Since open B particles so far have not been identified the corresponding conclusion for the Y (10.54) cannot yet be made. It is for example still possible that also $B^* B$ production contributes. Tuming the e⁺e⁻ storage rings to the ψ (3.772) or Y (10.54) provides $D\bar{D}$ and $B\bar{B}$ factories with very similar signal to background ratio of $\simeq 0.28$. A significant part of the background at the ψ (3.772) comes from the radiative tail of the nearby ψ' (3685). Since the integrated luminosities at SPEAR, DORIS II and CESR are known to be about $40 nb^{-1}$/day at ψ (3.772) and 250 nb^{-1}/ day at Y (10.54) the yields of new flavors are of the order

$$N = \Delta R \int L \, dt \cdot \frac{4\pi}{3} \frac{\alpha^2}{S} = 350 \text{ pairs/day}$$

Certainly an increase in luminosity would be very wellcome to make truely detailed studies of the lowest D, B states possible.

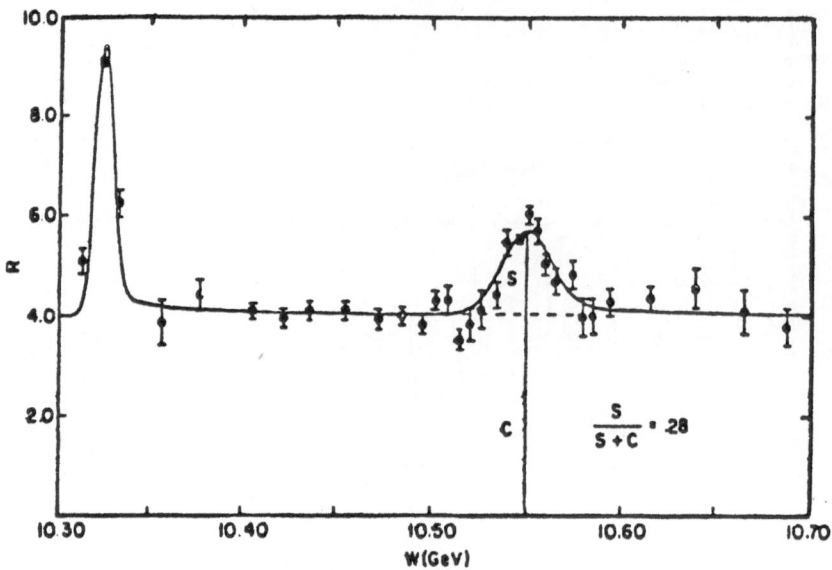

Fig. 5a from Ref. 8

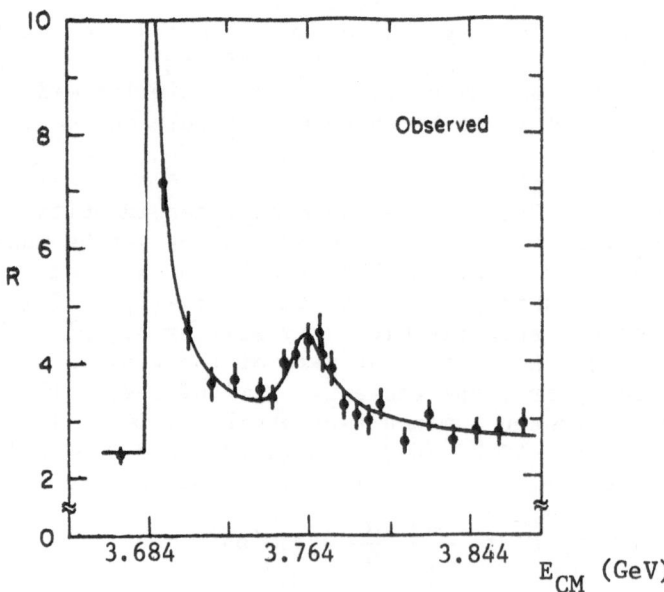

Fig. 5b from Ref. 9

Well above the threshold region the production cross-sections are given as $\sigma_{c\bar{c}} = 4/3 \cdot \sigma_{\mu\mu}$ and $\sigma_{b\bar{b}} = 1/3 \cdot \sigma_{\mu\mu}$. For PETRA with $\int Ldt \simeq 500$ nb^{-1}/day this yields

$$c\bar{c} + \text{hadrons} \quad \sim \quad 40/\text{day}$$
$$b\bar{b} + \text{hadrons} \quad \sim \quad 10/\text{day}$$

while at DORIS and CESR with $\int Ldt \sim 250$ nb^{-1}/day and at 11 GeV the corresponding rates are larger by a factor of 10. Since further the final states containing the heavy quarks particles are probably less complicated at 11 GeV than at 35 GeV (PETRA) it seems to be of great advantage to perform heavy flavor studies at DORIS II and CESR. Only two general purpose detectors, CLEO and ARGUS (not yet ready, start will be about end of 1982) can then be used to explore this interesting field.

The top quark t has been extensively searched for at PETRA. The narrow $(t\bar{t})$ resonances can be found by raising the energy of the storage ring in small steps comparable to the expected width of the resonances, it is of order 15 MeV at the higher PETRA energies. Secondly the threshold for open top production should show up as a step in R of the same size as for the charm threshold. The data from four experiments at PETRA have been averaged for this purpose to increase statistical precision and the result is shown in Fig.6.

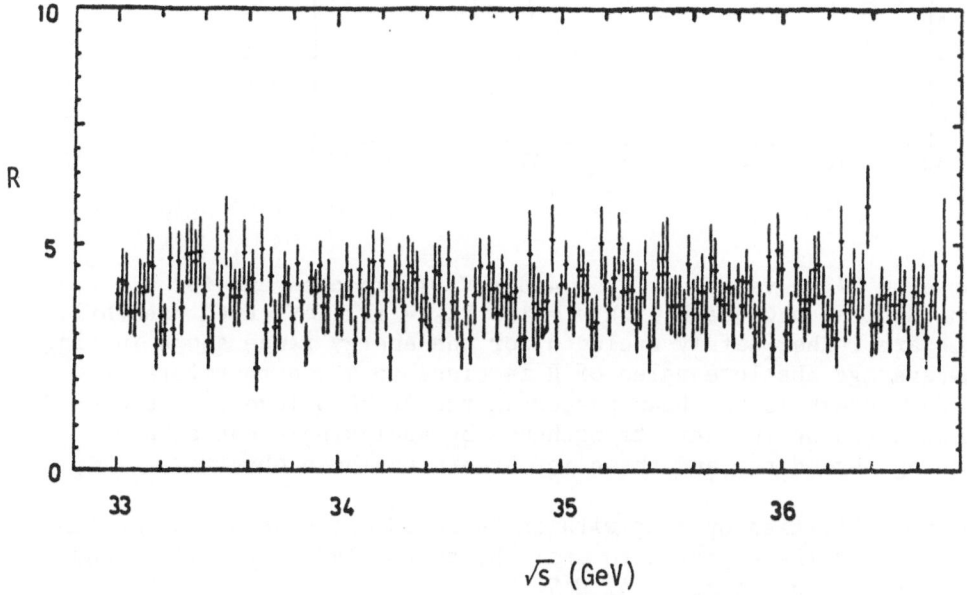

Fig. 6 from Ref. 10

Using the relation

$$\int \sigma dE = 6 \ \pi^2 \ / \ M^2 \cdot \Gamma_{ee} \cdot B_{had}$$

and assuming hadronic decays to dominate (B(had) near 100%) limits on Γ_{ee} the partiell width of a resonance for lepton pair decay can be deduced, the result is $\Gamma_{ee} <$ 0.61 KeV. The standard expectation for Γ_{ee} can be taken from a simple extrapolation of the data for the known resonances as shown in Fig. 7.

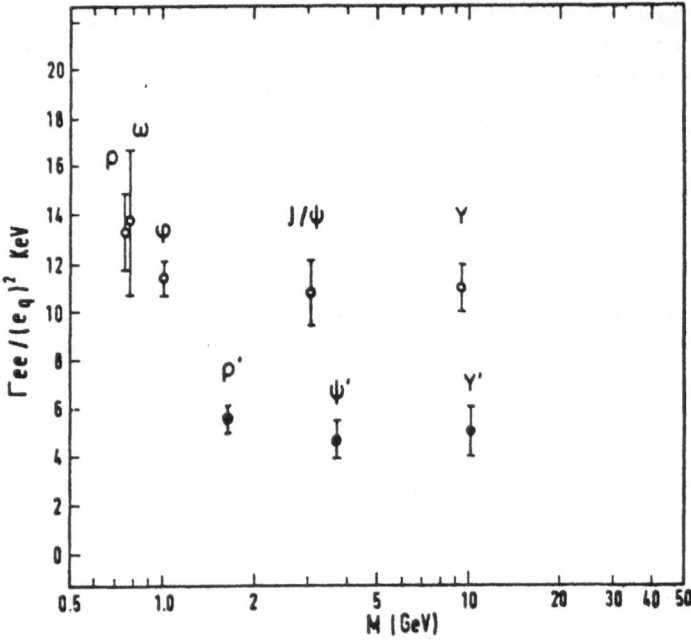

Fig. 7: The par-
tiell width of
($q\bar{q}$) resonances
(from Ref. 1)

For e_q = 2/3 a value of Γ_{ee} = 4.8 KeV is expected and for a hypothetical new quark with e_q= 1/3 Γ_{ee} = 1.2 KeV. Both possibili- ties are rather safely excluded for the energy range shown in Fig.6. The average absolute value of R is close to 4 and therefore also no new threshold has been passed at the $\Delta R \geqslant 0.5$ level. This con- clusion can be further strengthened by analysing event shapes; nothing beyond two and three jet events has been observed.

New possibilities open up with the planned increase of the maximum energy of PETRA expected to start by autumn 1982. It will extend the searches up to about 42 GeV.

RERERENCES

1. H. Meyer in Proceedings of the 1980 CERN School of Physics,
    Malente, CERN 81-04
2. J. Siegrist SLAC-Report No. 225, October 1979
3. DASP-Collaboration Nuclear Physics B 148, 189 (1979)
4. M.W. Coles LBL-Report 11513, 1980
5. E. Eichten et. al. Physical Review D 21, 203 (1980)
6. D. Wegener in International Conference on High Energy Physics,
    Lisbon, July 1981
7. Sh. Stone Cornell preprint, CLNS 81-514
8. D. Andrews et. al. Physical Review Letters 45, 219 (1980)
    G. Finocchiaro et.al.   "   "   "   45, 222 (1980)
    and also Ref. 7
9. R.H. Schindler et.al. Physical Review D 21, 2716 (1980)
10. J. Bürger in Proceedings of the 1981 International Symposium
    on Lepton and Photon Interactions at High Energies,
    Bonn 24.-29.August 1981
11. D. Schamberger in Ref. 10

MARK-J RESULTS ON HADRON AND INCLUSIVE MUON PRODUCTION

AND THE ELECTROWEAK INTERACTIONS OF QUARKS

Harvey B. Newman

Department of Physics
California Institute of Technology
Pasadena, California, USA

and

Deutsches Elektronen-Synchrotron DESY
Hamburg, Federal Republic of Germany

ABSTRACT

MARK-J results on $e^+e^- \rightarrow$ hadrons and inclusive muon production
are presented. The measured values of the total hadronic cross
section, the thrust distributions and the production rate and shape
of inclusive muon events are entirely consistent with the predictions of QCD with five quark flavors. The data show no evidence for
the continuum production of a new charge 2/3e or 1/3e quark.
Energy scans in the regions $29.9 \leq \sqrt{s} \leq 36.7$ GeV show no indication
of a hadron resonance signaling the existence of a new massive
quark-antiquark bound state. The energy dependence of the hadron
production cross section has also been used, in the context of the
standard electroweak theory, to determine $\sin^2\theta_w$ at high q^2 and to
show that quarks are pointlike up to an energy scale $\gtrsim 200$ GeV. The
inclusive muon data has been used to measure the semileptonic
branching ratio of b decays.

13

INTRODUCTION

The MARK-J[1,2,3] results on hadron production via one-photon annihilation

$$e^+e^- \rightarrow \text{hadrons} \tag{1}$$

and on inclusive muon production

$$e^+e^- \rightarrow \mu^\pm + \text{hadrons} \tag{2}$$

have been analyzed to search for new quark flavors, to study the electroweak parameters of quarks including c and b quarks, and to test the pointlike nature of quarks in the context of QCD and the standard electroweak theory. The data cover a wide range of center of mass energies 12 GeV $\leq \sqrt{s} \leq$ 36.7 GeV, including results obtained by extended periods of running at a fixed beam energy and by fine energy scans.

Details of the apparatus, event selection procedures, acceptance and radiative correction calculations are contained in Refs. 1-3.

HADRON PRODUCTION CROSS SECTION AND THE SEARCH FOR NEW FLAVORS

The results on the hadronic production cross section[F1] are expressed in terms of R:

$$R = \sigma(e^+e^- \rightarrow \text{hadrons}) \;/\; \sigma_{pt}(e^+e^- \rightarrow \mu^+\mu^-)$$

where σ_{pt} refers to the pointlike muon pair production cross section calculated according to QED. In the naive quark-parton model, the cross section for the hadron production process is simply given by the sum over flavors of the pointlike $q\bar{q}$ pair cross sections. Using this picture with spin 1/2 massless quarks and with three colors gives

$$R_o = 3\Sigma\, e_q^2$$

where e_q is the charge of the quark with flavor q. Considering the five known quarks (u,d,s,c and b), and correcting the naive model for gluon emission as predicted by QCD, one expects $R \sim 4$ over the entire PETRA energy range, with only a slight decrease in R with increasing beam energy.

Events from reaction (1) are selected by requiring that

(a) the total visible energy E_{vis} is more than 50% of the available c.m.-system energy $\sqrt{s}$;

(b) the energy is balanced within 50% of E_{vis} in both the longitudinal and the transverse directions with respect to the beam line; and

(c) the shape of the shower development in the layer structure of the detector is incompatible with a purely electromagnetic nature of the final state.

The acceptance for reaction (1) has been calculated with the the use of a Monte Carlo simulation[F2] to provide a phenomenological model of the hadron production process based on perturbative quantum chromodynamics (QCD) . The model, which incorporates q^2 evolution and the weak decays of heavy quarks[4] in the fragmentation process, also includes the effects of initial-state photon bremsstrahlung[5] correct in quantum electrodynamics (QED) to order α^3. After cuts the acceptance is 87% with little variation as a function of energy. Background contributions to the hadron sample from the two-photon process

$$e^+e^- \rightarrow e^+e^- + \text{hadrons} \qquad (3)$$

and

$$e^+e^- \rightarrow \tau^+\tau^- \qquad (4)$$

have also been calculated and subtracted by Monte Carlo techniques. The background subtraction for reaction (3) is $\Delta R_{\gamma\gamma} \lesssim 0.1$ and for reaction (4) is $\Delta R_{\tau\tau} = -0.18$ independent of beam energy.

Fig. 1 and Table I summarize the MARK-J measurements[2] of R. In addition to the statistical errors shown, we conservatively

TABLE   I

Results on   R

ALL DATA FROM 1978 → 1980

$\sqrt{s}$ (GeV)	R ± ΔR	N_{HAD}	$\int Ldt (nb^{-1})$
12.0	3.45 ± 0.24	233	98
13.0	3.40 ± 0.37	98	53
17.0	3.50 ± 0.45	68	60
22.0	4.31 ± 0.70	42	50
27.57	3.34 ± 0.24	216	508
30.0	3.90 ± 0.26	254	604
29.9 - 31.46	3.86 ± 0.14	919	2110
31.6	3.70 ± 0.42	88	243
33.0 - 34.0	3.63 ± 0.15	625	1793
34.0 - 35.0	3.54 ± 0.14	683	2132
35.0 - 36.0	3.90 ± 0.13	1065	3191
36.0 - 36.72	3.68 ± 0.17	560	1866
	3.70 ± 0.06	4851	12708

1981 DATA (up to mid-1981)

$\sqrt{s}$ (GeV)	R ± ΔR	N_{HAD}	$\int Ldt (nb^{-1})$
14.0	3.70 ± 0.07	2868	1526
22.0	3.60 ± 0.08	2415	3153
2.50	4.02 ± 0.21	411	615
32.986	3.86 ± 0.18	524	1384
33.915	3.76 ± 0.07	3480	9954
	3.71 ± 0.04	9698	16632

$$N_e = 262249 \ (1978-1980) + 528922 \ (1981)$$
$$= 791171$$

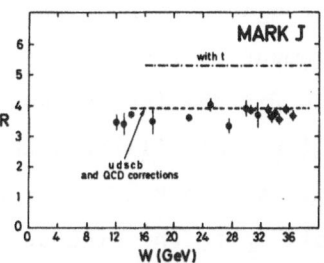

Fig. 1.   The total relative hadronic cross section R = σ(e$^+$e$^-$→
hadrons)/σ(e$^+$e$^-$ → μ$^+$μ$^-$)vs. center of mass energy.   The
dashed line is the expectation for the production of
u,d,s,c, and b quarks with QCD corrections.   The dot-
dashed line is the R value expected for production of
u,d,s,c,b quarks together with a charged 2/3 top quark.

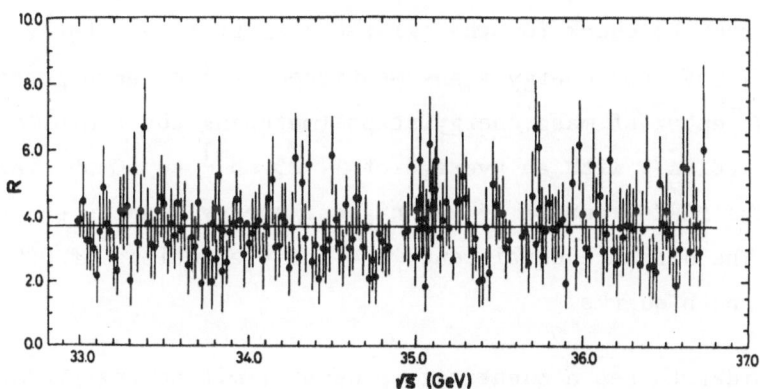

Fig. 2.   R values measured during the energy scan between 33.0 and
36.72 GeV.   The line is the mean R value of 3.75± 0.05.

estimate a systematic error of 10% due to model dependence of the
computed acceptance, event selection criteria, and the luminosity
measurement N_{HAD} in Table I refers to the accepted number of
hadronic events and N_e to the number of Bhabha events observed in the
central detector in the range $12^o \lesssim \theta \lesssim 168^o$ which are the basis
of the luminosity measurement[3]. The results in Fig. 1 are compared
to the predictions of QCD for the five known flavors u,d,s,c,b, as
well as for production of a sixth flavor with a quark charge of 2/3
in the continuum. No increase in R is seen corresponding to the
opening of a new threshold up to the highest energy of 36.72 GeV.
Other PETRA experiments have reported similar results[7].

In addition to the $t\bar{t}$ contribution to the hadronic continuum,
the toponium system should form one or more bound states. Interpretation of the vector mesons ρ, ω, ϕ, J, ψ', T, T', and T", as
non relativistic $q\bar{q}$ bround states, or "quarkonia" leads to the
prediction[6] that the gap between the lowest bound state and the
continuum is probably $\sim$ 1 GeV, and very likely $\leq$ 2 GeV.

In order to check for the existence of $t\bar{t}$ bound states lying
below 36.7 GeV, the energy scans mentioned earlier were performed
in 20 MeV center of mass energy steps (matching the r.m.s. energy
spread of PETRA), with an average of $\sim$ 25 nb^{-1} to 50 nb^{-1} per
point. The MARK-J result of the highest energy scan is shown in
Fig. 2. The data are consistent with the predictions of QCD for
u,d,s,c, and b quarks.

In order to set a quantitative upper limit on the production
of a narrow resonance, the data in Fig. 2 were fitted by a constant
plus a Gaussian distribution:

$$R = R_o + R_V \exp \left| -(w - M)^2 /2\Delta_w^2 \right| \qquad (w = \sqrt{s}), \qquad (5)$$

where R_o represents the nonresonant continuum, M is the mass of
the resonance, Δ_w is the r.m.s. machine energy width (Δ_w(GeV) =

2.2 x 10^{-5} s(GeV^2)), and R_v is the peak value of the resonant

constribution.   The largest value of R_v consistent with the data

was determined by trying fits with M, the center of the Gaussian

fixed at all center of mass energies at which data were taken.   The

largest value of R_v was obtained at 35.28 GeV, corresponding to an

upper limit on the resonance strength

$$\sigma_v \equiv \int (R - R_o) \sigma_{\mu\mu} dw \tag{6}$$

of 15.8 MeV nb at 90% confidence level.   Using the relation

between the resonance strength, the decay width into $e^+ e^-$ (Γ_{ee}), the

hadronic width (Γ_h), the total width (Γ), and the hadronic branching

ratio ($B_h \equiv \Gamma_h/\Gamma$),

$$\sigma_v \equiv \int \frac{3\pi}{M^2} \; \frac{\Gamma_{ee}\Gamma_h}{(w-M)^2 + \Gamma^2/4} \; dw = \frac{6\pi^2}{M^2} \, B_h \Gamma_{ee} \tag{7}$$

and taking radiative corrections into account we obtain

$$B_h \Gamma_{ee} < 0.85 \text{ keV (90\% confidence limit)}.$$

As shown in Fig. 3, this upper limit excludes the production

of a vector particle consisting of a $q\bar{q}$ bound state where the

quark has charge 2/3e.   On the basis of the experimental fact that

Γ_{ee}/e_q^2 is approximately constant for the vector-meson ground

states ρ, ω, ϕ, J, and Υ as predicted by duality arguments[6], one

expects $B_h \Gamma_{ee} \simeq 4$ keV for the lowest mass meson in the toponium

family.   Our results for the energy scan are in good agreement

with those obtained from other experiments[7].

JET ANALYSIS: NEW FLAVOR PRODUCTION AND EVENT SHAPES

A jet analysis of the hadronic events with $E_{vis} \geq 0.7 \sqrt{s}$ was

performed with use of the spatial distribution of the energy

deposited in the detector.   The jetlike appearance of the events

is parametrized in terms of the parameter thrust

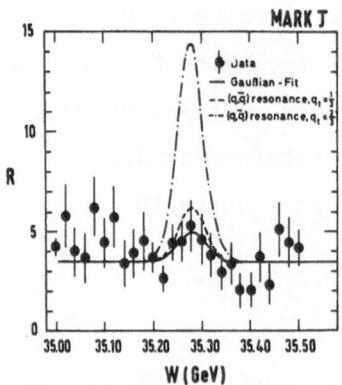

Fig. 3. Part of the energy scan data around 35.28 GeV. A fit using
         a Gaussian and a constant background is made to the R-
         values in order to search for the production of toponium.
         The solid line is the fit with position and width which
         minimizes the χ^2. The predictions for toponium production
         with charge 1/3 and 2/3 top quarks are also shown. In all
         cases the widths are those expected from the known machine
         energy resolution.

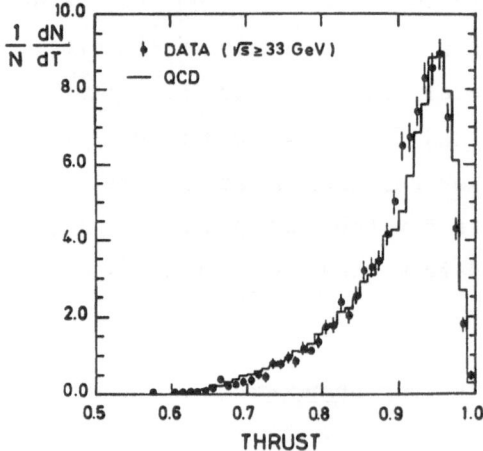

Fig. 4. The differential thrust distribution of hadronic events
         for c.m. energy $\geq$ 33 GeV compared with the QCD Monte Carlo
         (solid line).

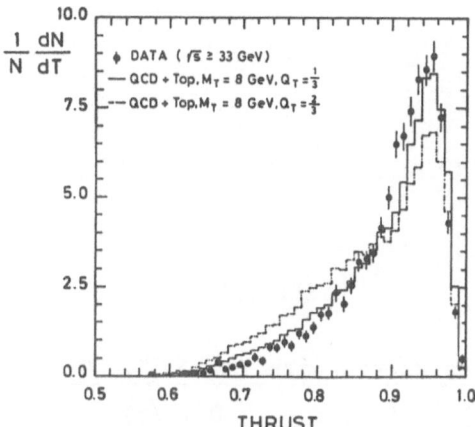

Fig. 5. The differential thrust distribution of hadronic events
for c.m. energy ≥ 33 GeV compared with the Monte Carlo
prediction which includes QCD and production of a hypothetical new quark with a mass of 15 GeV and a charge of
1/3e or 2/3e.

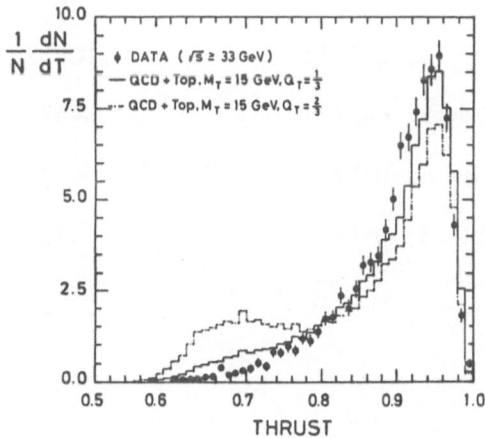

Fig. 6. A in Fig. 5 but with top quarks of mass 8 GeV.

TABLE II

Hadron Events with T < 0.75, T < 0.80

($\sqrt{s} \geq 33$ GeV, $E_{vis} \geq 0.7 \sqrt{s}$)

	Total Events	T < 0.75	T < 0.80
DATA	4762	216	479
QCD		268	513
QCD + TOP (M=15, Q=2/3)		1063	1416
QCD + TOP (M=15, Q=1/3)		467	774
QCD + TOP (M= 8, Q=2/3)		499	991
QCD + TOP (M= 8, Q=1/3)		326	667

$$T = \max \left| \Sigma_i |E^i_{//}| \ / \Sigma_i |E^i| \right| \qquad (8)$$

where E^i is an energy flow vector, whose direction is given by the position of a hit in a counter and magnitude by the corresponding deposited energy. $E_{//}^i$ is the parallel component of the energy flow vector along the axis which maximizes T and the sums are taken over all counter hits.

The normalized thrust distribution $N^{-1} \frac{dN}{dT}$ for the data taken with $\sqrt{s} \geq 33$ GeV is shown in Fig. 4. The data are compared to the predictions of the QCD Monte Carlo program of Ali et al.[4], which includes the effects of hard gluon bremsstrahlung as required by our data[1-3]. The data are consistent with the QCD predictions with the five known quark flavors, but rule out the continuum production of a sixth quark flavor with charge q = 2/3e or even q = 1/3e, if the mass of the new quark $m_q \gtrsim 8$ GeV. This is demonstrated in Figs. 5 and 6. Table II presents a quantitative comparison of the number of "spherical" events selected with two sets of cuts, T < 0.80 and T < 0.75, to the QCD predictions with and without new quark flavor production. The results in the table exclude q = 1/3 or q = 2/3e for all $m_q \gtrsim 8$ GeV.

Note that the results in Fig. 5, 6 and Table II are not sensitive to the uncertainties in the parameters used in the QCD model, which have been determined from the MARK-J data to be α_s = 0.17±0.02 at $\sqrt{s} \sim 35$ GeV and σ_q = 300±20 MeV from the data itself using jet analysis techniques described in Refs. 3 and 8.

Figs. 7 and 8 compare the thrust distributions dR/dT of the QCD model including new flavor production to the data. The areas under the curves in these figures are proportional to the total hadron production cross section. As the information on event rate and shape are included simultaneously

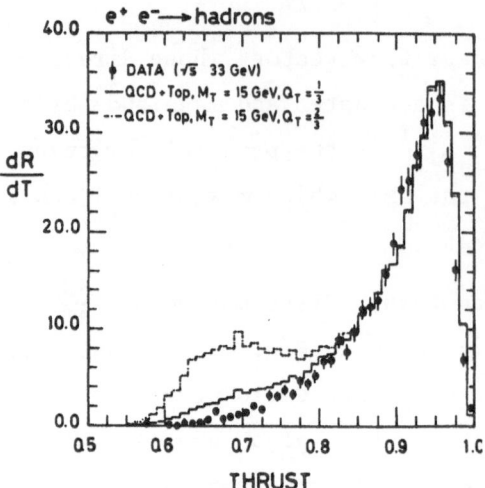

Fig. 7.   As in Fig. 5 but normalized according to production rate.

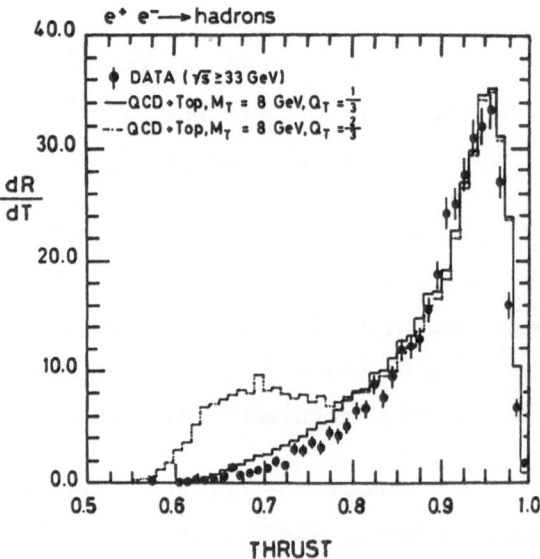

Fig. 8.   As in Fig. 6 but normalized according to production rate.

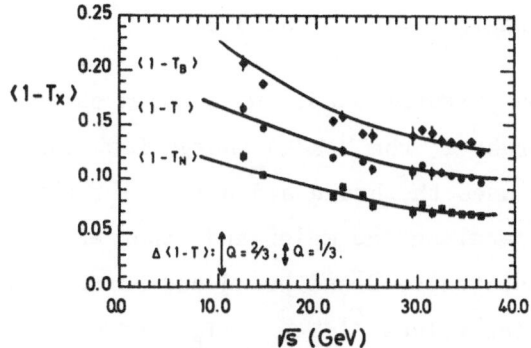

Fig. 9. The variation in the mean value $\langle 1-T_x \rangle$ as a function of the c.m., where T_x is respectively the broad jet thrust T_B, the total thrust, T, and the narrow jet thrust T_N. The solid lines are the predictions of the QCD Monte Carlo. There is no evidence for production of a new heavy quark, which could give a step in each distribution.

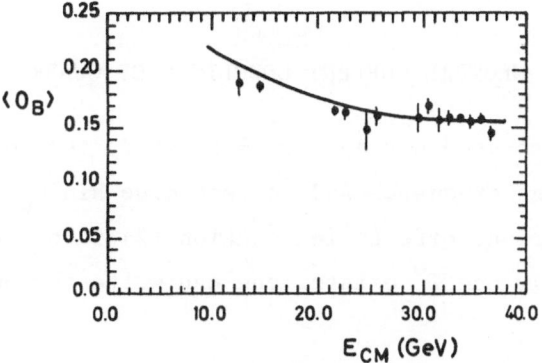

Fig. 10. The variation of the mean broad jet oblateness $\langle O_B \rangle$ as a function of the c.m. energy compared to the prediction of the QCD Monte Carlo.

in Figs. 7 and 8, one obtains even greater sensitivity to the
presence of a new quark flavor than one has in the comparisons
presented in the previous figures. Once again, even the difficult
case m_q = 8 GeV, charge 1/3e is ruled out.

The variation of event shapes with $\sqrt{s}$ is shown in Figs. 9-11.
In addition to thrust T, the MARK-J energy flow analysis described in Ref. 3 uses the broad and narrow jet thrusts T_B and
T_N, the energy flow along the major and minor axes F_{major} and
F_{minor}, and the oblateness 0 and broad and narrow jet oblateness
O_B and O_N. The mean values $<1-T>$, $<1-T_B>$ and $<1-T_N>$ are shown as
a function of $\sqrt{s}$ in Fig. 9. The data vary smoothly over the entire
range $12 \leq \sqrt{s} \leq 36.7$ GeV, and show no step indicative of a new
flavor threshold. The QCD predictions with u,d,s,c and b quarks
are in excellent agreement with the data. The step sizes expected
at $\sqrt{s}$ = 35 GeV from a quark threshold with charge 2/3e and 1/3e
are shown at the bottom of the figure. Figs. 10 and 11 show that
$<O_B>$ and $<F_{major}>$ also behave quite smoothly as a function of $\sqrt{s}$,
and the results are well described by the QCD predictions with
five quark flavors.

R AND THE WEAK NEUTRAL CURRENT COUPLINGS OF QUARKS

We have also used our data on R to study the weak neutral
current couplings of quarks and to determine $\sin^2\theta_w$. If one takes
weak neutral current effects in reaction (2) into account and
assumes that only one Z^o exists, one can write the quark cross
section as[9]

$$R_f = \frac{\sigma(e^+e^- \to \gamma, Z^o \to f\bar{f})}{\sigma_p} = Q_f^2 - 8s \, Q_f \cdot g_V \cdot g_{V_f} \cdot p(s)$$

$$+ 16s^2 g^2 (g_V^2 + g_A^2)(g_{V_f}^2 + g_{A_f}^2) \cdot p'(s),$$

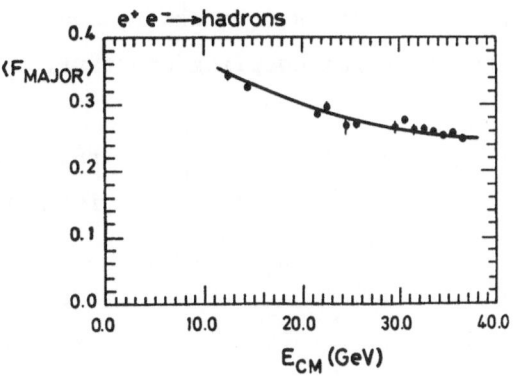

Fig. 11. As in Fig. 10 but for F_{major}, the fraction of energy flow
perpendicular to the thrust axis in the event plane.

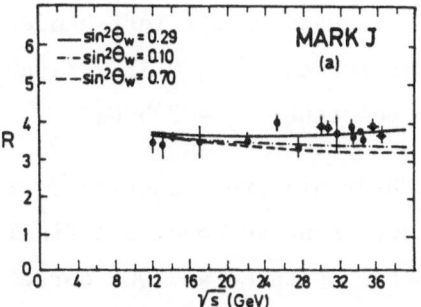

Fig. 12. Experimental results on R with statistical errors as a
function of the c.m. energy $\sqrt{s}$. The solid line shows the
theoretical prediction R_T, for the best fit to our data
($\sin^2\theta_w$ = 0.29 using α_s = 0.18 at $\sqrt{s}$ = 30 GeV). The
dependence of R_T on $\sin^2\theta_w$ is also shown for two extreme
values $\sin^2\theta_w$ = 0.70 and 0.10. The best fitted results
require that the normalization of the curves be scaled
down by 7.0% as shown (the normalization uncertainty is
10%).

where $\sigma_p = 4\pi\alpha^2/3s$ is the pointlike QED cross section, Q_f = charge
of the final state quark f, g_{V_f} and g_{A_f} are the weak vector and
axial vector coupling constants of quark f, and $g = 4.47 \times 10^{-5}$
GeV^{-2} is related to the Fermi coupling constant.

$$p(s) = \left[\left(\frac{s}{m_Z^2} - 1 \right) + \frac{\Gamma_Z^2}{s-m_Z^2} \right]^{-1} \quad \text{for } \gamma \text{ and } Z^0 \text{ interference,}$$

and

$$p'(s) = \left[\left(\frac{s}{m_Z^2} - 1 \right)^2 + \frac{\Gamma_Z^2}{m_Z^2} \right]^{-1} \quad \text{for pure } Z^0 \text{ exchange.}$$

In the framework of the Glashow-Weinberg-Salam (GWS) model[10],
assuming e-μ universality, the weak coupling constants are given
by

$$g_{V_f} = T_{3_f}^L - 2 Q_f \sin^2\theta_w$$

$$g_{A_f} = T_{3_f}^L \tag{10}$$

where $T_{3_f}^L$ is the weak isospin of the left handed quark. The mass
of the Z^0 can be expressed as $m_Z = 37.28/\sin\theta_w\cos\theta_w$ GeV and the
width is taken to be constant: $\Gamma_Z = 2.3$ GeV.

The electroweak hadronic cross section R_T is given in the
quark model by the incoherent sum over all final state quarks
including a color factor of three and QCD corrections[11]

$$R_T = 3 \cdot \Sigma_f R_f \left[1 + \frac{\alpha_s(s)}{\pi} + (1.98-0.115 \cdot N_f) \frac{\alpha_s^2(s)}{\pi^2} \right] \tag{11}$$

where α_s is the coupling constant of the strong interaction, which
has been measured at PETRA energies to be[12] $\alpha_s = 0.18$ for $\sqrt{s}$ in
the 30 to 36 GeV range, and which changes slowly with energy.
$N_f = 5$ is the number of different quark flavors.

The theoretical cross section R_T is compared with our

experimental result R for different values of $\sin^2\theta_w$ in Fig. 12.
A quantitative comparison is made by a χ^2-fit method where χ^2 is
defined as

$$\chi^2 = \frac{(F-1)^2}{\sigma^2_{syst}} + \sum_i \frac{(F\ R^i - R_T^i)^2}{\sigma_i^2} \tag{12}$$

where the sum runs over all measurements at different c.m. energies,
and where σ_i and σ_{syst} are the statistical and systematic errors
respectively. The systematic error σ_{syst} is estimated to be 10%.
The scaling factor F is used in order to take into account the
overall normalization uncertainties.

Taking all the MARK-J data in Table I into account, we obtain
the 95% and 68% confidence level contours in the $F-\sin^2\theta_w$ plane
displayed in Fig. 13. The results are:

$$0.20 \leq \sin^2\theta_w \leq 0.59 \qquad \text{(95\% confidence level limits)},$$

and

$$\sin^2\theta_w = 0.29 \begin{array}{c} + 0.26 \\ - 0.05 \end{array}$$

with a minimum χ^2 of 17.5 for 16 degrees of freedom.

The limits on $\sin^2\theta_w$ change by $\lesssim 0.02$ for relatively large
changes in α_s (0 to 0.24) and σ_{syst} (8% to 12%). The values for
the coupling constant α_s and the scaling factor F are correlated;
i.e. changes in α_s are compensated by an appropriate change in F.
This demonstrates that the method applied is sensitive to the
behaviour of R as a function of energy and not to the absolute R
values.

A combined analysis of MARK-J data for hadronic (reaction 1)
and leptonic final states ($e^+e^- \rightarrow e^+e^-$, $\mu^+\mu^-$, $\tau^+\tau^-$)[13] yields

$$0.19 \leq \sin^2\theta_w \leq 0.39 \qquad \text{(95\% confidence level limits)},$$

and

$$\sin^2\theta_w = 0.27 \begin{array}{c} + 0.06 \\ - 0.04 \end{array}$$

Similar results have been obtained by other groups at PETRA[7,15].

The determination of $\sin^2\theta_w$ is thus in good agreement with the value obtained from neutrino scattering experiments[14] which yield a value of $\sin^2\theta_w = 0.234\pm0.011$. The two types of experiments are done in entirely different kinematic regions. The good agreement in $\sin^2\theta_w$ obtained in two types of experiments gives important support to the validity of the GWS theory, and its applicability in the timelike region up to $q^2 = 1300$ GeV^2.

TEST OF THE POINTLIKE NATURE OF QUARKS

The MARK-J measurements of R have also been used to search for quark structure, by looking for a q^2 dependence of the total hadronic cross section beyond the effects predicted by QCD and the standard electroweak theory. The possible breakdown of point-like behaviour has been parametrized in terms of the form factors[16] which have traditionally been used to search for structure in leptons, and which depend on the cut off parameters Λ_+ and Λ_-:

$$\mathcal{F}_\pm \ (\Lambda_\pm) \ = 1 {- \atop +} \ \frac{q^2}{q^2 - \Lambda_\pm^2} \qquad (q^2 = s` \tag{13}$$

As shown in Fig. 14, a fit to the data using

$$R \ (\Lambda_\pm) \ \equiv R_T \mathcal{F}_\pm \ (\Lambda_\pm) \tag{14}$$

where R_T is given by Eq. 11, yields

$$\Lambda_+ > 190 \text{ GeV}, \quad \Lambda_- > 285 \text{ GeV} \quad (95\% \text{ confidence level}).$$

The fit is performed in a fashion similar to that used to determine $\sin^2\theta_w$ for quarks, except we now fix $\sin^2\theta_w = 0.23$ and replace R_T by $R(\Lambda_\pm) = R_T \mathcal{F}_\pm$ in Eq. 12. We then allow the scaling factor F and Λ_+, or F and Λ_-, to vary to obtain the 95% confidence level limits in the F-Λ plane.

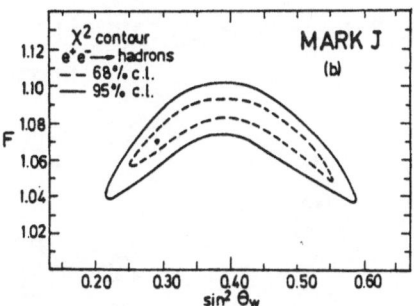

Fig. 13.   Scale factor F plotted against $\sin^2\theta_w$. The areas shown
indicated the 95% C.L. and 68% C.L. regions (inner curve).
The minimum χ^2 (the dot) is obtained for $\sin^2\theta_w$ = 0.29
and F = 1.07.

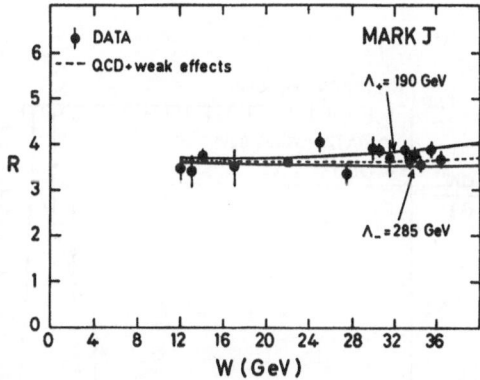

Fig. 14.   The total relative hardronic cross section R = σ(e^+e^- →
hadrons)/ σ(e^+e^- → $\mu^+\mu^-$)vs. center of mass energy W. The
dashed line is the value expected from inclusion of QCD
and weak effects as described in D.P. Barber et al., Phys.
Rev. Lett. 46, No.26, July 1981, page 1963, with $\sin^2\theta_w$ =
0.23 and α_s = 0.18. The full ines are fits at 95% confidence level lower limits to the cut off parameter Λ_+ and
Λ_- assuming form factor modifications of the form

$$\tilde{\tilde{j}} \ (\Lambda_{\pm}) = 1_-^- \frac{s}{s-\Lambda_{\pm}^{\ 2}}$$

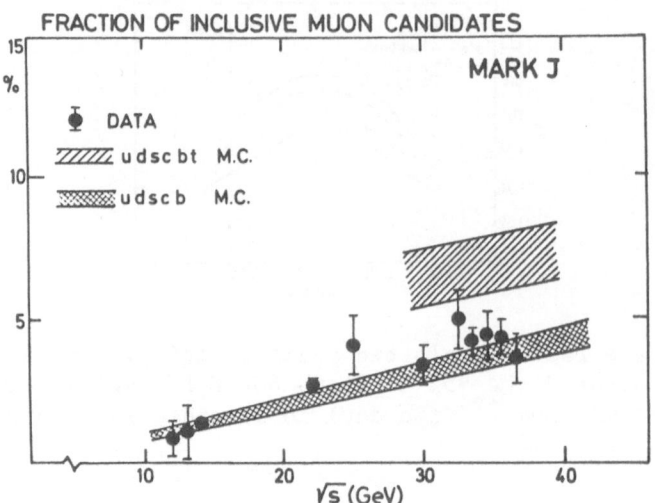

Fig. 15.   Relative production rate of hadronic events containing muon
            muon candidates as a function of center of mass energy
            $\sqrt{s}$.   The hatched areas are the QCD Monte Carlo predictions
            for five and six quark flavors.

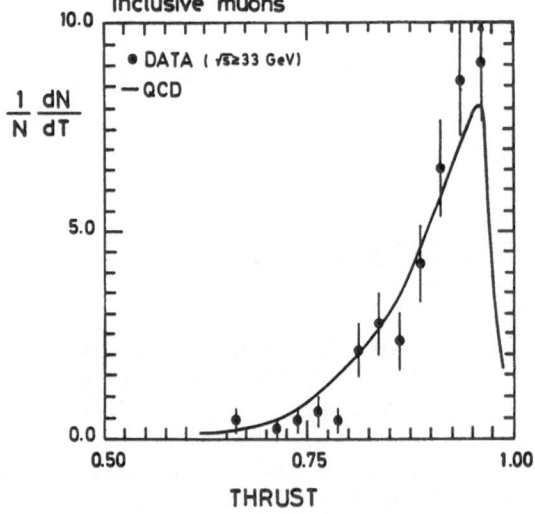

Fig. 16.   The thrust distribution for hadronic events with $\sqrt{s}$ > 33
            GeV which include a muon candidate.   The solid line is
            the prediction of the QCD Monte Carlo with the production
            of u,d,s,c, and b quarks.   The b → c branching ratio is
            assumed to be 100%.

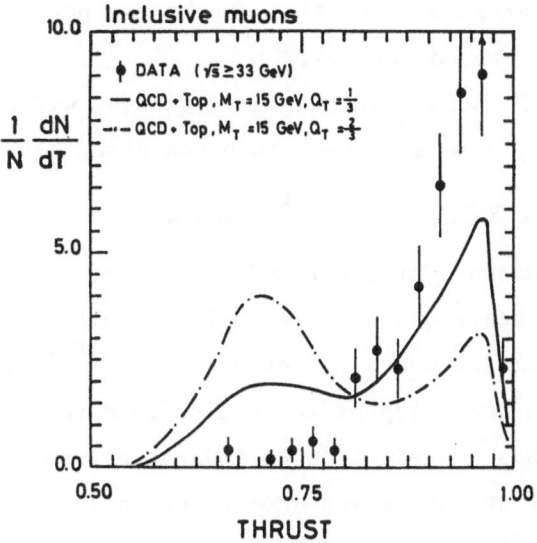

Fig. 17. As in Fig. 16 but with QCD Monte Carlo predictions which
include production of charged 1/3 or 2/3 top quarks of
mass 15 GeV.

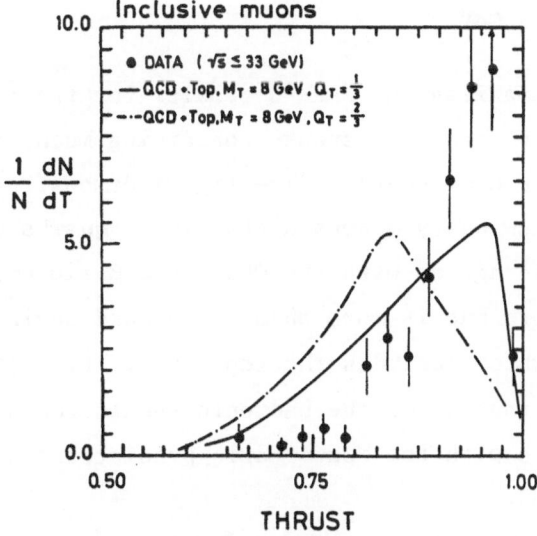

Fig. 18. As in Fig. 16 but with QCD Monte Carlo predictions which
include production of charged 1/3 and 2/3 top quarks of
mass 8 GeV.

We have therefore shown that quarks are pointlike, up to an energy scale of around 200 GeV.

Tests of the pointlike nature of quarks, without the inclusion of weak effects, have also been reported using TASSO data in Ref.17.

INCLUSIVE MUON EVENTS: $e^+ e^- \rightarrow \mu^\pm$ + hadrons

In the framework of the standard six quark model[18] for the weak decays of heavy quarks (c,b, and t) copious muon production is expected from the cascade decays $t \rightarrow b \rightarrow c$[18]. The onset of the production of a heavy lepton would also lead to an increase in muon production. Thus, in addition to indications based on R measurements and event shapes, a measurement of inclusive muon production in hadronic final states should provide a clear indication of the formation of top quarks or new leptons. All the hadron data for $\sqrt{s}$ from 12 to 36.7 GeV have therefore been analyzed and scanned to search for muons which are cleanly identified and momentum analyzed using the outer drift chambers of the MARK-J detector. As this implies that muon candidates must penetrate at least one meter of iron equivalent, the minimum detectable muon momentum is approximately 1.3 GeV.

Fig. 15 summarizes the MARK-J results for the relative production rate of hadronic events containing muon candidates as a function of the c.m. energy. The figure demonstrates once again the absence of new heavy mesons containing t-quarks up to 36.7 GeV. The observed rate agrees with the QCD Monte Carlo predictions for five quark flavors but is more than 5 standard deviations away from the prediction which includes the top quark. Fig. 16 shows that the thrust distribution of the hadronic events containing muon candidates agrees with the Monte Carlo predictions without new flavor production.

The scarcity of events at low thrust in Fig. 16 provides additional strong evidence which rules out new quark flavor production. This is demonstrated in Figs. 17 and 18, where one

TABLE III

Inclusive Muon Events with T < 0.75, T < 0.80

	Total Events	T < 0.75	T < 0.80
DATA	190	5	10
QCD		11.0	21.5
QCD + TOP (M=15,Q=2/3)		92.6	114.7
QCD + TOP (M=15,Q=1/3)		48.2	64.1
QCD + TOP (M=8, Q=2/3)		33.6	61.5
QCD + TOP (M=8, Q=1/3)		21.7	40.4

sees that new quarks of mass 8-15 GeV and charge 1/3e or 2/3e are clearly excluded. Table III gives a quantitative comparison of the number of events observed with T < 0.75 and T < 0.80 to the predictions of the QCD model including top quarks, with masses and charges corresponding to Figs. 17 and 18.

WEAK DECAYS OF HEAVY QUARKS (C AND B)

The main source of prompt muons in the hadronic events are decay products of bottom and charm quarks. Background contributions to the muon signal, arising from hadron punch through and decays in flight of pions and kaons[F3] have been calculated using Monte Carlo simulations[4] to be ∿ 2% of the hadronic events. The contribution of $\tau^+\tau^-$ events to the μ + hadron sample becomes negligible when the total energy cuts and energy balance cuts are applied.

Prompt muons in hadronic events can therefore be used to study the electroweak production of heavy quarks and their weak decays. The primary goal of the MARK-J study is to determine the forward-backward charge asymmetry in the production of heavy quark-anti-quark pairs, which is expected to be 50% larger for $c\bar{c}$ compared to $\mu^+\mu^-$ production, and a factor of three larger for $b\bar{b}$ production. This measurement is being carried out by looking at the charge asymmetry of the muons from semileptonic c and b decays.

A preliminary determination of the semileptonic branching ratio for bottom quarks has been performed in preparation for the asymmetry measurement. This measurement is not very sensitive to uncertainties in the fragmentation functions used in the Monte Carlo model[4], as the fragmentation parameters are tightly constrained by the detailed event shapes determined by our jet analysis (see Refs. 3,8, and 19). Fig. 19 shows that the inclusive muon momentum distribution predicted by the Monte Carlo

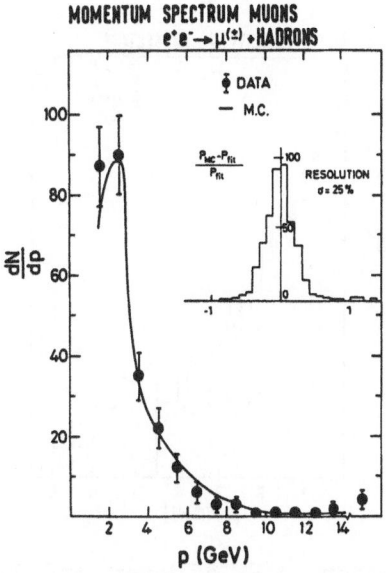

Fig. 19.   The momentum spectrum of muon condidates in inclusive
           muon events. The points are for the data. The solid
           line is the Monte Carlo prediction. The momentum re-
           solution curve deduced from the Monte Carlo is shown
           in the inset.

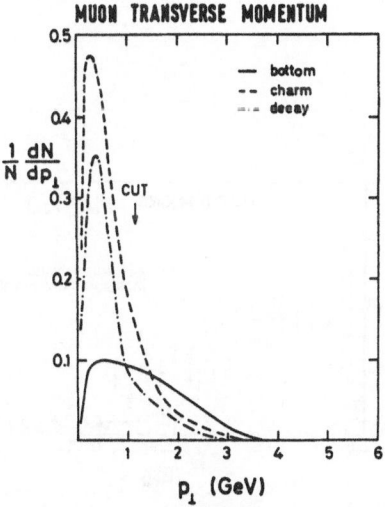

Fig. 20.   The distribution in P_T of muon candidates relative to the
           hadron jet axis in inclusive muon events a predicted by
           the Monte Carlo. The solid line is for muons resulting
           from bottom quarks, the dashed line is for charm decay,
           and the dot-dashed line is for muons from decay in flight.

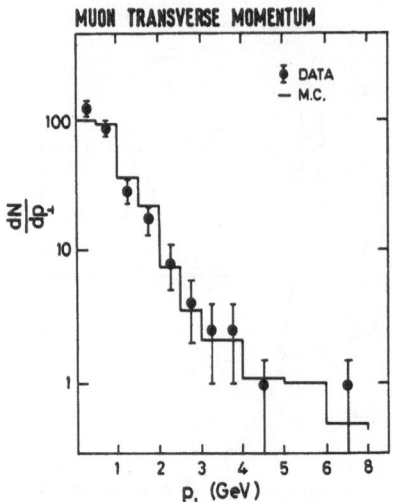

Fig. 21.   The distribution in P_T of muon candidates relative to the
           hadron axis in inclusive muon events. The solid line is
           the Monte Carlo prediction with u,d,s,c,b quarks including
           decay and punch through.

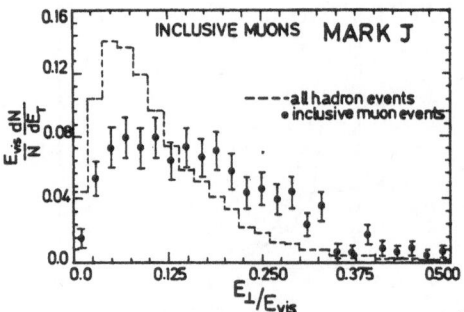

Fig. 22.   The transverse energy balance distribution for inclusive
           muon candidates. The dashed line is for all hadronic
           events with $\sqrt{s} > 27$ GeV and the points are for inclusive
           muon candidates.

model and our detector simulation is an excellent fit to the data.

The method used to obtain an enriched sample of hadronic events containing prompt muons from bottom decay is illustrated in Fig. 20. The figure shows the calculated P_T distribution for inclusive muons $\frac{1}{N}\frac{dN}{dP_T}$, where the P_T is transverse to the thrust axis of the jet not containing the muon. The P_T distribution for muons from c quark decay is peaked at relative low momenta, while the distribution for muons from b decay extends to higher values. Also indicated in this graph is the P_T distribution of the muons from π or K decay. By making a cut at $P_T \geq 1$ GeV one obtains a sample of events dominated by muons from the decay of the b-quark.

Fig. 21 shows the muon P_T distribution for 234 inclusive muon events. A cut at 1.2 GeV leaves 48 events of which ∿ 45% are estimated to be due to b-quark decay. The preliminary result for the semi-leptonic branching ratio for mesons containing a b-quark (denoted by B) is found to be:

BR (B → μ + x)= 8.0%±2.8% (statistical) ±2.0% (systematic).

The large systematic error is primarily due to the uncertainty of the punch through contribution which is still being analyzed. This result compares well with the results obtained at CESR[20] which are (for the CLEO group):

BR (B → μ + x) = 10.0%±1.3% (statistical) ±2.1% (systematic)
BR (B → e + x) = 13.6%±2.1% (statistical) ±1.7% (systematic)

The branching ratios quoted above refer only to the primary decay $b \rightarrow c\mu^-\bar{\nu}_\mu$ and not to the cascade muons from the subsequent decay $c \rightarrow s\mu^+\bar{\nu}_\mu$.

Fig. 22 shows the energy imbalance perpendicular to the beam axis for inclusive muon events compared to that for all hadron events. The inclusive muon events are substantially more imbalanced. This is primarily due to missing neutrino energy, with some additional contribution from muon energy escaping from the calorimeter.

SUMMARY AND CONCLUSIONS

(1)   The production rate and event shapes of hadron events and inclusive muon events show no sign of new flavor production up to $\sqrt{s}$ = 36.7 GeV, for new quark masses $\gtrsim$ 8 GeV and charge 1/3e or 2/3e.

(2)   The data are consistent with QCD and the standard electroweak theory (GWS). An analysis of R vs. $\sqrt{s}$ yields

$$\sin^2\theta_w = 0.29 \;^{+0.26}_{-0.05}.$$

(3)   Combining our data on $e^+e^- \to \ell^+\ell^-$ (ℓ= e,μ,τ) and hadron production we find

$$\sin^2\theta_w = 0.27 \;^{+0.06}_{-0.04}.$$

(4)   We have determined that quarks are pointlike to an energy scale $\gtrsim$ 200 GeV.

(5)   The production rate and the momentum, transverse momentum, and thrust distributions of inclusive muon events are consistent with the standard model of weak mixing.

(6)   Our preliminary result on the semileptonic branching ratio in b decays is:

BR (B $\to$ μ + x) = 8.0% $\pm$2.7%(statistical) $\pm$2.0%(systematic).

ACKNOWLEDGEMENTS

I wish to thank my colleagues of the MARK-J collaboration for help in the preparation of this talk. I have particularly benefitted from recent discussions with J.G. Branson, P. Duinker, D.P. Barber, M. Pohl,and H. Rykaczewski on matters relevant to the preparation of the manuscript. I am also grateful to Prof. Samuel C.C. Ting for his continuing support and encouragement.

REFERENCES

1. H. Newman, Proc. 1979 Intern. Symp. on Lepton and Photon Interactions at High Energy Physics (Fermilab, Batavia, Illinois, 1979);
    D.P. Barber et al., Phys. Rev. Lett. 43, 830 (1979);
    D.P. Barber et al., Phys. Lett. 89B, 139 (1979).

2. H. Rykaczewski, Aachen Report AC INTERN 81-05, August 1981.

3. D.P. Barber et al., Physics Reports 63, 337 (1980) and
    M.I.T./LNS Report No. 107 (1980).
    D.P. Barber et al., Phys. Rev. Lett. 44, 1722 (1980).

4. A. Ali, E. Pietarinen, G. Kramer and J. Willrodt, DESY Report 79/86 (1979), and Phys. Lett. 93B, 155 (1980).

5. F.A. Berends and R. Kleiss, DESY Reports 80/66 and 80/73 (1980).

6. J.L. Rosner, C. Quigg, H.B.Thacker, Phys. Lett. 74B, 350 (1978).
    C. Quigg, Contribution to the 1979 Intern. Symp. on Lepton and Photon Interactions, Fermilab.
    M. Greco, Phys. Lett. 77B, 84 (1978).
    T. Applequist and H. Georgi, Phys. Rev. D8, 4000 (1973).
    A. Zee, Phys. Rev. D8, 4038 (1973).
    M. Dine, J. Sapirstein, Phys. Lett. 43, 152 (1979).

7. Recent reviews of PETRA results on R and the search for new quark flavors include:
    P. Duinker, NIKHEF-H/81-30 (1981); Invited talk at the Intern. Conference on High Energy Physics, Lisbon, Portugal, July 1981, and at the Intern. School of Subnuclear Physics 19th Course, Erice, Sicily, August, 1981, and
    R. Felst, DESY Report No. 81/75; Invited talk at the 1981 Intern. Symposium on Lepton and Photon Interactions at High Energies, Bonn, Federal Republic of Germany, August, 1981.

8. H. Newman, Proceedings of the XXth Intern. Conference on High Energy Physics, Madison, Wisconsin, August, 1980.

9. J. Ellis and M.K. Gaillard, Physics With Very High Energy e^+e^-
    Colliding Beams, CERN 76-18, 21 (1976).

10. S.L. Glashow, Nucl. Phys. 22, 579 (1961);
    S. Weinberg, Phys. Rev. Lett. 19, 1264 (1967),and Phys. Rev.
    D5, 1412 (1972);
    A. Salam and J.C. Ward, Phys. Lett. 13, 168 (1964).

11. K.G. Chetyrkin et al., Phys. Lett. 85B, 277 (1979).
    M. Dine and J. Sapierstein, Phys. Rev. Lett. 43, 668 (1979).
    W.Celmaster and R. Gonsalves, Phys. Rev. Lett. 44, 560 (1980).

12. D.P. Barber et al., Phys. Rev. Lett. 43, 830 (1979).
    D.P. Barber et al., Phys. Lett. 89B, 139 (1979).
    W. Bartel et al., Phys. Lett. 91B, 142 (1980).
    Ch. Berger et al., Phys. Lett. 86B, 418 (1979).
    R. Brandelik et al.,Phys. Lett. 86B, 243 (1979).
    R. Brandelik et al.,DESY-Report 80/40 (1980).
    H.B. Newman, Proc. XXth Intern. Conf. on High Energy Physics,
    Madison (1980).

13. D.P. Barber et al., Phys. Rev. Lett. 46, 1663 (1981).
    D.P. Barber et al., Aachen Report PITHA 80/8 (1980).
    A. Böhm, Aachen Report PITHA 80/9 (1980) and Proc. XXth Intern.
    Conf. on High Energy Physics, Madison, Wisconsin, August, 1980.
    M. Pohl, Aachen Report PITHA 81/10 (1981).
    J.G. Branson, DESY-Report 81/73 (1981) and Proc. of the 1981
    Lepton-Photon Symposium, Bonn, 1981.

14. P. Langacker et al., Proc. Neutrino-79, Bergen Vol. 1, 276(1979).
    J.E. Kim et al., Pennsylvania Report UPR-158T (1980).
    I. Liede and M. Roos, Proc. Neutrino-79, Bergen, Vol. 1, 309
    (1979).
    J.J. Sakurai, ibid 267.
    L.M. Sehgal, Proc. Symposium on Lepton and Hadron Interactions,
    Visegard (1979).
    Ed.F. Csikor et al., (Budapest), Aachen Report PITHA 79/34, 29.

15. W. Bartel et al.,(JADE Collaboration), DESY Report 80/123(1980),
    and Phys. Lett. 101B, 361 (1981).
    A. Wagner, Proceedings of the XVIth Rencontre de Moriond,
    Les Arcs, France, 1981.
    Ch. Berger et al., (PLUTO Collaboration), DESY-Report 80/116
    (1980). Also see Branson, Ref. 13 above.

16. S.D. Drell, Ann. Phys. 4, 75 (1958).

17. P. Söding and G. Wolf, DESY Report 81/13 (1981).

18. M. Kobayashi and T. Maskawa, Prog. Theor. Phys. 49, 652 (1973).
    A. Ali et al., "Heavy Quarks in e^+e^- Annihilation", DESY
    Report 79/63 (1979).
    A. Ali, Z. Physik C, Particles and Fields 1, 25 (1979).

19. D. P. Barber et al., MIT-LNS Report No. 115 (1981).

20. J. Green et al., "Leptons from B Decay", in Cornell Report
     CLNS 81/513 and CLEO 81/06; also see
     A. Silverman, Proceedings of the 1981 Intern. Symposium on
     Lepton and Photon Interactions et High Energies, Bonn, 1981.

FOOTNOTES

F1. Further details on analysis methods, and previous MARK-J
     results on $e^+e^- \to$ hadrons, are presented in Ref. 2

F2. All Monte Carlo model calculations of QCD predictions, and
     acceptances for various processes, are performed by passing
     simulated events through a detailed representation of the
     MARK-J detector (see Ref. 3).

F3. Two detector simulations have been used in this determination.
     One simulation uses energy deposition distributions and
     punch through probabilities based on test beam results of
     our own and other groups, supplemented by published Monte
     Carlo data on hadron cascades. More recently, we have used
     a simulation which includes the generation of a complete
     hadronic cascade for each particle entering the detector.
     This program, developed by H.Fesefeldt of RWTH Aachen for
     general applications, is particularly designed to be accurate
     over the 0.1 - 10 GeV momentum range typical of particles
     produced in jets at PETRA.

RECENT RESULTS FROM JADE

Tomio Kobayashi

LICEPP
University of Tokyo

ABSTRACT

Recent results on top search, B-particle lifetime and on possible flavours changing neutral current decay of B-particles are reported. A measurement of charge asymmetry in $e^+e^- \to \mu^+\mu^-$ and the search for new particles have been also made.

1. INTRODUCTION

The JADE detector has been operating at PETRA e^+e^- storage ring since June 1979. We have accumulated the integrated luminosity of about 30 pb^{-1} so far at center of mass energies between 12 and 36.8 GeV. The apparatus, which has been described in ref.(1), is mainly composed of the cylindrical jet chamber operated in a solenoidal magnetic field of 4.8 kG, a cylindrical array of lead glass counters and muon chambers. It covers 97% of the full solid angle for charged particles and 90% for photons.

2. SEARCH FOR TOP QUARKS

The total cross section $\sigma(e^+e^- \to \text{hadrons})$ was measured for the center of mass energies between 12 and 36.8 GeV. Fig. 1 shows the recent results on the ratio R= $\sigma(e^+e^- \to \text{hadrons}) / \sigma(e^+e^- \to \mu^+\mu^-)$ together with data from other experiments[2]. The naive quark-parton model predicts

$$R_0 = 3\sum_q e_q^2 = 11/3$$

45

Table 1 :   Number of muons with P_T > 2 GeV/c in 4452 multihadronic
            events for $\sqrt{s}$ = 33 - 36.8 GeV with sphericity > 0.5.

observed	1
expected from u b	2
expected from t(Q=2/3)	104
expected from t(Q=1/3)	35
$(M_t = 16$ GeV/c^2, BR(t → μX) = 10%)	

where e_q is the charge of the quark with flavours q(q=u,d,s,c,b).
Including QCD corrections, one expects R ∿ 3.9 over the whole energy
range of PETRA. If a top quark with charge 2/3 exists in the PETRA
energy range, the step of R = 4/3 in the ratio R would be seen.
The data excludes this possibility.

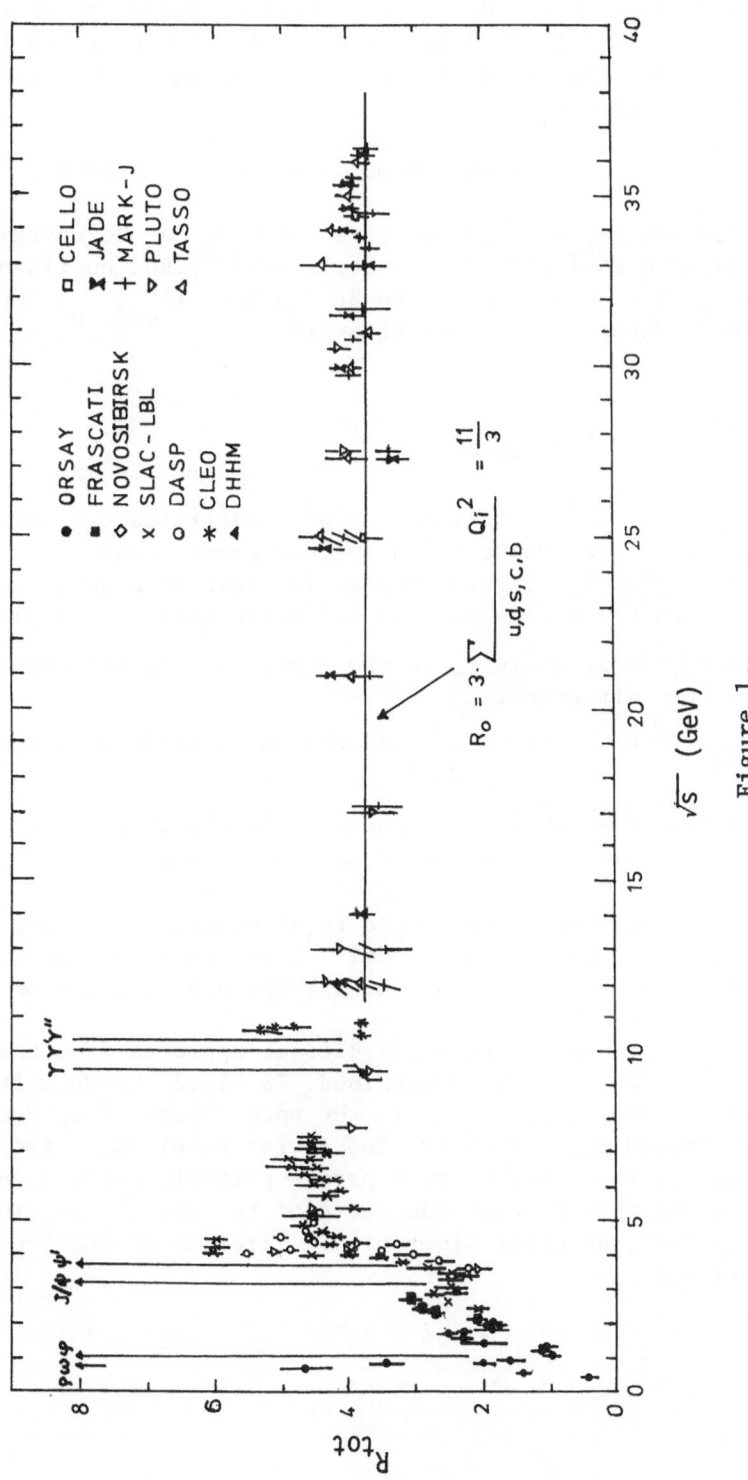

Figure 1.

We have also searched the production of prompt muons in the energy range 33 GeV to 36.7 GeV. The result is summarized in Table 1. It excludes the production of not only a top quark, but also a 1/3 charged heavy quark with

$$\sigma \cdot BR < 0.002 \cdot \sigma_{\mu^+\mu^-} \qquad (90 \% \text{ c.l.}).$$

The search for a $t\bar{t}$ bound state was made in the energy range 29.9 to 31.46 GeV[3] and 33.0 to 36.72 GeV[4]. 90% confidence upper limit on $\Gamma_{ee} \Gamma_h / \Gamma$ was measured to be 1.3 keV (Γ_{ee}, Γ_h: leptonic and hadronic decay widths, respectively).

3. LIFETIME OF B-PARTICLES[5]

An upper limit on the lifetime of B-particles was determined using multihadronic events which include prompt muons (p > 1.4 GeV/c). We have a sample of 349 such events. In order to enhance the fraction of events with heavy flavours the following cuts were applied:

i) $|\cos\theta_T| < 0.75$, where θ_T is the angle between the thrust axis and beam direction.

ii) $E_{vis} > 0.6 \sqrt{s}$, where E_{vis} is the total visible energy of the event.

iii) $T_1 = \sum |\vec{p_i} \cdot \vec{e_1}| / \sum |\vec{p_i}| > 0.2$, where $\vec{e_1}$ is the unit vector which is perpendicular to the thrust axis and minimizes $\sum |\vec{p_i} \cdot \vec{e_1}|$.

These cuts reduces the data sample to 31 events. According to the Monte Carlo based on the LUND model[6], one muon is expected to be originated from $u\bar{u}$, $d\bar{d}$ and $s\bar{s}$, and all the others are from $c\bar{c}$ or $b\bar{b}$ with the ratio of 1:1.2.

For each of these events the closest approach from the muon track to the beam (d_μ) was determined. In Fig.2 the quantity $\chi_\mu = d_\mu/\sigma_\mu$ is shown, where σ_μ is the uncertainty of d_μ determination, and is estimated to be 0.45 mm. The deviation of $<d_\mu>$ from 0 gives the measure of the lifetime of a parent particle which decays into a muon. Taking into account the ratio of $b\bar{b}$ and $c\bar{c}$ originating events (1.2:1), an upper limit of the lifetime of the B-particles can be derived:

$$\tau_b < 1.4 \cdot 10^{-12} \text{ sec} \qquad (95\% \text{ c.l.}).$$

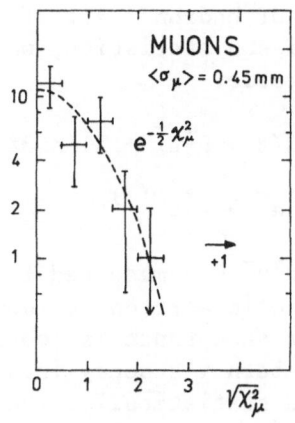

Figure 2.

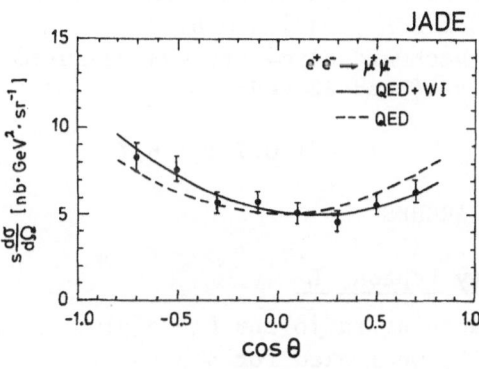

Figure 3.

4. LIMIT OF FLAVOUR CHANGING NEUTRAL CURRENTS

In the framework of "topless" theories $B \to \mu^+\mu^- + X$ should exist via flavour changing neutral currents. We have searched the dimuon events where both muons have opposite charges and are in the same jet. 8 candidates are found, whereas 6.9 events are expected from the standard decays of b- and c-quarks and background such as π,K decay, punch through of hadrons. All of our dimuon events have $M_{\mu\mu} < 5$ GeV/c^2. Using Poisson statistics, we can thus give an upper limit for the branching ratio:

$$BR(B \to \mu^+\mu^- + X) < 3\% \quad (90\ \%\ c.l.).$$

5. CHARGE ASYMMETRY IN $e^+e^- \to \mu^+\mu^-$ [7]

The process $e^+e^- \to \mu^+\mu^-$ was measured for the data until spring 1981, corresponding to an integrated luminosity of 19 pb^{-1} at $E_{cm} > 25$ GeV. The angular acceptance is $|\cos\theta| < 0.8$. Main background comes from τ pairs where both τ's decay into muons. It was calculated to be 1.3% and subtracted statistically. The final candidates were 778 events. Radiative corrections were applied to the data. Fig.3 shows the angular distribution for $e^+e^- \to \mu^+\mu^-$. The forward-backward asymmetry was calculated to be

$$A = -(11.8 \pm 3.8 \pm 1)\%$$

It agrees with the prediction of the standard Glashow-Salam-Weinberg model[8]: $A = -7.8\ \%$ for $|\cos\theta| < 0.8$.

The forward-backward asymmetry was measured also for the combined data at $E_{cm} = 14$ and 22 GeV:

$$A = (+\ 0.7 \pm 4.6)\%$$

6. NEW PARTICLE SEARCHES

Sequential Heavy Lepton, $L^{\pm}$

Since the branching ratio for $L \to e(\mu)\nu\nu$ is expected to be small for massive L, we looked for a heavy lepton through its hadronic decay mode. The signatures of the event are:
- high missing energy
- non back-to-back jet event shape.

We applied the following cuts:

    i) $|\cos\theta_T| < 0.75$

   ii) $E_{vis} > 0.33 \cdot E_{cm}$

  iii) $\psi_{ac} > 50^0$, where ψ_{ac} is the accoplanarity angle of the two jets,

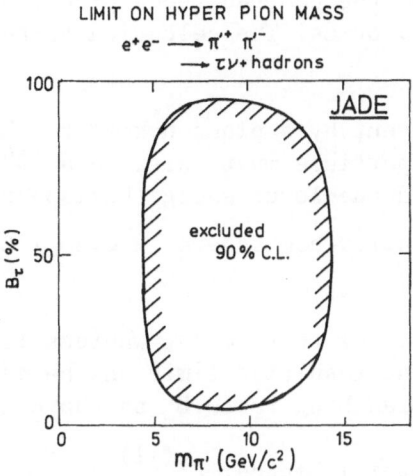

LIMIT ON HYPER PION MASS

$e^+e^- \longrightarrow \pi'^+ \pi'^-$

$\longrightarrow \tau\nu + $ hadrons

Figure 4.

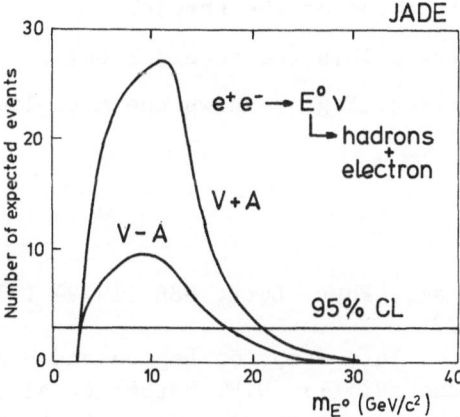

Figure 5.

and we found none, which excludes the heavy lepton with

$$6.8 < M_L < 18.1 \text{ GeV}/c^2 \text{ (95\% c.l.).}$$

Charged Technipion, π'^+[9]

We searched for the charged technipions, which are assumed to decay into either $\tau\nu$ or cs. The selection criteria are as follows:

   i) $|\cos\theta_T| < 0.7$

  ii) Divide the event by a plane normal to the thrust axis. One and only one particle must exist in a 70^0 cone around the thrust axis in the lower energy hemisphere.

 iii) E_{vis} (higher energy hemisphere) $> 0.63 \cdot E_{beam}$

  iv) $\psi_{ac} > 20^0$.

One event remained after these cuts, whereas 1.5 events are expected from $e^+e^- \rightarrow \tau^+\tau^-$. The resulting limit on the mass was determined as a function of the branching ratio B_τ as shown in Fig. 4[10].

Electron-type Neutral Lepton, E^0[11]

Heavy neutral electrons were searched through the process $e^+e^- \rightarrow E^0\nu \rightarrow e + \text{hadrons}$. Since the signature of such a reaction is highly unbalanced event shape, we applied the cuts:

   i) No charged tracks are contained in a 50^0 cone around the opposite direction of the thrust.

  ii) Shower energy within the cone < 2 GeV.

No event was observed. Fig. 5 shows the mass limits for V + A and V − A coupling of E^0.

REFERENCES

(1) V. Bartel et al., Phys. Lett. 88B (1979) 171, 92B (1980) 206, 99B (1981) 277.

(2) S. Orito, Proc. Int. Symp. on Lepton and Hadron Interactions 1979, p.52, and ref.(1), D.P. Barber et al., Phys. Rep. 63,7 (1980) 337, Ch. Berger et al., Phys. Lett. 81B (1979) 410, 86B (1979) 413, R. Brandelik et al., Phys. Lett. 83B (1979) 261, Z. Phys. C4 (1980) 87.

(3) W. Bartel et al., Phys. Lett. 91B (1980) 152, D.P. Barber et al., Phys. Rev. Lett. 44 (1980) 1722, Ch. Berger et al., Phys. Lett. 91B (1980) 148, R. Brandelik et al., Phys. Lett. 88B (1979) 199.

(4) W. Bartel et al., Phys. Lett. 100B (1981) 364.

(5) W. Bartel et al., DESY 82-014.

(6) B. Anderson et al., Phys. Lett. 94B (1980) 211.

(7) W. Bartel et al., Phys. Lett. 108B (1982) 140. See also the
     contribution to the LISBON Conference (1981) for Mark II results,
     M. Pohl, Review talk at the XVI-th Recontre du Moriond (1981)
     for other PETRA experiments.

(8) S.L. Glashow, Nucl. Phys. 22 (1961) 579, Rev. Mod. Phys. 52
     (1980) 539, A. Salam, Phys. Rev. 127 (1962) 331, Rev. Mod.
     Phys. 52 (1980) 525, S. Weinberg, Phys. Rev. Lett. 19 (1967)
     1264 , Rev. Mod. Phys. 52 (1980) 515.

(9) S. Weinberg, Phys. Rev. D19 (1979) 1277, L. Susskind, Phys. Rev.
     D20 (1979) 2619.

(10) More extensive analysis was recently made by JADE. See DESY 82-
     -023, submitted to Phys. Lett.

(11) F. Bletzacker and E. T. Nieh, Phys. Rev. D16 (1977) 2115.

DISCUSSION

H. Meyer: Have you studied the charge asymmetry of the $\mu^+\mu^-\nu$ events?

Answer: Yes. We observed the asymmetry which is consistent with QED
predictions

B. Foster: What branching ratio did you use for (in hyperpion search)
$\tau \rightarrow 5$ charged prongs? I do not think there is an experimental measurement of this quantity.

Answer: Branching ratio for $\tau \rightarrow 5$ charged prongs was assumed as
follows:

5 charged $\pi+\nu$	1.0 %
5 charged $\pi+\pi^0$'s$+\nu$	0.2 %

D. Antreasyan: You have an upper limit on the b-lifetime. The same
technique could be used to measure the τ lifetime. Do you have a
measurement ?

Answer: We are now working on that, and I hope we can get the result
in the near future.

RECENT RESULTS FROM TASSO[1]

B. Foster

Rutherford Appleton Laboratory, Chilton, Didcot
Oxon, OX11 0QX, UK

ABSTRACT

Results are presented on the inclusive production of π^0, K^0 and $\bar{K}^0$ and Λ and $\bar{\Lambda}$ in e^+e^- annihilation. These results, together with those on inclusive charged hadron production are used to obtain information on fragmentation mechanisms and the production of heavy quark flavours in e^+e^- annihilation.

INTRODUCTION

It is well-known that e^+e^- annihilation is a copious source of heavy quark production. In the absence of threshold effects

$$R = \frac{\sigma(e^+e^- \to \text{hadrons})}{\sigma(e^+e^- \to \mu^+\mu^-)}$$

is proportional to the sums of the charges squared of the quark types being produced. This means that about half of the hadronic events being produced by e^+e^- annihilation are initiated by heavy (charm or beauty) quarks. This is illustrated by figure 1, which shows the value of R measured by TASSO at high energies, together with the predicted values for different combinations of produced quarks. It is clear that, above W = 12 GeV (where W is the centre of mass energy) R is consistent with the production of udscb quarks and inconsistent with the production of a new, heavy, top quark.

The problem confronting experimentalists is to disentangle those events coming from heavy quarks (c and b) from those coming from light (u,d,s) quarks. There are several ways of approaching this difficult problem:

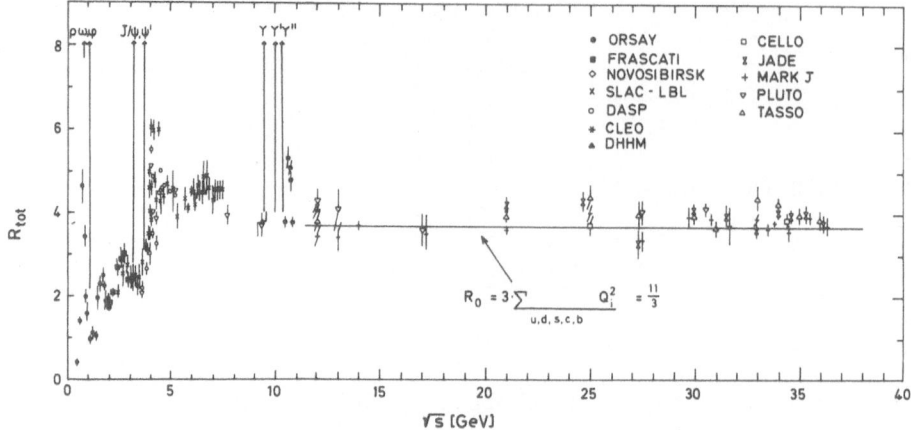

Figure 1.

1)   Since mesons containing heavy quarks have substantial weak
decay modes, many events have one or more leptons in the final
state. A study of inclusive leptons therefore gives information
on heavy quark properties.

2)   At or near threshold for heavy quark production, the shape of
e^+e^- events changes dramatically. A plot of sphericity against
aplanarity as shown in figure 2 shows hot top quark production
would lead to substantial numbers of "spherical" events as the
production threshold is passed.

3)   Mesons containing heavy quarks can have relatively long
lifetimes (approx 10^{-12} secs for charged D mesons). Thus charged
S's often decay after as much as 1 mm at the highest PETRA
energies. High precision drift chambers should be capable of
observing these distances. (I will discuss this possibility
later at this conference (2)).

4)   Decays of "leading" particles containing the primary produced
quark lead mostly to non-strange particles for light (u,d) quarks
while heavy (c,b) quarks lead to an enhanced strange particle
yield. Moreover, hadrons containing heavy quarks often lead to
larger multiplicities in the final state. Therefore the study of
inclusive distributions also gives information on heavy quark
properties.

     In this talk I will concentrate on method 4), since TASSO's
properties of high precision tracking, good neutral energy
resolution and particle identification make it particularly suited
for studies of inclusive particle production in e^+e^- annihilation.

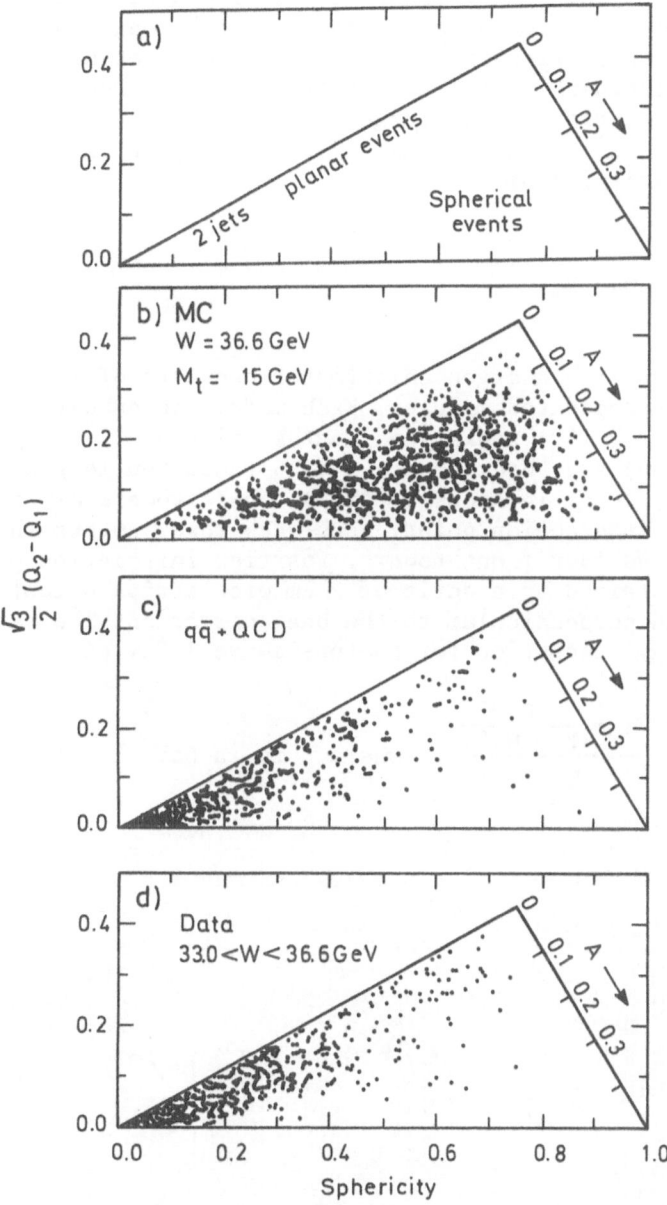

Fig. 2: Plot of sphericity against aplanarity showing
the population expected from b) top quark production,
c) light quarks and QCD effects and d) compared with
the data.

INCLUSIVE π^0 PRODUCTION

Figure 3 shows the TASSO detector viewed perpendicular to the direction of the incoming electrons and positrons. Full details of the detector, triggers and analysis procedures have been described elsewhere[3]. It is of particular relevance to note that the resolution of the large drift-chamber averaged over the whole driftcell is approximately 195 microns, which lead to a momentum resolution of

$$\frac{\sigma_{P_T}}{P_T} = 0.017 \; P_T \qquad 4)$$

The lead-liquid argon calorimeter consists of four mechanically separated modules. Each module is subdivided into "front towers," (6.1 radiation lengths thick and 7 cm by 7 cm in cross-section) and "back towers," (7.6 radiation lengths thick and 14 cm by 14 cm in cross-section). These towers are oriented towards the interaction point. Each back tower covers the same solid angle as four front towers. Position infromation on the shower is obtained by a seris of 2 cm wide strips oriented both parallel and perpendicular to the beam direction. The calorimeter has an energy resolution for photons above 1 GeV of

$$\frac{\sigma}{E} = \frac{0.11 + \frac{0.02}{(E - 0.5)}}{\sqrt{E}} \; , \quad \text{where E is in GeV.}$$

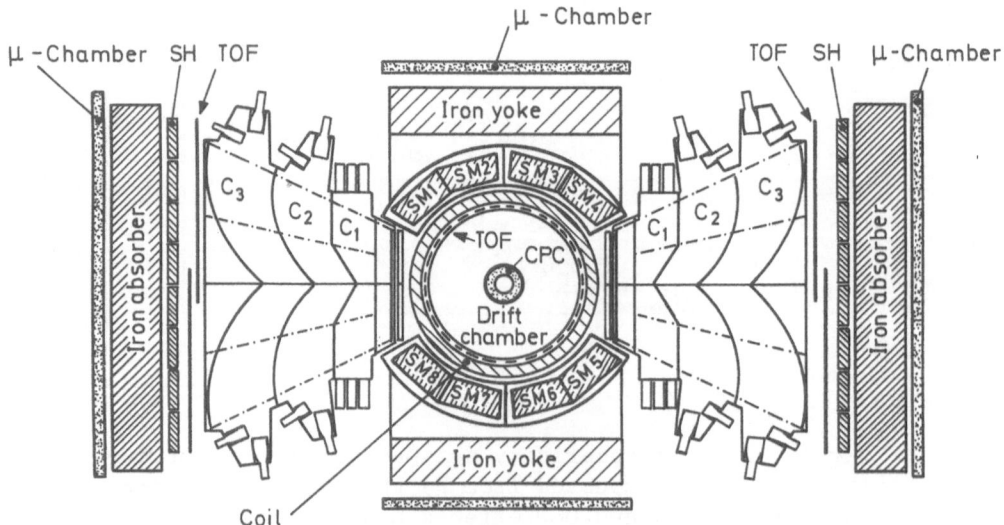

Fig. 3:   The TASSO experiment.

Further details on the construction and performance of the lead-
liquid argon calorimeter are contained in reference 5.

The data sample used to extract the π^o cross-section
consistent of 2173 events of the type $e^+e^-\to$ hadrons at W = 14 GeV
and 2797 events at W = 34 GeV. Photons were identified using a
clustering algorithm. A "cluster" was defined as any group of
front towers with an energy deposit of at least 20 MeV and separated
from any other cluster by at least one front tower with energy
less than 20 MeV. Precise localisation of the shower is possible
using the information from the strips. All charged tracks from
the event found in the central detector are extrapolated out to
the calorimeter. If any found cluster has a charged track pointing
to it to better than 50 mrad it is called a charged cluster and
removed from the analysis. All remaining clusters are called
neutral clusters. The result of taking all possible effective mass

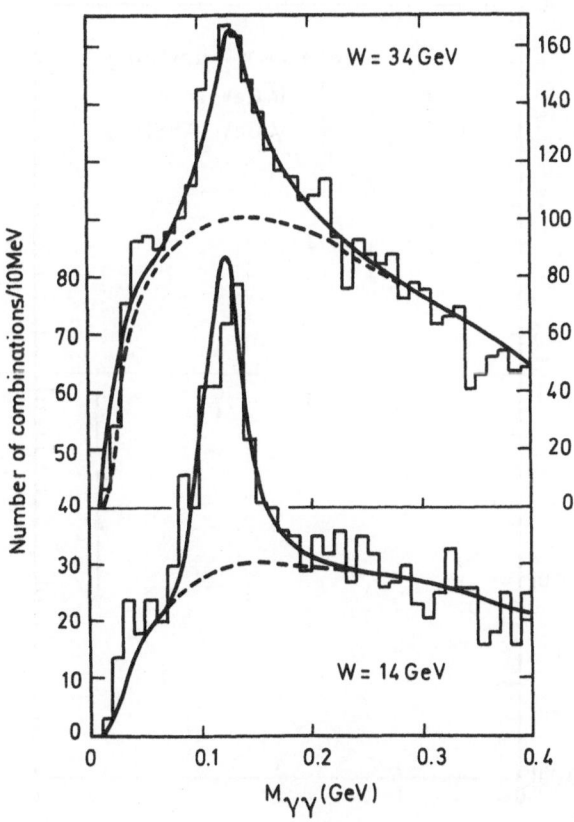

Fig. 4: Effective mass of pairs of γ's observed in
the TASSO liquid argon calorimeter at W=34 GeV and
W=14 GeV. The dotted line represents the combinatorial
background calculated from the Monte-Carlo.

combinations of these neutral clusters assuming them to be photons
is shown in figure 4. A clear signal is seen in both the high and
low energy data.

Calculation of the inclusive π^o cross-section requires a
knowledge of the detection efficiency and the shape of the
background. These were determined from Monte Carlo programs
simulating the passage of charged tracks in the detector including
the effects of multiple scattering, interactions, decays etc. The
reconstruction programs used for the Monte Carlo events were those
used for the real data. Electromagnetic showers were simulated
using the EGS program. The background underneath the π^o signal
was estimated by taking all effective mass combinations in the
Monte Carlo events which did not come from a π^o decay. The background, shown as the dotted lines in figure 4, was normalised to the
region $0.2 < M_{\gamma\gamma} < 0.4$ GeV, where the real π^o contribution is

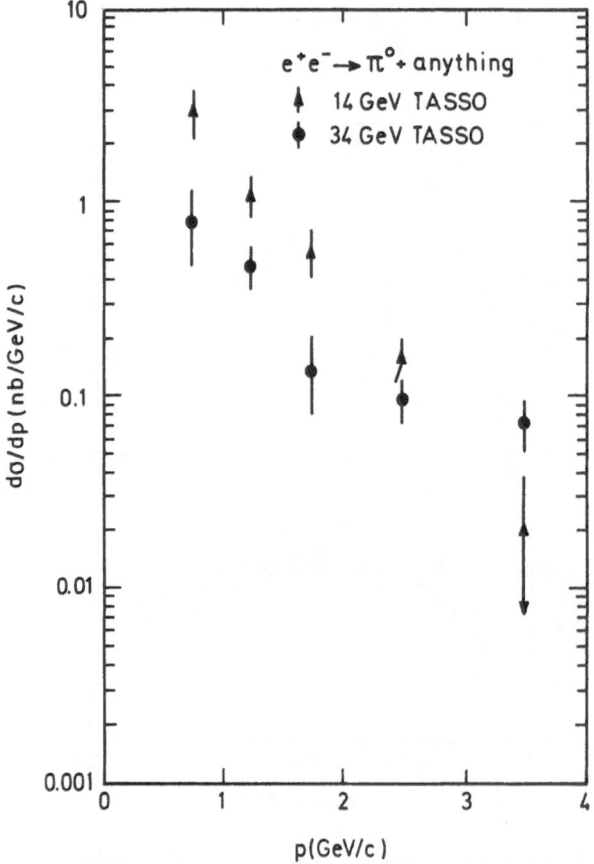

Fig. 5: The differential cross-section for π^o
production for W=14 and W=34 GeV.

expected to be negligible. The π^o detection efficiency estimated from this procedure varied between 5 % and 14 % over the π^o momentum range considered. Figure 5 shows the differential cross-section for both W = 14 GeV and W = 35 GeV calculated using the above procedure.

K^o, Λ PRODUCTION

The good pattern recognition properties and high precision of the TASSO charged particle tracking chambers allow good efficiency for detecting secondary vertices from K^o and Λ decay. The selection criteria used to detect these decays are outlined below.

Identification of a secondary vertex requires that the two tracks should intersect in three-dimensional space at a point well-separated from the primary vertex. In order to reduce the

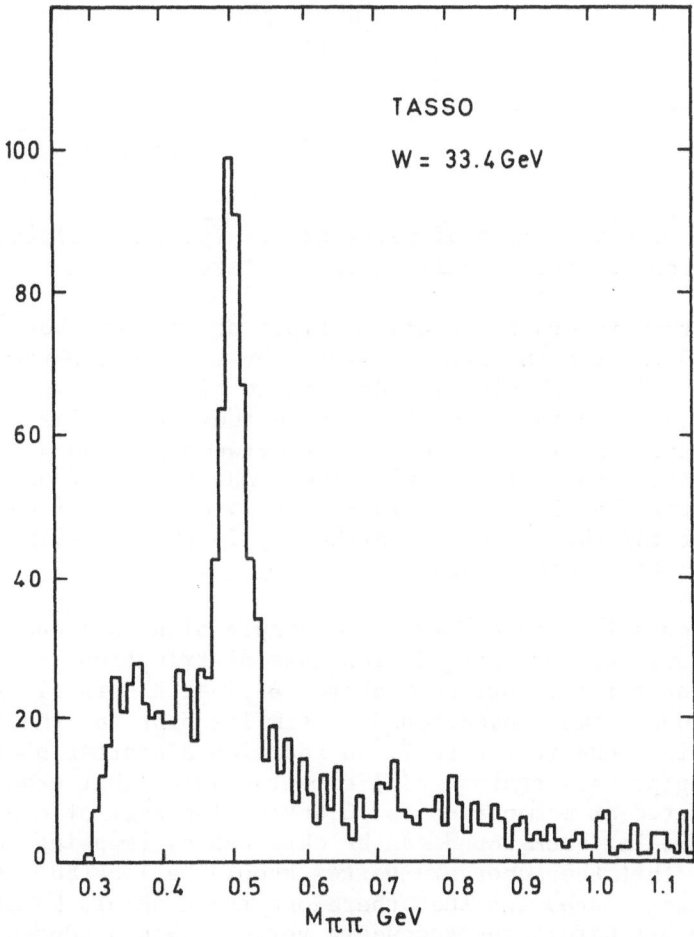

Fig. 6: The effective mass of all $\pi^+\pi^-$ combinations satisfying the K^o selection criteria described in the text.

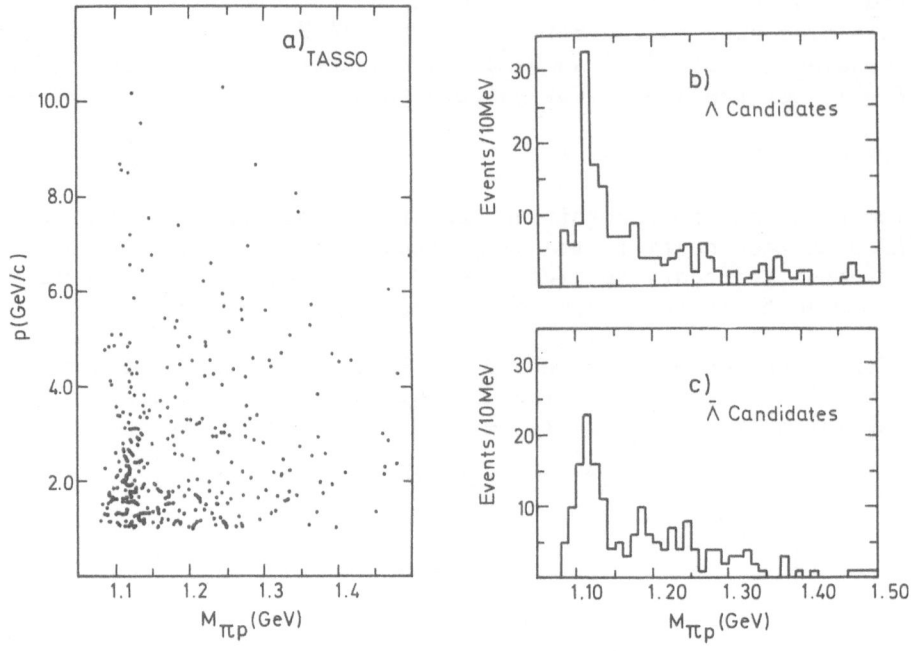

Fig. 7: The effective mass of pairs of pπ⁻ (p̄π⁺) satisfying the
Λ(Λ̄) selection criteria described in the text.

background from interactions and multiple scattering, both tracks
are required to miss the primary vertex by an amount which varies
for the K^O or Λ hypothesis. In addition no hit is allowed in a
tracking chamber on the line of flight between the primary and
secondary vertices. The line of flight between secondary and
primary vertices must line up with the momentum vector of the K^O
or Λ to better than 3 degrees. There are several additional cuts
which differ for the K^O and Λ hypotheses; further details are
contained in references 6 and 7.

These cuts lead to a very clean sample of $K^O(\overline{K}^O)$ and Λ(Λ̄).
Figure 6 shows the two charged pion mass distribution satisfying
the selection criteria outlined above. A clear K^O signal can be
observed above a small background. A similar plot for the Λ(Λ̄)
candidates is shown in figure 7. In addition a scatter plot of
the π⁻ (π⁺) p(p̄) mass against Λ(Λ̄) momentum shows that candidates
can be isolated at momenta up to 10 GeV/c. A sample of candidates
with even smaller background can be obtained by imposing the extra
requirement that the secondary vertex should be located after the
first tracking plane, and that therefore there should be at least
one missing hit before the secondary vertex. This produces a
signal with almost no background, at the cost of smaller efficiency.
This is illustrated in figure 8. It should also be noted that the
numbers of Λ and Λ̄ candidates are about equal on both methods.

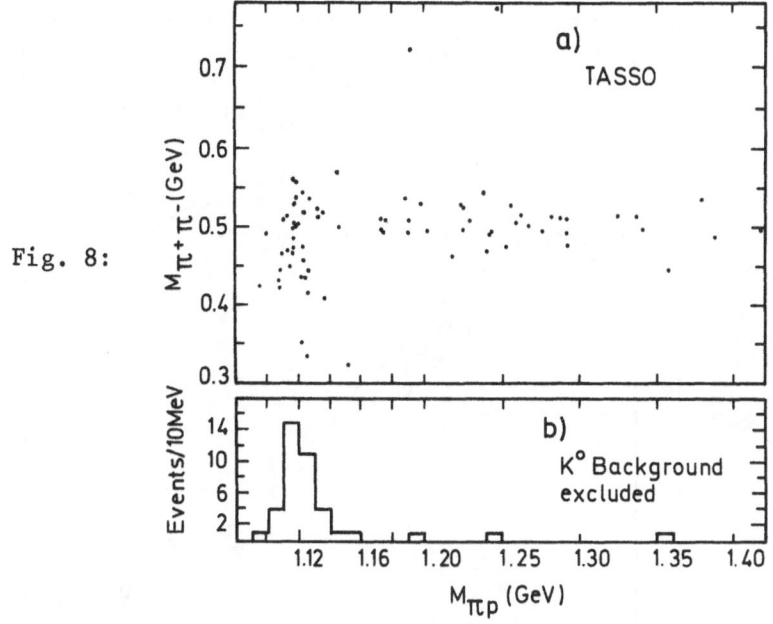

Fig. 8:

WHAT CAN WE LEARN FROM INCLUSIVE PRODUCTION?

I begin by discussing those features of inclusive production which can be obtained without particle identification, i.e. single charged hadron inclusive production. The naive quark model gives a prediction for the differential cross-section of inclusive hadrons as shown below:

$$\frac{d\sigma}{dx} (e^+e^- \to q\bar{q} \to hX) = \frac{8\pi\alpha^2}{s} \sum_q e_q^2 \, D_q^h(x) \qquad (1)$$

where $x = \dfrac{2E_h}{W}$ is the fractional energy of the charged hadron

$(x = \dfrac{P_h}{P_{beam}}$ if the mass of h is unknown), s is the centre of mass

energy squared, e_q is the charge on the primary quark and $D_q^h(x)$ is the fragmentation function, giving the probability for a primary quark to produce a hadron with fractional energy x.

Provided that no new quark thresholds are passed, such that $D_q^h(x) = D_{q'}^h(x)$, it can be seen from equation (1) that the quark-parton model predicts that $s \dfrac{d\sigma}{dx}$ is the same (i.e. scales) between different energies. Figure 9 shows $s \dfrac{d\sigma}{dx}$ measured by TASSO at three different energies, together with low energy data from the MARK II collaboration[8]. It can be seen that, for $x > 0.2$, $s \dfrac{d\sigma}{dx}$ approximately scales. However, significant variations can

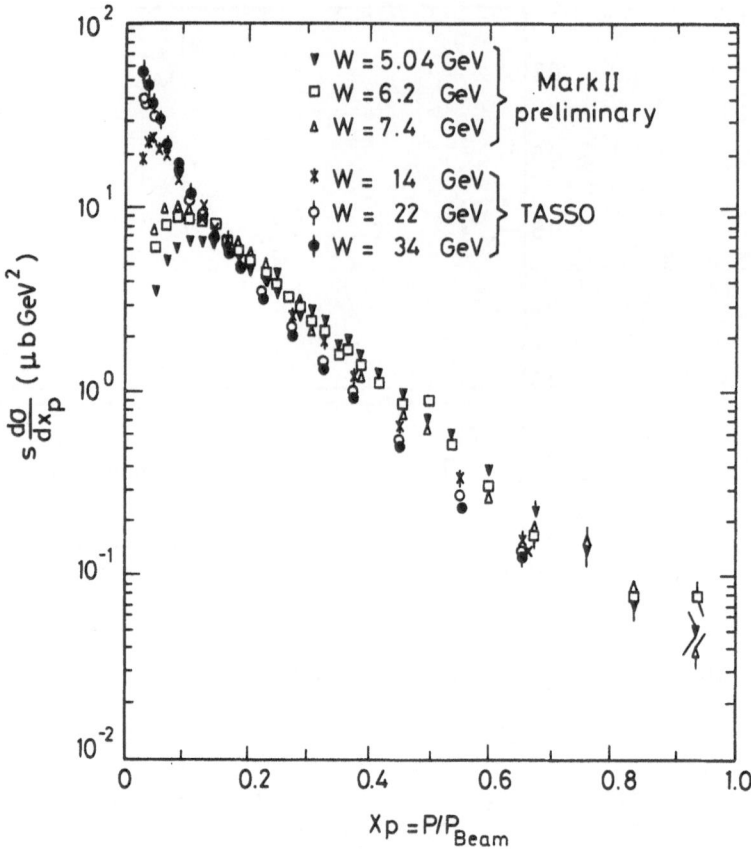

Fig. 9: The invariant cross-section $s \frac{d\sigma}{dx}$ at low energies from the
MARK II collaboration and at high energies from TASSO.

be seen in detail. The TASSO high energy data lie systematically
below both the MARK II and the low energy (W = 14 GeV) TASSO data.
Where the differences between the TASSO and MARK II data could be
due to the opening up of charm and bottom thresholds, the
differences between low and high energy TASSO data can clearly not
be explained by this mechanism. QCD predicts variations from the
simple parton model prediction, equation (1). A high energy
quark has a significant probability to radiate off a gluon, losing
energy and throwing the leading particle to a lower x-value.
Thus QCD predicts that at high energy, fewer high-x particles are
produced than at low energies. This is in qualitative agreement
with the data.

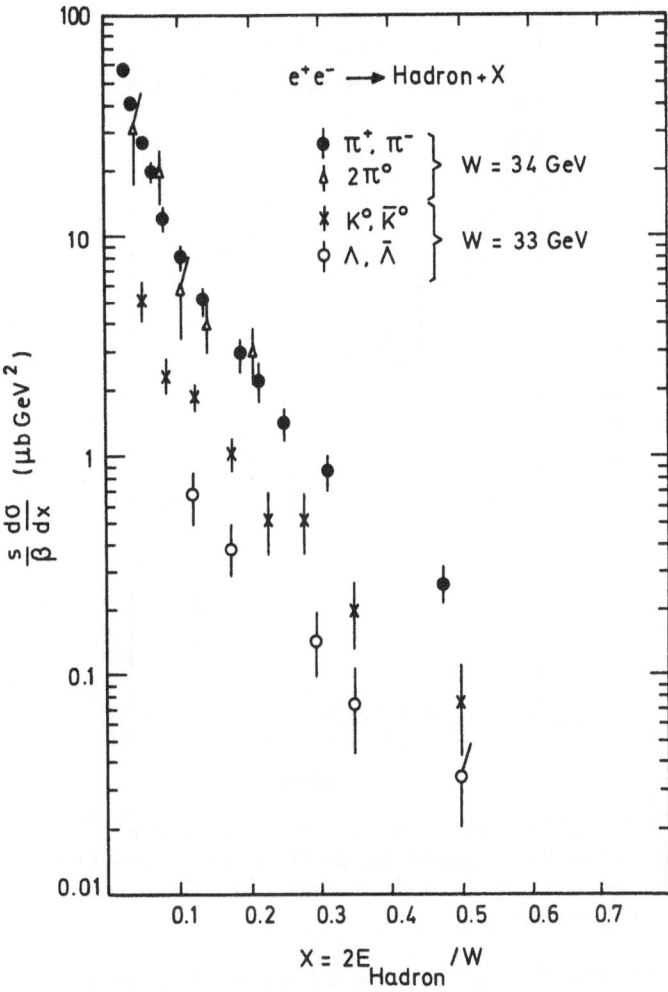

Fig. 10: The inclusive cross-section for $\pi^{\pm}$, π^{o}, K^{o} and $\Lambda(\bar{\Lambda})$ from TASSO at high energies.

Moving now to inclusive pion production, figure 10 shows the scaling cross-section $\frac{s}{\beta}\frac{d\sigma}{dx}$ for π^{o} production together with the average of charged pion production. The charged pion data was obtained from the Cerenkov counters in the TASSO hadron arms[9]. The agreement between the two is very good, so that on average charged and neutral pions carry away the same energy in $e^{+}e^{-}$ annihilations. This can also be seen in figure 11, which shows in addition lower energy π^{o} data from the Lead Glass Wall experiment at SPEAR[10], at W = 4.9 – 7.4 GeV, and TASSO data at 14 GeV. It is clear that the 14 GeV data lie systematically below the Lead Glass Wall data. Since both experiments have included systematic effects in the error bar, this is evidence for scaling violation of similar type to that observed in inclusive hadron production.

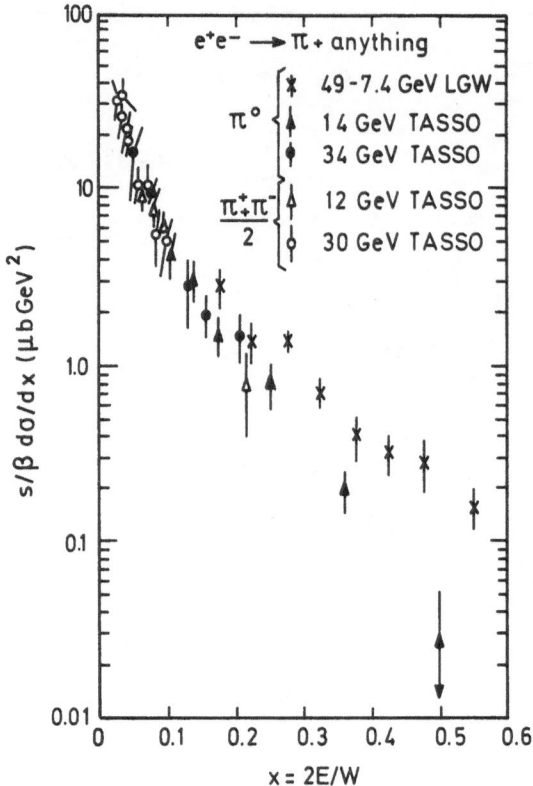

Fig. 11: The invariant cross-section for π^o
production from TASSO at high energy compared
with low energy data from the Lead Glass Wall
Collaboration[12].

    I now consider inclusive production of mesons containing strange
quarks. Measurements of K^o production in principle give information
both on fragmentation mechanisms and on heavy quark production,
since particles containing heavy quarks give rise to many strange
mesons in the final state. This is illustrated in figure 12, which
shows the scaling cross-section $\frac{s}{\beta}\frac{d\sigma}{dx}$ for $K^o(\overline{K}^o)$ production at
W = 33 GeV. Also shown are the relative contributions predicted
from a Field-Feynman model from a) primary strange quarks, b) primary
charm and bottom quarks and c) the $s\overline{s}$ sea which contributes to the
formation of particles in the fragmentation process. It can be seen
that the great majority of K^o_s at x < 0.5 come from the sea. However,
a sizeable proportion of the cross-section comes from c and b quarks
and it is therefore interesting to see the energy dependence of the
K^o $(\overline{K}^o)$ cross-section as the thresholds for b and c production are
passed. Figure 13 [11] shows data from the SLAC-LBL, PLUTO and TASSO
collaborations at energies between 4 and 33 GeV. The abscissa
show the quantity $R(K^o, \overline{K}^o)$ defined as

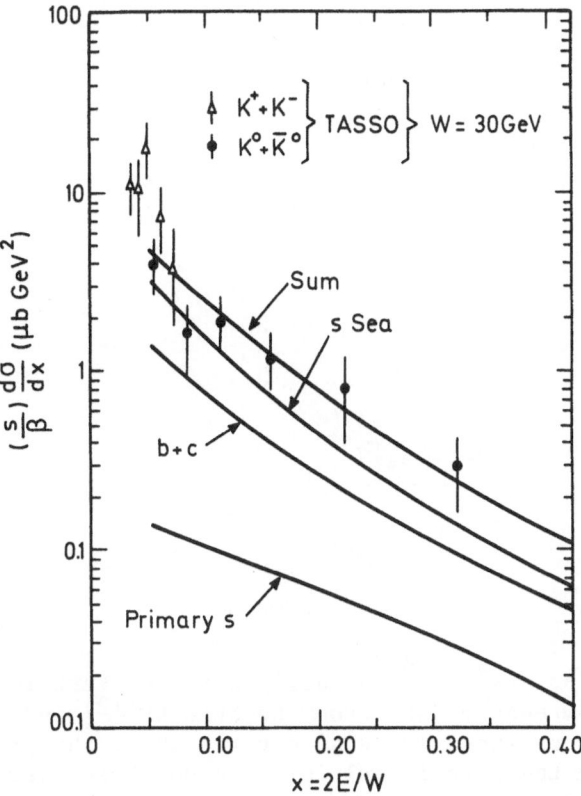

Fig. 12: The invariant cross-section for K^o production showing the expected contributions from primary strange quark production, from bottom and charm quarks and from the $s\bar{s}$ sea.

$$R(K^o, \bar{K}^o) = \frac{\sigma(e^+e^- \to K^o, \bar{K}^o + X)}{\sigma(e^+e^- \to \mu^+\mu^-)}$$

The low energy data show clear evidence for an increase in K^o yield above the charm resonance region. Above about 7 GeV, however, the rise in the $K^o(\bar{K}^o)$ cross-section follows very well the average increase in charged particle multiplicity over this energy range. In particular, there is no systematic difference below and above the b meson threshold. Clearly it is difficult to disentangle the properties of heavy quarks by looking at the inclusive production of strange mesons.

Finally, I examine the data on inclusive baryon production. Perhaps one of the greatest surprise at PETRA has been the large numbers of baryons produced in e^+e^- annihilations. At present no acceptable theoretical model is able to fully explain the

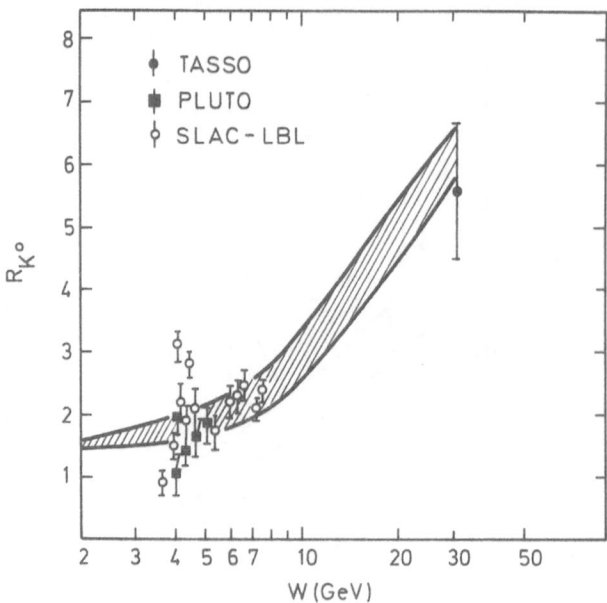

Fig. 13: R(K^O + $\overline{K}^O$) as defined in the text for TASSO
and low energy data from the SLAC-LBL[12) and PLUTO[13)
collaborations. The shaded curve shows the predicted
R from the rise in multiplicity observed over this
region.

inclusive baryon data. It might be expected that the heavy mesons
and baryons containing charm and beauty quarks would give rise to
many baryons simply from phase-space considerations. It would
seem more likely however that most of the baryons observed arise
from the fragmentation process.

Figure 14 shows the differential cross-section s $\frac{d\sigma}{dp}$ for Λ and
$\overline{\Lambda}$ from the TASSO experiment. Also shown are some data from the JADE
collaboration[12) on $\overline{\Lambda}$ production at low momentum and inclusive
(p,$\overline{p}$) data from TASSO from time-of-flight measurements. It can be
seen that all these particles have differential cross-sections
which can be described by an exponential with a common slope
suggesting a common production mechanism. We have seen that
production of heavy quarks can contribute to the increase in
baryon production, but we are forced to consider fragmentation
mechanisms also. I will consider two of the most popular models
below; extended Field-Feynman as implemented by T. Meyer[13) and
the LUND model[14).

a) In this modification of the Field-Feynman approach the
fragmentation scheme is extended by allowing the production of

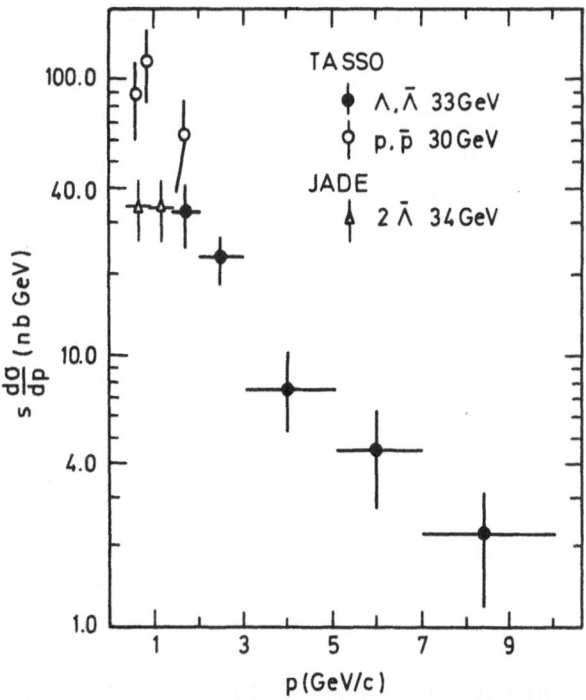

Fig. 14: The differential cross-section for $\Lambda(\bar{\Lambda})$ and $p(\bar{p})$ from TASSO and twice the $\bar{\Lambda}$ cross-section as determined by the JADE group[14].

"diquark lines" as well as the more normal quark lines. This is illustrated in figure 17a. The adjustable parameter used to control the amount of baryon production is

$$\frac{P(q \to q\bar{q} + baryon)}{P(q \to q\bar{q} + baryon) + P(q \to q + meson)} = R_b$$

where P( ) represents the probability for that process to occur. The value of R_b which fits the high energy data best is $R_b = 0.075$. The fit to the Λ ($\bar{\Lambda}$) and p ($\bar{p}$) data is shown in figure 18.

b)   The Lund model is essentially a string model, with the final state hadrons being produced when the string breaks, as shown in figure 17b. Baryon production is introduced by allowing the string to break into a $2q - 2\bar{q}$ diquark pair rather than a q - $\bar{q}$ pair. If one fixes the relative production rate

$$\frac{P(q\bar{q})}{P(q\bar{q}) + P(q)}$$

from low-energy SPEAR data, then the model falls well below the high energy data, as shown in figure 18. However, it should be

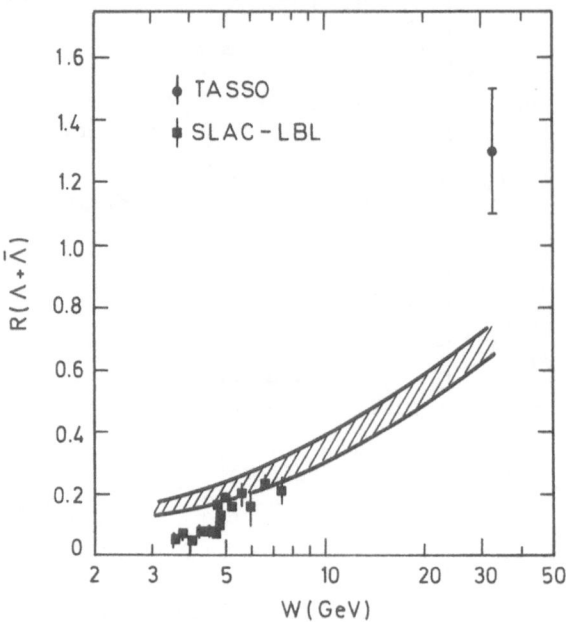

Fig. 15: R($\Lambda+\bar{\Lambda}$) for TASSO at high energies and SLAC-LBL
SLAC-LBL[15] at low energies. The shaded area shows
the R expected from the rise in multiplicity
observed over this energy range.

noted that if the requirement to fit the low-energy data is
removed the LUND model gives a relatively good fit to the proton
and lambda data at high energies.

CONCLUSIONS

Many open questions remain in the study of inclusive products
of e^+e^- annihilation, principally in baryon production. No model is
presently able to explain the baryon production rate over all
energies. An important open question is that of possible baryon-
antibaryon correltaions. More data should solve many of these
problems in the near future.

It is clear, however, that it is very difficult to study the
properties of heavy quarks from inclusive studies. More
information is necessary to achieve this aim. We believe that the
addition of a high precision drift chamber close to the event
vertex will enable us to see decays of short lived particles such
as $D^{\pm}$ and improve the chances of studying heavy quark properties.
I shall discuss this possibility in more depth later in this
conference.

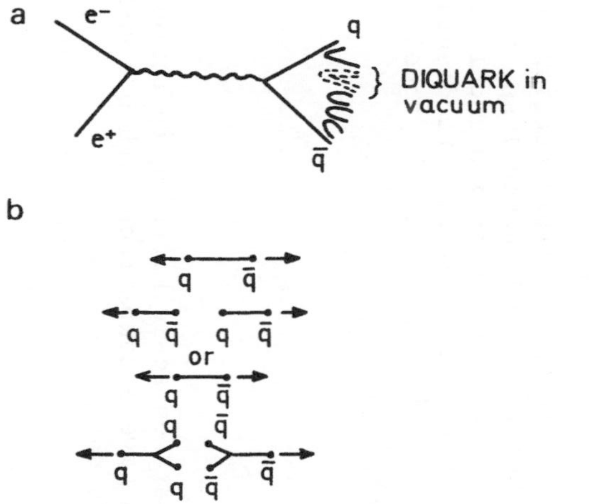

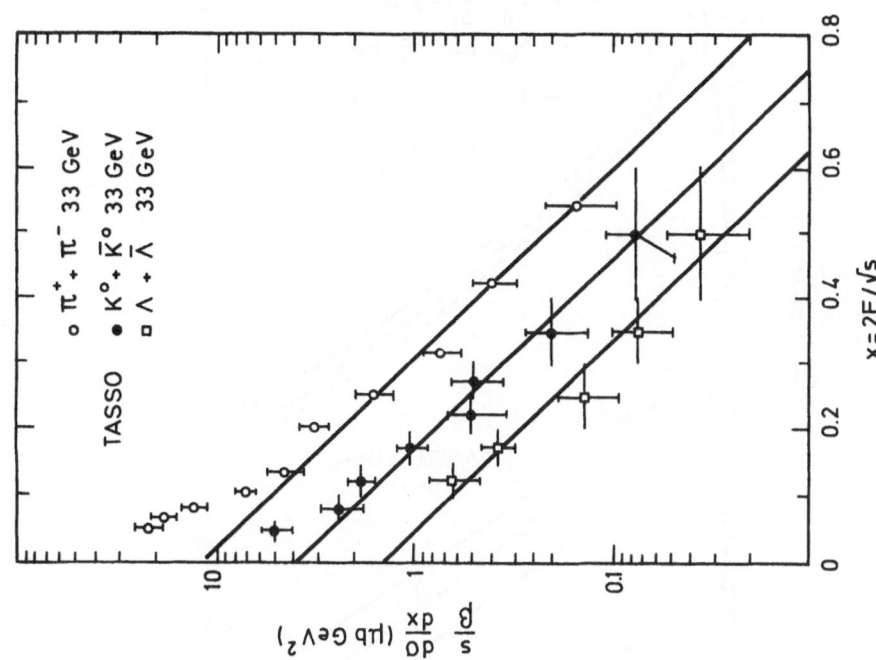

Fig. 16: The invariant cross-section for $\pi^{\pm}$, K^O and Λ production from TASSO. The straight lines are fits to e^{-8x} for $x > 0.1$.

Fig. 17: a) Baryon production proceeding via diquarks in the vacuum.
b) The LUND prescription for meson (upper diagram) and baryon (lower diagram) production via the breaking of coloured strings.

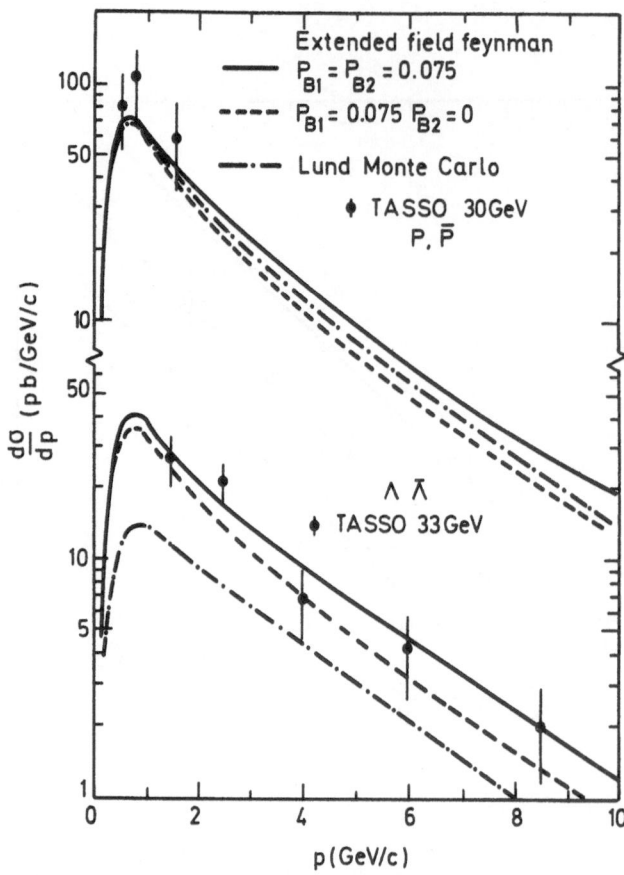

Fig. 18: The differential cross-sections for $p\bar{p}$ and $\Lambda\bar{\Lambda}$ from TASSO. The solid and dashed lines indicate the prediction from the Field-Feynman model with different fractions of baryon production. The dashed-dotted line shows the prediction from the LUND model normalised to low energy data.

DISCUSSION

H. Newman:

Since you showed data on scale breaking in inclusive charged particle production, do you have a comparison with QCD and a value of α_s extracted from this data?

Answer:

Yes, such a comparison exists. I did not show it since I did not think QCD comparisons terribly relevant to this conference.

G. Moneti:

The Lund Model reproduces well the p,p production, but it gives too small $\Lambda,\bar{\Lambda}$ production. This is due to their suppression of strange diquark-antidiquark pairs relative to strange quark-antiquark pairs. If that <u>extra</u> suppression is killed the Lund Model correctly describes the $\Lambda,\bar{\Lambda}$ production as seen in CLEO at ~10 GeV $\sqrt{s}$.

Answer:

Yes. If one ignores SPEAR data, Lund can also be made to fit the TASSO Λ and p data. However, the model does not at present seem able to describe both high and low energy data correctly.

B. Roe:

The distribution of momentum perpendicular to the jet ($p_\perp$) for p and Λ may help clarify some of the differences in production models. If the Λ come from decays of heavy Λ_b, Λ_c then the $p_\perp$ distribution should be significantly broadened. The distribution is also interesting to see if $p_\perp$ increases with mass as in hadron distributions.

Answer:

I agree. We are investigating these questions.

G. Moneti:

Did you see back to back or same side correlations of your baryon-antibaryon pairs?

Answer:

Yes. There is a small number of events with two baryons observed but too few to draw any conclusions as to how these baryons are produced. We must wait for more data.

Acknowledgements

I am very grateful to all my colleagues on TASSO and in particular to Drs. G. Wolf and S.L. Lloyd, for many useful and enjoyable discussions. I thank Frau E. Hell very much for her help with the manuscript.

REFERENCES

1)   The members of the TASSO collaboration are:
      R.Brandelik, W.Braunschweig, K.Gather, F.J.Kirschfink,
      K.Lübelsmeyer, H.-U.Martyn, G.Peise, J.Rimkus, H.G.Sander,
      D.Schmitz, D.Trines, W.Wallraff (Aachen). H.Boerner,
      H.M.Fischer, H.Hartmann, E.Hilger, W.Hillen, G.Knop, L.Köpke,
      H.Kolanoski, B.Löhr, R.Wedemeyer, N.Wermes, M.Wollstadt (Bonn).
      H.Burkhardt, S.Cooper, D.Heyland, H.Hultschig, P.Joos, W.Koch,
      U.Kötz, H.Kowalski, A.Ladage, D.Lüke, H.L.Lynch, P.Mättig,
      K.H.Mess, D.Notz, J.Pyrlik, D.R.Quarrie, R.Riethmüller,
      A.Shapira, P.Söding, G.Wolf (DESY). R.Fohrmann, M.Holder,
      H.L.Krasemann, P.Leu, D.Pandoulas, G.Poelz, O.Römer, P.Schmüser,
      B.H.Wiik (U. Hamburg). I. Al-Agil, R.Beuselinck, D.M.Binnie,
      A.J.Campbell, P.J.Dornan, D.A.Garbutt, T.D.Jones, W.G.Jones,
      S.L.Lloyd, J.K.Sedgbeer, R.A.Stern, S.Yarker (Imperial College
      London). K.W.Bell, M.G.Bowler, I.C.Brock, R.J.Cashmore, R.Carnegie
      R.Devenish, P.Grossmann, J.Illingworth, M.Ogg, G.L.Salmon,
      J.Thomas, T.R.Wyatt, C.Youngman (Oxford). B.Foster, J.C.Hart,
      J.Harvey, J.Proudfoot, D.H.Saxon, P.L.Woodworth(Rutherford).
      E.Duchovni, Y.Eisenberg, U.Karshon, G.Mikenberg, D.Revel,
      E.Ronat (Weizmann). T.Barklow, J.Freeman, P.Lecomte, T.Meyer,
      G.Rudolph, E.Wicklund, Sau Lan Wu, G.Zobernig (Wisconsin).

2)   B. Foster, "The TASSO High Precision Vertex Detector Project",
      these proceedings.

3)   R. Brandelik et al., Phys.Lett. 83B, 261 (1979).
      R. Brandelik et al., Z.Phys. C4, 87 (1980).

4)   H. Boerner, Ph. D. Thesis, University of Bonn, BONN-IR-81-27
      (1981), (unpublished).

5)   R. Brandelik et al., Phys.Lett. 108B, 71 (1982).

6)   R. Brandelik et al., Phys.Lett. 94B, 92 (1980).

7)   R. Brandelik et al., Phys.Lett. 105B, 75 (1981).

8)   R.J. Hollebeek, Proceedings of International Symposium on Lepton
      and Photon interactions at High Energies, Bonn (1981).

9)   R. Brandelik et al., DESY 82-009, submitted to Phys.Lett. B.

10)   D.L. Scharre et al., Phys.Rev.Lett. 41, 1005 (1978).

11)   From G. Wolf, "High Energy e^+e^- Interactions", DESY 81-086 (1981).

12)   W. Bartel et al., Phys.Lett. 104B, 325 (1981).

13)   T. Meyer, Z. Phys. C12, 77 (1982).

14)   B. Anderson et al., University of Lund preprint LU-TP 81-03.

RESULTS FROM CESR

G.C. Moneti

Syracuse University, Syracuse, NY and
CERN, Geneva, Switzerland
On behalf of the CLEO [1] and CUSB [2] Collaborations

INTRODUCTION

I shall report results obtained by the CLEO and CUSB collaboration on the decay of B mesons from T(4S) decay. The characteristics
of decay final states with leptons and/or kaons are consistent with
dominance of the b → c decay mechanism. Some exotic decay models can
be excluded. A summary table of data about T spectroscopy is also
shown.

THE APPARATUS

In CESR, the Cornell Electron Storage Ring, there is only one
bunch of electrons and one bunch of positrons: consequently only two
interaction regions. The South interaction region, in the large,
former Synchrotron experiment hall, houses the CLEO general purpose
magnetic detector. The North interaction region is in a hole more than
a hall, and the Columbia - Stony Brook collaboration (CUSB) was very
smart in designing a calorimetric detector that could fit there.

The design energy of CESR is 16 GeV but for obvious physics
reasons we have been running only at energies between 9.4 and 11.5 GeV.

The CLEO detector has been built by the Cornell, Harvard,
Rochester, Rutgers, Syracuse, Vanderbilt collaboration. Fig.1 shows a
longitudinal view of the detector. Starting from the beam pipe we have
a triplet of proportional chambers with cathode strip readout, followed
by the inner cylindrical chamber made of 17 cylindrical layers of drift
cells. This is enclosed in the magnet coil, formerly a 0.5 T Al coil,
now a 1.0 T superconducting coil. Outside of the coil we have a triplet

of planar drift chambers followed by either an ionization measuring
device (dE/dx counter) or atmospheric pressure Cerenkov counters. We
started running with mostly Cerenkov counters, but last summer all
octants have been equipped with dE/dx counters. After the ionization
measuring device we have time of flight scintillators (TOF) followed by
shower counters made of a sandwich of lead and proportional tubes.
There are other shower counters at the ends of the inner drift chamber
and of the octants, in order to have complete coverage of the solid
angle. Finally we have the iron, partly as magnet yoke, and all of it
as a filter to identify muons that are detected by drift chambers all
the way around and also in slots within the iron.

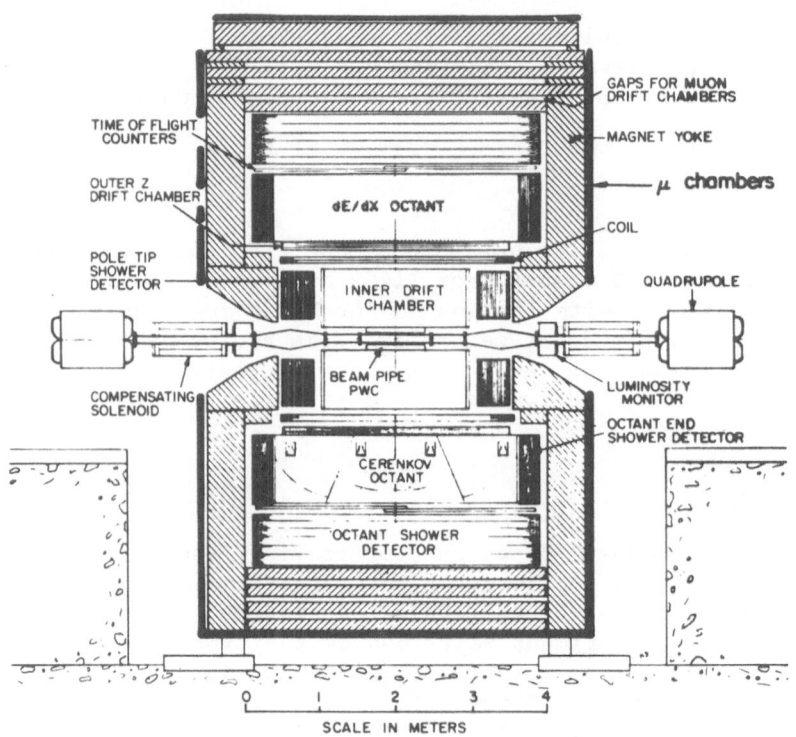

Fig. 1 Cut of CLEO detector along the beam.

Fig.2 shows a reconstructed event in a beam-eye view of CLEO
showing the octagonal symmetry of the detector outside of the coil.

Fig.3 shows a schematic view of the calorimetic CUSB detector
built by the Columbia, Stony Brook, Louisiana State, Max Plank Institute collaboration. There are five layers of NaI detectors, four thin
ones and a thick one, followed by a layer of Pb-glass detectors. The

detector has a square symmetry around the beam and is split in two halves along the beam. End caps calorimters have recently been added as well as a shield of magnetized iron to identify muons and their charge.

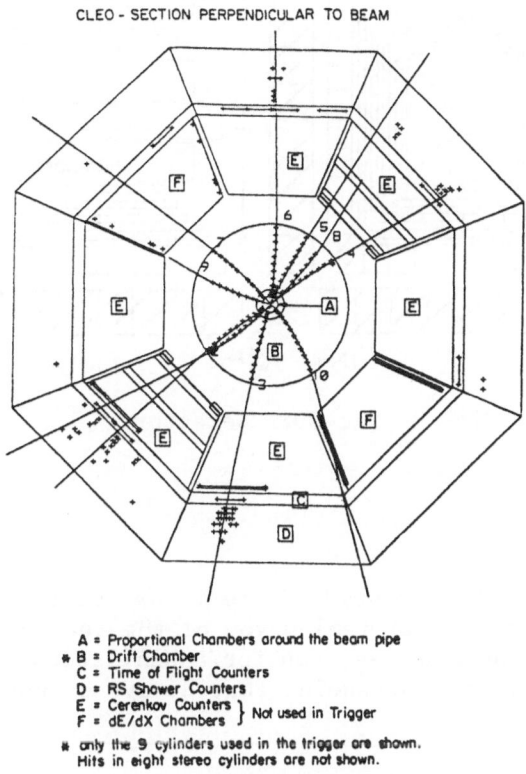

CLEO - SECTION PERPENDICULAR TO BEAM

A = Proportional Chambers around the beam pipe
* B = Drift Chamber
C = Time of Flight Counters
D = RS Shower Counters
E = Cerenkov Counters } Not used in Trigger
F = dE/dX Chambers

* only the 9 cylinders used in the trigger are shown.
Hits in eight stereo cylinders are not shown.

Fig. 2   Beam-eye view of CLEO with reconstructed event.

Around the beam pipe there are 12 planes of drift chambers to detect the direction of all the charged particles and proportional chambers with cathode strip readout are located in between in the NaI chrystals, in order to better determine the direction of the showers.

We can split our experimental results in two categories. One is properties of the bottom flovored (B) meson, the other one is spectroscopy of the T family. I shall talk only of the results in the first category, mainly B meson decay properties, but at the end, I shall try to summarize in one table the essential data we have collected on the spectroscopy of the T family, data that show the validity of the flavor-independent non-relativistic potential model and, more important, provide very useful information and checks for QCD.

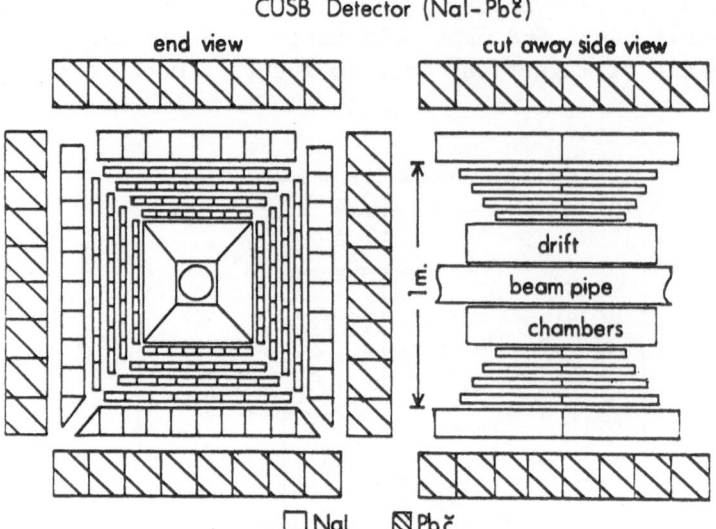

Fig. 3 The CUSB detector.

THE B FACTORY

Fig.4 shows our "B factory". Since it has been already discussed by H. Meyer [3], let me just remind you of the reasons why we believe that this bump in the cross section for annihilation into hadrons is indeed a resonance above threshold, that decays essentially 100% into B̄B.

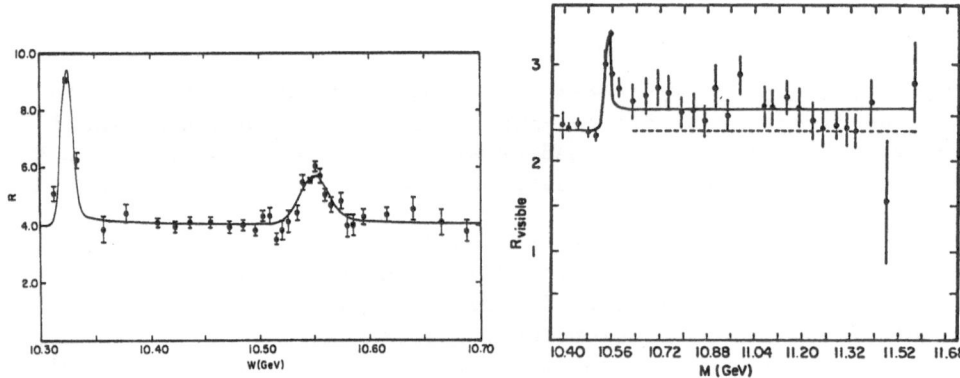

Fig. 4 R_h vs. c.m. energy in the T(3S) and T(4S) region (CLEO)

Fig. 5 R_{vis} vs. c.m. energy above the T(4S) (CUSB)

The first reason is that, after unfolding the instrumental width
due to the beam energy resolution of CESR (FWHM $\simeq$ 10.7 $\pm$ 0.8 MeV),
its total natural width comes out to be 19.1 $\pm$ 1.8 MeV, definitely a
strong decay uninhibited by the OZI rule.

The second one is the increase in charged multiplicity. In the
continuum immediately below resonance the charged multiplicity is
8.2 $\pm$ 0.4 while, after continuum subtraction, the multiplicity of
events coming from the bump is 11.6 $\pm$ 0.4 as expected for the decay
of two objects of mass just above 5 GeV.

The third reason is that the continuum-subtracted shape
distributions (sphericity, thrust, H_2/H_0, π_1, etc..) are
all drastically different from the ones in the continuum and well
fitted by isotropic, phase-space like distributions, again as expected
for the decay of two large mass objects nearly at rest.

More recently we have run at energies just above the $T(4S)$ (10.6
to 11.6 GeV) and, just measuring directly the cross section for
annihilation into hadrons, we have found a step in the ratio $R_h =
\sigma_h/\sigma_{\mu\mu}$. The CLEO value is 0.20 $\pm$ 0.08, while CUSB (Fig.5)
finds 0.29 $\pm$ 0.06. That means that somewhere close to the $T(4S)$
there is the threshold for production of pairs of bottom-flavored
mesons (the expected step is $\Delta R_h \simeq$ 0.38 asymptotically). In fact I
should remark that CLEO sees also a step both in $e^{\pm}$ and $\mu^{\pm}$ production. Attributing such a step to production and semileptonic decay of
B meson and dividing the step in the inclusive lepton production cross
section by the respective B meson branching ratios (to be discussed
later) CLEO gets, from $e^{\pm}$ production ΔR_h = 0.35 $\pm$ 0.20, and from $\mu^{\pm}$
production ΔR_h = 0.49 $\pm$ 0.25, in agreement with the asymptotic expectation.

Finally the bump we are discussing is just where the $T(4S)$ is
expected by the flavor-independent non-relativistic potential model,
that so successfully predicts the masses of the other members of the
T family.

Of course all the results that I shall discuss in what follows
are also a confirmation that at the $T(4S)$ we are observing the
decay of an object into a pair of bottom-flavored mesons.

So, at the $T(4S)$ we have our small "B factory". It can be compared
to the "D factory" at the ψ'' : H. Meyer already pointed out that the
smallness of the $B\bar{B}$ production cross section is compensated by the
higher luminosity of CESR, so that the number of $B\bar{B}$ events/day one can
collect at CESR is comparable to the number of $D\bar{D}$ events/day that one
can collect at SPEAR or DORIS. I would like to add that the "B factory" is substantially inferior to the "D factory" because of signal
to background (i.e. continuum) ratio: that is 1:3 at the $T(4S)$ while
it is 1:1 at the ψ'', thus requiring considerable higher statistics

at the T(4S) to obtain statistical errors comparable to those obtained
at the ψ".

On the subject of statistics and number of events, I would like
to make a remark. During last spring we have been collecting data at
the rate of about 70 cm^{-2} day^{-1}. After the change to the minibeta
configuration we have already reached a rate of about 250 cm^{-2}
day^{-1}, four times the previous one, with hopes of further improvement. Consider that roughly half the data that I shall discuss were
collected in about three months : those three months will be equivalent to three weeks from now on, so that we should be able to
improve our statistics fairly rapidly and make more quantitative many
of the statements that are somewhat qualitative or semi-quantitative
at this stage.

THE B MASS

What do we want to know about the B meson? Obviously the mass,
the spin, the lifetime. Some branching ratios are of specific interest : the semileptonic branching ratios and the inclusive branching
ratio in kaons. They have a direct bearing on the weak interaction
properties of the b-quark and specifically on the outstanding question of the ratio between the (b → c) and (b → u) decay rates.

Furthermore there is a substantial number of non-standard models
of b decay, some of them related to the problem of the existence of
the top-quark, there is also the possibility of Flavor-Changing
Neutral Currents (FCNC) in the b sector, there are the models that
introduce multiplicative quantum numbers, etc..

Finally there is the question : can we observe B^0 – $\bar{B}^0$ mixing?

About the lifetime we have nothing better than what we said a
year ago [4], i.e. that τB ≲ 10^{-10}s, as determined by comparing the
circle of confusion at the vertex for events in the continuum and
events from the T(4S).

On the mass of the B meson we have two handles : (i) the observed
width of the T(4S) tells us that we are at least a FWHM above threshold, and (ii) the lack of its decay into B$\bar{B}^*$ + $\bar{B}$B* tells us that we
are no more than ∿ 50 MeV above the B$\bar{B}$ threshold.

Let us discuss the second point. A simple scaling argument tells
us that

$$M(B^*) - M(B) = \frac{m_c}{m_b} [M(D^*) - M(D)] \simeq 50 \text{ MeV}.$$

This argument has been refined by Martin [5] that sets the limits 52 MeV < M(B*) - M(B) < 57 MeV. Because of this small mass difference, the only possible decay of the B* is B* → B + γ and the γ loboratory energy will be ≃ 50 MeV, since the B* decays nearly at rest. These photons can be detected in the CUSB detector. Fig.6 shows the continuum-substracted photon spectrum from the T(4S). Monte Carlo simulation says that, if the T(4S) decayed 100% into BB* a peak should be observed, like the shaded area in the figure. From the absence of such a peak CUSB can set an upper limit to the branching ratio :

Br [T(4S) → BB* → BB̄γ] < 20%   (90% C.L.).

The detailed calculations of Eichten [6] of the dependence of the T(4S) on its distances from the BB̄ and BB* decay thresholds give then the following limits for the B mass :

20 < M[T(4S)] - 2M(B) < 50 MeV

or

5,249 < M(B) < 5,264 MeV

that can also be expressed as M(B) = 5,257 ± 8 MeV, where the error is not Gaussian, but rather the half width of an allowed interval.

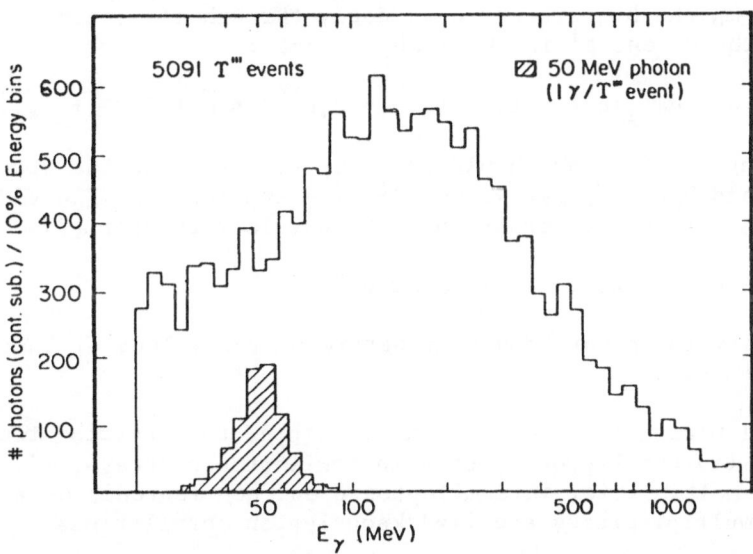

Fig. 6   Inclusive photon spectrum on the T(4S) after continuum subtraction. The shaded area represents the photon spectrum expected for one B* → B + γ decay per T(4S) event (CUSB).

I should remind you that there is also an estimated error of 30
MeV in the absolute c.m. energy scale of CESR. We have just installed
an electron polarimeter that should allow us to eliminate this uncertainty in the near future.

B DECAY PROPERTIES

Let me start the discussion of the B decay properties with a
brief review of the basics of the weak interaction of the b quark
[7]. If we believe in the standard model, the left handed quarks are
in three SU_2 doublets :

$$\begin{pmatrix} u \\ d' \end{pmatrix}_L \quad \begin{pmatrix} c \\ s' \end{pmatrix}_L \quad \begin{pmatrix} t \\ b' \end{pmatrix}_L$$

where d', s', b' are obtained applying the Kobayashi-Maskawa matrix
to the strong interaction eigenstates d, s, b :

$$\begin{pmatrix} d' \\ s' \\ b' \end{pmatrix} = V \begin{pmatrix} d \\ s \\ b \end{pmatrix}$$

Since $m_t \gg m_b$ the b quark can only decay through its contributions to the d' and s' in the weak current :

$$J_\lambda^+ = s_1 s_3 < \bar{u}|\Gamma_\lambda|b > + (c_1 c_2 s_3 + s_2 c_3 e^{i\delta}) < \bar{c}|\Gamma_\lambda|b > + \ldots\ldots$$

Studying the B decay should allow us to determine the two K-M matr
$V_{ub} = s_1 s_3$ and $V_{cb} = c_1 c_2 s_3 + s_2 c_3 e^{i\delta}$. The knowledge of V_{ub} and
V_{cb} based on information other than B decay is very poor [7] :

$$|V_{ub}| = 0.06^{+0.03}_{-0.06} , \qquad |V_{cb}| \leq 0.5 .$$

So, it is very important to better determine the values of these
matrix elements.

Several measurements allow, in principle, to determine the ratio
$|V_{ub}/V_{cb}|$: (i) the lepton spectra in semileptonic decays, (ii)
the charged multiplicity in semileptonic and non leptonic decays,
(iii) kaon multiplicities and (iv) kaon lepton correlations.

In the following I shall assume the validity of the spectator
model for B meson decay. This is shown in Fig.7 : the b quark is
assumed to decay freely and the only function of the other (anti)-
quark is to combine with the decay products of the b quark to produce
the final hadronic system. Based on fairly sound theoretical arguments, the spectator model is supposed to be much more valid for B
decay than for D decay, and our results agree with such arguments.

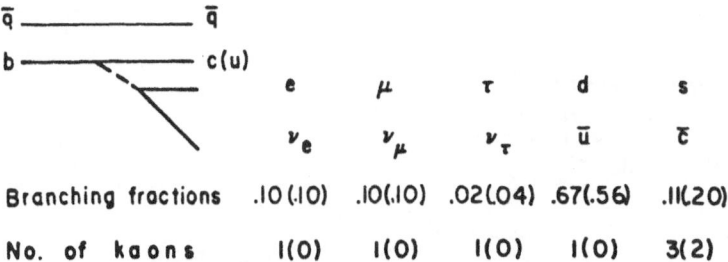

Fig. 7 The spectator model of B meson decay.

SEMILEPTONIC BRANCHING RATIO

In order to measure the semileptonic branching ratio it is
necessary to identify the charged lepton produced directly in the
decay of the b quark, excluding those produced in the subsequent
decay of the c quark or τ lepton.

The direct leptons are selected by requiring that their momentum
be greater than 1 GeV/c. According to (V-A) theory, we expect 77% of
the direct electrons to be above this limit, while only 8% of the
leptons from secondary D or τ decay are so energetic; the contamination due to them is obtained by combining this small fraction with
the respective branching ratios (8.2% for D and 17% for τ) and
comes out very small.

Direct electrons can be detected by CLEO and by CUSB. CLEO has
found about 450 electron at or just below the $\Upsilon(4S)$ energy. They
are identified requiring an electromagnetic shower of the appropriate
energy in correspondence with a track in the inner detector. Furthermore, depending on the type of octant traversed by the particle, we
require either a hit in the Čerenkov counter or an ionization pattern
in the dE/dx detector, consistent with that of an electron. The over-
all acceptance x detection efficiency is 0.17. CUSB selected about
300 electrons requiring the special coincidence of track in the
central drift chamber with a shower, having the characteristic longitudinal energy deposition, in the NaI detectors. Their acceptance x
efficiency is 0.15.

The inclusive visible electron cross section, corrected for contamination but not for detection efficiency, is shown in Fig.8 and 9
and compared with the cross section for annihilation into hadrons.
Both experiments show a strong relative enhancement of the electron
over the hadron yield at the $\Upsilon(4S)$.

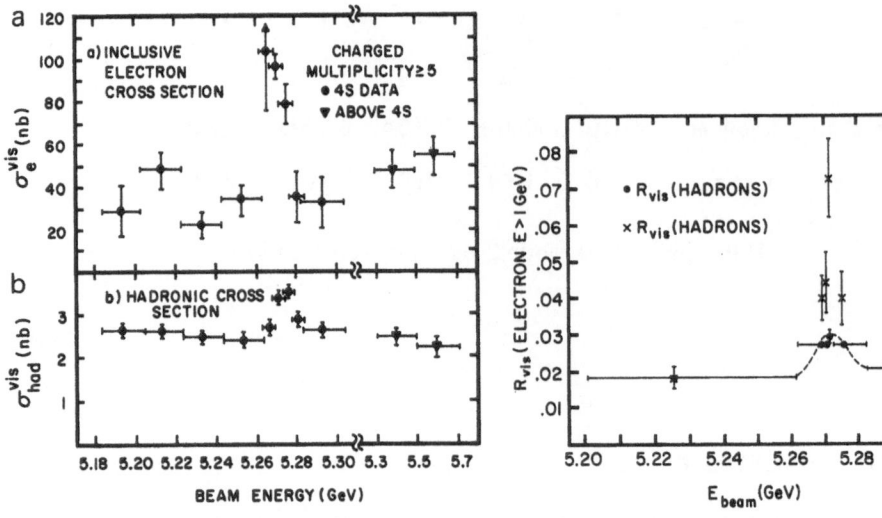

Fig. 8   Inclusive electron (a) and   Fig. 9   Inclusive electron and
         hadronic (b) visible cross   hadronic visible cross
         section vs. c.m. energy   section vs. c.m. energy
         (CLEO)   (CUSB)

Using these data and the estimated electron detection efficiency one can readily obtain the B semileptonic branching ratio into electrons. The CUSB result is

$$Br(B \rightarrow e^{\pm} + X) = (13.6 \pm 2.5 \pm 3.0)\%.\qquad (CUSB)$$

CLEO has chosen to analyse the more recent results seperately from those already published; the old published result [8] is :

$$Br(B \rightarrow e^{\pm} + X) = (13.0 \pm 3 \pm 3)\%,\qquad (CLEO, previous)$$

while the statistically independent new one is :

$$Br(B \rightarrow e^{\pm} + X) = (13.6 \pm 2.1 \pm 1.7)\%\qquad (CLEO, 1981 only)$$

The first error is statistical and the second systematic, as usual.

CLEO can also identify direct muons requiring a spatial coincidence between a hit in the muon chambers outside of the iron and the projected passage of an inner track. 400 muons have been selected and the corresponding inclusive cross section is shown in Fig.10. From this we obtain :

$$Br(B \rightarrow \mu^{\pm} + X) = (10 \pm 1.3 \pm 2.1)\% \qquad (CLEO).$$

The electron and muon branching ratios are statistically compatible with electron universality and can be combined to obtain the branching ratio for decay in one kind of charged lepton ($e^{\pm}$ or $\mu^{\pm}$) :

$$Br(B \rightarrow \ell^{\pm} + X) = (11.8 \pm 1.8)\%$$

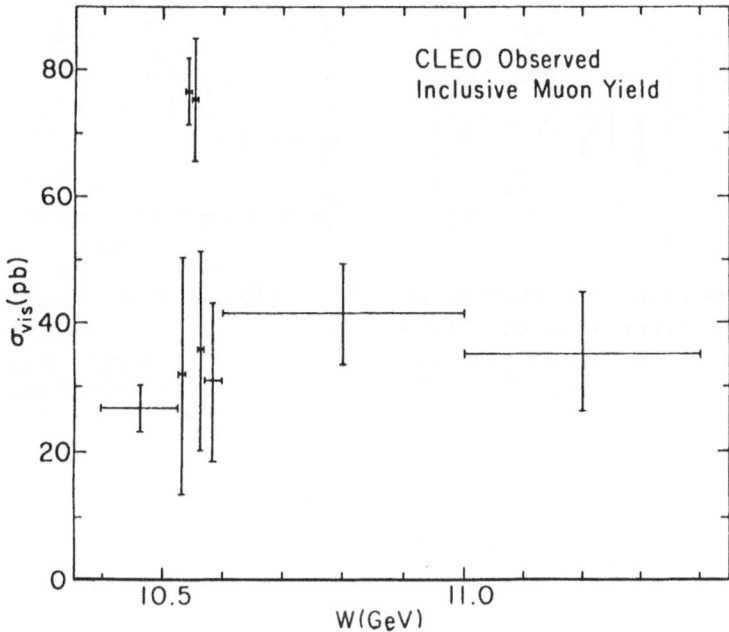

Fig. 10 Inclusive visible muon cross section vs. c.m. energy (CLEO)

LEPTON SPECTRA

The direct lepton spectra are an important tool to distinguish the (b → c) from the (b → u) decay chain. The CLEO and CUSB electron spectra are shown in Fig.11 and the CLEO muon spectra in Fig.12. In all three cases we see the Monte Carlo calculated spectra according to (V–A) decay of the b quark and for various values of the mass of the hadronic system. One can see that a hadronic mass of 1.0 GeV (that expected for the b → u decay chain) is excluded by the data.

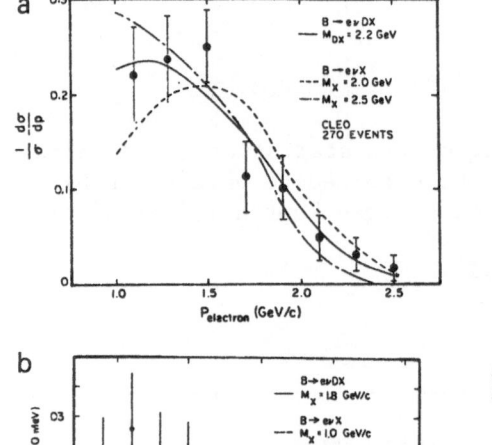

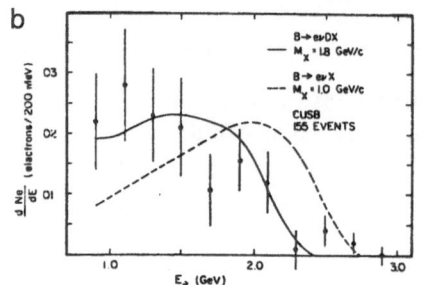

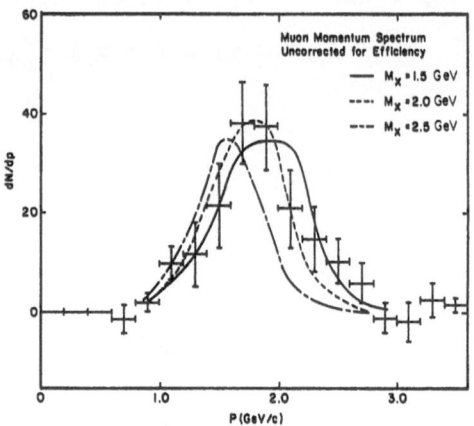

Fig. 11 Momentum distribution of
electrons from the $\Upsilon(4S)$
after continuum subtrac-
tion (a) CLEO, (b) CUSB

Fig. 12 Momentum distribution
of muons from the
$\Upsilon(4S)$ after continuum
subtraction (CLEO)

The best fit to the spectra is obtained with the hypothesis B
$\rightarrow \ell\nu X$ with $1.8 < M_X < 2.2$ GeV or equivalently taking X to be a
mixture of D, D^*, $D\pi$ and $D\pi\pi$: this is just what is expected from the
$(b \rightarrow c)$ decay chain.

A confirmation of the dominance of the $b \rightarrow c$ decay mode comes
from the seperate measurements (CLEO) of the charged multiplicities
in the semileptonic and nonleptonic decays of the B. They are :

$< n_{ch} >/B = 3.5 \pm 0.35$ (semileptonic decays)
$= 6.3 \pm 0.35$ (non leptonic decays)

Subtracting the charged lepton track from the first number, one is
left with 2.5 ± 0.35 i.e. the multiplicity expected from a 50-50

mixture of D and D*. The nonleptonic charged multiplicity is
consistent with an average decay into D + $4\pi^{\pm}$ + $2\pi^0$.

DECAY INTO KAONS

A more quantitative estimate of the relative rate $(b \rightarrow u)/(b \rightarrow c)$
can be obtained from the study of the kaons from B decay.

Charged kaons are identified in CLEO by time of flight, in the
momentum interval 0.5 < p < 1.0 GeV/c; the r.m.s. time resolution is
0.45 ns over a base of $\sim$ 2.4 m. The efficiency varies from 0.11 to
0.04 with increasing momentum; the pion contamination varies, with
the relative population of K$^{\pm}$ and $\pi^{\pm}$, from 10% at the $\Upsilon(4S)$ to 26%
in the continuum. In the most recent data half the useful solid angle
was covered also by dE/dx counters, providing an identification check
on part of the sample of 1400 K$^{\pm}$. Presently all octants are equipped
with dE/dx counters and there will be a double check on all identi-
fications and a reduction of background from hadron interactions in the
coil.

K^0's are identified through their decay into $\pi^+\pi^-$ both in CLEO
and in CUSB. CLEO has selected about 1500 K$^0 \rightarrow \pi^+\pi^-$ with p > 0.3 GeV/c
by requiring vertices more than 7 mm away from the beam line and
that the $\pi^+\pi^-$ effective mass be in the interval m(K^0_s) ± 20 MeV
(see Fig.13). The efficiency reaches 0.23 above p = 0.8 GeV/c and the

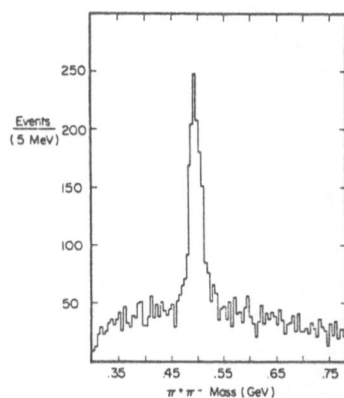

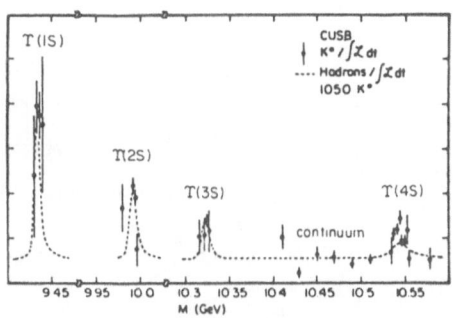

Fig. 13 Mass distribution for
$\pi^+\pi^-$ pairs (CLEO)

Fig. 14 Inclusive visible K^0
and hadrons production
cross sections vs. c.m.
energy (CUSB).

contamination is 27%. CUSB has found ∿ 1300 K⁰ → π⁺π⁻ requiring a
secondary vertex 10 to 60 mm from the beam line and imposing various
cuts on coplanarity and opening angle (in order to avoid lambda contamination). Through Monte Carlo simulation they estimate a contamination of 23% and an efficiency that peaks at 0.09 at p = 0.6 GeV/c.
Fig.14 shows the visible cross section for $K^{\pm}$ and K^0 from CLEO (for
the Υ(4S) and vicinity) and for K^0 from CUSB (at all four Υ resonances
and in the continuum). While at the bound resonances and in the continuum σ_{vis}^K follows σ_{had}, at the Υ(4S) there is a clear excess of
kaons.

Fig.15 illustrates the naïve expectation of the no. of kaons
per annihilation event in the continuum, in $B\bar{B}$ decay through the b
→ c chain, in $B\bar{B}$ decays through the b → u chain. A strong enhancement
of kaon production is expected in the second case. These are naïve
expectations that have been refined by Monte Carlo calculations that,

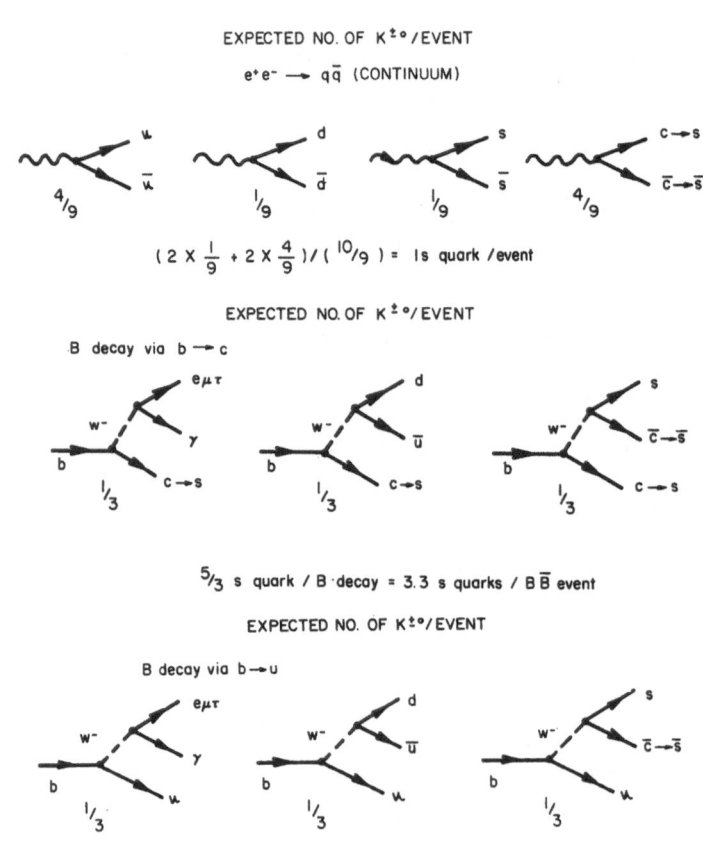

Fig. 15 Expected no of K per event expected per continuum event,
        per B decay via b → c, per B decay via b → u, according to
        the naive spectator mode.

that, among other things, take into account phase space effects. Because of the model dependence of these calculations, we believe that the ratio of expected no. of kaons in $B\bar{B}$ decay to that expected from continuum annihilation can be trusted more than the absolute numbers.

Table I gives the experimental results and compares them with the Monte Carlo expectations.

Focusing on the ratio between $\Upsilon(4S)$ and continuum, one can see that the experimental result is in excess of the Monte Carlo expectation for ($b \to c$) decay. If we then try to describe the data as a superposition of ($b \to c$) and ($b \to u$) decays :

$$R_{DATA} = f\ R_{MC}(b \to c) + (1 - f)\ R_{MC}(b \to u),$$

we get $f = 1.12 \pm 0.25$.

Of course it is necessarily $f < 1$ and from this we can conclude that $|V_{ub}/V_{cb}|^2 < 0.4$ (90% C.L.) the best value being zero.

TABLE I   Measured number of charged and neutral kaons per hadronic event for continuum annihilation and for $\Upsilon(4S)$ decays, and their ratio, compared with Monte Carlo expectations. The errors are statistical only.

	Charge	0.5<p<1.0	0.3<p<3.0	all p	Detector
Continuum DATA	$K^\pm$	0.41 ± 0.07	—	1.12 ± 0.19	CLEO
	K^0	—	0.65 ± 0.04	0.73 ± 0.05	CLEO
	K^0	—	—	0.82 ± 0.05	CUSB
Y(4S) DATA	$K^\pm$	0.82 ± 0.10	—	2.02 ± 0.24	CLEO
	K^0	—	1.13 ± 0.20	1.43 ± 0.25	CLEO
	K^0	—	—	1.52 ± 0.20	CUSB
$R=\dfrac{T(4S)}{continum}$ DATA	$K^\pm$	—	—	1.80 ± 0.32	CLEO
	K^0	—	—	1.96 ± 0.34	CLEO
	K^0	—	—	1.85 ± 0.30	CUSB
Two-jet MC	$(K^\pm+K^0)/2$	0.33	0.80	0.90	
$\bar{B}B(b \to c)$MC	$(K^\pm+K^0)/2$	0.65	1.26	1.60	
$\bar{B}B(b \to u)$MC	$(K^\pm+K^0)/2$	0.30	0.65	0.90	
$R=\dfrac{\bar{B}B(b \to c)}{Two\text{-}jet}$ MC	$(K^\pm+K^0)/2$	—	—	1.78	
$R=\dfrac{\bar{B}B(b \to c)}{Two\text{-}jet}$ MC	$(K^\pm+K^0)/2$	—	—	1.00	

A similar conclusion, of V_{cb} dominance, is reached from the
CLEO analysis of 212 double-kaon events. Out of them 55 ± 16 events
are estimated to come from T(4S) decay. In Table II the experimental results are again compared to Monte Carlo expectations for the
two b-decay hypotheses and once more the b → c channel is strongly
favored.

The same pattern of comparison leading to the normal conclusion
of b → c dominance, comes from the analysis of 107 CLEO events with
a lepton and at least one kaon detected. Table III shows this number
after contamination and continuum subtraction. Combining it with the
kaon multiplicity in T(4S) decays with a lepton, it is possible to
estimate the kaon multiplicities seperately for semileptonic and nonleptonic B-meson decays. The prediction for the semileptonic decays
is quite safe; we hope to improve rapidly the statistical accuracy
and give a reliable limit for $|V_{ub}|$.

TABLE II   Comparison of the number of events with two kaons from
            T(4S) decay, with Monte Carlo predictions for BB decay in
            the hypotheses of b → u and b → c decay mechanism. For
            the second case also the kaon multiplicity has been varied.

	$K_s^0 K_s^0$	$K^{\pm}K^0$	$K^{\pm}K^{\pm}$	TOTAL
DATA (CLEO)	14.6 ± 7.9	25.4 ± 11.4	15.2 ± 8.0	55 ± 16
M.C., b → u, 1.8 K/evt	7.3	11.2	4.3	22.8
M.C., b → c, 3.2 K/evt	11.7	18.0	6.9	36.6
M.C., b → c, 3.9 K/evt	19.0	29.3	11.2	59.5

TABLE III   Comparison of observed kaon-lepton correlations with
             Monte Carlo expectations for the b → c and b → u
             decay mechanisms.

	DATA (CLEO)	M.C. expectation b → c	b → u
No. of K-lepton events in about 2750 B$\bar{\text{B}}$ events	39 ± 11	39	11
$\langle n_K \rangle$ in B$\bar{\text{B}}$ decays with visible lepton	2.5 ± 0.7	2.6	1.0
$\langle n_K \rangle$ /B semi-leptonic decay	0.76 ± 0.74	1.0	0
$\langle n_K \rangle$ /B nonleptonic decay	2.0 ± 0.31	–	–

EXOTIC DECAY MODELS:

Up to this point the results have been interpreted in terms of the standard model. It is now time to dicuss other, more or less exotic, models and see if our results can prove or disaprove their validity. I shall consider the following possibilities :

1) $b \rightarrow q\ell^+\ell^-$ via Flavor Changing Neutral Currents, in connection with "topless" models [9,10] or not.

2) $b \rightarrow q\ell\ell'$, $b \rightarrow q\ell\ell$, $b \rightarrow \overline{q}\ell\ell$, i.e. non conservation of some lepton number (for example via introduction of a multiplicative quantum number [11]).

3) $b \rightarrow qq\ell$, $b \rightarrow \overline{qq}\ell$ as proposed in some "topless" models [12].

4) $b \rightarrow qa^- \rightarrow q + \{ \begin{smallmatrix} \tau & \nu \\ c & s \end{smallmatrix}$ where a^- is a charged Higgs boson with mass $M_\tau < M_a < M_b$ [13].

We can compare the predictions of these non-standard models with three kinds of measurements : (i) the charged energy fraction, (ii) baryon multiplicity and (iii) dilepton production in $B\overline{B}$ decays.

We define the charged energy fraction ρ_c in $\Upsilon(4S)$ decays as the ratio of the sum of the energies of all observed charged particles (considered as $\pi^\pm$) to the c.m. energy : $\rho_c = E_{ch}^{vis}/\sqrt{s}$. We have then produced detailed Monte Carlo simulations of $B\overline{B}$ decays, according to the models, in the CLEO detector, with the following criteria : (i) the models are constrained to produce the semileptonic branching ratios that we have measured, (ii) vary all other available parameters in such a way as to maximize ρ_c.

TABLE IV   Comparison of charged energy fraction predicted, via Monte Carlo simulation, by some exotic models, and the same quantity measured in $\Upsilon(4S)$ decay. The stated uncertainty includes both statistical and systematic errors (roughly 50/50).

Monte Carlo, type 2. exotic B decays	: $\rho_c \leq 0.49$
Monte Carlo, type 3. exotic B decays	: $\rho_c \leq 0.54$
Monte Carlo, standard model	: $0.54 \leq \rho_c \leq 0.62$
DATA, $\Upsilon(4S)$ decay, continuum subtracted	: $\rho_c = 0.60 \pm 0.02$

Preliminary results of these measurements and analyses are
summarized in Table IV; they are in good agreement with the standard
model and they disagree with models of type 2 by five standard deviation and with type 3 by three.

Regarding the decay via charged Higgs boson, there is a correlation between semileptonic decay and charged energy fraction. In
Fig.16 we can see that the measured point is well outside the region
allowed by this model.

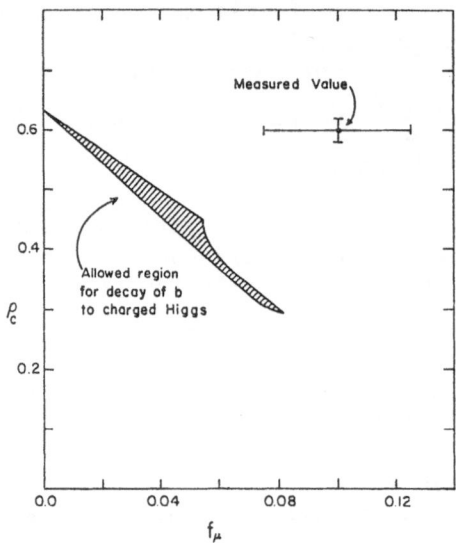

Fig. 16 Charged energy fraction ρ_c vs. apparent semileptonic decay
        branching fraction f_μ, as calculated for the b → ua
        and as measured (CLEO).

Going back to the "topless" model (3. above) that predicts that
all B meson decay into baryons, we have measured the following $\bar{p}$ and
Λ, $\bar{\Lambda}$ baryon multiplicities :

    $2n(\bar{p})/(B\bar{B}$ event$) = 0.29 \pm 0.19 \pm 0.05;$
    $n(\Lambda$ and $\bar{\Lambda})/(B\bar{B}$ event$) = -0.07 \pm 0.10 \pm 0.07.$

These numbers are consistent with zero, and they are far from
the naïve expectations for model type 3.; we are however performing
more Monte Carlo because of the complication due to possible
abundant production of neutrons.

The third type of measurement I mentioned is the number of
$\Upsilon(4S)$ decays with two detected leptons in the final state. CLEO
has observed only 10 such events in the $\Upsilon(4S)$ energy region
($5e^+e^-$ and $5e^\pm\mu^\pm$) (and 2 in the continuum below the $\Upsilon(4S)$). Despite the paucity of the statistics, we can say that this number

agrees with what is expected from the independent semileptonic decay of the B and $\bar{B}$ and we can set a limit to the branching ratio into $\ell^+\ell^-$:

$$Br(B \to \ell^+\ell^-X) < 7.4 \times 10^{-3} \quad (90\% \text{ C.L., CLEO}).$$

From this we can set also the following limit

$$r = \frac{Br(B \to \ell^+\ell^-X)}{Br(B \to \ell\nu X)} < 0.08 \quad (90\% \text{ C.L., CLEO})$$

that is enough to exclude some "topless" models of type 1 [9,10] that require $r \gtrsim 0.12$ if b is a left handed singlet. The possibility remains however of c, b forming a right handed doublet with (V+A) interaction [14].

SUMMARY OF T SPECTROSCOPY RESULTS

As promised at the beginning I shall limit myself to presenting a summary of the many results in Table V and Fig.17.

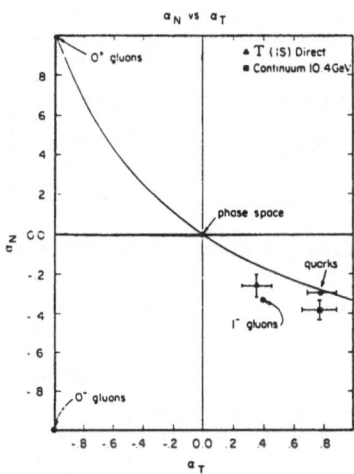

Fig. 17 The predictions for α_T and α_N for $J^P = 1^-$, 0^-, 0^+ gluons and quark jets. The curve represents the case of a normal randomized around a two-jet direction (CLEO).

I shall only remark that the measurement of branching ratios $Br(T(nS) \to \mu^+\mu^-)$ allows us to calculate the total and gluonic width of these bound resonances. We have thus been able to get a fairly accurate determination of the strong interaction coupling constant α_s at about the mass of the T.

Furthermore two different measurements, $Br(T(2s) \to T(1S) \pi^+\pi^-)$ and the angular distributions in the $T(1S) \to 3$ gluon decay agree with the vector gluon hypothesis and disagree with that of scalar or pseudo-scalar gluons.

All the research discussed here was supported in part by various grants of the U.S. Departement of Energy and National Science Foundation. The speaker wishes to express his gratitude to CERN for the hospitality extended to him while this talk was prepared and written.

TABLE V   Summary of results (new or updated) on T spectroscopy

		T(1S)	T(2S)	T(3S)	
M(T(nS)) - M(T(1S)), MeV		-	559.5±0.3	890.7±0.5	CLEO
Γ_{ee} , keV		1.32±0.02±0.10	0.55±0.02±0.04	0.39±0.02±0.03	CLEO
Br(T(nS)→$\mu^+\mu^-$) , %		3.6±0.9 (1)	1.6±1.0 (2)	∿ 1.9	CLEO
		-	18.2±3.2	3.9±1.9	CLEO
Br(T(nS)→$\pi^+\pi^-$T(1S)), %		-	19±7	9.7±4.6	CUSB
	Continuum				
2 < n($\bar{p}$) >	0.22±0.04	0.32±0.07	0.41±0.08	0.38±0.10	(±15%)CLEO
< n(Λ and $\bar{\Lambda}$) >	0.14±0.025	0.23±0.05	0.31±0.03	0.17±0.06	(±20%)CLEO
< n(K) >	1.8±0.2	1.7±0.2	1.7±0.2	2.0±0.2	CLEO
	1.6±0.3	Fig.14	Fig.14	Fig.14	CUSB

Angular distributions in T(1S) → 3 gluon decay, with respect to beam :

	DATA	Theory with | without hadronisation	
Thrust axis	$[1+(0.35\pm0.11)\cos^2\sigma_T]$	0.39	0.30
Normal to plane :	$[1-(0.29\pm0.06)\cos^2\sigma_N]$	-0.33	-0.20

(1) Allows the following determination :

$$\alpha_s(0.48\ M_T)_{\overline{MS}} = 0.152\ ^{+0.020}_{-0.015}\ ,\qquad \Lambda_{\overline{MS}} = 83^{+58}_{-33}\ MeV.$$

(2) Allows determination of :

$$\Gamma[T(2S) \to T(1S)\pi\pi]/\Gamma[\psi(2S) \to \psi(1S)\pi\pi] = \frac{9.3^{+5.6}_{-3.6}}{109} = 0.085 \pm 0.058$$

in agreement with vector gluon prediction (0.10), not with scalar gluons (1.0).

DISCUSSION :

H. Meyer (Wuppertal) : – The data that you showed on ΔR_h above the
T(4S) and below have a rather sharp step just at the position of the
T(4S). Spin 1/2 quarks have a threshold factor $1/2(3\beta -\beta^3)$ with a
rather slow turn on. From a comparison to ΔR_h above charm threshold
the T(4S) appears to be more similar to the $\psi(4.03)$ even more so,
since ΔR at the T(4S) seems to be rather large for a quark with e_q =
1/3. One then becomes suspicious about the behaviour of R_h just above
the 4S: it should have more structure.

Answer :   – You are perfectly right. I showed the cross section
for annihilation into hadrons vs. center of mass energy as measured
by CUSB. The sharp step you mentioned may well be an optical illusion
due to the superimposed fitted line. In fact you can compare the CUSB
results with those of CLEO (Fig.18) and you will see that a sharp
step does not seem to be there at all.

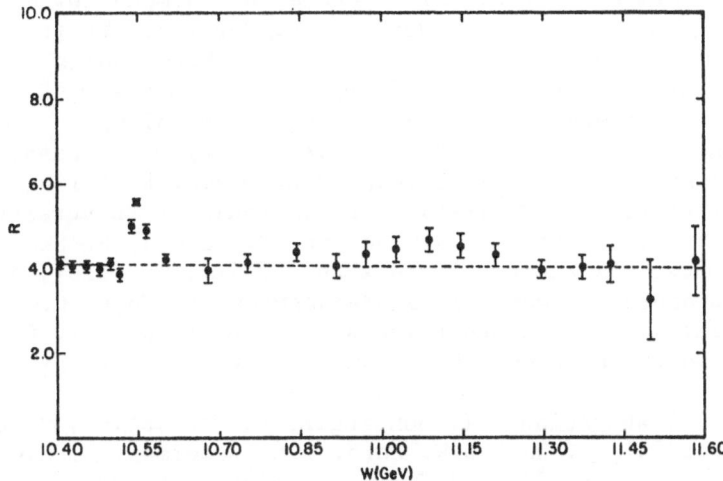

Fig. 18 R_h in the energy range T(3S) to T(4S) (CLEO).

H. Newman (DESY) : – How large are the non-perturbative effects in
the determination of α_s from the branching ratio of $\psi(1S) \to \mu^+\mu^-$?

Answer :   – High order perturbative corrections to the lowest
order expression for the ratio $\Gamma(\psi(1S) \to 3$ gluons$)/\Gamma(\psi(1S) \to \mu^+\mu^-)$
have been calculated in the MS scheme. Neglecting corrections of or
α_s^2/π^2, one can choose to calculate α_s for M = 0.48(T(1S)) : in this
case the first order correction vanishes. The ratio $\Gamma[T(1S) \to$
3 gluons$]/\Gamma[T(1S) \to \mu^+\mu^-]$, that we measure, should not be affected
non perturbative effects.

B. Foster (Rutherford) : – What proportion of (B $\to$ e,μ) branching
ratio come from B $\to$ τ decays? Are there big acceptance differences
betwen these channels?

Answer :   - Our Monte Carlo modelling shows that only 8% of the
electrons coming from secondary D and τ decays have a momentum above
1.0 GeV/c. In the case of the τ this fraction should be multiplied by
the product of branching ratios Br(B → τνX) Br(τ → eνν) that we estimat
to be 0.04 x 0.17 = 0.007. This small contamination has anyway been
subtracted (together with that due to secondary D decays) before
arriving at the quoted B semileptonic branching ratio.

REFERENCES

[1] CLEO Collaboration : D. Andrews, P. Avery, K. Berkelman,
    R. Cabenda, D.G. Cassel, J.S. DeWire, R. Ehrlich, T. Ferguson,
    M.G.D. Gilchriese, B. Gittelman, D.L. Hartill, D. Herrup,
    M. Herzlinger, D.L. Kreinick, N.B. Mistry, E. Nordberg,
    P. Perchonok, R. Plunkett, K.A. Shinsky, R.H. Siemann,
    A. Silverman, P.C. Stein, S. Stone, R. Talmann, D. Weber,
    R. Wilcke, - Cornell University; C. Bebek, J. Haggerty,
    M. Hempstead, J.. Izen, C. Longuemare, W.A. Loomis,
    W.W. MacKay, F.M. Pipkin, J. Rohlf, W. Tanenbaum, R. Wilson -
    Harvard University; A.J. Sadoff - Ithaca College; K. Chadwick,
    J. Chauveau, P. Ganci, T. Gentile, H. Kagan, R. Kass,
    F. Lobkowicz, A.C. Melissinos, S.L. Olsen, R. Poling,
    C. Rosenfeld, G. Rucinski, E.H. Thorndike - University of
    Rochester; J. Green, J.J. Mueller, F. Sannes, P. Skubic,
    A. Snyder, R. Stone - Rutgers University; A. Brody, A. Chen,
    M. Goldberg, N. Horwitz, J. Kandaswamy, H. Kooy, G.C. Moneti,
    P. Pistilli - Syracuse University; M.S. Alam, S.E. Csorna,
    A. Fridman, R. Hicks, R.S. Panvini - Vanderbilt University.

[2] CUSB Collaboration : T. Böhringer, F. Costantini, J. Dobbins,
    P. Franzini, K. Han, S.W. Herb, L.M. Lederman, G. Mageras,
    D. Peterson, E. Rice, J.K. Yoh - Columbia University;
    G. Finocchiaro, J. Lee-Franzini, G. Giannini, R.D. Schamberger
    Jr., M. Sivertz, L.J. Spencer, P.M. Tuts - The State University
    of New York at Stony Brook; R. Imlay, G. Levman, W. Metcalf,
    V. Sreedhar - Louisiana State University; G. Blanar, H. Dietl,
    G. Eigen, E. Lorenz, F. Pauss, H. Vogel - Max Planck Institute
    of Physics.

[3] H. Meyer, this Conference.

[4] K. Berkelman, in "Proceedings of the XXth International
    Conference on High Energy Physics", Madison, WI (July 1980)

[5] A. Martin, Phys. Lett. 103B (1981) 51.

[6] E. Eichten, Phys. Rev. D22 (1980) 1819.

[7] For review of the theory of B decays see, e.g. M.S. Chanowitz
     in "Proceedings of the International Symposium on High Energy
     e^+e^- Interactions", Vanderbilt University, May 1-3 1980;
     D. Hitlin in Proceeding of SLAC Summer Institute, August 1980;
     J.P. Leveille, University of Michigan Preprint UM HE 81-18,
     March 1981.

[8] C. Bebek et al., Phys. Rev. Lett. 46 (1981) 84;
     K. Chadwick et al., Phys. Rev. Lett. 46 (1981) 88.

[9] V. Barger and S. Pakvasa, Phys. Lett 81B (1979) 195,
     G.L. Kane and M.E. Peskin, University of Michigan Preprint UM
     HE 81-51.

[10] G.L. Kane, University of Michigan Preprint UM HE 80-18.

[11] E. Derman, Phys. Rev. D19 (1979) 317.

[12] H. Georgi, S. Glashow, Nucl. Phys. B167 (1980) 173. See also :
     H. Georgi, M. Machacek, Phys. Rev. Lett. 43 (1979) 1639.

[13] E. Golowich, T.C. Yang, Phys. Lett. 80B (1979) 245;
     L.N. Chang, J.E. Kim, Phys. Lett. 81B (1979) 233;
     C.H. Albright, J. Smith, H. Tye, Phys. Rev. D21 (1980) 711.

[14] H. Tye, M.E. Peskin, Private communication.

CHARM AND BEAUTY PHOTOPRODUCTION AT FERMILAB

Jeffrey A. Appel

Fermi National Accelerator Laboratory
Batavia, Illinois 60510

INTRODUCTION

Four manifestations of Charm have been observed in photoproduction at Fermilab so far. These four are (1) multimuon indications of the total Charm cross section and observations of (2) Ψ and Ψ', (3) D^O and D^* and (4) Λ_c. The relevent photoproduction experiments in the search for Charm at Fermilab are the broad band neutral beam experiments by a Columbia-Fermilab-Illinois (CFI) collaboration[1], the Tagged Photon Beam experiment by the TPS collaboration[2] and the muon beam experiment with an active iron target by the Berkeley-Fermilab-Princeton (BFP) collaboration.[3]

The photon beam experiments have similar forward multiparticle spectrometers. There are important differences among the two experiments, however. These include differences in the beams (hadron contamination, energy and flux) and detectors (solid angle acceptance of the forward spectrometers, target system instrumentation and trigger capability). Both experiments include a two-magnet forward spectrometer system with two Cerenkov counters for charged particle identification. Wire chamber systems are installed as far upstream as possible allowing large angle acceptance for those particles which pass through the first magnet, but do not continue all the way through the downstream spectrometers. The Tagged Photon beam energies used during last year's run by the TPS collaboration range from 50 to 150 GeV with no hadron contamination. The energy of each incident photon was known to a few percent. In the broad band neutral beam, the energy and intensities were higher. However, the 1% hadron contamination in the beam resulted in half the event triggers and serious backgrounds for some of the interesting physics. On the other hand, the average energy for observed charm events was typically around 150 GeV. In measuring Ψ photoproduction, the Fermilab-Illinois part of the CFI collaboration used liquid hydrogen and liquid deuterium targets while for the charmed particle production the full CFI group used a segmented scintillator target. The TPS experiment used a one and a half meter long liquid hydrogen target for all measurements.

In the TPS, the acceptance reaches out to approximately ± 150 milliradians in the verticle and horizontal directions, while in the broad band neutral beam, acceptances are just a little more than half this size. In order to make use of the larger angular acceptance, the TPS Cerenkov counters have 20 cells in each, both upstream and downstream detectors, while the CFI chambers had 12 and 16 cells respectively. The steel absorber in the CFI spectrometer is divided longitudinally to allow insertion of hodoscopes. These were used in identifying muons and crudely projecting them back to the target for trigger purposes. A typical event in the TPS is shown schematically in Figure 1. A rather clean multiparticle event is seen in the forward spectrometer and a single recoil is seen in the system of

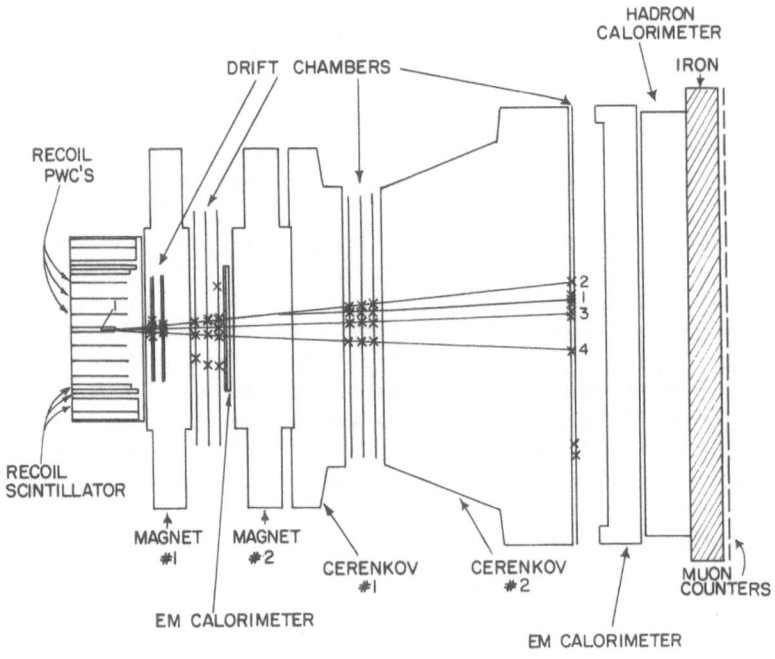

Figure 1. Schematic plan view of the tagged photon spectrometer showing a multiparticle event.

cylindrical proportional wire chambers and scintillators surrounding the target. This system allows identification of protons and measurement of the kinetic energy up to approximately 0.5 GeV. The only directly measured photoproduction data presented here comes from the broad band neutral beam experiments. Results from the TPS collaboration should start appearing soon. The multimuon spectrometer of the BFP collaboration has been discussed in detail elsewhere[4] and will not be reviewed here.

CHARM TOTAL CROSS SECTION

In the Berkeley-Fermilab-Princeton muon experiment, the measurements are extrapolated to zero in q^2 of the virtual photon. This procedure provides equivalent photoproduction results from the muon scattering experiment. In the BFP experiment, the muon beam strikes a magnetized active iron target and the relevent charmed photoproduction results are derived from multi-muon events. The muons, other than the fast forward beam sign muons, are assumed to come primarily from decays of Charm in the dense target. We leave the bulk of the discussion of the total cross section measurements to later talks at the Conference. In summary, however, the measurements by the Berkeley-Fermilab-Princeton group show a slightly rising Charm production (approaching 1 μb per nucleon) by photons as the energy rises. As the experimenters note, this rise does not saturate the rise in the total photon cross section.[5]

CHARMONIUM

The Ψ and Ψ' measurements reported here were made by the Fermilab-Illinois part of the CFI group. The Ψ' results are new.

The experimental observation of rather clean dimuon and dielectron signals centered at the appropriate mass for the J/Ψ allows the group to make measurements as a function of energy. The energy is obtained from the observed Ψ energy. Only two track events were used, thereby excluding only 5% of events seen with extra tracks. Additional multiparticle events were vetoed by wide angle counters just downstream of the target and a requirement of ≤ 6 forward particles in the trigger. In the diffractive events, the incident photon energy is equal to the dilepton energy. The assumptions used in arriving at the final results are given in Table I. The decay to electron pairs and muon pairs is observed at the same rates ($\sigma_\Psi B_{ee}/\sigma_\Psi B_{\mu\mu}$ = .98 ± .06). The hydrogen and deuterium σB data both show a linear increase for photon energy between 60 and 300 GeV with a slope of (.007 ± .0014) nb per GeV. Most of the data is on deuterium. Yet enough hydrogen data was taken to see that except for the coherent deuterium peak near t=0, the t slopes are consistent with each other. The deuterium data have t shapes which are essentially energy independent.

TABLE I

CROSS SECTIONS WERE COMPUTED USING A MONTE CARLO PROGRAM
TO CALCULATE THE SPECTROMETER ACCEPTANCE.

ASSUMPTIONS:

1. A $1+\cos^2\theta$ DECAY ANGLE DISTRIBUTION CONSISTANT WITH
   THE DATA.

2. AN EXPONENTIAL DEPENDENCE ON THE FOUR MOMENTUM TRANSFER
   SQUARED, t, WITH A SLOPE B OF -4 GeV^{-2}. VARIATION
   OF B FROM 60 TO 1 CHANGED OUR ACCEPTANCE BY LESS THAN
   20% AND PRODUCED NO APPRECIABLE E DEPENDENCE.

PROGRAM INCLUDED EFFECTS OF:

1. BEAM SIZE (2″ SQUARE).
2. TARGET LENGTH (5% OF INTERACTION).
3. CHAMBER INEFFICIENCIES (2% PER TRACK).
4. TRIGGER INEFFICIENCIES.
5. ELECTRON BREMSSTRAHLUNG.
6. GEOMETRIC ACCEPTANCE.

YIELDS CORRECTED FOR:

1. ELECTRONICS DEAD TIME (17%).
2. ACCIDENTAL MUON HALO VETOES (10%).
3. TRIGGER COUNTER INEFFICIENCIES (3%).
4. BETHE-HEITLER BACKGROUND (5%).
5. BRANCHING RATIOS.

The experimenters studied the character of the target recoil for the 95% of the Ψ events which appeared to be diffractively produced as seen in the forward spectrometer. Two layers of scintillation counters and one layer of lucite Cerenkov counter were placed around the target for this purpose. As with photoproduction of rho's, the Fermilab-Illinois group found 70% of the events consistent with elastic scattering where a proton and only a proton appeared where it should or was absent where it shouldn't have gotten out of the target. These events are called elastic. The other 30% of the events, called quasi-elastic, either had additional hits or a recoil particle in a non-coplaner recoil element. The elastic events were produced with dσ/dt proportional to exp(-2.8t) while the quasi-elastic cross section was proportional to exp(-1.3t). All these results refer to the diffractively produced ψ's. The Berkeley-Fermilab-Princeton and EMC muon experiments report significant non-diffractive production of Ψ's. They both observe a Ψ total cross section roughly consistent with that seen by the Fermilab-Illinois group. However, defining non-diffractive events (called inelastic by these groups) as those which have greater than 4-1/2 to 5 GeV of visible energy in the calorimeter target, the BFP and EMC groups see half the production as non-diffractive.[6]

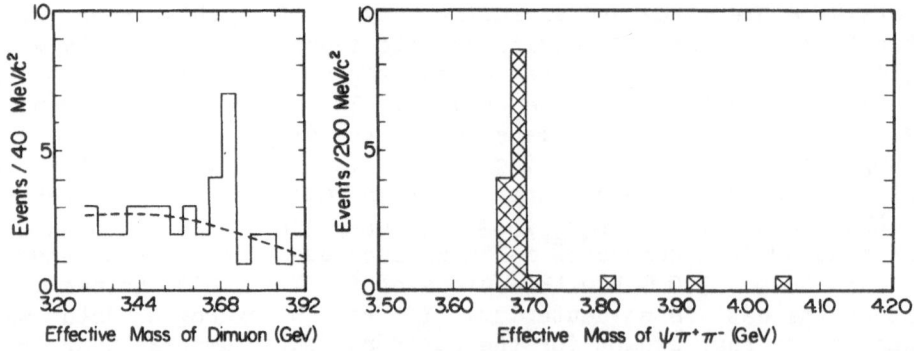

Figure 2. Effective mass distributions showing ψ' events from the Fermilab-Illinois collaboration.

The Fermilab Illinois group sees Ψ' to two muon and to two electron peaks at 3.68 GeV/c^2 sitting on the tails of dileptons from the Ψ. An even cleaner sample of events consistent with the Ψ' decaying to $\Psi\pi\pi$ is shown in Figure 2. Notice that there are only a few events consistent with non-diffractive Ψ production. The Ψ' results are summarized in Table II. Note, too, that the Ψ' photoproduction cross section at an average energy of 160 GeV is about 20% of that of the Ψ at a similar energy. More significantly, using the value $d\sigma/dt$ at $t = 0$ in a vector dominance model, $\sigma_{\Psi'p}$ and $\sigma_{\Psi p}$ are about the same size. This makes it reasonable to consider the Ψ', like the Ψ, in the family of hadrons as required to consider them related to the open Charm which is discussed next.

TABLE II

Ψ' RESULTS

$\sigma_{\gamma p \to \Psi' p} \ (\Psi' \to e^+e^-) = 6.8 \pm 3.4$ nb

9 EVENTS

$\sigma_{\gamma p \to \Psi' p} \ (\Psi' \to \mu^+\mu^-) = 5.2 \pm 2.8$ nb

7 EVENTS

$\sigma_{\gamma p \to \Psi' p} \ (\Psi' \to \Psi\pi^+\pi^-) = 6.0 \pm 1.5$ nb
$\phantom{\sigma_{\gamma p \to \Psi' p} \ (\Psi' } \hookrightarrow \mu^+\mu^-$

14 EVENTS

AVERAGE $= 6.0 \pm 1.3$ nb

AT AN AVERAGE ENERGY OF 160 GeV

$\sigma_{\gamma p \to \Psi' p} \simeq 30$ nb AT SAME ENERGY

$\dfrac{\sigma_{\Psi'p}}{\sigma_{\Psi p}} = .7 \pm .32$

CHARMED MESONS

We will concentrate here on the recent D* production data and then look at two unusual branching modes of the D^0.

In order to see the D^0, it is necessary to look at the D^0's which result from the decay of the D*. Figure 3 shows two decay modes which have consistent and correct value D^0 mass peaks. In order to achieve these, a cut has been made on the low D*-D mass difference. Figure 4 shows the enhancement at the appropriate Δ ($\Delta = M_{K\pi\pi} - M_{K\pi}$) value in the D* $\to$ Dπ decay for those events which have the correct D^0 mass. It was by going back and cutting on this Δ value that gave the peaks in the previous figure.

The D* data are consistent with a pair production mechanism. Equal

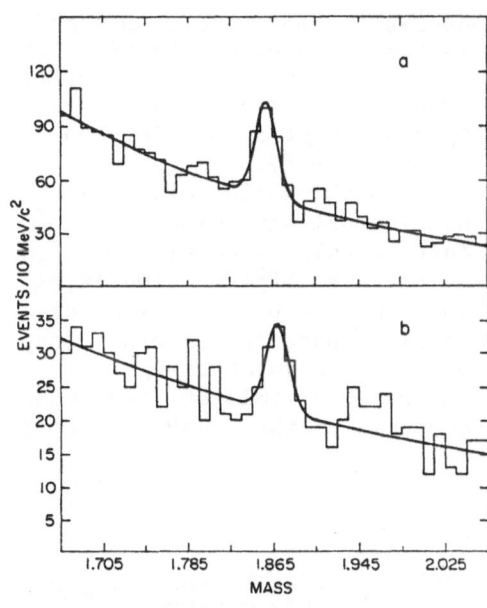

Figure 3. The (a) $\overline{K}^{\mp}\pi^{\pm}$ and (b) $K^0\pi^+\pi^-$ invariant mass distributions obtained by the Columbia-Fermilab-Illinois Collaboration for combinations within the D*$^{\pm}$ mass difference peak shown in Fig. 4b.

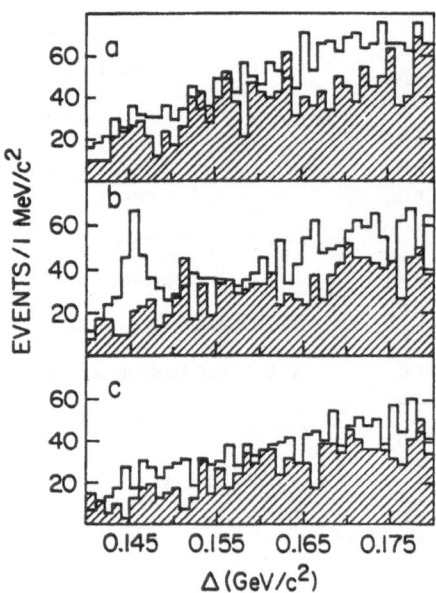

numbers of D*+ and D*- were seen in the experiment. The energy, x, P_t and multiplicity distributions are all consistent with diffractive production. Furthermore, 35 ± 9% of the D*'s are produced with an additional charged K, consistent with an additional decaying D. From a study of the spectrum of π's produced with D*'s, the CFI group concludes that 45 ± 25% of the D*'s are produced with another D*. Both of these statistically poorly measured results are very suggestive of nearly pure diffractive production of the D*'s.

Finally, the CFI collaboration reports results on the Cabbibo structure of D decays. They observe $D \rightarrow K^+K^-$ with a branching fraction of 20± 9% (compared to the SPEAR result of 11 ± 3%). The prediction of simple Cabbibo theory is 4.6%. This collaboration, therefore, is not inconsistent with the surprisingly large value observed at SPEAR. Finally, the limit on D^0 $\bar{D}^0$ mixing (or doubly suppressed Cabbibo decay mode of irregular charge

Figure 4. Mass difference distributions ($\Delta \equiv M_{K\pi\pi}-M_{K\pi}$) obtained by the Columbia-Fermilab-Illinois Collaboration for combinations with a Kπ mass (a) below, (b) straddling, and (c) above the known mass of the D^0. Both charm and anticharm states are included in this plot. The shaded distributions show the appropriately normalized contributions from hadronic contamination in the photon beam.

combinations of Kπ) gives a branching fraction limit of less than 11% at the 90% confidence level (compared to the SPEAR result of <16%).

CHARMED BARYONS

Average properties of the Λ_c events observed by the CFI group are listed in Table III. The properties they observed for the charmed baryons are listed in Table IV. Notice the approximately equal numbers of both charge states in the decay modes pK and p̄K, of Figure 5. This again suggests diffractive pair production.

Limits on other decay modes of the Λ_c are given in Table V as derived from the data shown in Figure 6. The shaded areas in Figure 6 represent the events produced by the hadronic contamination in the broad band beam. This background is clearly

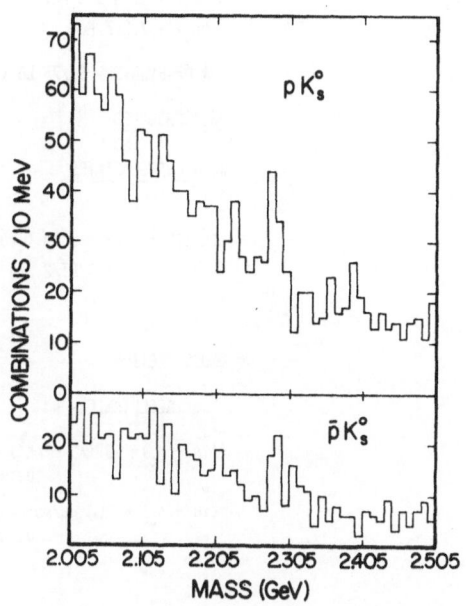

Figure 5. Λ_c search mass distributions.

limiting the results and one may hope for better information from
the tagged photon beams in the not too distant future.

TABLE III

AVERAGE PROPERTIES OF Λ_c EVENTS

(SIGNAL/NOISE $\simeq$ 2/3)

$<E_\gamma>$ = 165 GeV

$<t>$ = .48 GeV^2

$<E_{\Lambda_c}>$ = 83 GeV

$<P_{\perp,\Lambda_c}^2>$ = .49 GeV^2

$<E_{K_s}>$ = 40 GeV

$<\# \text{ PARTICLES}>$ = 5.2

(CUT 4-7)

$<E_{\Lambda_c}/E_\gamma>$ = 0.52

$<\text{TOTAL MASS}>$ = 5.1 GeV

$<\text{RECOIL MASS}>$ = 2.5 GeV

TABLE IV

CHARMED BARYON PROPERTIES

$M(pK_s)$ = 2284±1±5 MeV

$\Gamma(pK_s)$ = 7.2±3 MeV

OF EVENTS = 55/75 BG (~6σ, > 8σ FROM FIT)

PHOTOPRODUCED

BOTH CHARGE STATES $\dfrac{p^+K_s}{p^-K_s} \simeq 1$

LIFETIME: $\gamma c\tau$ < 3 CM (WOULD HAVE SEEN DECAY)

$<\gamma> \simeq 83/2.284 = 36$

$\tau < 3\times10^{-12}$ SEC

CROSS SECTION

ACCEPTANCE $\simeq$ 4%

$\sigma(\gamma N \rightarrow \Lambda_c) \times BR(\Lambda_c \rightarrow pK^o)$ = 3.0 ±1 nb/NUCLEON
USING LINEAR A DEPENDENCE

$\sigma(\gamma N \rightarrow \Lambda_c)$ = 200 nb/NUCLEON
USING BR = 1.5%

TABLE V

BRANCHING RATE LIMITS
90% C.L.

$$\frac{BR(\Lambda_c \rightarrow \Lambda\pi^+)}{BR(\Lambda_c \rightarrow pK^o)} < 0.3$$

$$\frac{BR(\Lambda_c \rightarrow \Lambda\pi^+\pi^+\pi^-)}{BR(\Lambda_c \rightarrow pK^o)} < 3.1$$

$$\frac{BR(\Lambda_c \rightarrow pK^-\pi^+)}{BR(\Lambda_c \rightarrow pK^o)} < 1.4$$

$$\frac{BR(\Lambda_c \rightarrow pK^o\pi^+\pi^-)}{BR(\Lambda_c \rightarrow pK^o)} < 3.3$$

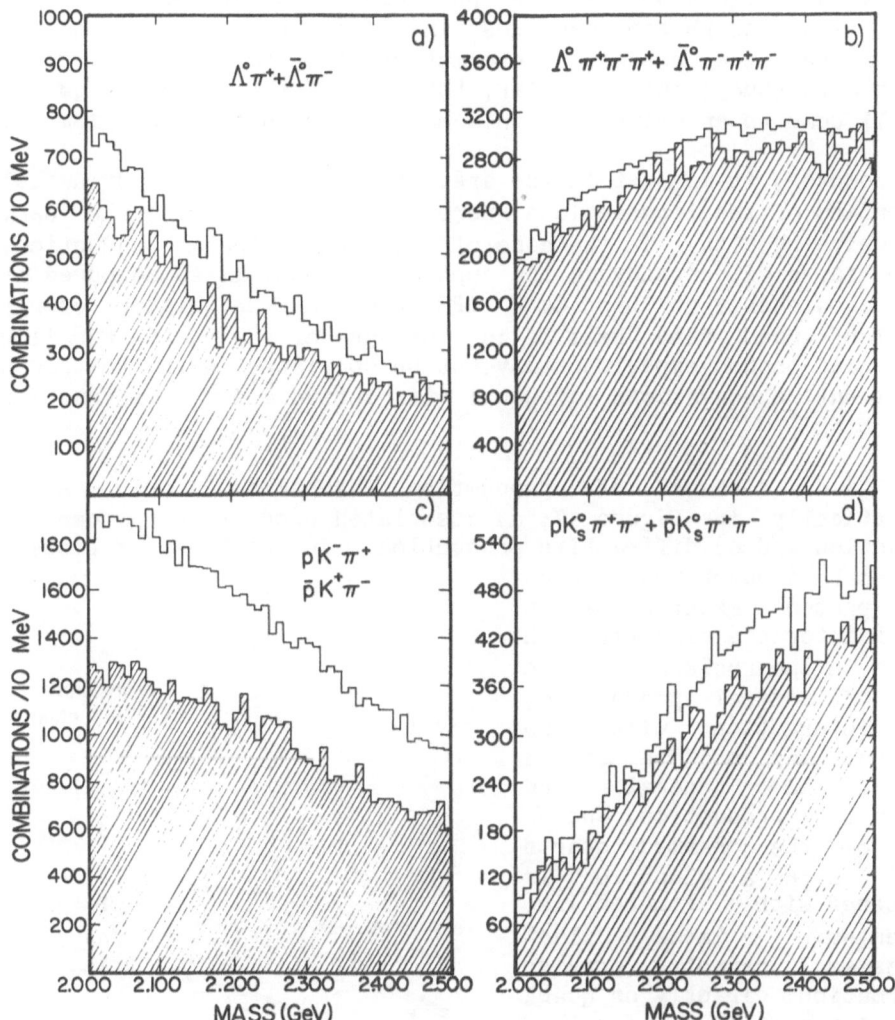

Figure 6. Effective mass distributions for various particle combinations used in the search for charmed baryons. The shaded distributions show the appropriately normalized contributions from hadronic contamination in the photon beam.

LESSONS

What do we learn about heavy quarks from these photoproduction results? First and perhaps most fundamentally, the Ψ and Ψ' particles behave like hadrons. In addition to this simple identification, it is possible to study the details of quark, gluon and photon dynamics involving the charmed quark. Furthermore, the fundamental parameters of dynamical theories are amenable to study; in particular, the mass of the charmed quark. We will come to an example of this mass determination in a moment.

However, it is just in the area of determining the dynamical properties of the interactions where experiments provide the least conclusive results. In the case of the open Charm production, there is a discrepancy in the apparent mechanisms as measured at low photon energy at the CERN SPS and in the higher energy Fermilab broad band beam. In the case of the Ψ, the Fermilab broad band photon experiment may not be consistent with the muon experiment observations of equal diffractive and non-diffractive production.[6]

Three classes of production mechanisms are shown schematically in Figure 7; a) associated production, b) central production and c) diffractive production. In the first of these, the charmed quark produced at the photon materialization vertex interacts directly with quarks in the nucleon. In the case of central production, the interaction with the nucleon may not be via a quark. In central and diffractive production, one or more of the charmed quarks fuses with a gluon or is scattered with a Pomeron-like mechanism. These last processes cannot provide information directly on quark quark interactions.

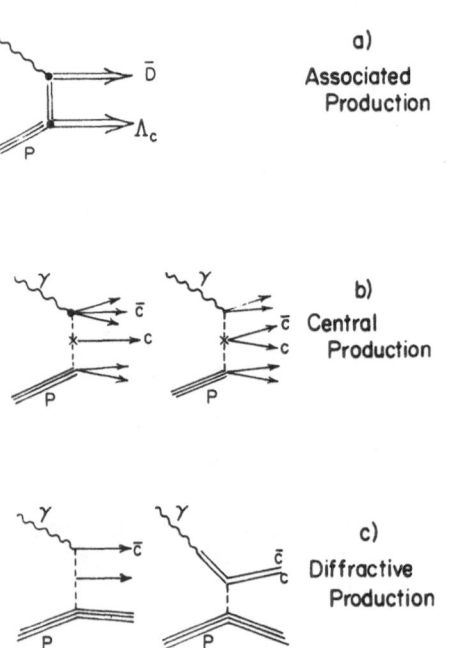

a) Associated Production

b) Central Production

c) Diffractive Production

Figure 7. Charm Production Models.

In order to determine physically interesting results, we expect to compare measurements to one or another of the processes represented in Figure 7. However, the experimental discrepancies get in the way. The first discrepancy may well be accounted for by the differences in the energies at which the the Charm production was observed. Figure 8 shows the Ψ production cross section threshold behavior as observed in experiments. In addition to the appropriate energy for these measurements, other energy scales are indicated on the abcissa. They are scaled by M^2, the mass squared of the indicated states. It is clear that within the statistics of the actual measurements, the threshold behavior of producing the heavier $D\bar{D}$ state and heavier still $\Lambda_c\bar{\Lambda}_c$ states may account for the apparent discrepancies in the production mechanisms. Not only would this explain the behaviors

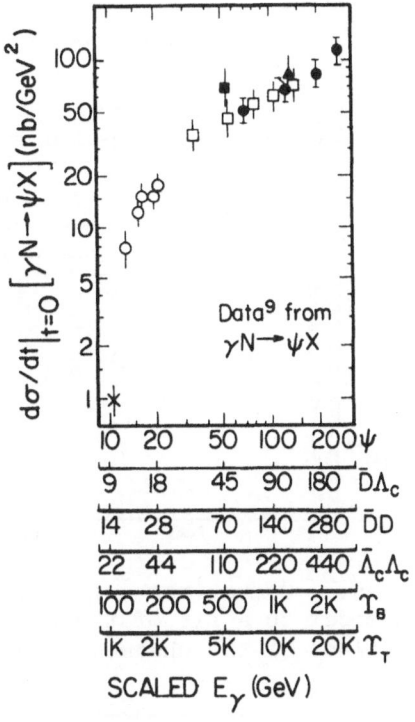

Figure 8. Ψ photoproduction threshold data shown with various energy scales.

observed, but it may allow us to obtain additional information from photoproduction by examining the behavior at different energies.

Once descrepancies are resolved, interesting physical quantities can be determined. As an example, the photoproduction of Ψ's is sensitive to interesting physical parameters. Figure 9 provides the threshold and higher cross sections for various charmed quark mass values in the photon gluon fusion model[7] of Figure 7c. As another example, Figure 10 gives the p_t^2 distribution of D*'s in the CFI data and compares it to the predictions of the photon gluon fusion model for three different values of $c\bar{c}$ primordial p_t. Additional information required to achieve the agreement is a soft gluon distribution function of the form:

$$F_g = \frac{(1-x)^5}{x}$$

and a dressing function of the form:

$$D_c(z) = \exp(-5.5z)$$

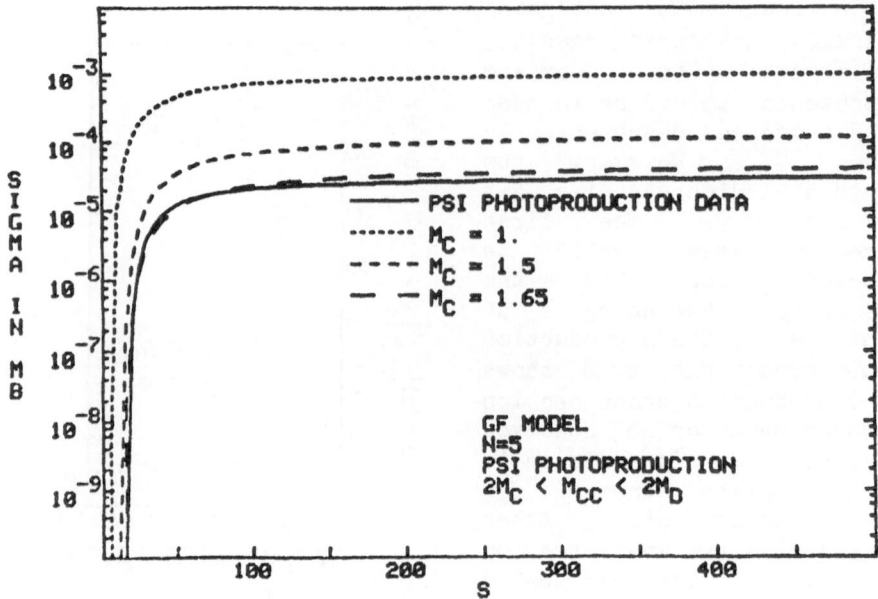

Figure 9.   Comparison of the predictions of the photon-gluon
fusion model to a fit of the world's ψ photoproduction data for
three values of the charmed quark mass.

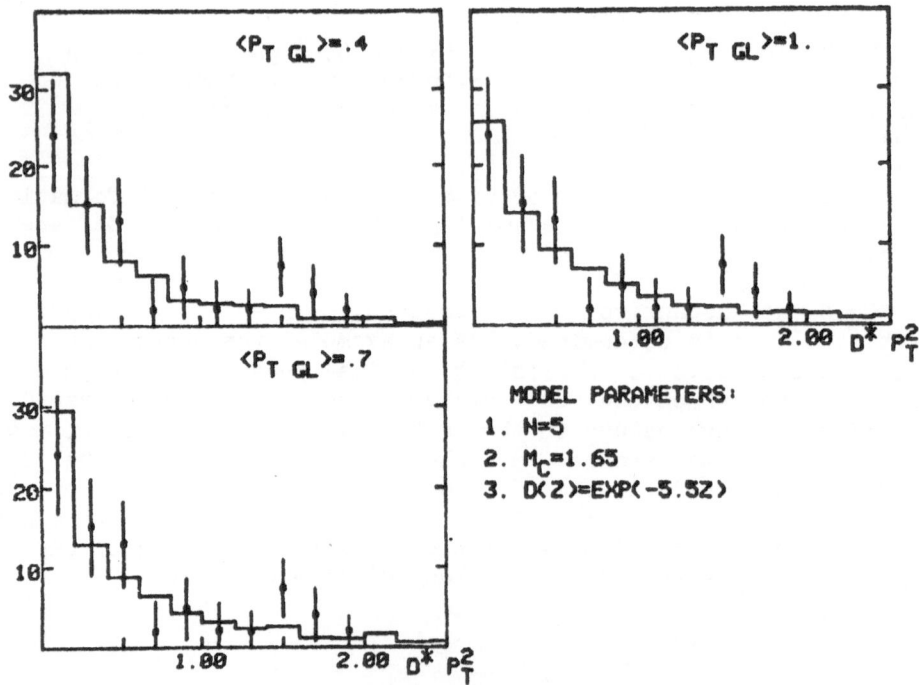

Figure 10.   Comparisons of the $p_T{}^2$ distributions of D#+'s to those
predicted by a photon-gluon fusion model.

Clearly, with so much theoretical input, no single experimental result will provide a conclusive measurement of parameters. However, we may hope that a series of measurements will result in a coherent picture.

SCALING FOR TRUTH AND BEAUTY

The gluon fusion model of the last section can also be used to predict upsilon and open beauty cross sections. The parameters in the theory are crucial for predicting the threshold and asymptotic production levels. However, the threshold behavior can also be predicted by simply scaling the Ψ production threshold behavior measured by previous experiments. In Figure 8, the solid curve is the scaled prediction for bottomonium in the gluon fusion model of Reference 8. The curve is drawn by adjusting the magnitude of the cross section and scaling the energy by the square of the mass produced in the interaction. As can be seen from the curve and lower scale, Truth lies somewhere in the future for photoproduction.

ACKNOWLEDGEMENTS

The data presented in this talk are due to the Columbia-Fermilab-Illinois collaboration and the Berkeley-Fermilab-Princeton collaboration. I am especially indebted to Joel Butler, John Cumalat, Irwin Gaines and Marshall Mugge for providing me with the data and discussing their significance with me. Excellent and more detailed discussions of the data are available in References 9, 10, and 11. I am also indebted to the INFN for partial financial support of my participation in this Conference at Erice.

FOOTNOTES AND REFERENCES

1. M. S. Atiya, S. Holmes, B. Knapp, W. Lee, W. J. Wisniewski (Columbia), M. Binkley, C. Bohler, J. Butler, J. Cumalat, I. Gaines, M. Gormley, D. Harding, R. L. Loveless, J. Peoples (Fermilab), P. Avery, P. Callahan, G. Gladding, M. Goodman, T. O'Halloran, C. Olszewski, J. J. Russel, A. Wattenberg, J. Wiss (Illinois).

2. V. Bharadwaj, B. Denby, A. Eisner, R. Kennett, A. Lu, R. Morrison, D. Summers, S. Yellin, M. Witherell (University of California, Santa Barbara), P. Estabrooks, M. Losty, J. Pinfold (Carleton University), S. Bhadra, A. Duncan, J. Elliott, U. Nauenberg (University of Colorado), J. A. Appel, J. Biel, D. Bintinger, J. Bronstein, P. Mantsch, T. Nash, K. Stanfield, S. Willis (Fermilab), G. Kalbfleisch, M. Robertson (Oklahoma), D. Blodgett, S. Bracker, G. Hartner, R. Kumar, G. Luste, J. Martin, K. Shahbazian, J. Spalding, C. J. Zorn (Toronto).

3. A. R. Clark, K. J. Johnson, L. T. Kerth, S. C. Loken,
    T. W. Markiewicz, P. D. Meyers, W. H. Smith, M. Strovink,
    W. A. Wenzel (Berkeley), R. P. Johnson, C. Moore,
    M. Mugge, R. E. Shafer (Fermilab), G. D. Gollin,
    F. C. Shoemaker, P. Surko (Princeton).

4. G. Gollin, et al., IEEE Trans. Nucl. Sci., 26, 59 (1979).

5. M. Strovink, Proceedings of the 1981 Int. Symp. on Lepton
    and Photon Interactions at High Energies, Bonn, August
    24-29, 1981.

6. In the version of this paper presented at the conference,
    particular emphasis was placed on the possible
    discrepancy in the non-diffractive production of ψ. If
    half the ψ production is non-diffractive, it is difficult
    to understand how so little of it would have survived the
    Fermilab-Illinois trigger requirements. Unfortunately,
    it is impossible to put quantitative limits on this since
    the data required to study this question were not
    recorded.

7. A. R. Clark, et al., PRL 43, 187 (1979).

8. L. M. Jones and M. W. Wyld, Jr. , Phys. Rev. D, 17, 2332
    (1978).

9. M. Binkley, et al., Phys. Rev. Lett. 48, 73 (1982).

10. J. Wiss, AIP Conference Proceedings, Sixth International
    Conference on Experimental Meson Spectroscopy-1980,
    Brookhaven National Laboratory, New York, p. 257.

11. J. Butler, Baryon 1980, Proceedings of the IVth
    International Conference on Baryon Resonances, Toronto,
    Canada, July 14-16, 1980, p. 329.

DISCUSSION

L. Montanet:

In the broad band photon experiment, the D signal is enhanced
by selecting events with low Δ-values of $D^{*} \to D\pi$. Have they
tried the same selection to enhance the Λ_c, i.e., using the
low Δ-value of $\Sigma_c \to \Lambda_c \pi$?

Answer:

Yes, the Columbia-Fermilab-Illinois group looked for the low-Δ

enhancement for Λ_c's. However, they see no dramatic excess of events with Δ about 170 MeV. They interpret this to indicate that less than half the observed Λ_c come from the decay of Σ_c.

B. Margolis:

Do you have any information on the ratio of photoproduction of D* to photoproduction of D mesons?

Answer:

Using a 3σ excess of 660 ± 230 events in the $K^-\pi^+$ mass distribution, correcting for relative efficiencies and removing D^0's from D*'s, the CFI group quotes $.4^{+.22}_{-.10}$ for the D*/D^0 ratio in their data.

PROMPT NEUTRINO EXPERIMENTS ON CHARM AND OTHER FLAVOURS

Gianni Conforto

Istituto Nazionale di Fisica Nucleare
Firenze, Italy

INTRODUCTION

Prompt neutrinos are those originating from the (semi)leptonic decays of particles whose decay length is much shorter than their interaction length in matter.

They are produced by letting a high energy beam from a proton accelerator interact in a target, the "dump", many interaction lengths thick. Because of their long decay lengths, the copiously produced long-lived particles such as pions and kaons have in the dump a decay probability of 10^{-4} - 10^{-5}. This reduction of the yield of ordinary neutrinos allows the detectors placed downstream of the dump to be sensitive to the much weaker sources of prompt neutrinos.

EARLY RESULTS

The first evidence for prompt neutrinos was obtained at CERN in 1977-78.[1-3]

Figure 1 shows the lay-out of the experimental arrangement. A 2 m long copper dump was used in place of the beryllium target for ordinary neutrino operation.

The main characteristics of the detectors employed are summarized in Table I. In neutrino detectors it is not always convenient or even possible to have a posi-

117

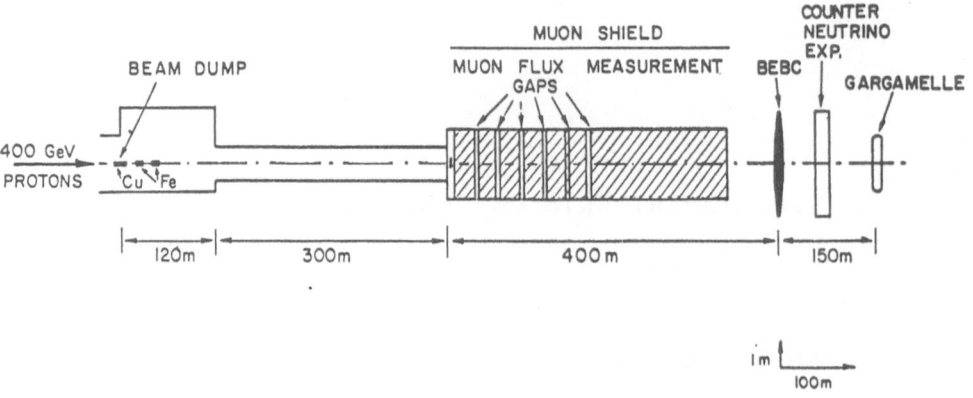

Fig. 1. Lay-out of the CERN 1977-78 beam dump experi-
         ments.

tive identification of ν_e or $\overline{\nu_e}$ charged current events.
In this case this class of events is obtained statisti-
cally from the muonless event sample by subtracting out
the contribution due to the ν_μ, $\overline{\nu_\mu}$, ν_e and $\overline{\nu_e}$ neutral
current events calculated on the basis of the observed
ν_μ and $\overline{\nu_\mu}$ charged current signal.

Figure 2 shows, as an example, the results obtained
by BEBC.[3] The histograms are the event energy distributions experimentally obtained, the smooth curves those
calculated under the hypothesis that all neutrinos interacting in the detectors originate from pion and kaon decays. The integrals of the calculated distributions are
21, 1.3, 3.4 and 0.3 events respectively. It is therefore evident that a statistically significant "excess" of
e^+ and e^- events is observed.

From the very beginning, the possibility of explaining this excess in terms of the production and subsequent
semileptonic decays of charmed particles looked very attractive. Specifically, D-mesons are assumed to be centrally produced according to the invariant differential
cross section

$$E \frac{d^3\sigma}{dp^3} \sim (1 - x_F)^n e^{-bp_\perp},$$

where $x_F \equiv P_\parallel^{cm}/P_{max}^{cm}$ and, typically, $n \approx 3$ and $b \approx 2$. The
D-meson semileptonic branching ratios are known to be

Table 1.   Characteristics of the detectors used at CERN
           in the 1977 and 1979 prompt neutrino runs

BEBC 1977 and 1979	Hydrogen bubble chamber filled with heavy Ne/H$_2$ mixture External muon identifier Electron identification (with sign) Fiducial mass $\sim$12 t
CDHS 1977 and 1979	Magnetized iron-scintillator calorime- ter interspersed with drift chambers Muon detection Fiducial mass $\sim$500 t
CHARM 1979	Fine grain marble-scintillator calorim- eter followed by toroids Electron identification (without sign) Fiducial mass $\sim$100 t
GARGAMELLE 1977	Heavy liquid bubble chamber filled with heavy Freon External muon identifier Electron identification (without sign) Fiducial mass $\sim$10 t

about 10% and therefore total charm production cross sections can be derived from the observed data.

The assumptions made clearly introduce a fair amount
of arbitrariness. In addition, the very small solid
angles subtended by the detectors imply a detection efficiency at $x_F = 0$ which is typically two to three orders
of magnitude smaller than that at $x_F = 1$. Thus, all
total cross section determinations are extremely sensitive to the assumed longitudinal form of the invariant
cross section.

The values of the total charm productions cross section on a free nucleon determined by the experiments of
references 1-3 assuming an $A^{2/3}$ dependence are shown by
the points marked a) in figure 3.

These results, ranging from about 30 to a few hundred microbarn, were however found to be in substantial
disagreement with the upper limit of 1.5 microbarn ob-

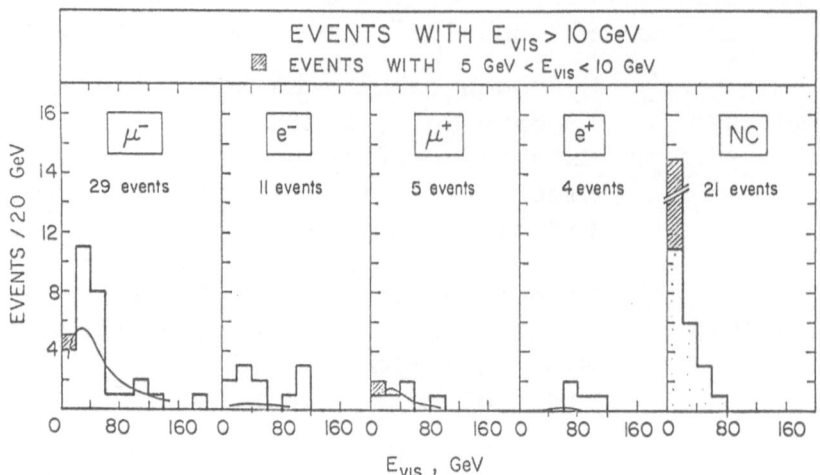

Fig. 2.   Event energy distributions experimentally ob-
          tained by BEBC[3] in the 1977-78 CERN Beam Dump
          experiment.   The smooth curves are calculated
          under the hypothesis that all observed neutrinos
          originate from pion and kaon decays.

tained using the same A dependence in an emulsion experiment at 300 GeV.[4]

    This conflict is nowadays, if not completely re-
solved, at least largely attenuated by taking into account the emulsion experiments detection efficiency which
is a rather strong function of the charmed particles
lifetime.[5]   The 90% confidence level limits to the total
charm production cross section derived from the results
of references 4-6 using the theoretically more palatable
A^1 dependence are shown as a function of the charmed particles lifetimes in figure 4.   It is interesting to note
that the two curves of figure 4 should coincide on the
right-hand side, as the two experiments they refer to
have very similar sensitivities in that range of lifetimes.   The difference between the two curves can therefore be taken as an estimate of the error introduced by
the different assumptions made in the efficiency calculations.   Thus, for lifetimes of the order of several times
10^{-13} s, the upper limit derived from the emulsion experiments could be as large as about 10 μb.

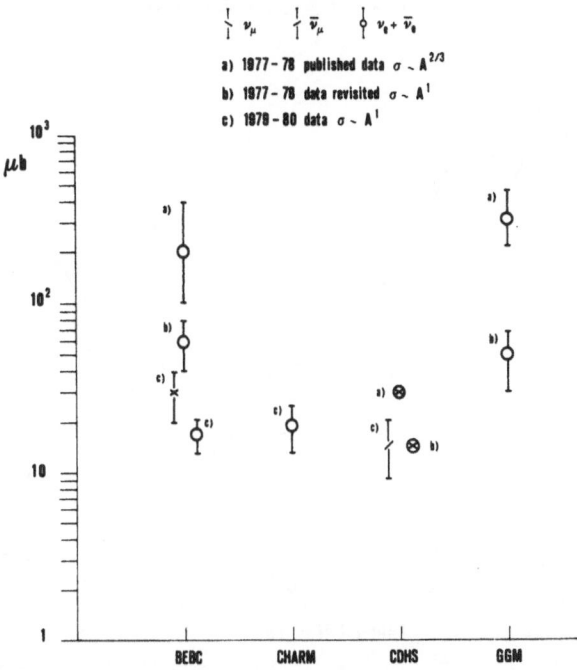

Fig. 3.   Summary of the total charm production cross sections obtained in beam dump experiments at CERN
in 1977-78 and 1979-80. Different symbols are
used to represent results obtained from μ^-, μ^+
and e$^+$ events or combinations thereof. Points
marked a) are the values originally published
(references 1-3) assuming an $A^{2/3}$ dependence,
those marked b) are new calculations (reference
7) based on the same data but using an A^1 dependence, the same form of the invariant differential cross section for all experiments and
various other minor corrections which became
available after publication, those marked c) are
the values obtained in the 1979-80 experiments
(references 8-10).

    The new values of the total charm production cross
section calculated from the beam dump data of references
1-3 using an A^1 dependence, the same form of the invariant differential cross section for all the experiments
and various other minor corrections which became availa-

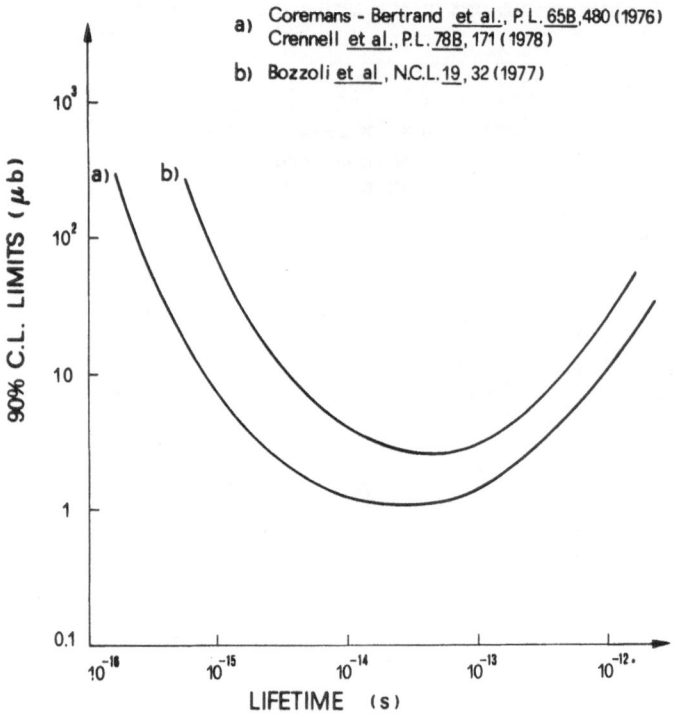

Fig. 4. 90% confidence level limits to the total charm
         production cross section obtained from the emul-
         sion experiments of reference 4 (supplemented by
         the efficiency calculation of reference 6) and 5
         using an A^1 dependence as a function of the
         charmed particles lifetime.

ble after publication[7] are shown in figure 3 by the
points marked b). They are in the region of a few tens
microbarn.

     In conclusion, considering the different beam ener-
gies and the many theoretical and experimental uncertainties, no major conflict can be claimed to exist at present between the emulsion upper limits and the beam dump
data interpreted in terms of charm.

SECOND GENERATION RESULTS

     In order to definitely establish the promptness of
the signals observed in 1977-78, a new series of experiments was carried out at CERN in 1979-80.[8-10]

     The main novelty introduced was the use of the two

dumps shown in figure 5. The non-prompt component due to
long-lived particle decays is obviously three times as
large in the ς = 1/3 than in the ς = 1 data. Thus,
contrary to the "subtraction" method used in 1977-78,
this approach allows a direct determination of the prompt
signal by means of the so-called "infinite density extrapolation". Figure 6, in which the μ^+ and μ^- event
rates observed for the two dump densities by the three
experiments are shown, illustrates this procedure using
the CDHS results. The prompt signal is simply given by
the intercept of a line through the data points at the
two densities.

Total charm production cross sections can obviously
be derived from the data following the same procedure
outlined in the preceding paragraph. The data, however,
indicate some deviations from that simple picture.

The first is illustrated in figure 7, which shows
the energy spectra of the prompt μ^+ and μ^- events obtained by CDHS.[10] It can be seen that, at least when the
presumably more reliable extrapolation method is used, no
significant prompt μ^+ signal is observed. This corresponds to an about two standard deviation significant

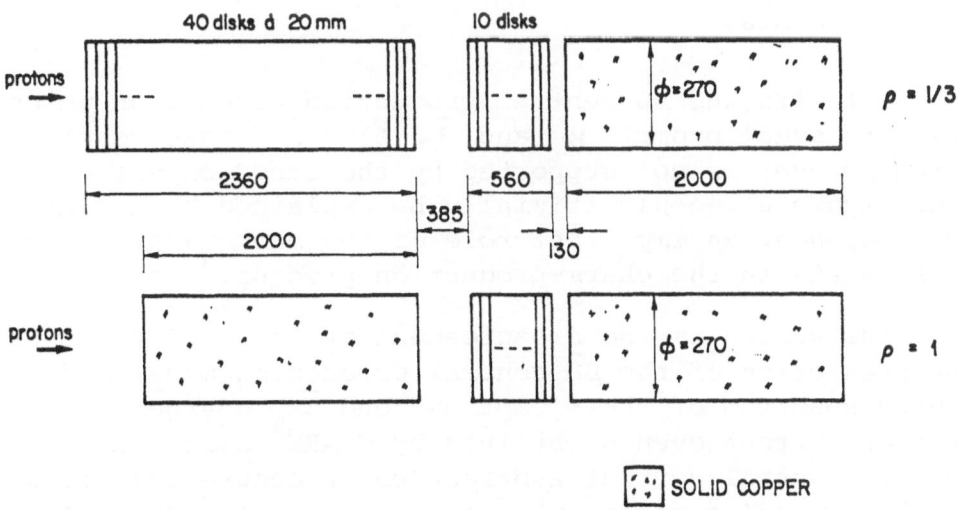

Fig. 5. The ς = 1 (full density) and the ς = 1/3 copper dumps used at CERN in 1979. Units are mil-
limeters.

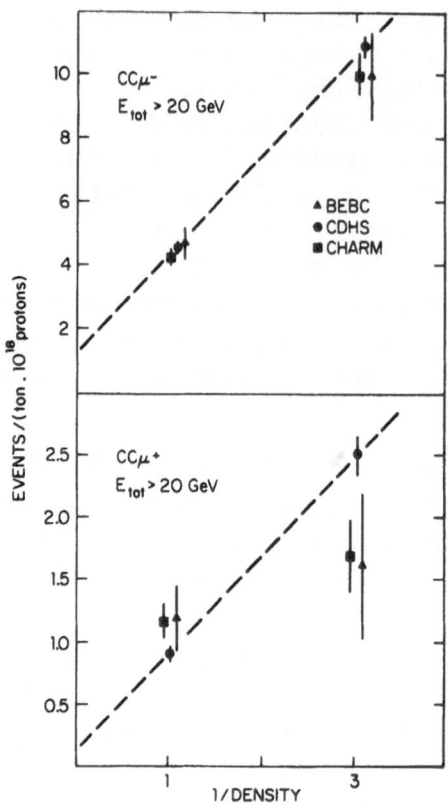

Fig. 6. μ^+ and μ^- event rates observed for the two
          dump densities by the three 1979-80 CERN experi-
          ments.

contrast with the $D\bar{D}$ central production assumption which
predicts equal prompt ν_μ and $\bar{\nu}_\mu$ fluxes. This result,
which however is not supported by the other experiments
(see figure 6), could trivially be explained by assuming,
for instance, an important role of the baryon-antimeson
final state in the charm-production process.[7-10]

   The second is the disagreement at low energy between
the prediction of the $D\bar{D}$ central production model and the
energy spectrum of the prompt ν_e and $\bar{\nu}_e$ charged and
neutral current events obtained by CHARM[9] using the ν_μ
and $\bar{\nu}_\mu$ neutral current subtraction procedure described
in the preceding paragraph. This is visualized in figure
8. No satisfactory explanation exists at present for
this effect, which also has a significance of about two
standard deviations and has not been confirmed (or

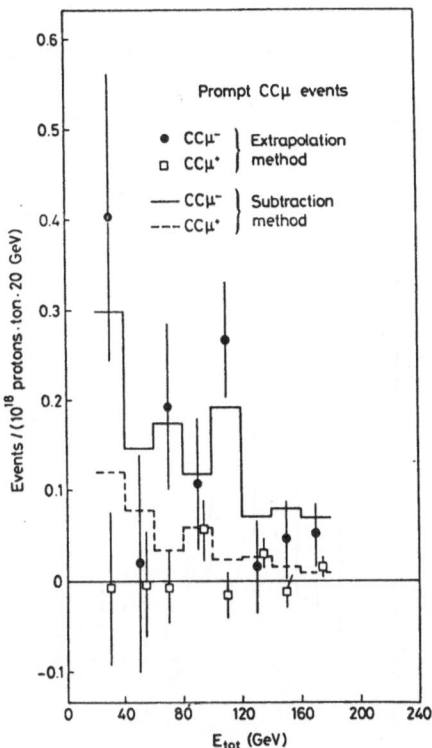

Fig. 7. Energy spectra of the prompt μ^+ and μ^- events
obtained by CDHS in 1979–80.

denied) by any other experiment.

The third is the most puzzling result, on the small-
ness of R, the ratio of the prompt ν_e + $\bar{\nu}_e$ and ν_μ +
$\bar{\nu}_\mu$ fluxes, consistently reported by the three 1979–80
CERN experiments. The measured values of R are shown in
figure 9. Different data or procedures are used to obtain the prompt electron signals. The agreement is really quite remarkable. Even the old 1977–78 CDHS value[2]
(open circle in the figure) is below unity and completely
consistent with the new results.

On the basis of electron muon universality, the
branching ratios for the electronic and muonic decay
modes of charmed particles must be very similar, implying
an R very close to unity regardless of the production dynamics. Thus, there is no way the values of R obtained
can be understood in terms of conventional charm production and decay only.

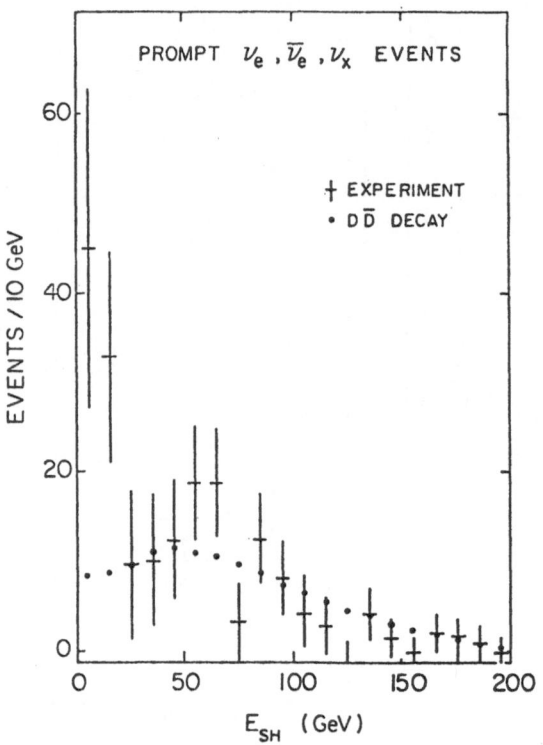

Fig. 8. The energy spectrum of the prompt ν_e and $\bar{\nu}_e$
         events obtained by CHARM in 1979-80.

At any rate, if these still open questions are ignored, the cross section values marked c) in figure 3 are
obtained. The difference between the two BEBC entries
reflects the smallness of R. Because of the inconsisten-
cy between the $D\bar{D}$ central production model and the data,
the CDHS point is calculated using μ^+ events only, averaging the values obtained by means of the subtraction and
extrapolation procedures.

TODAY'S SPECIAL

An experiment, specifically designed to study prompt
neutrinos, has been set up in recent years at Fermilab by
a Firenze, Michigan, Ohio, Washington, Wisconsin (FMOWW)
collaboration.

The lay-out is shown in figure 10. The detector, an
about 150 t lead-scintillator calorimeter followed by a
muon spectrometer, is only at approximately 60 m from the
dump. As a result, the angular acceptance extends up to

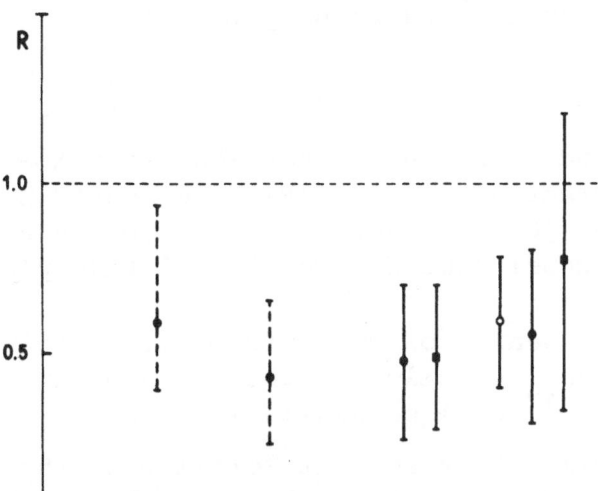

Fig. 9.   The ratio R = prompt ν_e + $\overline{\nu}_e$ flux/prompt ν_μ + $\overline{\nu}_\mu$ flux as measured, using various procedures for the prompt electron signal determination, in the three 1979-80 CERN experiments. The open circle point represents the value obtained for R by CDHS in the 1977-78 experiment.

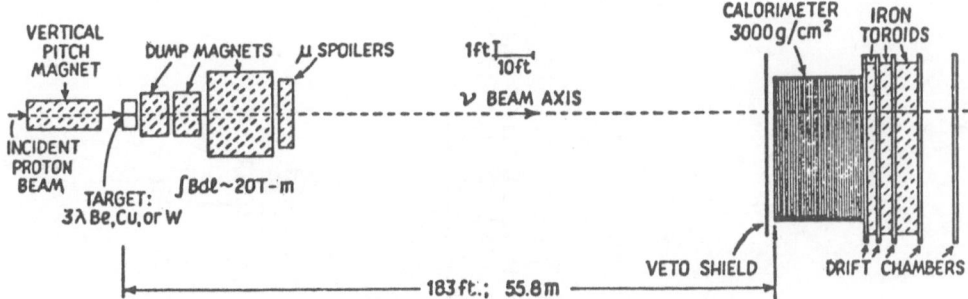

Fig. 10. The lay-out of the Firenze, Michigan, Ohio, Washington, Wisconsin prompt neutrino experiment at Fermilab.

about 40 mrad. The large flux of muons from the dump is
swept up and down by a series of magnets providing a
transverse momentum of about 7 GeV/c.

In the first half of 1981 the experiment had its
first data run and preliminary results are now beginning
to appear.[11]

FUTURE PLANS

A second data run of the FMOWW experiment at Fermilab and another series of CERN experiments with the dump
placed about 400 m closer to the detectors at the end of
the decay tunnel (see figure 1) will take place in early
1982.

Thus, a wealth of new information, including perhaps the first detection of the tau neutrino from F-
meson decays,[12] will be available soon.

Work will start soon at Fermilab on the construction of the prompt neutrino beam which will become oper-
ational at Tevatron energies in a few years.

The schematic lay-out is shown in figure 11. The
thick lines represent the new additions to the existing
neutrino area. The full energy, full intensity proton
beam from the Tevatron is transported, by means of conventional magnets, to a variable density beryllium,
copper or tungsten dump located about 200 m upstream of
the 15' bubble chamber. The beam can be steered in the
range 0-40 mrad. The magnetized muon shield downstream

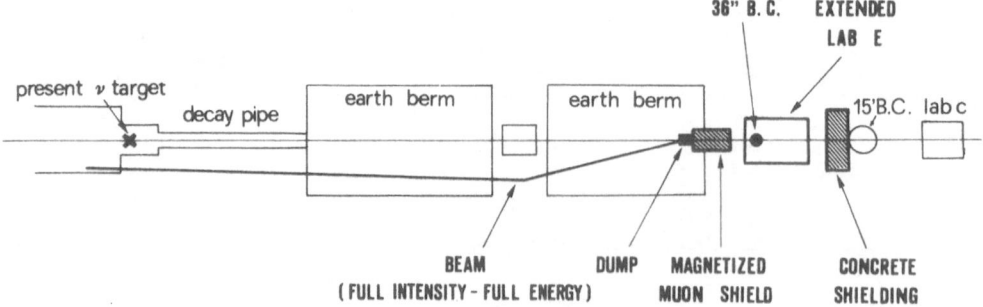

Fig. 11. Schematic lay-out of the prompt neutrino beam
          at the Fermilab Tevatron.

of the dump (see preceding paragraph) is designed to
reduce the muon flux to the low levels required for
bubble chamber operation.

Two experiments have already been approved to run in
this beam. One [13] uses a 36" freon high resolution bubble
chamber supplemented by an external particle identifier,
the other [14] the 15' bubble chamber filled with heavy
neon. By detecting hundreds of ν_τ interactions and visible τ decays and thousands of ν_e induced events they will
be able to explore extremely interesting new areas of
neutrino physics.

DISCUSSION

TREILLE : Assuming that the excess of prompt ν over
prompt $\bar{\nu}$ in CDHS experiment is due to Λ_c (or charmed
baryon) production, can one get from their data some in-
dication on the Λ_c production (x distribution, $(\sigma \cdot B)_{\Lambda_c}$
for a given x distribution)?

Answer: The small solid angle subtended by the detectors introduces such a severe variation of the acceptance
as a function of x that any x distribution determination
is, in my opinion, hopeless.

In principle, assuming the production x distributions
and the semileptonic branching ratios for charmed mesons
and baryons, the acceptances of the apparatus for
the $D\bar{D}$ and $\Lambda_c\bar{D}$ reactions can be calculated. Indeed, the
$15 \pm 6 \mu b$ point based on the μ^+ rate shown in figure 3
refers to the inclusive $\bar{D}$ production, i.e. the sum of the
two reactions. Similarly, from the observed prompt ratio
$\bar{\nu}_\mu / \nu_\mu = 0.46 \pm 0.18$ a number for the $\Lambda_c\bar{D}$ production
cross section can be derived. In practice, however, the
assumptions one has to introduce, in view also of the
difference between the branching ratios for semileptonic
decays of charged and neutral D's, are so unjustified
that I do not believe one can go beyond the qualitative
statement that the data can be explained by a sustantial
$\Lambda_c\bar{D}$ contribution.

ANTREASYAN : What are the measurements for the "prompt"
neutral current events ? Isn't this relevant to the
ν_e / ν_μ ratio you mentioned ?

Answer: Neutral current events, prompt or non-prompt, can be singled out only in detectors capable of positive electron and muon identification (see for instance figure 2). When this is not experimentally possible, the neutral current contribution in the muon-less sample is calculated from the observed ν_μ charged current events and the known (~ 0.32) ratio of the neutral to charged current cross sections. A further, somewhat less justified, assumption is made by taking the same number to be valid also for ν_e interactions.

NEWMAN: What is the current thinking on the relevance of the data on the ratio of ν_e to ν_μ events to possible neutrino oscillations?

Answer: After about twenty years of experimentation in ν_μ beams, there is no evidence that the ν_μ oscillates into anything else. The next best possibility is that the ν_e oscillates into ν_τ and indeed this possibility has been proposed as an explanation for the smaller-than-unity ν_e / ν_μ ratio observed in beam dump experiments. However, this interpretation has also some difficulties. Due to the fact that ν_τ interactions are expected to be muonless most of the time, one would expect to observe a smaller ν_e / ν_μ ratio when electron events are positively identified than when they are arrived at after neutral current subtraction from the muonless events. This is not observed in the data (see figure 9).

MUSSET: Apart from possible physical effects to explain $R = \frac{e}{\mu}$ in the beam dump, one could also worry about possible systematics. Indeed, it is difficult to separate electrons (ν_e), muons (ν_μ) and neutral currents in bubble chambers in which the grain is very small compared to X_o. That should be even more difficult with grains larger by order of magnitude. I wonder if the same experiments could also "normalize" themselves with ν_e contained in ordinary ν_μ beams.

Answer: This is a suggestion to pass on to all groups who have experiments set up where ordinary as well as prompt neutrino beams are available. I suspect however some difficulties due to the smallness and poor knowledge of the ν_e content of ν_μ beams. Not much ν_e physics has been achieved in ν_μ beams so far.

ROE: 1.   The BEBC group set a limit on ν_τ in their data of $\sim 10\%$, $\sim 2\sigma$ lower than required if the ν_e/ν_μ ratio observed by the CHARM group was due to $\nu_e - \nu_\tau$ oscillations.

2.   The BEBC group reported at the ν 79 Bergen conference that they did obtain a prompt neutral current signal. From that they were able to obtain the very important result that the nc/cc ratio for ν_e was the same as for ν_μ within a factor of two. This was the first observation of the ν_e neutral current.

Answer: 1)   I have already commented on the fact that $\nu_e \leftrightarrow \nu_\tau$ oscillations is not a very likely explanation for the smallness of the ν_e/ν_μ ratio.

2)   I believe in their final paper,[8] their neutral current signal is quoted to be consistent with equal $\overset{(-)}{\nu_e}$ and $\overset{(-)}{\nu_\mu}$ neutral current cross section, but the $\overset{(-)}{\nu_e}$ neutral current signal is also consistent with zero.

REFERENCES

1.   P. Alibran et al., Phys. Letters 74B, 134 (1978).
2.   T. Hansl et al., Phys. Letters 74B, 139 (1978).
3.   P.C. Bosetti et al., Phys. Letters 74B, 143 (1978).
4.   G. Coremans-Bertrand et al., Phys. Letters 65B, 480 (1976).
5.   W. Bozzoli et al., Nuovo Cimento Lettere 19, 32 (1977).
6.   D. Crennell et al., Phys. Letters, 78B, 171 (1978).
7.   G. Conforto, Cosmic Rays and Particle Physics - 1978, T.K. Gaisser Editor, American Institute of Physics, New York 1979, p. 221.
8.   P. Fritze et al., Phys. Letters 96B, 427 (1980).
9.   M. Jonker et al., Phys. Letters 96B, 435 (1980). See also: P. Monacelli, these Proceedings.
10. CDHS Collaboration, to be published. See also: F. Dydak, Proceedings of the ν-80 Conference, Erice, Italy; G. Conforto, Gauge Theories, Massive Neutrinos and Proton Decay, A. Perlmutter Editor, Plenum

        Press, New York, London 1981, p. 281.
11. B. Roe, these Proceedings.
12. G. Myatt et al., CERN SPSC/P143.
13. I.A. Pless et al., Fermilab E-636.
14. C. Baltay et al., Fermilab E-646.

PROMPT PRODUCTION OF NEUTRINOS BY 400 GeV PROTONS ON TUNGSTEN:

FIRST RESULTS FROM FERMILAB E613

Presented by B. Roe

R.C. Ball[2], S. Childress[4], C.T. Coffin[2], G. Conforto[1,2],
M.B. Crisler[3], M.E. Duffy[5], G.K. Fanourakis[5], H.R.
Gustafson[2], J.S. Hoftun[3], L.W. Jones[2], T.Y. Ling[3], M.J.
Longo[2], R.J. Loveless[5], D.D. Reeder[5], T.J. Roberts[2],
B.P. Roe[2], T.A. Romanowski[3], D.L. Schumann[5], E.S. Smith[5],
J.T. Volk[3] and E. Wang[2].

[1] Istituto Nazionale di Fisica Nucleare, Firenze, Italy
[2] Department of Physics, University of Michigan, Ann
    Arbor, MI 48109 U.S.A.
[3] Department of Physics, Ohio State University, Columbus,
    OH 43210 U.S.A.
[4] Department of Physics, University of Washington,
    Seattle, WA 98195 U.S.A.
[5] Department of Physics, University of Wisconsin, Madison,
    WI 53706 U.S.A.

ABSTRACT

Data from Experiment E613, a Fermilab experiment to study
prompt neutrino production, are now being analyzed. The experiment
was specifically designed to study neutrinos from the decay of
charmed (or other massive) particles. The detector, an 80 metric
ton (fiducial volume) lead-scintillator calorimeter, is approximately 60 m from the target. As a result, the angular acceptance
extends up to 40 mrad. We have run with a full density tungsten
and 1/3 density tungsten target. We report on a sample of 938
neutrino 1μ events. If we parametrize the $D\overline{D}$ production cross-
-section as proportional to $(1-|x|)^3 e^{-2p\perp}$ we find that our prompt
neutrino 1μ events if interpreted as due to $D\overline{D}$ production imply
$\sigma_{D\overline{D}} = 17.2 \pm 3.2$ (stat.) ± 3.4 (sys) μb. For E_ν greater than 25 GeV,
the $\overline{\nu}$ flux/ν flux = .8$\pm$.35 and the $\overline{\nu}$ flux is 2.9σ from 0.

We report on the production of prompt neutrinos by 400 GeV protons incident on a tungsten beam dump target. Figure 1 shows the basic layout of the experiment. A 150 ton calorimeter detector is located 60 meters from the target. A beam dump of 11 m of magnetized iron followed by 11 m of massive iron shields the detector. Strongly interacting particles are absorbed and most muons are ranged out or bent away from the calorimeter by the magnetic field. A residual flux of about $3x10^5$ muons per typical beam pulse of $2x10^{12}$ protons was measured at the front of the detector.

The detector, shown schematically in Fig.2, consists of a 150 ton calorimeter made up of 30 modules. Each module contains 12 teflon-coated lead plates 6.3 mm thick, to give a total of 14.4 rad. lengths, 0.5 hadron absorption length, and 106 gm/cm^2 per module. Light from the liquid scintillator surrounding the plates is detected by 10 photomultiplier tubes per module. Each module is followed by two PWC planes, one with horizontal and one with vertical wires on 2.54 cm centers (see inset to Figure 2). These are operated in a proportional mode with analog readouts of the pulse height. The PWC's have a wire-to-wire gain uniformity of 10% without software corrections and the readout is linear for showers with as many as 100 particles per wire. The calorimeter is followed by a muon spectrometer with drift chambers and solid iron magnets. For a trigger we require pulses of sufficient amplitude from appropriate calorimeter phototubes as described later and no pulse from the triple wall of veto counters in front of the detector. The beam was operated in the slow extraction mode with a spill time of about 1 s, corresponding to a live-time fraction for the apparatus of about 70%.

The transverse dimensions of the sensitive region are 3 m wide by 1.5 m high. Beam center is 0.75 meters from one side horizontally and centered vertically. The horizontal asymmetry allows us to record neutrino interactions out to a 40 mr production angle.

A set of six calorimeter modules was calibrated in a test beam to determine the energy scale. The calorimeter energy response was found to be quite linear and to have a resolution

$$\sigma/E = .55/\sqrt{E} \quad \text{(GeV) for hadrons and}$$
$$.27/\sqrt{E} \quad \text{(GeV) for electrons}$$

To form a trigger the signals from the modules are summed in 24 overlapping groups of 8 longitudinally and 2 vertically. The signals on the two sides are kept separate. Pulse height requirements are them imposed on the summed signals (see Figure 3). This together with light attenuation in the liquid scintillator results in a trigger threshold which varies with position in a known and calibrated manner. The threshold varies between 5 and 12 GeV (equivalent hadronic energy) for most of the detector. The variation

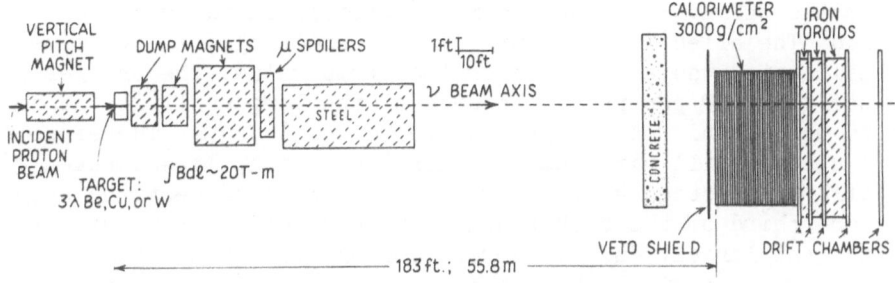

Fig. 1. Layout of Fermilab Beam Experiment E613.

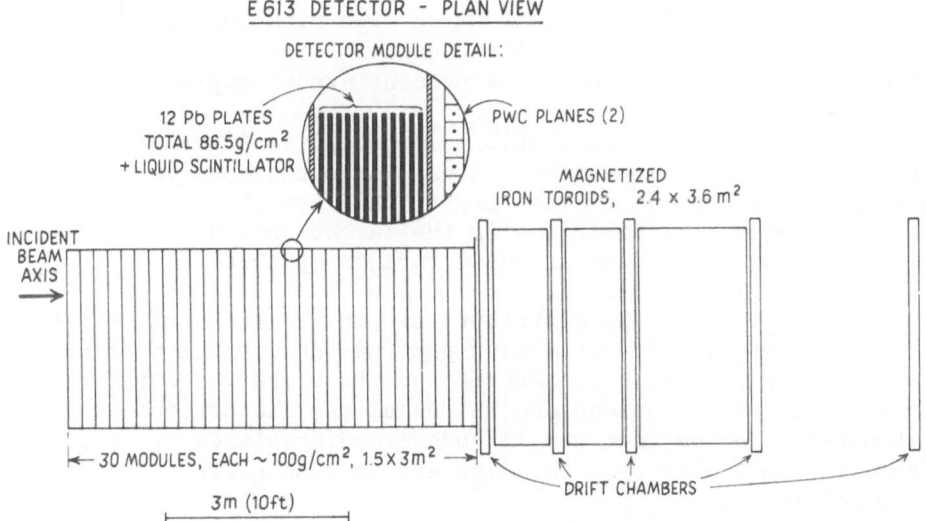

Fig. 2. Fermilab E613 Calorimeter.

of the average threshold as a function of incoming neutrino angle
(θ_ν) is very weak. The effective threshold for 50% acceptance averages around 6 GeV as shown in Figure 4.

In the Spring 1981 run 1.87×10^{17} protons on target were
obtained. The effective exposure after correcting for live time,
bad spills etc. was a little over half this value. The triggering
rate was about 30 per pulse. Because of the long spill it was important to understand the background due to cosmic rays. Therefore for
each 1 second spill with beam another 1 second spill was taken with
beam off. We recorded about 1000 triggers per good neutrino event.
About one third of the background was due to cosmic ray triggers
and the rest largely due to showers coming from muon interactions
in the floor or roof blocks and impinging on the apparatus.

Because of their unambiguous signature in the detector, muon
neutrino charged current events have been studied initially.
Comparison of two independent analysis methods indicates that,
within the fiducial volume, we have 92% efficiency for keeping
events in which the muon goes through the drift chambers. Our fiducial volume boundaries are 6" in from the lead plate edges. We
include only events starting in modules 3 through 25.

The results presented here are based upon about 80% of the muon
neutrino charged current data from our run with the tungsten target.
The raw distribution of 1μ event energies is shown in Figure 5 for
ν's and $\bar{\nu}$'s. Table I details the present event sample.

The raw spatial distributions are shown in Figure 6. Note that
the events cluster about beam center. Since the beam is pitched up
by a magnet just ahead of the target, the distrubution of events at
the calorimeter from neutrinos in the target should be 25 cm higher
than that from background produced further upstream.

We must correct the distributions for the threshold effects
and for the probability of a muon passing through the toroids.
These effects distort the observed Bjorken x and y distributions.
However, the distributions are known and the distortions calculable
although further refinements in understanding biases are required.
Figure 7-9 indicates that the data are in reasonable agreement with
the predicted shape.

Some of the muon neutrino events are due to neutrinos from
pions and kaons that decay before absorption in the target. We have
determined this fraction by using tungsten targets of two different
densities and extrapolating to infinite density. About 21% of our
tungsten running was done with a segmented tungsten target whose
average density was one-third the normal density of tungsten. We
note that because we detect events over a large solid angle our
correction is expected to be smaller than that for the CERN CDHS

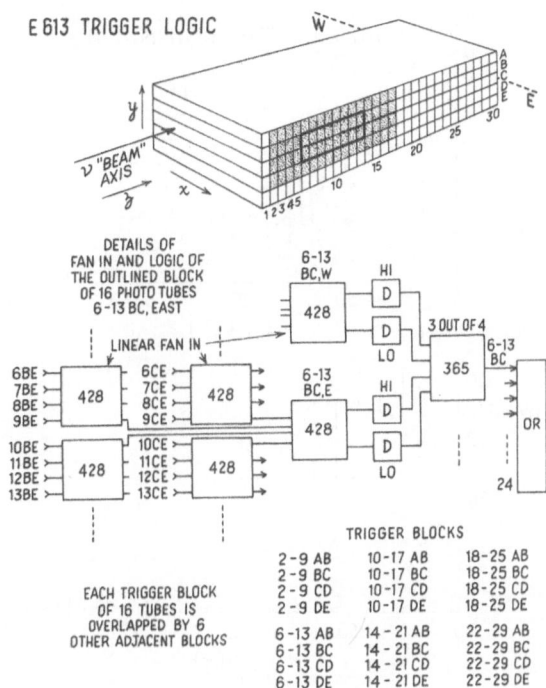

Fig. 3. Fermilab E613 Neutrino Interaction Energy Trigger.

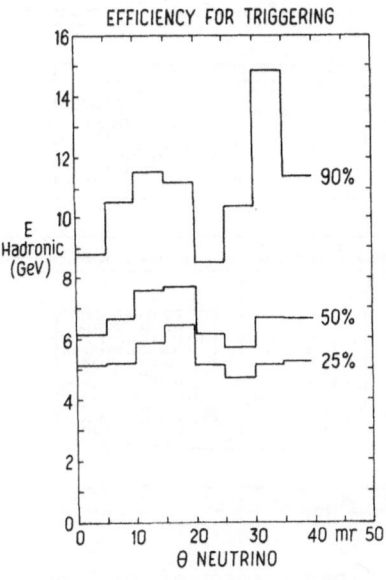

Fig. 4. Variation of Threshold with Neutrino Angle.

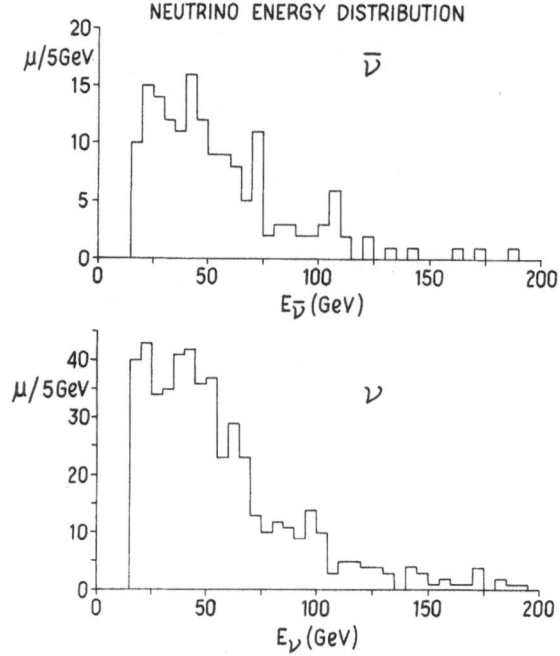

Fig. 5. Raw Energy Distribution of 1μ Events.

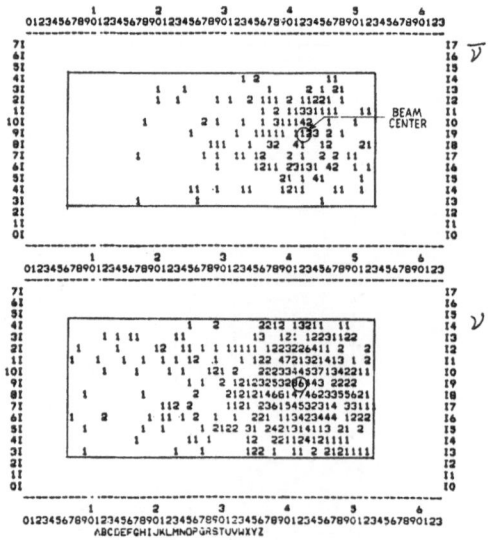

Fig. 6. Raw Spatial Distribution of 1μ Events in the Calorimeter.

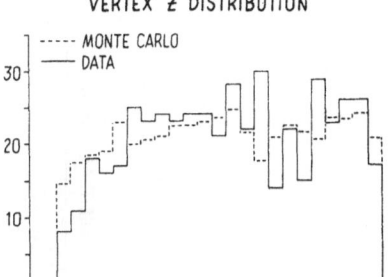

Fig. 7. Distribution of Events as a Function of Module Number (Z Distribution).

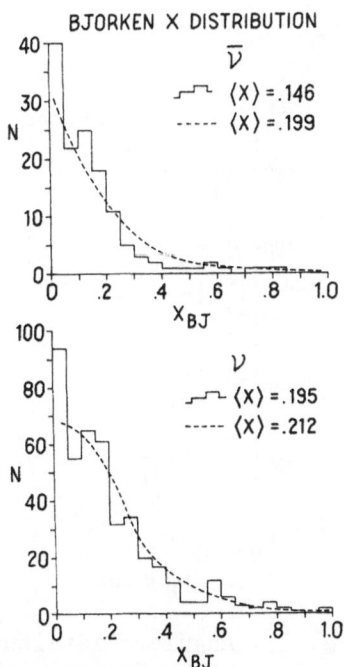

Fig. 8. Raw Bjorken x Distribution of 1μ Events.

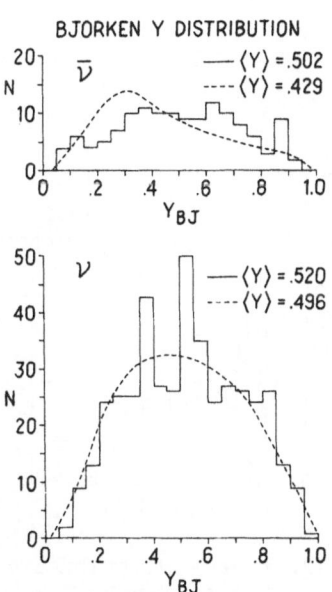

Fig. 9. Raw Bjorken y Distribution of 1μ Events.

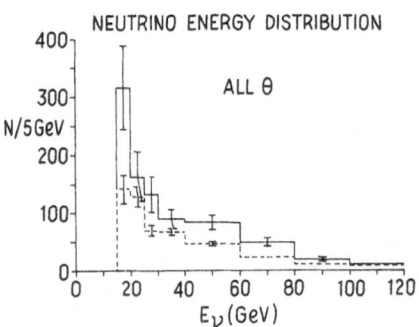

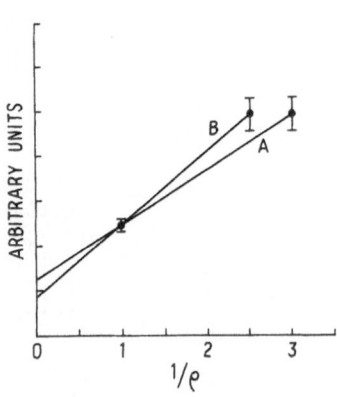

Fig. 10. Overall Extrapolation
of Neutrino Plus Anti-
-Neutrino Events to an
Infinite Density
Target.

Fig. 11. Energy Distribution of
all 1μ Events for Full
and One-Third Density
W Targets.

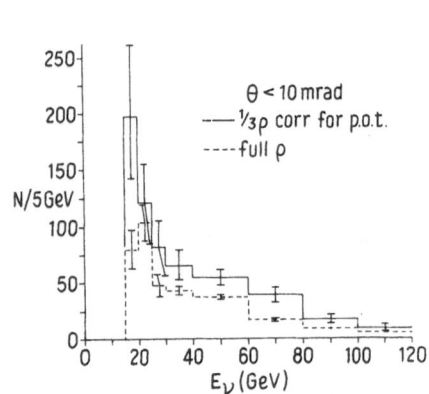

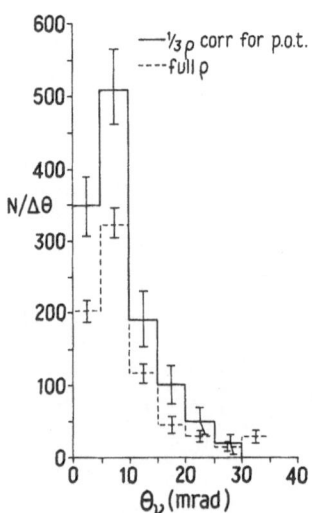

Fig. 12. Energy Distribution of 1μ
Events for Full and One-
-Third Density W Targets.

Fig. 13. Angular Distribution
of 1μ Events with
E>15 GeV for Full
and One-Third
Density W Targets.

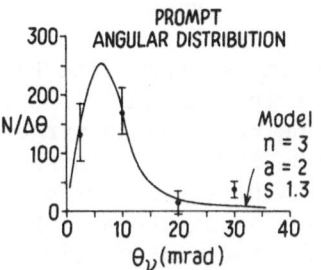
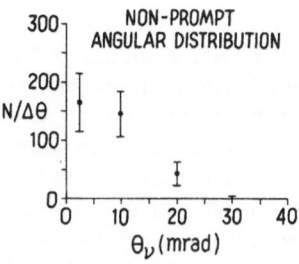

Fig. 14. Energy Distribution of Weighted 1μ Prompt and Non-Prompt
Neutrino Events.

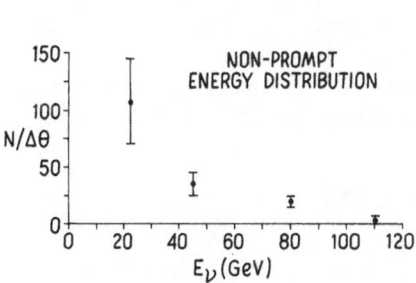
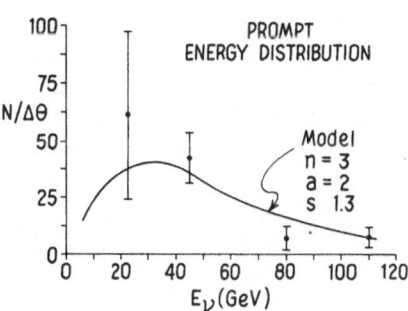

Fig. 15. The Angular Distribution of Prompt and Non-Prompt Neutrino
Events.

group. This is because the pion decay neutrinos are concentrated
at smaller angles than the prompt neutrinos and because we use a
tungsten target which has a higher atomic weight and a higher density than the CERN copper target.

In a prompt neutrino experiment control of upstream background
is essential. We continually monitored upstream scraping by a
series of monitors around the beam pipe. These were calibrated by
introducing known amounts of materials in the beam. The background
was estimated from these monitors to be less than 1% of the non-
-prompt signal inherent in our high density tungsten target. Material just upstream of the target (SWICS, SEM, etc.) introduced a
larger but calculable background of 8.3% of the non-prompt high
density tungsten value. A correction averaging around 15% was introduced by hadrons punching through the three absorption length
targets into the essentially infinite iron and copper stop in back
of the target. The net effect of these corrections on the extrapolation is to reduce the effective density ratio of the two targets
from 3 to about 2.5, as shown in Figure 10. In Figure 10 these
corrections are ignored in extrapolation A and included in extrapolation B.

In order to determine the prompt cross-section as a function
of energy and angle, we have divided the events into bins of energy
and angle and done the extrapolation on a bin-by-bin basis using
background corrections as a function of energy.

The energy and angular dependence of the neutrino events are
shown in Figures 11 through 15. If we assume that the major production mode is via $D\bar{D}$ production we can go further. We assume the
cross-section is proportional to $A^{1.0}$. (We hope to test this hypothesis by using copper and beryllium targets. In the present run
14% of the data was with copper targets). The semi-leptonic branching
ratio is taken as the mean of the D^0 and D^+ branching ratios obtained
from e^+e^- data, i.e. 16.4±2.4%. We parametrize the $D\bar{D}$ cross-section
as

$$E \frac{d^3\sigma}{dp^3} \propto (1-|x|)^n e^{-ap\perp}$$

where x is Feynman x. We assume that the protons cascade in the
target with an average elasticity (ε) of 2/3 and that the charm
cross-section varies as s^k. The CERN BEBC group[1] used k=.5.
J. Leveille[2] suggests k=1.3 is a more appropriate value.

The results for n, a, σ are then:

n	a	ϵ	k	$\sigma_{D\overline{D}}(\mu b)$	stat. error	syst. error
3	2	.67	.5	15.6	± 3.1	± 3.1 (BEBC model)
3	2	.67	1.3	17.2	± 3.4	± 3.4
4	2	.67	1.3	21.4	± 4.3	± 4.3
3	1.6	.67	1.3	18.6	± 3.7	± 3.7
3	2	.3	1.3	24.4	± 4.9	± 4.9

The BEBC group has obtained a value of $\sigma_{D\overline{D}}$(n=3, a=2, k=0.5) of 30±10 μb from $\nu_\mu+\overline{\nu}_\mu$ and 17±10 μb from $\nu_e+\overline{\nu}_e$ events[1]. The CHARM group reported (n=4, a=2, no cascading) 19±6 μb [3].

The E613 data was separated into μ^+ and μ^- events. From the relative event rates the ratio R of $\overline{\nu}_\mu/\nu_\mu$ fluxes was derived. We find that for $E_\nu > 25$ GeV, R=.8±.35. The ν flux is 2.9σ from 0. R is expected to be one if the flux is dominated by neutrinos from DD production while DY_c production should give R<1. Unlike the CDHS result summarized by G. Conforto at this conference we do have positive evidence for a $\overline{\nu}$ signal.

REFERENCES

1. P. Fritze et al. Phys. Lett. 96B (1980).
2. J. Leveille, private communication.
3. M. Jonker et al. Phys. Lett. 96B, 435 (1980).

DISCUSSION

J. Appel
Question 1: When you showed the ν and $\bar{\nu}$ y distributions, you suggested
that the $\bar{\nu}$ discrepancy from expectation was due to inaccuracies in
the energy acceptance affect the ν distribution similarly?

Question 2: In the cross section determination, does the high even
rate at large angle contribute disproportionally? This might occur
if the cross section was summed from bin by bin averages. Or, is
the cross section from the total event rate (i.e., the fit to the
full distribution)?

Answer 1: Errors in energy acceptance affect both distributions by
the same percentage at a given y value. Since relatively there are
many more low y antineutrino events than neutrino events the discre
pancy appears larger in the anti neutrino y distribution.

Answer 2: The total event rate was used to determine cross-sections,
the data were not fit to theory as a function of θ_ν

RECENT RESULTS FROM THE EXPERIMENT NA16

(LEBC-EHS Collaboration)

G. Zumerle[(*)]

CERN
EP-Division
CH-1211 Geneva 23

INTRODUCTION

Hadronic production of charm is attractive to study both the decay properties of the charm particles and the production mechanism of the heavy charm quarks.

The use of a visual, high resolution detector allows a high purity sample of charm events to be selected on the basis of their characteristic topology. This type of detector also allows the decay particles to be distinguished from these coming from the primary vertex.

APPARATUS

The layout of the experimental set up is shown in fig. 1 and is described in detail in[1]. The experimental configuration may be separated into (a) vertex detection; (b) charged particle momentum analysis; (c) gamma detection and measurement; (d) particle identification; (e) trigger.

VERTEX DETECTION

The bubble chamber, LEBC, 20 cm in diameter and 4 cm in depth, gives bubbles of $\sim$ 50 μm diameter with density of $\sim$ 80 bubbles/cm. A detailed description of the construction and operation can be found in[2].

(*) On leave of absence from INFN, Padova, Italy.

CHARGED PARTICLE MOMENTUM ANALYSIS

The momentum precision, using both the measurement in the bubble chamber[*] and in the spectrometer, is $\Delta P/p = 1.5\%$ over the full momentum range. This has been checked using a sample of strange particle decays. Fig. 2 shows the effective mass distributions for the K^0 and Λ decays. The width of the distribution agrees with the previous figure on the momentum resolution and implies a resolution of 10-20 MeV for charm particles with all the tracks "hybridized" (i.e. tracks which have been reconstructed in both the bubble chamber and the spectrometer).

Using a sample of strange particle decays, the hybridisation efficiency for tracks above 2 GeV, entering the acceptance region, has been measured to be better than 90%.

GAMMA DETECTION AND MEASURMENT

A full description of the gamma detection system can be found in[3]. Table 1 gives the main parameters of the two lead-glass calorimeters.

Table 1

		IGD	FGD
Acceptance	$\Delta\varphi$	$\pm$ 80 mr	$\pm$ 30 mr
	$\Delta\lambda$	$\pm$ 80 mr	$\pm$ 17 mr
Energy resolution $\Delta E/E$(FWHM)		$(\dfrac{15.}{\sqrt{E(GeV)}} + 2.)\%$	$(\dfrac{10.}{\sqrt{E(GeV)}} + 2.)\%$
Angular resolution $\sigma_\varphi = \sigma_\lambda$		0.4 mr	0.1 mr

Fig. 3 shows typical γ-γ effective mass distributions.

PARTICLE INDENTIFICATION

Since only a prototype of the pictorial drift chamber ISIS was available during the NA16 experiment, only limited ionisation precision, $\Delta I/I = 18\%$ FWHM[4] was obtained. In any case, however, this detector provides valuable spatial information which is most useful during pattern recognition.

(*) Track residuals $\sim$ 7 μm, fiducial errors $\Delta\varphi$ (vertical, bending plane) = 0.1 mr, $\Delta\lambda$ (horizontal) = 0.7 mr.

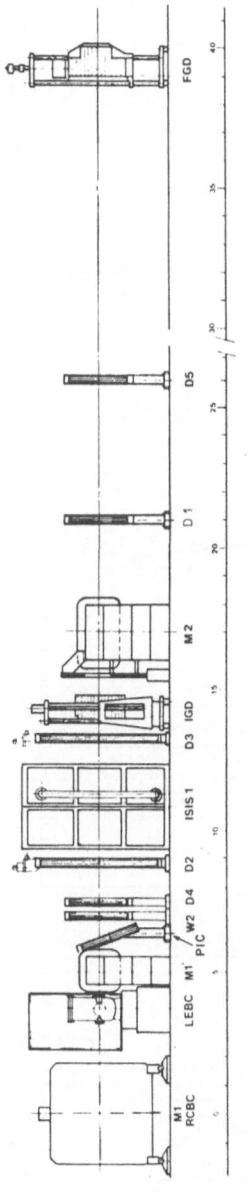

Fig. 1. EHS lay out. LEBC: bubble chamber; M1, M2: bending magnets; PIC, W2, D1-D5: wire chambers; ISISI: particle identification; IGD, FGD: electromagnetic calorimetres.

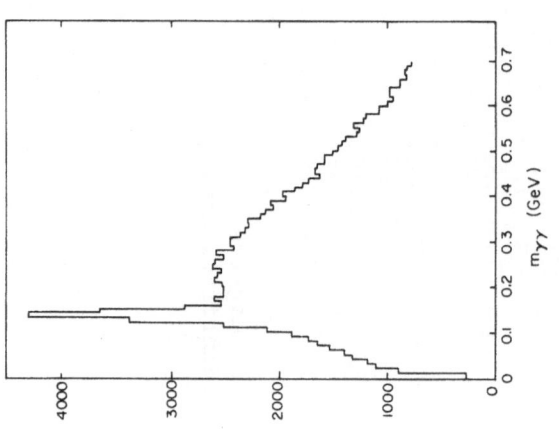

Fig. 3.   Effective mass distribution
of γγ combinations.

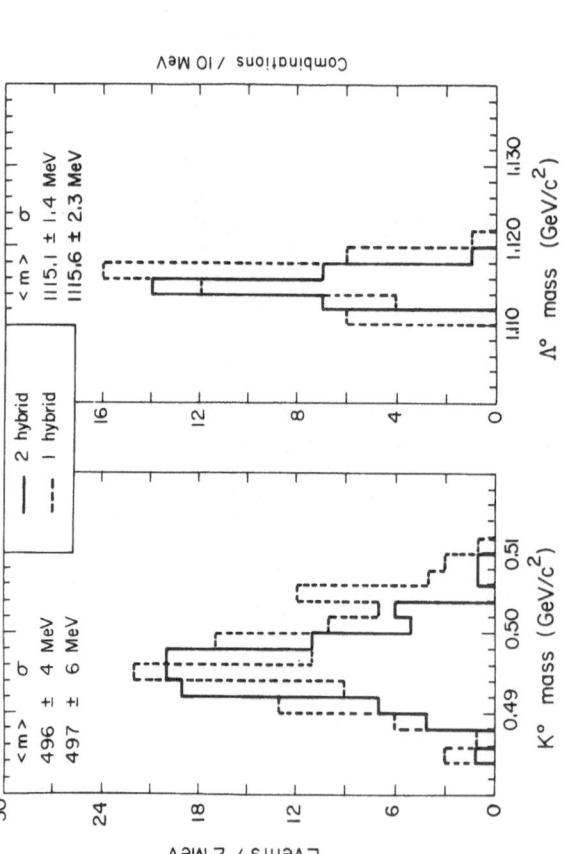

Fig. 2.   Effective mass distribution for a sample of K⁰, Λ
decays.

TRIGGER

   A simple interaction trigger was used, requiring at least
one particle on both sides of a narrow vertical aperture
($\Delta\lambda$ = 12 mr) placed downstream of the bubble chamber.

THE DATA

   In order to extract the charm candidates, two independent
scans of the film are made, followed by a check done by a
physicist. Events with decays are then measured and analysed
only if the decay vertex falls into a cylindre of 600 μm
radius around the beam - the so called "charm box".

   The event statistics of the experiment is given in table 2.

   The data from a limited sample, $\sim$ 200 K pictures with π^-
beam (5 ev/μb) and $\sim$ 100 K pictures with p beam (1.7 ev/μb) will
now be presented.

   These data only concern topologies with a very low non charm
background, i.e. decays with more than 2-prongs or events with 2
decays in the charm box. The sample statistics is given in
table 3, where C_n (V_n) means a charged (neutral) decay in n
prongs.

Table 2

pictures taken	$\sim 10^6$ 650 K p-beam (360 GeV) 400 K π^--beam (360 GeV)
events in fiducial volume (13 cm of H_2)	4.5×10^5
events with $\gtrsim$ 1 decay measurements	5×10^4 2×10^4
events in "charm box" not obviously strange	$\sim$ 2000 (analysis in progress)

Table 3

Topology	Identified not charm	Identified charm	Under study
C_3	21	10	17
C_5, C_7	–	–	2
V_2	11	4	4
V_4, V_6	–	2	5
C_1	2	0	6

In this sample a decay is identified as a charm decay if:
(a) there is no fit to a strange particle decay with a
probability $> 10^{-4}(4\sigma)$; (b) there is a constrained fit to
a charm decay.

Table 4 gives the relevant quantities for the identified
charm decays.

LIFETIMES

To avoid biases in the lifetime determination as much as
possible, only decays with all tracks completely reconstructed,
giving 3 constraint fits, have been used.

To take into account scanning inefficiencies for short
decays, the lifetime is computed as

$$\tau = \frac{1}{N} \frac{m}{c} \Sigma \frac{\ell_i - \ell_i^{min}}{P_i} \tag{1}$$

where ℓ_i^{min} is the minimum distance from the primary vertex at
which the decay would have been detected. Eq. (1) is the
maximum likelihood estimate of the lifetime if the scanning and
reconstruction efficiency can be approximated by a step function
at $\ell_i = \ell_i^{min}$.

For charged D mesons, the distance visibility interval is
the one in which in the observed decay, at least one track
misses the primary vertex by more than 100 µm; ℓ_i^{min} is then

$$\ell_i^{min} = \ell_i * \frac{100 \ (\mu m)}{y_i^{max}(\mu m)} ,$$

y_i^{max} being the maximum impact distance of the decay tracks.

The lifetime is found to be

$$\tau_{D^{\pm}} = (6.7 \ _{-2.4}^{+4.9}) \times 10^{-13} \ s .$$

For the neutral D mesons it is found that a decay is
topologically clear if its distance from the primary vertex is
greater than 1 mm. ℓ_i^{min} therefore is set equal to 1 mm. Using
eq. (1) the lifetime is evaluated to be

$$\tau_{D^0} = (2.3 \ _{-0.9}^{+1.7}) \times 10^{-13} \ s .$$

Table 4.   NDF is the number of degrees of freedom in the fit of the decay. P, ℓ, t, y_{max} are the momentum, the decay length, the time-of-flight and the maximum impact parameter (projected on the film plane) of the charm particle.

Decay	NDF	Mass (MeV)	P (GeV)	ℓ (cm)	t (10^{-13} s)	y_{max} (μm)	Comments
π^- beam							
$D^- \to K^+\pi^-\pi^-$ C_1 not ident.	3	1867±7	209	2.76	8.2	205	amb F
$D^- \to K^+\pi^-\pi^-\pi^0\pi^0$ C_1 not ident.	3	1863±20	43	0.87	12.6	230	
$D^+ \to K^-\pi^+\pi^+\pi^0$	1	1867±20	8.5	0.21	15.4	197	amb F
$D^- \to K^+\pi^-\pi^-$	2	1850±27	27	1.24	28.8	120	amb F/Λc
$D^0 \to K^-\pi^+\pi^0\pi^0$	3	1857±22	119	0.41	2.1	70	
$\bar{D}^0 \to K^+\pi^+\pi^-\pi^-$	3	1862±9	79	0.75	5.9	206	
$D^- \to K^+\pi^-\pi^-$	3	1840±12	188	0.25	0.9	20	amb F
$D^- \to K^+\pi^-\pi^-\pi^0$	2	1858±31	119	4.51	24.1	1153	amb F
$D^0 \to K^-\pi^+\pi^0\pi^0$	2	1880±33	76	0.32	2.6	165	
$\bar{D}^0 \to K^+\pi^+\pi^-\pi^-$	3	1850±14	82	0.35	2.6	111	
$D^0 \to K^-\pi^+\pi^0$	2	1865±35	44	.09	1.3	74	
$D^- \to K^+\pi^-\pi^-$	3	1865±9	36	0.30	5.3	324	
$D^0 \to K^-\pi^+\pi^0\pi^0$	3	1847±20	43	0.15	2.1	100	
$D^+ \to K^-\pi^+\pi^+\pi^0$	3	1861±12	70	1.34	11.8	140	amb F
p beam							
$F^- \to K^-K^+\pi^-$	3	2025±11	43	0.27	5.5	136	amb Λc
$D^- \to K^+\pi^-\pi^-\pi^0$	3	1861±19	251	0.50	1.2	37	amb F
$D^- \to K^+\pi^-\pi^-$	3	1859±7	78	2.07	16.7	580	amb Λc

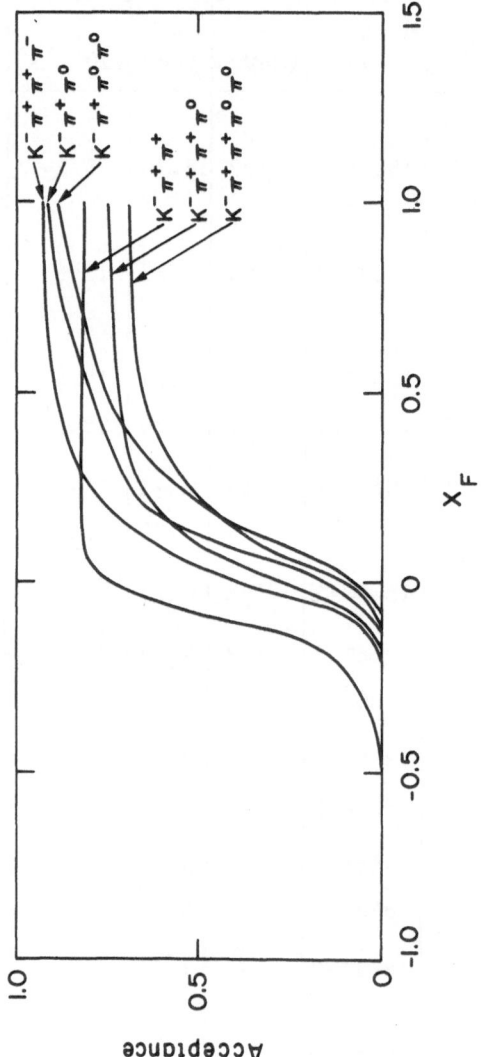

Fig. 4.   Acceptance curves for D meson decay products.

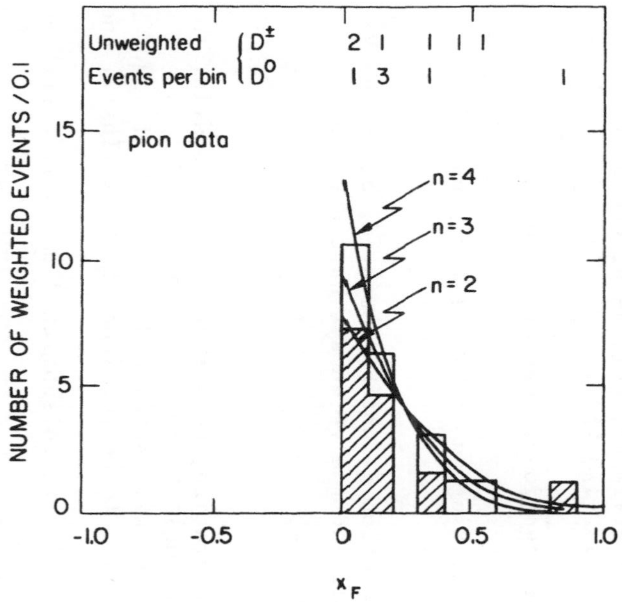

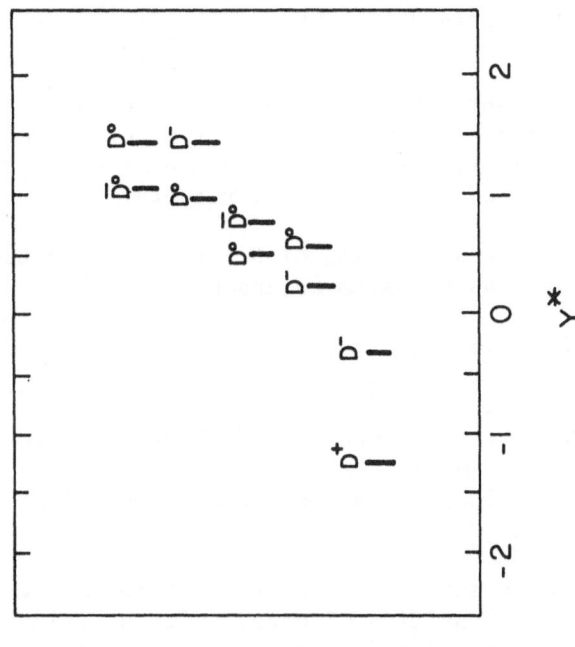

Fig. 5. Weighted x_F distribution for D meson decays from the $\pi^- p$ sample. The curves represent $(1-x_F)^n$, $n = 2,3,4$.

Fig. 6. Centre of mass rapidity for pair events.

PRODUCTION

Fig. 4 gives the experimental acceptance for the decay products of D mesons as a function of $x_F = 2 P_{11}^{cm}/\sqrt{s}$.

From these curves only the main production characteristics in the forward emisphere can be studied. From now on only the π^- beam sample will be discussed since the p beam data is still too scarce to give any significant results.

Fig. 5 shows to x_F distribution of D mesons, corrected for acceptance. The data can be parametrized by

$$\frac{dN}{dx_F} = (1-x_F)^{3.2 \pm 1.0}$$

The average transverse momentum of the D mesons with respect to the beam direction is 0.78 GeV/c. The five events where both charm decays have been identified, show a close correlation in rapidity (fig. 6).

The inclusive cross section for $D^{\pm}$ production can be computed from the six identified C3 decays with positive x_F. Since the sensitivity is 5.0 ev/μb, the seen inclusive cross section is 1.2 μb. An overall correction factor must then be applied, to take into account: (a) the events not recognized as strange but still not understood as charm (x 1.9); (b) the acceptance (x 2.1 in average); (c) the branching ratio $D^{\pm} \rightarrow$ 3 prongs/$D^{\pm} \rightarrow$ all $(1/0.45)$[5].

Thus, the cross section is estimated to be

$$\sigma(\pi^- p \rightarrow D^{\pm} + \text{anything}, x_F \gtrsim 0) = 11. \pm 5 \ \mu b$$

The D^0 cross section cannot be computed, since the V4 sample is too small and the V2 decays have not been completely analyzed.

REFERENCES

1. LEBC-EHS Collaboration, NA16 experimental details, to be submitted to Nucl. Instr. and methods.
2. J.L. Benichou, et al., submitted to Nucl. Instr. and Methods.
3. B. Powell, et al., to be published in Nucl. Instr. and Methods.
4. W. Allison, et al., to be submitted to Nucl. Instr. and Methods.
5. R. Schindler, et al., to be submitted to Nucl. Instr. and Methods.

DISCUSSION

Diambrini: How does the sharp cut hypothesis affects the lifetime evaluation?

Answer: The statistical error on the lifetime is still much more important than effects of this type.

Newman: Is the strong correlation in y betyeen D-$\overline{\text{D}}$ pairs simply a consequence of the tendency for all heavy particles to be produced near y = 0? This analogous to old results at lower energies on p and Λ production, which show relatively narrow, gaussian rapidity distributions centred at y = 0.

Answer: I do not think so, since the inclusive y distribution is much wider than the Δy distribution.

D'Ali: My question refers to the y correlation. How does the D-$\overline{\text{D}}$ invariant mass distribution look?

Answer: Since there is a correlation, the invariant mass must be small. It is difficult to say more with the few events that we have.

Websdale: Could you take advantage of the factor x 10 improvement in charm/background ratio possible using a photon beam in EHS?

Answer: There are plans to check the behaviour of the EHS spectrometer in the presence of the heavy c.m. background produced by a photon beam. The acceptance however, for charm particles produced in the central region, using a beam of the order of 100 GeV, could be small without modifications to the first part of the spectrometer.

PHOTO-PRODUCTION OF CHARMED PARTICLES BY 19.5 GeV GAMMAS

SLAC BC 72/73 Collaboration

T. Kitagaki
Tohoku University
Sendai, Japan

1. EXPERIMENT

The photo-production experiment, BC 72/73, has been periodically running at SLAC since summer 1980. In this experiment, monochromatic gammas of 19.5 GeV are produced by the backward scattering of ultra violet laser beam of 0.266 μm into the 30 GeV SLED electron beam. Fig. 1 shows the measured energy spectrum of the gammas. It is seen that 85% of the beam are in E_γ > 15 GeV. The detection system, the SLAC Hybrid Facility is shown in Fig. 2. A high resolution camera (HRO) is installed in addition to the three normal cameras. The HRO has a spacial resolution of approximately 55 μm in the depth of ± 6 mm. The triggering of HRO for hadronic events are made by either the PWC-168E algorithm (delay 150 μs) or the lead glass wall signal (delay 1 μs). The triggering efficiency is estimated to be 90% for hadronic events and 80% for charm events.

2. SEARCH FOR CHARMED PARTICLES

Charmed particles are searched using the HRO pictures taken at the bubble growth of about 60 μm in diameter. Short decay vertices or off primary vertex tracks are the most important signature of charmed particles and every events are examined looking for anomaly near the vertex on high magnification scanning tables. For example, at Tohoku all anomalous events found on scanning tables are photo-recorded and carefully examined by physicists. Candidates for charm events are measured on normal 3 views and HRO films and the impact distances of tracks against primary vertices are examined.

In order to collect clean samples, we require that at least one of the tracks from the short decay vertex has the impact parameter greater than 110 μm, about twice of bubble diameter. This cut is useful to remove ambiguous events. Further, we require that the anomalous event be inconsistent with the decay of known strange

Table I.   Selected charm decays for life time analysis

Charged

Event	d^{MAX} μm	d_2 μm	l mm	l_{eff} mm	T_{OBS} $(\times10^{-13}$s)	$\bar{T}_{eff}$ $(\times10^{-13}$s)	TYPE	$P_{CONST.}$	T $(\times10^{-13}$s)	Constrained $(\times10^{-13}$s)
596 – 237	470	230	2.0	1.5	12.8	9.6	$D^+/(F^+)$	9.95	9.4	
916 – 016	150	50	1.7	0.34	9.3	1.8	$D^-/(F^-)$	8.7	2.4	
1293 – 147	1020	225	3.2	2.63	32.4	26.7	$F^+/D^+/\Lambda_c$			
1415 – 225	1440	260	1.75	1.25	10.8	7.8	D^+/F^+	7.5	10.4	
1002 – 194	175	44	1.4	0.13	10.0	0.9	D^-/F^-			
1216 – 085	450	50	0.53	0.03	7.9	0.45	D^+/F^+	4.2	0.45	
781 – 275	380	345	1.38	0.88	8.7	5.6	$D^-/(F^-)$	8.1	6.7	
1562 – 192	190	70	0.64	0.14	4.2	0.95	$D^-(F^-)$	7.9	1.1	
1187 – 229	185	170	0.86	0.36	6.3	2.6	$D^+/F^+/\Lambda_c$			
1528 – 833	550	300	3.14	2.57	15.0	12.3	D^-/F^-	13.0	12.3	
Mean	501	174	1.66	0.98	11.8	6.9		8.5	6.1	$6.1^{+4.0}_{-2.0}$

Neutral

Event	d^{MAX} μm	d_2 μm	l mm	l_{eff} mm	T_{OBS} $(\times10^{-13}$s)	$\bar{T}_{eff}$ $(\times10^{-13}$s)	TYPE	$P_{CONST.}$	T $(\times10^{-13}$s)	Constrained $(\times10^{-13}$s)
944 – 015	250	180	1.6	0.9	13.6	7.7	$D^0/\bar{D}^0$	9.4	5.9	
435 – 714	220	193	1.5	0.75	18.7	9.3	$D^0/\bar{D}^0$			
1256 – 003	780	230	2.4	1.90	22.4	17.7	$D^0/\bar{D}^0$			
1216 – 085	124	86	0.62	0.07	2.2	0.45	$\bar{D}^0$	9.3	0.5	
1404 – 883	604	374	1.56	1.06	13.0	8.8	$D^0/\bar{D}^0$			
1187 – 229	888	43	1.8	0.13	22.7	1.7	$D^0/\bar{D}^0$			
1528 – 833	580	210	0.98	0.48	19.4	9.5	$D^0/\bar{D}^0$			
636 – 188	403	50	0.93	0.19	5.4	1.1	$D^0/\bar{D}^0$			
1285 – 919	707	327	4.79	4.04	23.4	19.7	$\bar{D}^0$	11.5	21.8	
1079 – 160	156	103	0.99	0.29	7.3	2.1	$D^0/\bar{D}^0$			
Mean	471	180	1.72	0.98	14.8	7.8		10.1	9.4	$9.4^{+13.}_{-3.4}$

Table II.   Life time of Charmed Particles.

Method	(i) $<\tau^{eff}>$	(ii) $<l><l^{eff}><\tau>$		(iii) $<d^{max}>$	MEAN
Charged $D^\pm/F^\pm/\Lambda^+$	6.9	7.3 7.3 6.7 7.1		9.1	$7.7^{+4.0}_{-2.0}$
Neutral $D^0/\bar{D}^0$	7.8	7.3 7.3 9.0 7.9		8.7	$8.1^{+4.0}_{-2.0}$

particles by examining masses and P_T. For the selection of charm events, the fit of invariant mass to the charm mass is not used in present analysis, since the fake rate of forming charmed particle mass combinations is considerably large. One prong decays are ignored, because of the difficulty of analysis. So far, 30 events with one or two clearly identified charmed particles were collected from these selections in a sample of 205,198 hadronic events. Table I. Figs. 3 and 4 show typical clean examples of charm events.

3. PRODUCTION CROSS SECTION

The correction for the fiducial volume for the HRO pictures (x 0.85) and the fraction of gammas above 15 GeV (x 0.85) yield a total hadronic events of 148 k events (= 205 k x 0.85 x 0.85) corresponding to the 30 charm events.
The requirements,

$$d_{max} > 110 \quad \mu m$$

$$d_{2nd} > 40 \quad \mu m \quad and \qquad (1)$$

$$l > 0.5 \ mm,$$

reduce the number of charm events to 20 from 30. A Monte Carlo simulation gives the efficiency for selecting the charm events to be $\epsilon = 0.35 \pm 0.15$ from these cuts. Therefore, the charm production cross section is estimated to be

$$\sigma_{charm} = \frac{110^{\mu b} \ x \ (20/0.8/0.9)}{(148,000/0.9) \ x \ \epsilon} = 53 \ nb,$$

where 0.9 and 0.8 are the triggering efficiency of pictures for hadron events and charm events. The 0.9 for 20 ev is the scanning efficiency of charm search. Our experiment is still continuing. Therefore, at this conference, I do not change our previous figure which we reported at Lisbon Conference, 1981, (Fig.5)

$$\sigma_{charm} = 40^{+40}_{-20} \ nb. \ (19.5 \ GeV \ \gamma p)$$

4. CHARM LIFETIME

For the study of charm lifetime, we impose the cuts (1). The two cuts on the impact parameter is to ensure that they are clear short decay with multiplicity greater than two. The cut on the decay length reduces those events ambiguous between a neutral decay or a charged decay. Thus, 10 unambiguous charged decays and 10 unambiguous neutral decays are collected (Table I).

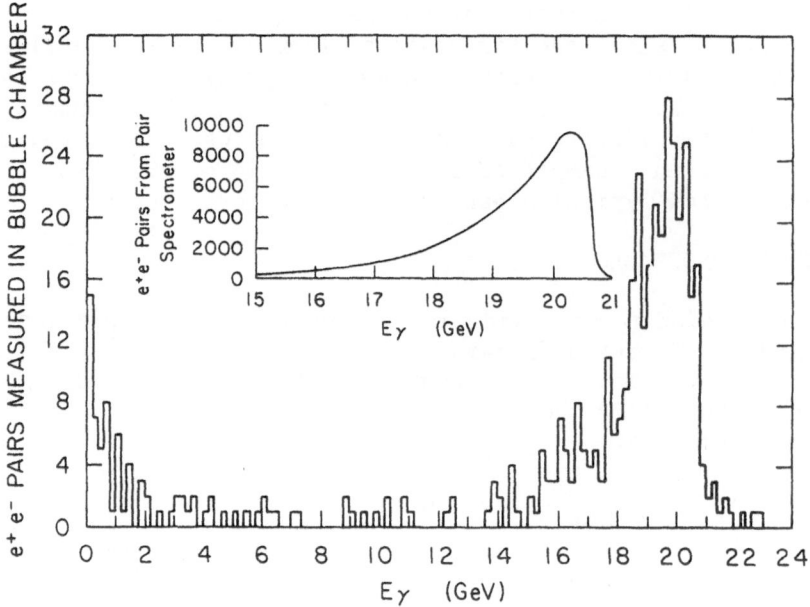

Fig. 1. The photon energy spectrum at the bubble chamber.

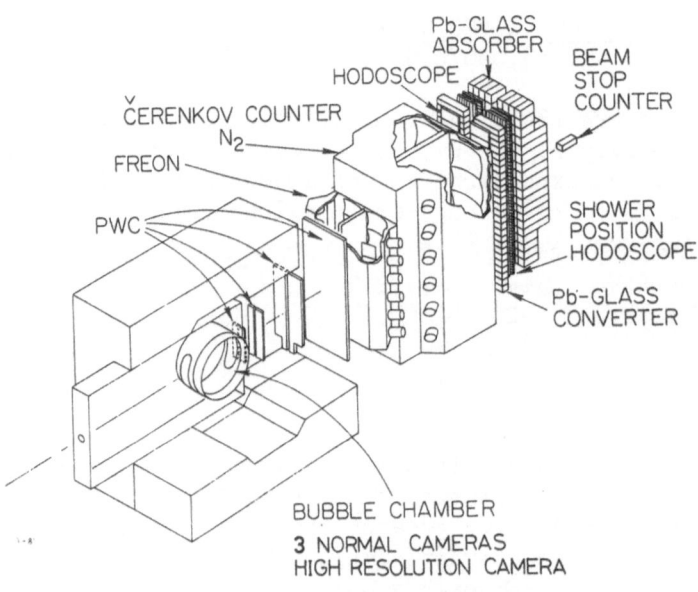

Fig. 2. The SLAC Hybrid Facility.

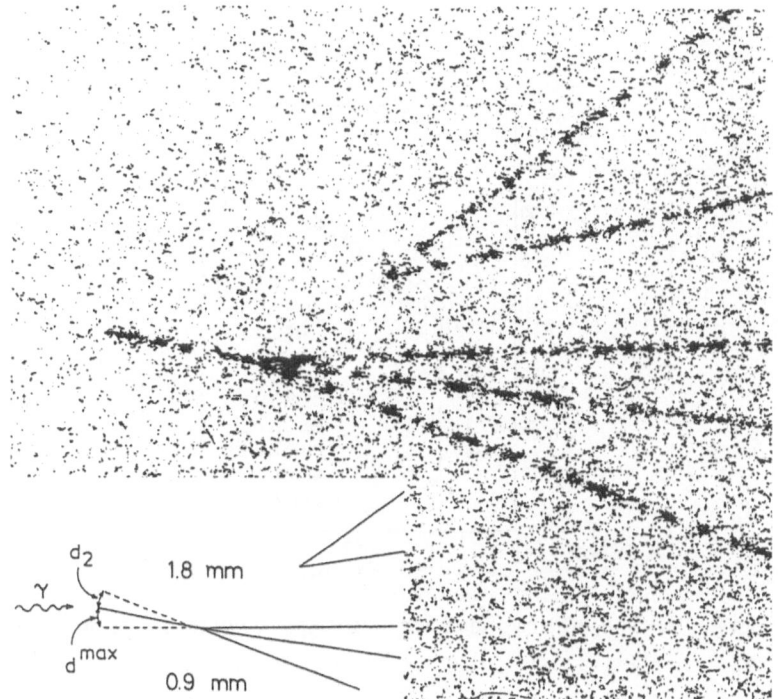

Fig. 3. Neutral V + 3 prong decay, 1187 - 229.

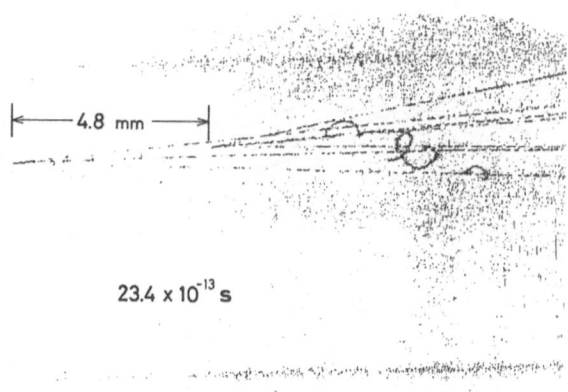

Fig. 4. Neutral 4 prong decay, 1285 - 919.

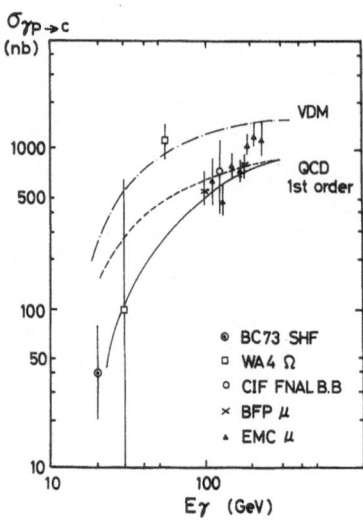

Fig. 5. The charm photoproduction cross section.

DEFINITION OF EFECTIVE DECAY LENGTH

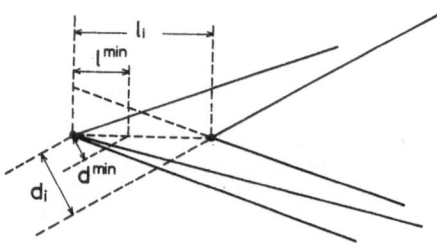

$L_{eff} = l_i - l_{min}$

Fig. 6. Effective decay length, $(l_i - l_i^{min})$.

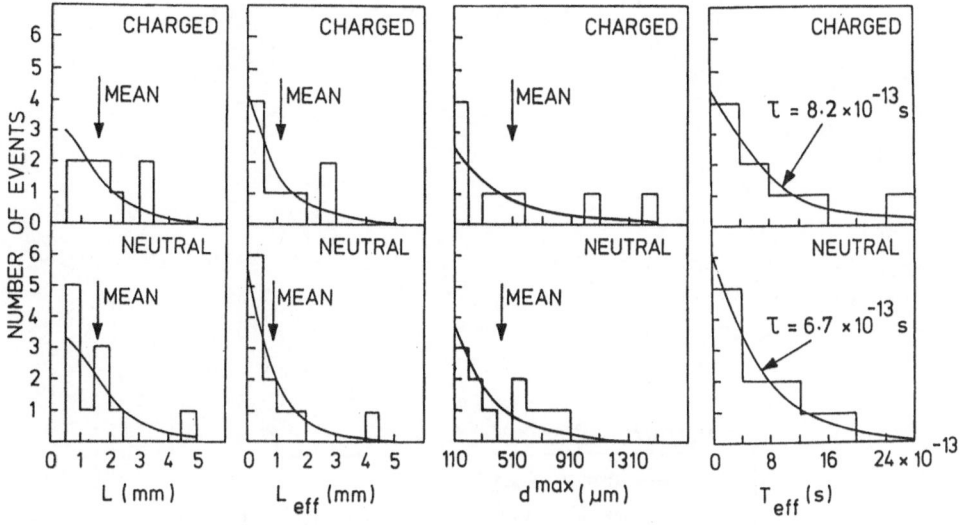

Fig. 7(a). Mean values of L, L_{eff}, d^{max} and T_{eff}, (Data).

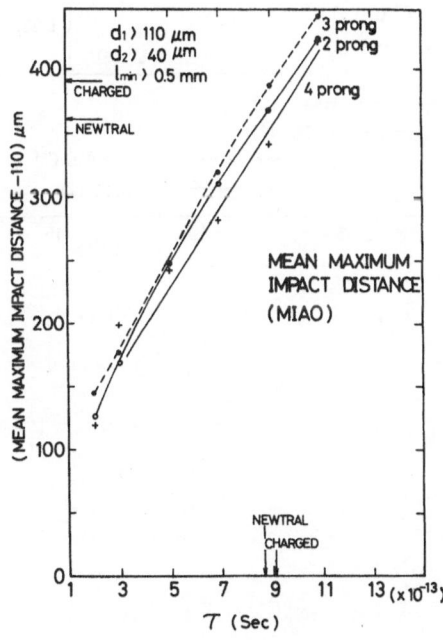

Fig. 7(b). An example of Monte Carlo simulation, for the mean maximum impact distance.

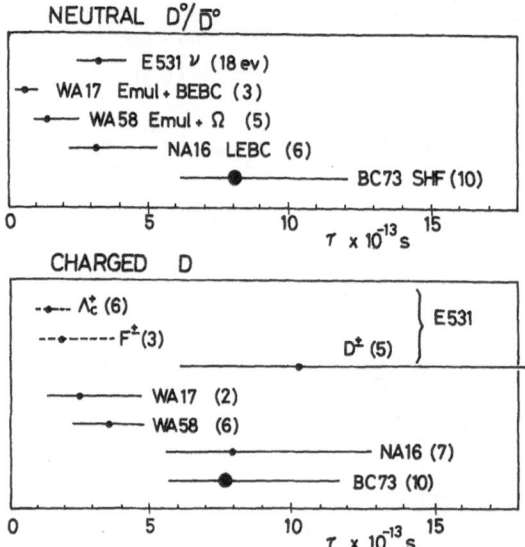

Fig. 8. Charm lifetime.

Estimation of lifetime is first based on (i) the maximum like-lihood method for the effective decay length. The effective decay length is $(l_i - l_i^{min})$, where l_i is the observed decay length and $l_i min$ is the minimum decay length under our cuts for the particular decay kinematics (Fig.6).

$$l_i^{min} = l_i \frac{d^{min}}{d_i}$$

The mean lifetime is given by

$$<\tau> = \frac{1}{N} \sum_{}^{N} \frac{(l_i - l_i^{min})}{P_i c/m}$$

where m and P_i are the mass and momentum of D particle, respectively. For the constrained 7 charged and 3 neutral events, we obtained

$$\tau(\pm) = 6.1^{+4.0}_{-2.0} \times 10^{-13}s$$

$$\tau(0) = 9.4^{+13.}_{-3.4} \times 10^{-13}s$$

To use the other unconstrained events for the life estimation, (ii) we compare the data with a Monte Carlo simulation on the mean length, mean effective length and mean proper time. (iii) also the mean maximum impact parameters are compared with a Monte Carlo simula-tion (Fig. 7 (a) and (b)). Table II summarizes the estimations of the lifetimes. The mean of (i), (ii) and (iii) gives

$$\tau(\pm) = 7.7^{+4.0}_{-2.0} \times 10^{-13}s$$

$$\tau(0) = 8.1^{+4.0}_{-2.0} \times 10^{-13}s$$

and

$$\frac{\tau(\pm)}{\tau(0)} = 0.95^{+1.0}_{-0.5}$$

Fig. 8 shows a comparison of the present data with other experiments. At the present level of the statistics, it is still premature to discuss a detailed comparison of the lifetime data.

DISCUSSIONS

Sacton: Could you tell us your mass resolution? I am quite surprised by the fact that most of your charged charmed particles are ambiguous between D^- and F^- meson.

Answer: Our mass resolution is typically ± 15 MeV. The ambiguity between $D^{\pm}$ and $F^{\pm}$ primarily come from the lack of particle identification. In the present analysis, the mass value is not used as a good signature of the charm particle.

Diambrini: A length of 100 μm, i.e. twice the bubble diameter, seen from the main vertex subtend an angle of $\sim 10°$. Is the l = 0.5 mm cut sufficient in order to avoid neutral-charged decay contamination?

Answer: Our data indicated that this 0.5 mm cut removed half of charge/neutral ambiguous events and none of the unambiguous charged or neutral decays.

Bellini: How many of your 20 candidates are found in associated production events?

Answer: In the original candidate sample of 30 events 9 events have two identified vertices. However, we have only two examples of charmed pair production in the 20 charms left for the lifetime analysis under the three cuts.

Reay: Comment. Your $D°$ lifetime is quite long, and there are two possibilities for this. One is the possibility that every experiments are wrong. There is a second possibility: if there are two lifetimes, with the short one $\sim 2 \times 10^{-13}$ sec, $\tau°_{short}$ events would all lie inside your 1/2 mm cut. You then would be sampling only the $\tau°_{long}$ lifetime.

Baltay: How many strange particle decays do you have in your $D°$ sample?

Answer: $2K°$ are observed in the 18 events containing the 20 charms for lifetime analysis. One K^+ and one K^+K^- are identified by the Cerenkov counter.

Bizzarri: For the events not constrained how can you determine the proper time, since you do not know the number of missing particles?

Answer: For the events not constrained we handled them statistically using a Monte Carlo simulation.

DISCUSSION SESSION ON: VISUAL VERTEX DETECTORS

A. Bettini

Istituto di Fisica
Via Marzolo, 8
35100 Padova (Italy)

Visual detectors of sufficient spatial resolution allow direct observation of charmed or beautiful particles decay vertices at least for proper life-times $\gtrsim 10^{-14}$ s. Information from a downstream spectrometer is needed to distinguish the signal from topologically similar backgrounds (important even in hydrogen targets), to identify the parent hadron without combinatorial confusion and obviously to determine its momentum. Identification of the charged decay particles and π^0 detection of a large solid angle are also necessary for a complete reconstruction.

The discussion session aimed to analyse the advantages and limitation (biasses, trigger rates, analysis times, etc.) of the present high resolution visual techniques (mainly emulsions and bubble chambers) with a look on future developments (emulsions scanning automation and holographic readout of bubble chambers).

Jean Sacton gave an introduction with a comparison between the emulsion and bubble chamber techniques, Romano Bizzarri discussed classical optics for high resolution bubble chambers and Paul Lecoq the tests made for the holographic technique. Charles Baltay (that did not submit a written version of his talk) and T. Kitagaki presented the current ideas on holography in big bubble chambers expecially for ν_τ physics; Sergio Natali and V.Z. Peterson discussed the scanning and measurements of the holograms. Lucien Montanet talked on the physics programs at the EHS (no written version included in these proceedings) and G. Van der Hage (no written version) and G. Romano on advances in photographic emulsion technique.

From the discussion the conclusion can be drawn that high resolution (classical or holographic bubble chambers) are well adapted for the study of charmed hadrons (lifetimes and production mechanisms) when coupled to a spectrometer, while emulsions seem to be better suited to search for beautiful particles, provided selective enough triggers are found.

PHOTOGRAPHIC EMULSION VERSUS BUBBLE

CHAMBERS IN CHARM AND BEAUTY SEARCHES

Jean Sacton

Université Libre de Bruxelles

INTRODUCTION

This talk is intended to serve as an introduction to the discussion session on the use of visual detectors in the search for charm and other flavours. I have limited it to the presentation of some general considerations and a few elements of comparison between the photographic emulsion and the bubble chamber techniques. In this field of research both these techniques are currently used as high spatial resolution vertex detectors incorporated in large and complex counter set-ups (*).

The main difficulties encountered in searching for charmed and beautiful hadrons are related to the short lifetimes of these particles ($\lesssim 10^{-12}$ s) and to their small production cross-sections (see Table 1), even at SPS energies. In addition, most of the charmed hadrons present a large variety of decay modes of which only a fraction has been identified to date. First results from CESR indicate that the average charged particle multiplicity in the hadronic decay of beautiful hadrons is as high as 6.31 ± 0.35[1]; no B meson decay has yet been kinematically reconstructed.

Most of the figures quoted in this report are indicative only.

(*) Note that large bubble chambers are successfully used for charm studies in neutrino interactions and for beam dump experiments. This aspect will not be discussed here.

RESOLUTION AND VISIBILITY

In Table II, I have displayed typical values of some parameters measuring the spatial resolution of visual detectors such as photographic emulsion and the small bubble chambers in use at CERN for short-lived particle studies. These parameters are (i) $\underline{d}$, the size of the track element (diameter of a grain or bubble), $\overline{(ii)}$ $\underline{n}$, the number of elements per unit of track length (density) and $\overline{(iii)}$ $\underline{x}$, the distance over which observation around the vertex is possible, in the same view field. They are to be compared with the physical quantities characterizing the decay of unstable particles (see fig. 1) : (i) the flight path $l = \beta\gamma \, c \, \tau \sim \frac{P}{m} x$ 30 μm for a lifetime $\tau = 10^{-13}$ s and (ii) the impact parameter $y \sim c \, \tau$. For $\underline{visual}$ detection it is generally agreed that y should be about three times the size of the track element. The figures of Table II concerning the holographic chambers should be taken with caution because this technique is still in the development phase. It should be noticed, however, that the presently achieved bubble density is quite low. Smaller mean gap lengths are required for an efficient and unambiguous resolution of the production and decay processes at high energies.

The relative resolution power of the methods under discussion is provocatively illustrated in fig. 2 taken from reference (2). It shows an example of the associated photoproduction of charmed particles actually observed in emulsion[3] (fig. 2.a) as it would appear in an holographic bubble chamber with d = 5 μm, n = 500/cm (fig. 2.b) and in LEBC in running conditions similar to those of the NA16 experiment, d $\approx$ 40 μm, n $\approx$ 70/cm (fig. 2.c).

From these considerations, it can be concluded that in the near future it is unlikely that the bubble chamber technique will seriously compete with photographic emulsion in the detection of particles with lifetime smaller than 5×10^{-14} s. It is worth mentioning that the empirical limit quoted above for the y-parameter can be significantly reduced in emulsion at the price, however, of systematic measurements of the coordinates of individual grains along each track. S. Petrera and G. Romano[4] have shown that in good experimental conditions y distances of the order of a few tenths of a micron can be resolved which would allow the observation of particles with lifetimes as short as a few times 10^{-15} s.

DATA ANALYSIS RATE AND EXPOSURE TIME

Let us consider the case of hadronic charmed particle production at SPS energies. A "standard" (i.e. large) bubble chamber Collaboration can handle (scan, rescan and check) some 500.000

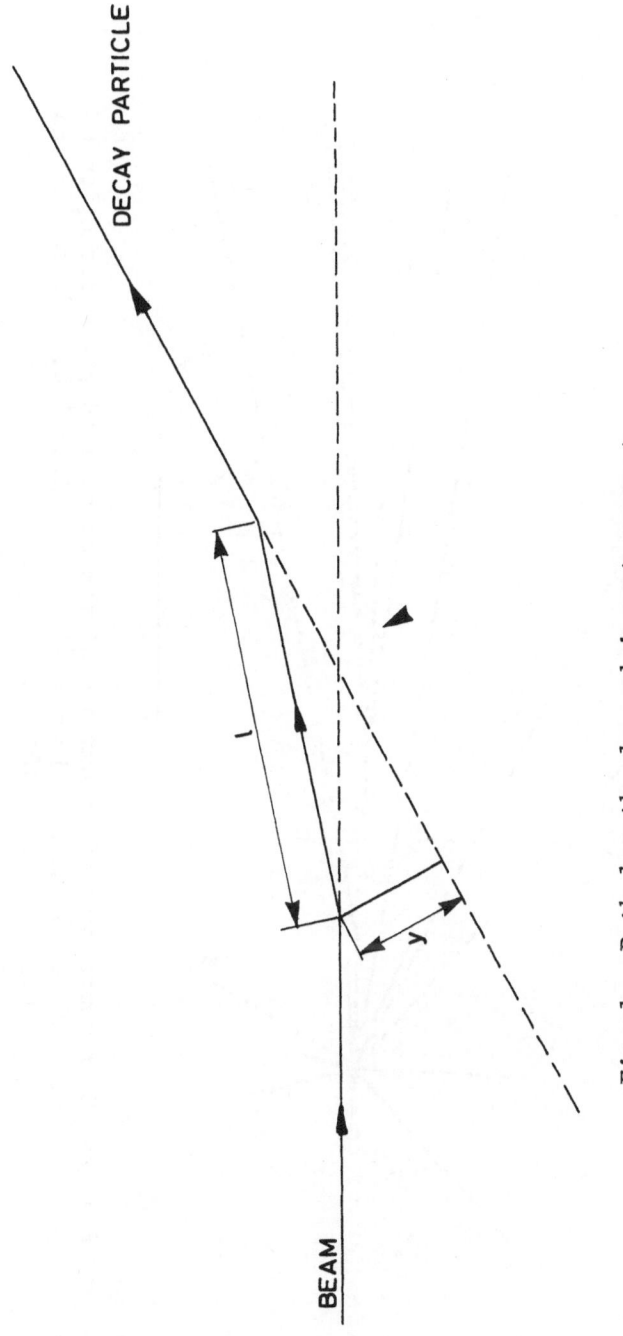

Fig. 1. Path length -l- and impact parameter y.

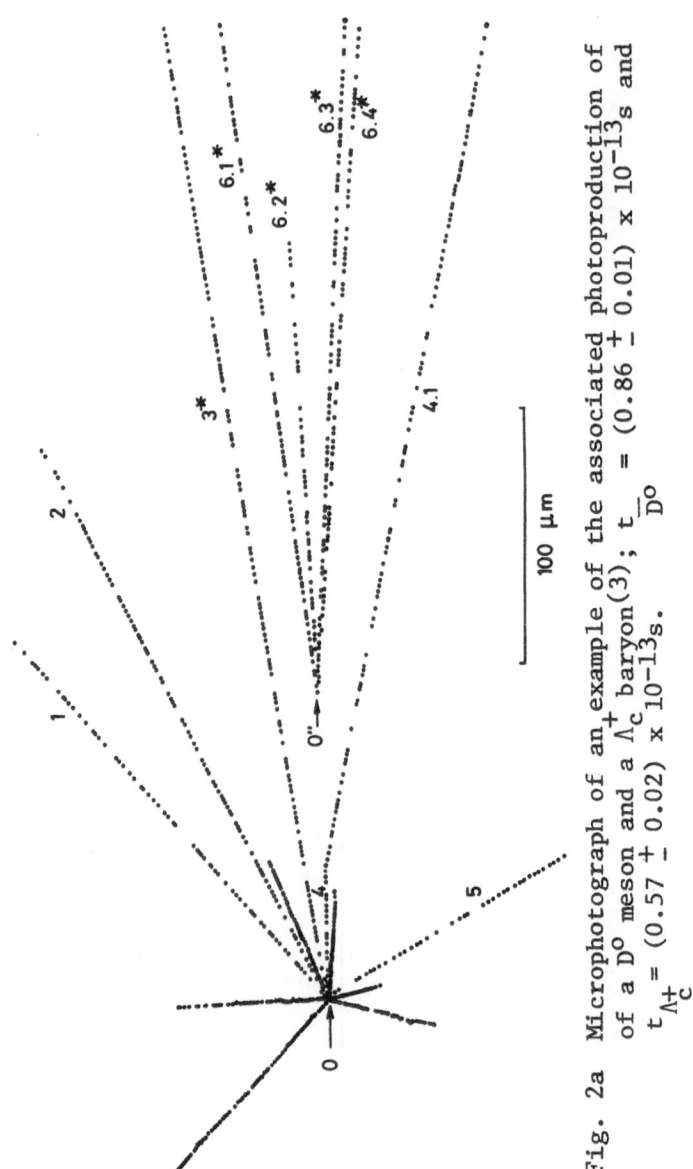

Fig. 2a Microphotograph of an example of the associated photoproduction of
of a D^0 meson and a Λ_c^+ baryon(3); $t_{\overline{D}^0} = (0.86 \pm 0.01) \times 10^{-13}$s and
$t_{\Lambda_c^+} = (0.57 \pm 0.02) \times 10^{-13}$s.

Holographic LEBC $d = 5 \, \mu m$
                          $n = 500 \, / \, cm$

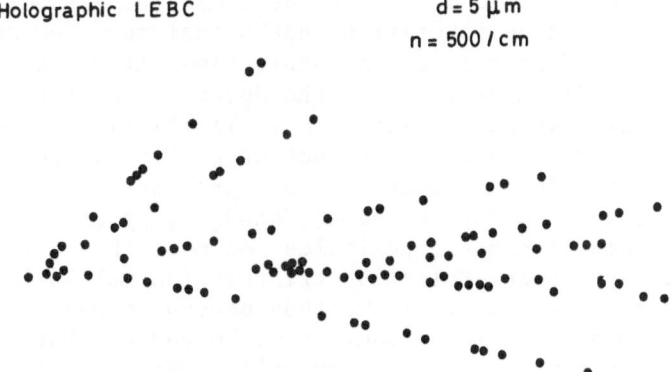

Fig. 2c   The same event as it would be seen in LEBC (NA16 running
          conditions). This figure is taken from ref. (2).

50 μm 40 μm bubbles
                                $n = 70 / cm$

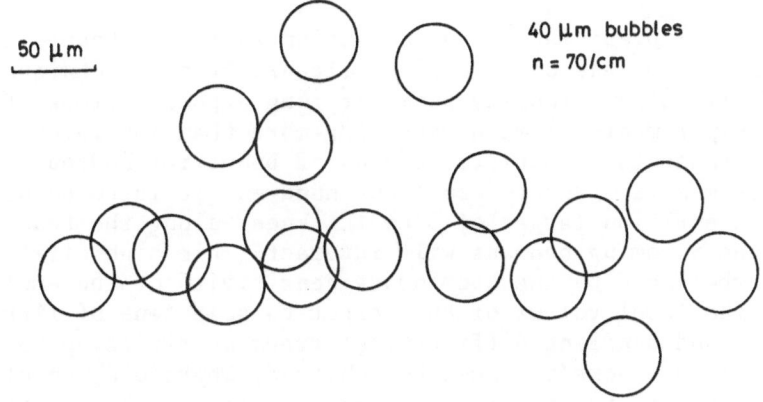

Fig. 2b.   The same event as it would be seen in a holographic
           bubble chamber.

Table 1

Projectile	$\dfrac{\sigma_{charm}}{\sigma_{tot}}$	$\dfrac{\sigma_{beauty}}{\sigma_{tot}}$ (expected)
Hadrons	$\sim 10^{-3}$	$\sim 10^{-6}$
photons	$\sim 10^{-2}$	$\sim 10^{-5}$
neutrinos$^{(*)}$	$\sim 10^{-1}$	$\sim 10^{-3}$

(*) Note that σ_{tot} for neutrinos is small

interactions in 12 to 15 months. Using a minimum bias or interaction trigger such a collaboration should analyse a few hundred charmed particle decays in a reasonable time. At variance, the small available field of view for the observation of emulsion pellicles under high magnification (see Table II) implies tedious and time consuming track following to detect particles decaying after flight paths of a few millimeters. This procedure reduces the scan speed, as compared to bubble chamber studies, by at least a factor 50 for horizontally exposed pellicles and more than a factor 10 for vertical exposures. Selective triggers (on multiplicity, presence of muons or kaons, ...) are thus needed to reduce significantly the number of interactions to be looked at. Unfortunately, such triggers often preclude or make difficult reliable estimates of absolute decay rates and cross sections. It is thus not surprizing that there are presently no dedicated charmed particle hadro-production studies running with photographic emulsions.

The small predicted hadro-production rates for beautiful particles impose the use of selective triggers in all techniques, combined with high intensity beams or long exposure times. Four emulsion experiments[5] each requiring more than 10^8 interactions are now planned to accumulate samples of beautiful hadron decays ranging from a few tens to about one hundred. It is to be noted that in an emulsion target of 3 cm thickness along the beam about 10 % of the incoming hadrons will interact. This high efficiency is counterbalanced by the continuous sensitivity of the emulsion. To limit the total volume of the target to some tens of litres (high cost and handling difficulties) track densities up to or more than 1000 tracks/mm^2 must be achieved. Therefore, an efficient location of the interesting events -selected by the triggering system- implies the use of accurate beam hodoscopes and vertex locaters. Silicon microstrip detectors, high precision drift chambers and CCD devices are expected to provide vertex predictions with accuracies of about 20 μm across the beam and about 100 μm along the beam.

In this field of research the high resolution bubble chambers are severely handicaped by their short sensitive time, their low repetition rate and their small fiducial volume (the use of high resolution optics limits the depth of field to a few millimeters). Typically, they can record a maximum of 10 to 15 interactions per burst of $\sim 5 \times 10^5$ particles ! Holographic bubble chambers should improve this situation by at best a factor 10 to 20, still leaving beauty hadro-production studies questionable. It could be worthwhile to investigate the feasibility of photo-production experiments where one expects a more favourable signal to noise ratio.

Table 2

Technique	Grain or bubble diameter (d) (μm)	Grain or bubble density (n)	Track length in view field (x)
Nuclear emulsion	0.5 - 1.0	$20\text{-}30/_{100}$ μm	~ 100 μm (horizontal exposure) ~ 300 μm (vertical exposure)
Small bubble chambers with high resolution optics e.g. LEBC (*)	40 - 50	$\sim 60/_{cm}$	a few cm
e.g. HOLEBC (**)	20 - 25	$\sim 80/_{cm}$	a few cm
Small bubble chambers with holographic optics e.g. HOBC (***)	~ 10	$200 - ?/_{cm}$	a few mm (TV scan) a few cm (scan table)

(*)   Lexan Bubble Chamber used in the CERN NA16 experiment

(**) Hydrogen Holographic Lexan Bubble Chamber tested recently at CERN with both high resolution and holographic optics.

(***) Holographic Bubble Chamber : heavy liquid bubble chamber tested recently at CERN.

CONCLUSIONS

i) The bubble chamber technique (high resolution or holographic optics) is well suited to the study of charmed hadrons with lifetimes in the range 10^{-13} to 10^{-12} s.

i) Searches for beautiful hadrons remain presently a domain for triggered emulsion experiments due to the smallness of the production cross-sections (provided the lifetime is not much shorter than 10^{-14} s).

iii) For particles of lifetimes shorter than a few times 10^{-14} s the emulsion technique is still without competitor.

REFERENCES

(1) M.S. Alam et al., contributed paper to the 1981 International Symposium on Lepton and Photon Interactions at High Energies held at Bonn.

(2) J.H. Mulvey; invited talk at the 1980 SLAC Summer Institute Conference - Oxford University preprint 84/80.

(3) A. Fiorino et al.; Lett. al Nuovo Cimento 30 - 166 - 1981.

(4) S. Petrera and G. Romano; Nucl. Instr. 174 - 61 - 1980.

(5) E653 at FNAL (B. Reay), WA71 (G. Diambrini-Pallazzi), P166 (P. Musset) and P167 (G. di Caporiacco) at CERN.

HIGH RESOLUTION BUBBLE CHAMBERS

R. Bizzarri[*]

CERN
EP-Division
CH-1211 Geneva 23

I shall briefly discuss the performances obtained or to be expected from small bubble chambers with "classical" optics (i.e. no holography). Examples of such chambers are the heavy liquid chamber BIBC and the two hydrogen chambers LEBC and HOLEBC. I shall specifically focus my attention to these last two.

Limits on the accessible cross sections

Let it be n = of particles in the picture ($\lesssim$ 20)
ν = expansion frequency of the bubble chamber ($\sim 30s^{-1}$),
the useful incident flux during the accelerator flat top is then
$n\nu$ = 600 particles/s. The effective flux f is obtained by multipling by the accelerator duty cycle. For SPS the duty cycle is 2/12 and $f \sim$ 100 particles/s; a low flux indeed.

For a given cross section σ and a data taking time T it is straightforward to compute the number of events $n_{ev} = f.\sigma.\ell.\rho.N.T$, where

ℓ = useful length of the bubble chamber $\sim$ 10 cm.
ρ = density of the liquid hydrogen = 5.6×10^{-2} g/cm^3.
N = Avogadro's number = 6×10^{23}.

A reasonable assumption (also in terms of film to be scanned) is $T \sim 10^6$s ($\approx$ 12 days at full efficiency), giving a sensitivity

_____

[*] On leave from the Rome University, Rome, Italy.

of 35 ev/µb: this technique is suitable to study events with a
cross section of at least a few µb.

Limits on the Accessible Life Times

Due to the confusion introduced by other tracks coming from
the primary vertex, the best signal of the presence of a short
lived decay is given by the fact that its decay products do not
point to the production vertex. A reasonable criterion is to ask
that the "impact parameter" y (fig. 1) is bigger than ∿ 2 times
the minimum resolved bubble size. It is

$$y = L \sin\theta_d = \beta\gamma ct \ \frac{p^* \sin\theta^*}{\gamma p^* \cos\theta^* + \beta\gamma E^*} = ct \left(\frac{\beta\beta^* \sin\theta^*}{\beta^* \cos\theta^* + \beta} \right)$$

where the quantities

$$
\begin{aligned}
L &= \text{path length} = \beta\gamma ct \\
t &= \text{life time} \\
\beta c &= \text{velocity} \\
\gamma &= 1/\sqrt{1-\beta^2}
\end{aligned}
$$

refer to the parent particle, and

$$
\begin{aligned}
\theta_d &= \text{emission angle in the laboratory.} \\
\theta^* &= \text{emission angle in the c.m.s.} \\
p^*, E^*, \beta^* c &= \text{momentum, energy and velocity in the c.m.s.}
\end{aligned}
$$

refer to the secondary particle – the factor inbetween
parenthesis varies strongly with θ^*, but its average value is
of the order of magnitude of 1. A particle of mean life τ can
therefore be detected with a reasonable efficiency if it is

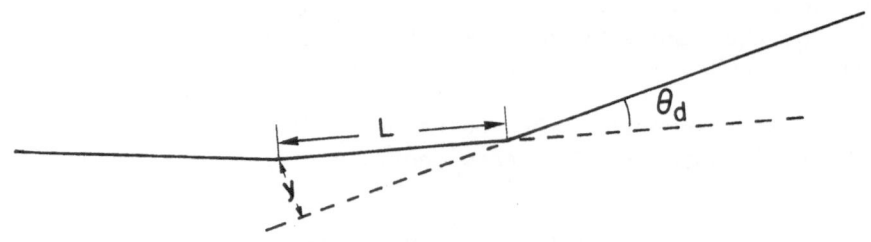

Fig. 1: The "impact parameter" $y = L \sin\theta_d$ of the decay
         product of a charmed particle travelling a distance of L.

$$c\tau \gtrsim 2 \times R$$

R being the minimum distance between resolved bubbles. As we shall see, it is $R \simeq 20 \div 40$ μm which requires $\tau \gtrsim 2 \times 10^{-13}$ s for the technique to be useful.

Limits on the Resolution

For a perfect lens, the minimum distance R between point objects resolved according to the Rayleigh criterion is related to the half-depth of field δ (defined by the Rayleigh criterion) by the relation

$$R = 0.61 \sqrt{2\lambda\delta}$$

which is illustrated in fig. 2 for a wave length λ = .5 μm.

The depth of field must be large enough so as to contain a large fraction of the incident beam. For instance the H2 beam at the CERN-SPS has an r.m.s. spread at its horizontal focus of .7 mm; fig. 3 shows the fraction of the primary beam within ± δ. A reasonable working point could be at δ = 1 mm (80% of the beam in focus), which gives R = 20 μm.

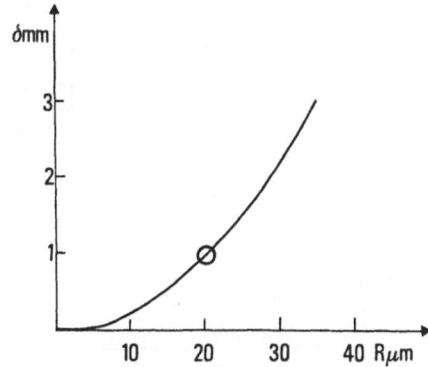

Fig. 2: The minimum resolved distance between bubbles R vs. the half depth of field δ. A possible working point is indicated.

In the experiment NA16 with the bubble chamber LEBC, two
lenses were used of focal length f = 180 mm, open at f/11 and
with a space to film demagnification m = 3.2. The half depth of
field was δ = 2 mm and the theoretically expected resolution
R' = 27 μm. Experimentally it was found R ≃ 40 μm. One could not
do better because of:

- Effect of the bubble chamber and vacuum tank windows (mainly
  astigmatism and field curvature).

- Limits on the lens quality at larger openings.

- Effect of the limited film resolution (because of the 3.2
  demagnification).

In the next proposed experiment with HOLEBEC, one could use
lenses of f = 300 mm at f/17 with a demagnification m = .9. The
half-depth of field should then be δ = 1 mm and one can expect
a resolution very near to the theoretical value R = 20 μm, since

- The lenses will be corrected for the presence of the windows.

- The lenses will be diffraction limited up to f/6.5.

- The film resolution will be much less important because of the
  small demagnification.

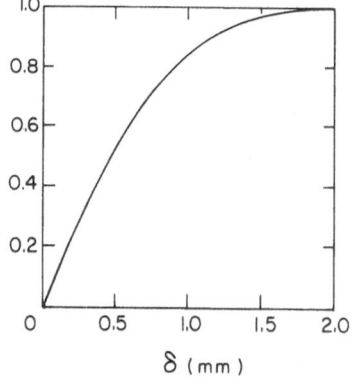

Fig. 3: Fraction of the incident beam in focus vs. the half depth
        of field σ.

Bubble Density

   The density of bubbles along the track is also an important
parameter for the human ability to detect a short decay. A
bubble density of 150 bubbles/cm should be possible,
corresponding to an average distance between bubbles of 67 μm.
This is a confortable situation, but a much smaller bubble
density would be likely to reduce the detection efficiency.

Comment by N. Reay

      There is a difference between finding decays and fitting decays.
Impact parameter techniques are good for finding, but many classes
of events require a separation between production and decay many
times the instrument resolution if you wish to fit the event. For
example, $D^{\pm}$ and Σ_c^{++} decays would give a pion track from the production vertex which is very close to the decay tracks, many times with-
in a few milliradians. One needs either very good spatial resolution and/or exceptional mass resolution to facilitate fitting.

HOLOGRAPHY IN SMALL BUBBLE CHAMBERS

P. Lecoq

CERN-EP
1211-Geneva-23

INTRODUCTION

Several authors have mentioned that the holographic technique could be used to register particle tracks produced in bubble chambers [1-3]. The striking advantage over classical optics lies in achieving an excellent spatial resolution without sacrificing the depth of field.

Interest in spatial resolutions of a few micrometers is motivated by the short lifetime of charm particles, typically a few 10^{-13} s. For a detailed study of the production and decay mechanisms it is a great advantage to register the decay of these particles by direct observations.

The first holographic recording of bubble chamber tracks was done at the CERN-SPS in 1981 [4]. In a small bubble chamber filled with freon one obtained a spatial resolution of 8 μm over a depth of field of around 10 cm. The bubble density was around 300 per centimetre.

The small, heavy liquid bubble chamber HOBC was built for the NA25 experiment [5], the main aim of which is to determine the total charm cross section at different incident momenta. The charm signal is very much enhanced by using a muon trigger, which will result in a relatively small number of holograms to study.

Holography in liquid hydrogen was also tested using the holographic lexan bubble chamber HOLEBC with the aim to prepare a future holographic experiment in hydrogen [6].

FIRST RESULTS FROM THE NA25 EXPERIMENT

The freon bubble chamber HOBC

The holographic bubble chamber, HOBC, is a small (2ℓ) heavy liquid rapid cycling bubble chamber designed for the use of in-line holography. It is a see-through chamber with two high quality optics windows in BK7.

The film plane is situated 12.5 cm from the centre of the fiducial volume. It is sucked onto a five faces capstan, which is fixed on the vacuum tank. The useful part of the film has the same size as the cross section of the bubble chamber liquid (11 x 5 cm^2).

More details on the bubble chamber are given in Ref. 12.

The first holographic arrangement

The bubble chamber was designed to allow the maximum flexibility for the optics. Direct illumination is possible through two optical windows covering an optical volume of 11 x 5 cm^2 by 6 cm in depth.

After some tests on an optical bench, where both the in-line and the two-beam geometries have been tested, the in-line (or Gabor type) holographic set-up was preferred for the following reasons:

   i) It is very simple to adapt to a small bubble chamber of the HOBC type;
  ii) The set-up for the scanning machine is also very simple;
 iii) The required laser power is a minimum for both the recording (about 2 mJ for a 50 cm^2 hologram) and the replay of the holograms (1mW with an optical magnification of 5, which gives a total magnification of about 100 on a TV screen);
  iv) The resolution in the HOBC geometry at 12.5 cm distance from the high-resolution target (USAF 51 Chart) was better than 4 μm. This distance between the film plane and the HOBC centre was chosen as small as possible because previous tests showed a dramatic deterioration in resolution at increasing distances [7];
   v) The contrast is very high, better, for instance, than for conventional pictures. In addition, it can easily be enhanced on the replay machine by very simple suppression of the background with a spatial filter.

The laser is an excimer laser (the amplifier medium is the excimer gas XeCℓ) producing 10 ns ultraviolet pulses (308 nm)

with an energy of 200 mJ at a frequency up to 30 Hz.
This laser pumps a dye laser filled with Coumarin-307 in order to
select the wavelength of 514 nm fitting the argon line for the
replay system, with a maximum output energy of 15 mJ. In fact,
the dye amplifier was obscured to limit the output energy to 3 mJ
only, which was enough to expose the film at a density of 1.3.
The film, Agfa 10E56 emulsion on a 170 µm polyester base with
antihalo coating, was sucked onto a metallic capstan. Exposed at
a density of 1.3 the noise was quite similar to that of 10E56
holographic plates and the signal-to-noise ratio was excellent.

The improved holographic set-up

    The first holograms showed that the picture quality was very
dependent on the turbulence in the chamber. As it is practically
impossible to obtain pictures with a few microns resolution
through a turbulent medium with any optical system (classical or
holographic), extensive work was done to define operating
conditions for the bubble chamber with the maximum liquid
stability. In addition, the effect of the turbulences was
analysed in order to define an optical set-up minimizing their
influence. The turbulences affect the illuminating wave as it
travels to the object, causing a non-uniform illumination of the
optical field owing to amplitude variations. The object wave is
also distorted, both in amplitude and in phase. Finally the
reference wave going through the medium also suffers from phase
and amplitude variations, causing an unpleasant background and
additional sources of noise at the reconstruction.

    However, a possible compensation of the variations of the
object and of the reference beam exists under certain conditions.
A given bubble diffracts useful information in a forward cone,
which defines an area on the hologram the diameter of which
depends on the distance of the film to the bubble. This area also
gives the useful section of the reference beam for this bubble.
If this area can be made small compared to the mean turbulence
size, both the object and the reference rays are affected by the
same phase shifts and the interference pattern is not modified by
the turbulences. More details are given in the literature [8,9].
The obvious solution is to reduce the distance of the object to
the hologram. As this was not possible in our case due to
mechanical constraints, we decided to use a relay lens of very
good quality and to take a hologram of the image of the bubble
rather than of the bubbles themselves. In order to keep the
reference beam parallel a field lens system was added
approximately in the plane of image formation of the first lens;
with this configuration the constraint on the optical quality of
the field lens is a minimum as only a small part of it is used for

a given bubble. In fact the field lens system was made of two
converging lenses with relatively large radius of curvature in
order to minimize aberrations due to prismatic effects on the
edges of the field. The idea to use an afocal system was already
mentioned in previous publications [10,11]. The new optical
set-up is shown in Fig. 1. With this system the distance from the
centre of HOBC to the hologram has been artificially reduced from
12.5 cm to 4 cm. This system of lenses had a magnification of 1.1.
Data taking and results

Data taking and results

The use of the afocal system dramatically improved the
picture quality, especially when the chamber liquid was not
perfectly stable. Very good images with sharp bubbles and an
impressive contrast have been obtained. With a laser delay of
3 µs, the reconstructed bubble size measured with a microscope
was around 12 µm (full width above the background). The
intrinsic resolution of the system is probably better than 10
µm, as the bubbles appear round-shaped and sharp (Fig. 2).
Fig. 3(a) shows a bubble profile projected on a CCD camera. One
division corresponds to 12 µm in space. Two bubbles are easily
separated at a distance of about 10 µm (Fig. 3(b)). The
contrast is better than 70%. The slope of the profiles suggests
that the limiting resolution is better than the bubble size as the
top is flat and the edges are sharp (Fig. 3(c)). With a delay of
1 µs, 7 µm bubbles were seen, but not so sharp and with much
less contrast; at this level the quality was probably limited by
the relay lens which has a frequency cut-off of about 6 µm.

The 3 µs delay was finally chosen for the data taking.
Under these conditions more than 10000 holograms with an event
trigger and 1100 holograms with a muon trigger have been taken
with a bubble density of 200 bubbles cm^{-1} for the beam tracks.
These holograms will be used for the first physics study of
particle interactions using the holographic technique.

Unfortunately the expansion rate of the bubble chamber had to
be limited to 1.4 Hz, because of the time needed to evacuate the
heat left in the liquid after the bubble recompression, which was
causing turbulent channels. There is some hope to improve the
situation by using a modified expansion system and working at
higher temperature. Tests are now in preparation. A series of
holograms were taken at a high particle flux in order to test the
storage capability of holograms. As we were working in a slow
extracted beam a kicker magnet was installed (risetime
∿ 15 µsec) which could remove the beam before and after the
useful bubble chamber sensitive time for a total of 5 µsec, but
even when the beam was kicked out, some particles from the beam
halo went through the chamber. At fluxes in excess of 10^6

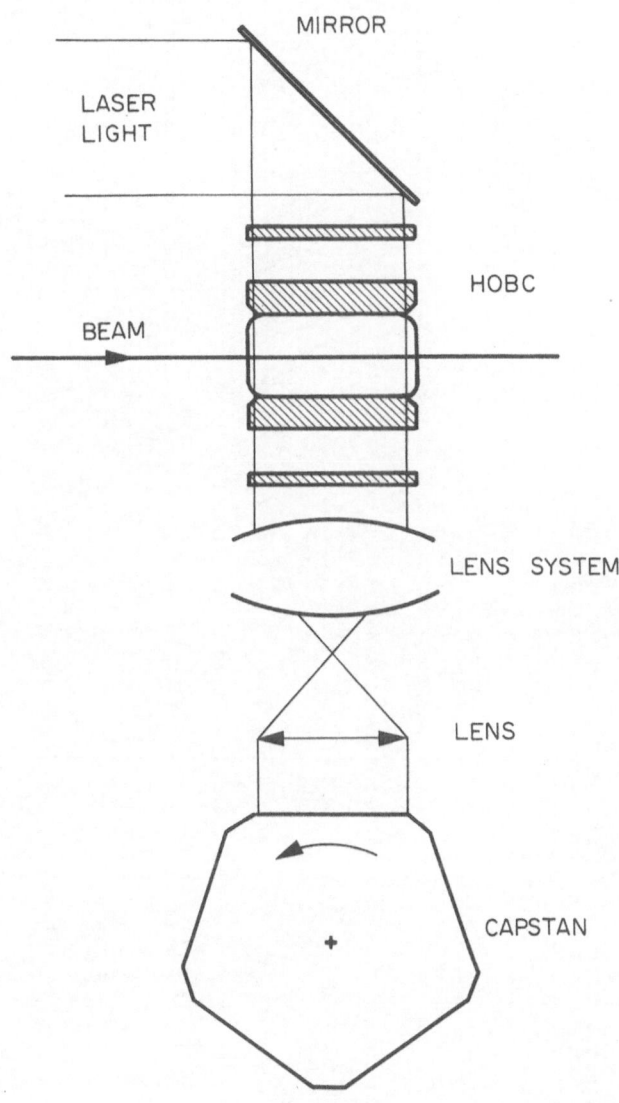

Figure 1

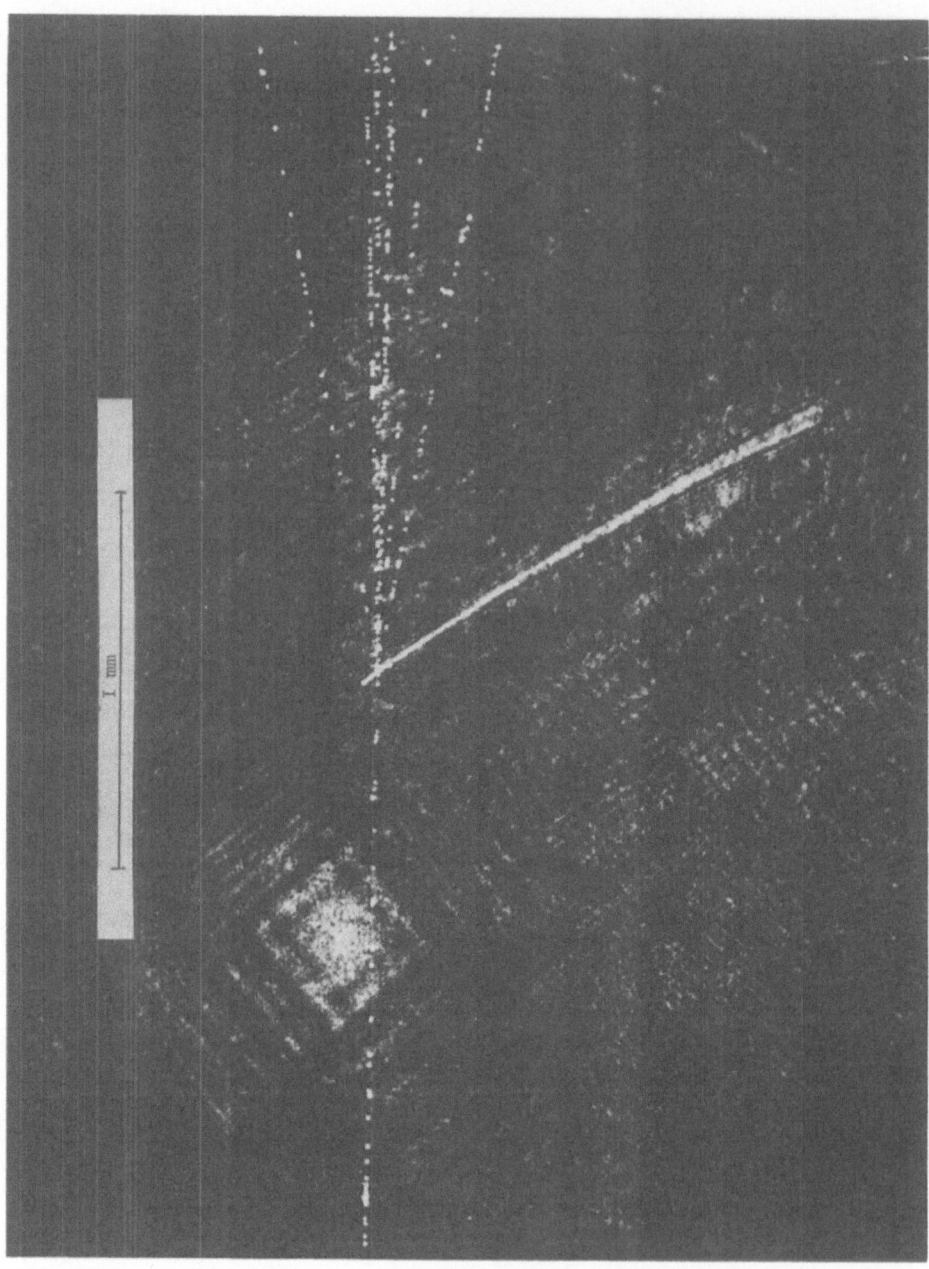

Figure 2

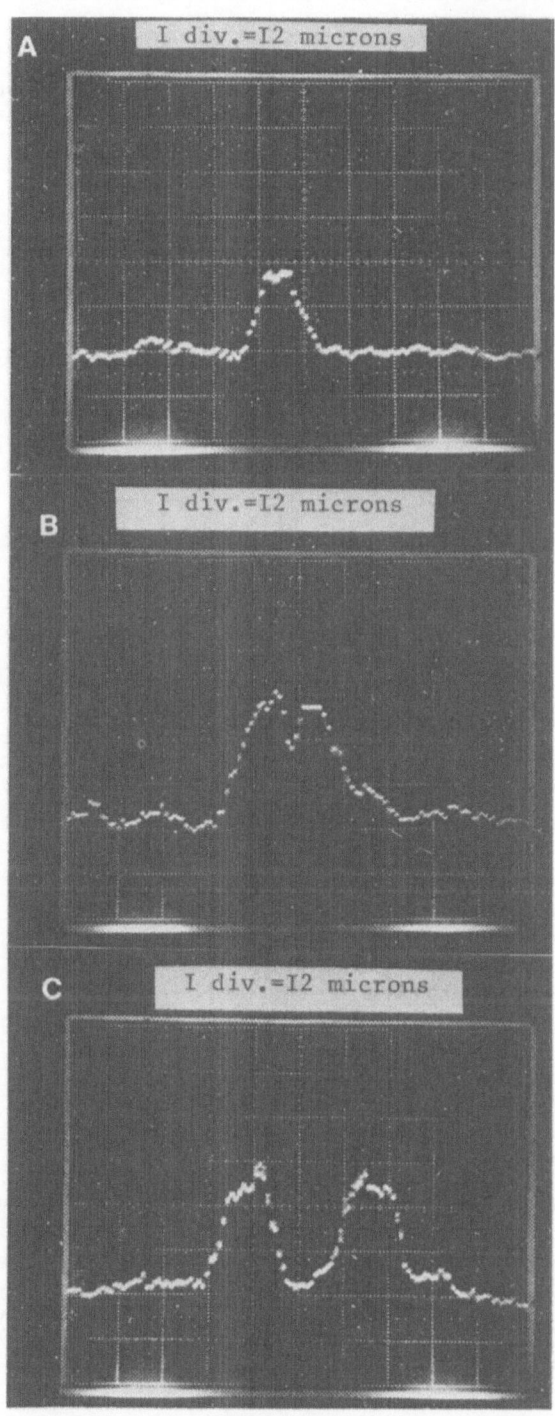

Figure 3

particles per second, the probability was high of having a few
very old tracks (1-2 ms). In the vicinity of these tracks the

image quality becomes poor. However, more than 100 tracks have
been stored on some holograms. In this case the image quality is
still very good if all the tracks have bubbles smaller than
50 μm. We just noticed a slight degradation of the
signal-to-noise ratio.

PRELIMINARY TEST IN HOLEBC

Experimental set-up

    HOLEBC is a name for the hydrogen holographic lexan bubble
chamber which is described in details in ref. [6]. It is also a
see-through bubble chamber with about the same dimensions as
HOBC. The differences with HOBC as far as the optics is concerned
are :

- The main window is made out of lexan which is not a very good
  optical material.
- Due to the vacuum tank, it was not possible to place the film
  at a distance smaller than 350 mm from the centre of the
  chamber.
- The chamber was running at 30 Hz, but the liquid was very
  turbulent, which caused some deterioration to the image quality
  even for classical optics pictures (either dark-field or
  bright-field). Since the time of the test the situation has
  been improved by a large factor.

It was decided to try two different holographic set-ups:

- The in-line combined with the afocal system (the same which was
  used with HOBC) in order to compensate for the distance effect
  and to be less sensitive to the turbulences.
- A two-beam set-up where the reference beam is not going through
  the chamber liquid in order to test the possible advantage of
  having a clean reference wave.

Results

    Results with the in-line set-up were very encouraging. Of
course the turbulences caused an unpleasant background and some
distortions on the tracks but not worse than for classical optics
bright-field images for objects having the same size. Very good
15 μm tracks were seen with a high contrast (fig. 4). Due to
the level of turbulence the quality of 10 μm tracks was just at
the limit for an efficient scanning.

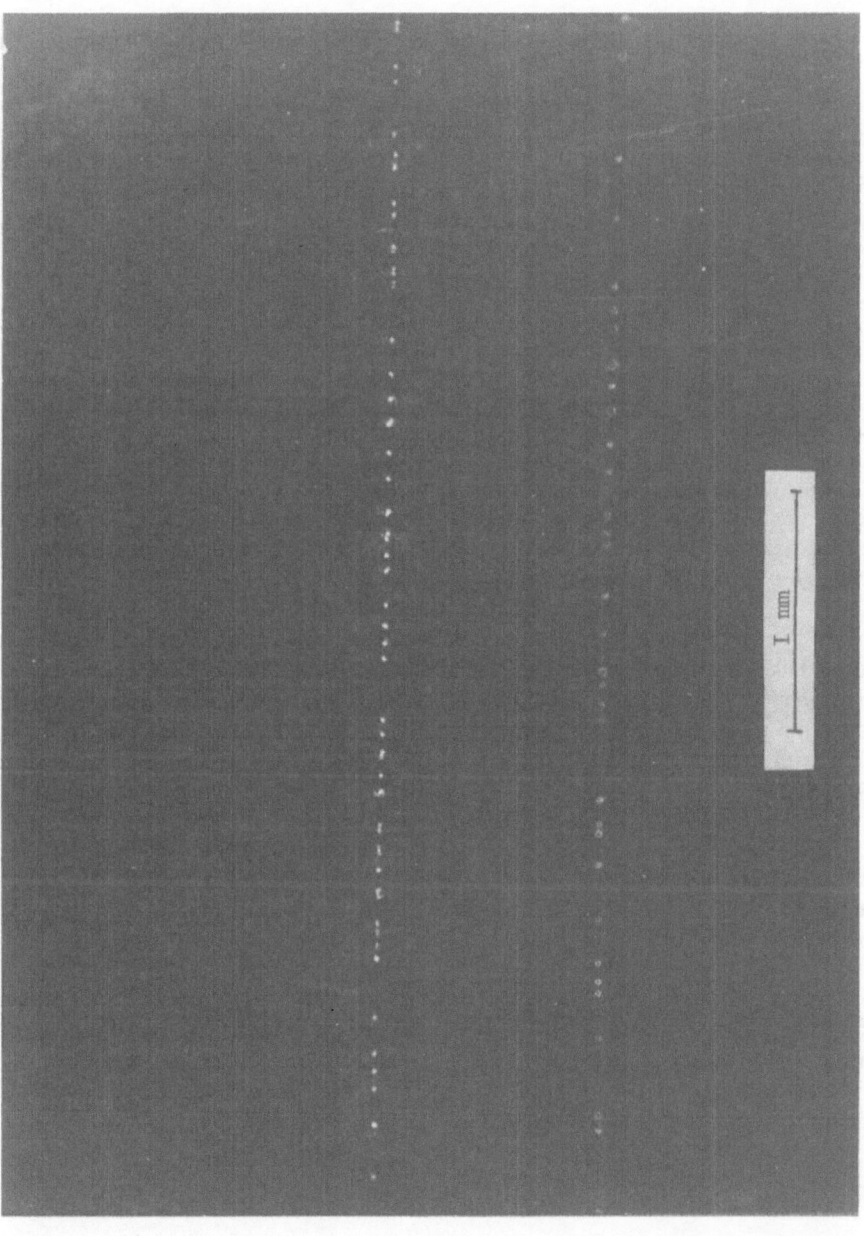

Figure 4

The high intensity tests gave very interesting results as more than 100 incident tracks per hologram do not cause a dramatic effect on the picture quality. The situation here is more favourable for hydrogen than for freon as the bubble growth is much slower in hydrogen.

The tests with the two beam set-ups were rather disappointing:

30 μm bubbles were seen but with an important speckle having a frequency of about 10 μm. In these conditions it was not possible to see the bubbles round-shaped with sharp edges as it is on in-line holograms. In addition the set-up is more complex to align and the requirements for the laser power are more severe for both the recording and the replay system.

CONCLUSIONS

An advantage of holography is to have the maximum resolution in the full volume of the bubble chamber, which allows a gain in sensitivity by a factor of 10 compared to classical optics as 100 tracks per hologram look reasonable.

The experience with the analysis of NA25 holograms indicates that holograms are not more difficult to analyse than classical optics high-resolution pictures. The optics of the scanning and measuring machine must, however, be carefully set up in order to exploit the maximum quality of the holograms, which is in general very good.

The in-line set-up with an afocal lens system seems to be the best for small very high resolution chambers, although the conclusion could be different for other applications (large chambers).

The results show that holography is a very powerful technique which can be used in very high resolution particle physics experiments. If the highest possible resolution is to be reached ($\sim$ 5 μm) a large effort must be initiated in order to reduce as much as possible the turbulences in the chamber.

REFERENCES

[1]   W.T. Welford, Appl. Opt. 5 (1966) 872.

[2]   J.H. Ward and B.J. Thompson, J. Opt., Soc. Am. 57 (1967)
        275.

[3]   F.R. Eisler, Nucl. Instr. and Meth., 163 (1979) 105.

[4]   M. Dykes, P. Lecoq, D. Güsewell, A. Hervé, H. Wenninger,
        H. Royer, B. Hahn, E. Hugentobler, G. Ramseyer and
        M. Boratav, Nucl. Instr. and Meth. 179 (1981) 487.

[5]   Bari-Brussels-CERN-Mons-Paris-Strasbourg-Vienna
        Collaboration, Proposal to study the production of charm
        and bottom particles using a holographic bubble chamber,
        CERN/SPSC/80-120, SPSC/P 155.

[6]   Brussels-CERN-Oxford-Padova-Rome-Rutherford-Stockholm-
        Trieste-Vienna, A prototype experiment to study charmed
        production and decay using a holographic high resolution
        bubble chamber (HOLEBC) and the European Hybrid
        Spectrometer, CERN/SPSC 80-116, SPSC/P 154.

[7]   P. Lecoq and P. Olivier, Proc. Meeting on the Application
        of Holographic Techniques to Bubble Chamber Physics,
        Rutherford Laboratory, 1981, RL 81-042 (1981 94.

[8]   H.T. Yura, Appl. Opt. 12 (1973) 1188.

[9]   J.W. Goodman, w.H. Huntley, Jr. D.W. Jackson and
        M. Lehmann, App. Phys. Lett. 8 (1966) 311.

[10]   P. Lecoq, Proc. Meeting on the Application of Holographic
        Techniques to Bubble Chamber Physics, Rutherford
        Laboratory, 1981, RL 81-042 (1981) 175.

[11]   R. Bizzarri, The use of lan afocal optical system to image
        holographic bubble chambers, CERN/EP/EHS/PH 81-21.

[12]   A. Hervé, K.E. Johansson, P. Lecoq, P. Olivier,
        J. Pothier, L. Veillet, G. Waurick, S. Tavernier,
        CERN/EP 82-28, submitted to NIM.

Discussion

C. Baltay: Did you use film or glass plates in the case where you
             had a separate reference beam? I ask this question
             because film flatness may be a problem here.

Answer: For the HOLEBC tests we used plates. It was clear that
             as the fringe separation is given by $\lambda/\sin\theta$ where θ is
             the angle reference-object beam, all the constraints on
             the film positioning and flatness are probably much more
             severe in the case of two beam holograms than for in
             line holography. On the other hand, our colleagues from
             Stockholm have shown that for a resolution of about 15μm
             the results are quite similar on film and on plates.

J. Sacton: What bubble density did you achieve in both HOBC and
             HOLEBC?

Answer: The bubble density in HOBC was 200 bubble/cm. However,
             it was clear that it was limited by parasitic boiling
             and some work is on the way to try to cure it. For
             instance the Bonn people obtained 400 bubbles/cm in
             BIBC and we do not see any reason to not reach this value.
             For HOLEBC the maximum bubble density was 150/cm. This
             will be probably more difficult to improve.

G. Vanderhaeghe: What is the resolution in depth (z) of your system?
             It should depend obviously on the magnification under
             which you look at your holograms.

Answer: It is clear that the z resolution depends on the magni-
             fication of the replay system and more generally on the
             optical quality of this system. On an optical bench
             with a magnification of 250 I got a RMS of 20 μm for
             10 μm bubbles, but this value could easily go
             to 50 μm on an "industrial" scanning machine.

H. Newman: Is it feasible to swith a multiple array of small bubble
             chambers sequentially into the beam? This could help
             overcome the problems of heating in the freon case, and
             could also overcome the limitation in the repetition
             rate to about 30 Hz.

Answer: Why not!

A. Bettini: Have you any feeling on the limit of resolution for
             holography?

Answer: Holography in itself is a very powerful method. On a
             bench there is no problem to get a resolution of 2 μm
             or even less. In a bubble chamber, as I said the
             problem of turbulences is very important as far as the

resolution is concerned. I would say that a resolution
of 5 mm looks a reasonable limit in the near future.
Beyond that is just a question of optimism.

B. Kitagaki: Comment on turbulence. The High Resolution Optics of
SLAC 40" BC has a resolution of about 55 μm. In our
photon experiment, the heat pile up due to old track
bubbles along the 3 mm diameter beam path started to kill
the H.R.O. when the number of incident photons beyonded
30 per pulse in the 10 Hz operation. This optical
turbulence on HRO limited the number of incident photons
per pulse and the repetition rate of bubble chamber in
our experiment.

S.O. Holmgren: Bench tests done in Stockholm show, contrary to our
expectation that: when doing tests of 2-beam holography
we observe the same quality of holograms on film as on
plates. The geometry used was the same as for the 2-beam
test with converging beam used in the HOLEBC tests (nov.
'81).

Answer: On a hologram we record fringes with a fringe distance of
$\lambda/\sin\theta$ where θ is the angle between the object and the
reference beam. Therefore it is clear that the constraints
on the film resolution and positioning are more severe for
two beam holograms then for in-line if the maximum resolution is required. On the other hand, for 10 μm bubble we
are still far from the theoretical limit of resolution of
the optics and the film does not seem to limit the quality
at least for angle not larger than 20°.

R. Bizzarri: A brief comment on the results with HOLEBC. Paul
Lecoq has only shown an hologram with 10 μm bubbles but
at 20 μm diameter the quality of the bubble images is
much better and comparable with that obtained with
classical optics. Of course we sufferedfrom the turbolences of the bubble chamber but one should point out
that this was the first long run of this chamber and
that it was equipped with very large optic windows
since many tests were to be done with different illumination schemes both with holograms and classical optics.
I believe therefore that we should not anticipate major
difficulties in bringing the turbolences down to a
tolerable level.

SCANNING HOLOGRAMS

Sergio Natali

University of Bari and I.N.F.N.
Via Amendola,173
Bari, 70126 (ITALY)

Due to the short time I will not review all is being done on
the problem of scanning holograms. I will only report on my experience, made scanning some of those 10000 holograms taken in
H.O.B.C., at CERN, for experiment NA25, as just reported by Dr. P.
Lecocque in the previous talk. Each hologram is triggered by an interaction in the chamber, the primary particles being pions at 340
GeV/c and the aim of the experiment being the study of charm production.

The holograms, recorded on 50 mm film with the "in line" technique, can be analyzed simply shining a parallel expanded laser
beam through the film, obtaining immediately above it the real image
of the chamber which can then be scanned and measured with a technique half way between emulsions and bubble chambers. In fact, the
bubble size being $\sim$ 10 μ, one needs a magnification at least of the
order of 100 X which restricts the field of view to order of 1 $\div$ 2
mm.

At our energies, most of the events show a narrow forward cone formed by the majority of the particles produced in the interaction. In this cone, the various tracks cannot be resolved before
1 mm. In conclusion, displaying the interaction vertex in a field
of 1 $\div$ 2 mm is generally not enough to spot a Charm decay which
could be strongly collimated forward, in the direction of all the
other hadrons produced in the interaction. Most of the information
about this event lies outside this zone.

One could then enlarge the field of view at the same magnification, playing well known tricks like aiming at the vertex back
from the end point of each track, or looking for tracks crossing.
Our feeling, already discouraged by the need of a very powerful la-

ser, has been that for events in space one could not play such tricks
with the same efficiency as in the usual bubble chamber picture projection. We adopted the well known scheme of imaging an $\sim$ 1 mm field,
magnified few times, on the vidicon of a TV camera which then displays the event on a TV set, giving the required magnification. A
lot is nowadays being done to improve the resolution of such systems.
I will mention only that the holograms were of such a good quality
that the cheapest commercial equipment was enough to give images
satisfactory for the equipment.

To overcome the difficulties due to the small field, we decided to scan the events along lines originating from the primary vertex, searching for deviations of the outgoing tracks from these techniques.   To experiment the method we built a special apparatus, which
I will shortly describe, (Fig. 1).

In Fig. 1 the optical axis of the TV camera is pointing to the
vertex of a four prong interaction, displayed on an x-y moving stage, where the holographic film is clamped. The film drive is not
shown for simplicity and the laser beam is reaching the film from
below.

The x-y stage is digitized and used to locate the interaction
point according to the predictions from beam chambers. Focussing in
z is achieved moving up and down a lens mounted on the same support
as the TV camera. On the monitor screen, connected to the TV camera,
a cross with arms of apparent size of 1 μ is marking the position of
the optical axis and is superimposed to the primary vertex.

The turntable on the "swing" stage, supporting the x-y stage,
is used to rotate each track in turn in the direction of motion of
the swing axis. If the track is straight, can be perfectly allined
with the swing axis; swinging the picture back and forth one will
see on the monitor screen the track moving up and down perfectly
stable with respect to the reference cross.

Any deviation from this behaviour is interpreted as "something
happening" to the track. Incidentally, the movement of the image has
the additional effect of reducing the holographic bakground while
the track appears intensified and continuous since it persists in
its position on the screen. Experience shows that in these conditions even a small defocussing can be tolerated, adding speed to
the analysis.

The sensitivity of the technique depends on the stability of
the center of rotation of the turntable, on the size of the distortions introduced in the hologram and the bubble size. With our apparatus and film, assuming a potential path of 2 cm for each secondary
track, we were able to detect kinks of projected angle α (in degrees)
if happening at a distance d (in mm) from the vertex
$$d \geq 1/2 \, \alpha.$$
The apparatus is now used by the scanners, instructed to follow
each track from an interaction for 3 cm and:
a) To ceck for compatibility of all tracks with production at the
   primary vertex.
b) To identify neutrals (γ's, V$^\circ$) and check theyr compatibility with

production at a given vertex.

Every event showing something unusual according to criteria a) and b) is sent for further study, which is done by physicists. Is is found very useful recordingthe scanning on tape which is easily replyed to identify the reasons of the selection.

The average time necessary to analyze an interaction in this way, overheads included, is 15 minutes.

In conclusion our experience indicates that holograms can be analysed as quickly and reliably as in other visual techniques and that to them is open the same order of magnitude of large scale experiments.

DISCUSSION

B. Roe: Why not use differential X and Y magnification in TV screens to make kinks, etc. visible?

Answer: I do not know how much the use of anamorphism could really speed up the scanning, being limited by the restricted field where it operates, in particular if we will go down to 5 micron bubbles. And it is expensive. CERN is now trying out this technique extensively.

A MEASURING DEVICE FOR HOLOGRAMS FROM LARGE VOLUME BUBBLE CHAMBERS

R.J. Cence, M.D. Jones, S.I. Parker and
V.Z. Peterson (presented by. V.Z. Peterson)

Physics Department
University of Hawaii at Manoa
Honolulu, Hawaii 96822

ABSTRACT

Measuring full-size reconstructed holographic images of
bubbles from a large bubble chamber presents problems of
size, rapid location of the desired event, and precise
measurement of a local region. We present a conceptual
design of a practical system using computer-controlled
mirrors to align the desired vertex region along the axis
of motion of a vidicon whose field of view includes the
local region. Precise measurements of short-decay paths
are then made by fine adjustments of the vidicon position.
This method requires normal stereo photography for scanning and preliminary reconstruction of the event.

INTRODUCTION

The advantages of high resolution in detecting short-lifetime
decays, and the advantages of holographic photography in preserving
depth of field with high resolution, are well known. The use of
small bubble chambers with holographic photography has been demonstrated in hadron beams; e.g., by the BIBC group. In this area of
small or medium-sized bubble chambers for use with hadrons or photon
beams, where the volume to be photographed is small, holography has
been pushed with the greatest vigor and with promise of early
success.

In neutrino physics the beam-target volume cannot be made
small; neutrino beams are difficult to focus or collimate. Yet

201

neutrino interactions are one of the cleanest and best sources of charmed particles, and "beautiful" particles are expected at higher energies; Charm plus Beauty yields at the Tevatron should constitute at least 10% of neutrino events, as compared to 1% of photon interactions and 0.1% for hadron beams. Thus, for bubble chambers, which can accommodate only a few interactions per picture, a large bubble chamber in an intense neutrino beam may be more useful for studying charm and beauty events in detail than a small chamber rapidly cycled in a hadron or photon beam.

The advantages of holography are clearly indicated for Tevatron neutrino beams in bubble chambers: with the present 15-foot conventional optics, the bubble size and resolution are about 500 microns. (The resolution can be improved to ~ 200 microns by sacrificing depth-of-field.) Due to the short lifetime of the tau, only a small fraction of tau decays are expected to be visible decays with conventional optics. With holography one can hope to achieve resolution better than 50 microns over the full-chamber volume. Since the median decay distance for the tau-lepton in the Tevatron beam-dump experiment is about 800 microns, holography can improve both the yield and quality of the detectable signal for taus. It may also make it possible to detect "beauty" decays, which are expected to have even shorter decay distances.

PROBLEMS WITH LARGE VOLUME HOLOGRAPHY

Holographic photography over a large volume poses difficulties not encountered with small volumes where "direct-view" holography (i.e., film area comparable to cross section being photographed) is practicable. The Welford scheme of two-beam holography in Scotchlight-illuminated bubble chambers has been tested at RHEL[1] and at BEBC[2] and appears to be feasible. A different method has been proposed by Baltay[3] for the 15-foot chamber: the laser beam enters at the bottom of the chamber via a diverging lens, and the direct and scattered amplitudes are recorded on the film without any focusing lenses. Preliminary bench tests of this idea at Columbia appear encouraging.

A problem with large-volume holography is connected with viewing and measuring the holographic image. When the hologram is projected with CW laser light, the full-size bubble images will be produced in space over the original volume of the bubble chamber. The 20 m^3 fiducial volume of the 15-foot chamber presents quite a problem to scan for small bubbles. Furthermore, conventional measuring methods would require moving a high-precision measuring device over large regions of space. This is obviously impractical.

We have devised a scheme for avoiding these problems, which

appears to us to be practical. First of all, one does <u>not</u> scan the
<u>holographic</u> image but relies on conventional three-view pictures
(the three "normal" stereo cameras of low resolution) to locate
interesting events. Stereo reconstruction determines their spatial
coordinates with the usual precision of bubble-chamber event reconstruction. Secondly, one uses this information to establish which
specific <u>region</u> of holographic image space should be viewed at high
magnification. Mirrors are adjusted (by computer control) to view
the vertex (or other restricted region of interest). Precise
measurements can then be made of decay distances on the <u>full-size</u>
bubble image.

A schematic drawing of the proposed system is shown in Fig. 1.
The hologram is illuminated with a CW laser of nearly the same wavelength as the pulsed laser used in the initial exposure. This will
reproduce the real bubble image in space by the interference of
coherent light from all parts of the hologram. The total image
will occupy a volume in space relative to the hologram corresponding
to the bubble-chamber volume, including angles up to 40 degrees
from the optic axis. However, by using plane mirrors tilted at the
proper angle, the image in question can be brought into the "measuring volume" where the vidicon is located. This "measuring volume"
is cylindrical with its axis located above a one-dimensional track
upon which the vidicon cart rides. This cart can be moved to bring
the image within the field of view of the vidicon, and the magnified image can be displayed on a TV monitor.

This method has the advantage that the mirror directs the
narrow cone of interference rays from the hologram in the direction
of the vidicon, so that only a slight swiveling of the vidicon is
required to align it for optimum viewing. A lens can be used
between the bubble's real image and the vidicon for greater magnification. Measurement of the decay distances can be made on the TV
monitor, or directly in space by moving the vidicon with the points
in question centered in the field of view.

Computer-controlled motion is essential for this scheme to
work, and the spatial coordinates of the localized region must be
known reasonably accurately in advance from the stereo photographs.

SOME CRITICAL ELEMENTS OF THE MEASURING MACHINE

The mirrors involved must be very flat, and their motion
smoothly and accurately controlled. The angular accuracy of mirror
setting required to position the image of a bubble well within the
aperture of a vidicon (2.5 cm) at a typical distance of 400 cm is
only $\sim$ 1 milliradian; such mechanical motion (two angles) is easily
made and measured. Incremental shaft encoders with accuracies of
better than 1 part in 10^5 per revolution are now available. Thus

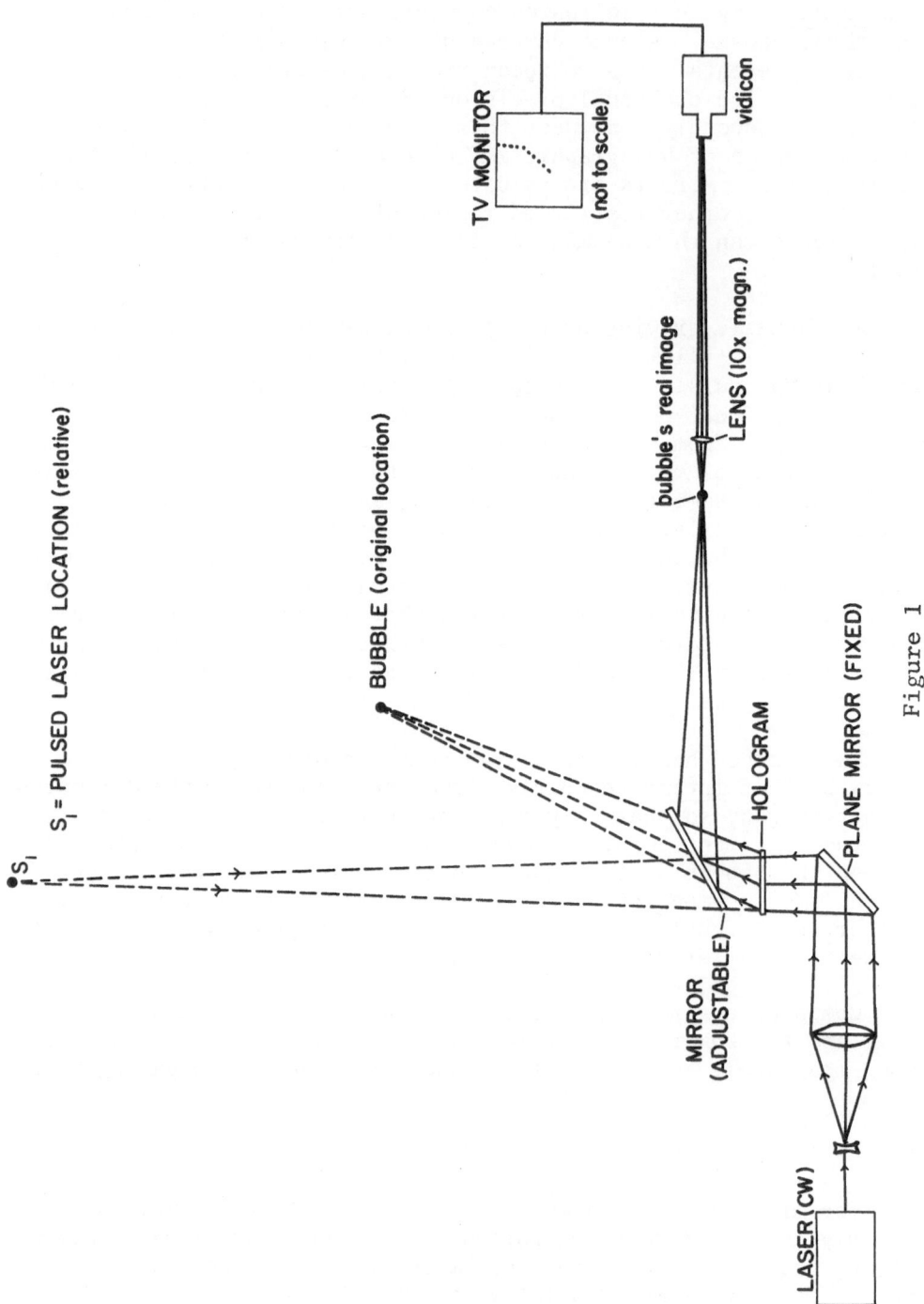

Figure 1

with computer control of mirror position the vertex can be positioned accurately within the field of view and moved about with computer control.

The measurements on the reprojected image can be made either (1) on the magnified CRT image or (2) by moving the vidicon by precision x-y stage to bring several points to the center of the vidicon (cross hair). The former method is more convenient, but the linearity and CRT are not involved. In the latter method, the non-linearities of the vidicon movement are the only limitation. In either case the image of bubble on the vidicon may be magnified: typically, x10 at the vidicon and x200 on the CRT. Thus a 50-micron diameter bubble image would appear as 500 microns at the vidicon and 10 mm on the CRT. This is a convenient scale for viewing and making measurements. Short-decay distances would be magnified to a scale where their detection would be readily done on the CRT; angles could be enhanced by making the transverse magnification greater than the longitudinal magnification.

STATUS OF THE PROPOSED MEASURING MACHINE

At the present time this is just a proposal, although development funds have been sought and a preliminary design has been sketched. It is our intention to compare this proposal with other methods of measuring large-chamber holograms during this conference, and then choose the best method for further development for the 15-foot chamber holograms.

ACKNOWLEDGMENTS

This work was supported in part by the U.S. Department of Energy (High Energy Physics Program) under DOE Contract DE-AM03-76SF00235.

REFERENCES

1) C.M. Fisher, E. Miranda, V. Peterson, and R. Sekulin, "Laboratory Tests of Holography for Large Bubble Chambers," Proceedings of the Rutherford-Appleton Lab. Bubble Chamber Holography Meeting (January 19-20, 1981), RL-81-042, p. 181.
2) F. Pouyat and H. Royer, "Preliminary Test of Holography for BEBC," BEBC BUG Newsletter (July 29, 1981).
3) C. Baltay, "Improved Optics for the 15-foot Chamber," C. Baltay remarks at Fermilab meeting (April 1981).

DISCUSSION

B. Roe: Are you planning to take conventional pictures of the event as well as the laser image?

Answer: Yes, we plan to also utilize classical optics (normal stereo trial), taken at a later time with larger bubbles.

TOPOLOGICAL BACKGROUND ON CHARMED AND BEAUTY PARTICLE PAIRS
PRODUCED IN HIGH ENERGY HADRON INTERACTIONS IN NUCLEAR
EMULSIONS

Giorgio Romano

Istituto di Fisica "G.Marconi" - Università di Roma
Istituto Nazionale di Fisica Nucleare - Sezione di Roma
Piazzale A.Moro, 2 00185 Roma (Italy)

INTRODUCTION

Making use of the fact that new flavours must be produced
in pairs in strong interactions and that beauty particles are
expected to decay often into charmed particles, the contribution of background simulating decays will be computed from a
pure topological point of view. Of course, the addition of
further requirements (triggers) could clean the signal to a
larger extent.

It will be assumed that in the interaction of (350-400) GeV
hadrons in emulsion the production rate of charmed particle
pairs is ~ 5×10^{-3}/interaction ($\sigma \approx 30 \, \mu$ b/N and σ ~ A); the
corresponding figures for $B\bar{B}$ production are currently estimated
to be ~ 10^3 times smaller.

EMULSION DATA

A primary high energy interaction produces, on average

- 16 charged high energy hadrons λ(inter.) $\simeq$ 400 mm
- 16 γ's (from π^o) λ(convers.) $\simeq$ 40 mm
- 1 high energy neutron λ(inter.) $\simeq$ 400 mm
- 0.5 K^o ($\beta\gamma$ ~ 30) λ (decay) $= \beta\gamma c\tau \simeq$ 810 mm for K^o_S
- 0.5 Λ^o ($\beta\gamma$ ~ 20) λ (decay) $\simeq$ 1600 mm

However, charged hadrons and γ's likely to give rise to interactions simulating the decay of a heavy particle are those
emitted in a forward cone including about half of them. More-

over, the requirements of the right parity (odd for charged primary, even for neutral primary) and of the absence of evaporation prongs further reduce to 1 in 20 secondary interactions those simulating a decay. Elastic scatterings without visible recoil (kinks) at an angle larger than a suitable cut-off value (typically of the order of 1°) would contribute at most to an equal amount as the interactions.

Therefore, the secondary interactions and the strange particle decays simulating heavy particle decays will have the following probabilities (L = measured path, in mm)

$$P^{ch} \simeq 2 \times 10^{-3} \times L \; ; \quad P^{o} = P^{o}_{inter.} + P^{o}_{K^{o}_{S}} + P^{o}_{\Lambda^{o}} + P^{o}_{\gamma} \simeq 0.9 \times 10^{-3} \times L$$

In computing P^{o}_{γ}, the probability that an electron pair simulates a decay (opening angle $\alpha \lesssim 0.5^{\circ}$ and no obvious low energy electron) has been estimated to be $\sim 10^{-3}$.

It should be noted that some neutral decay topology, like 4 or more charged prongs, are much less affected by background.

SEARCH FOR CHARMED PARTICLES

The amount of background obviously increases with the scanning length and will then depend on the scanning scheme. No loss is expected if the scanning is confined within paths corresponding to five times the currently accepted maximum values of the mean life-times:

charged C's $\tau \sim 8 \times 10^{-13}$ s $\beta\gamma \sim 20$ $L_{max} = 24$ mm

neutral C's $\tau \sim 2.5 \times 10^{-13}$ s $\beta\gamma \sim 20$ $L_{max} = 7.5$ mm

Of course, by allowing a small loss of events, more efficient scanning schemes can be envisaged (for instance: scanning for a first vertex within a shorter path; further extensive search for a second vertex only if the first is found).

Assuming only D's are produced, $D^{*}/D = 3$ and by requiring the presence of two secondary vertices within the scanning volume, one obtains (R = signal/signal+background) the rates shown in Table 1. It is easily seen that in no case a single secondary vertex can be considered as a clean decay, and even double charged topologies are dominated by background if no other selection criterion can be applied. Neutral topoligies are much cleaner if primary interaction and secondary vertices can be univocally linked (it is usually done by means of external detectors in the current hybrid experiments), otherwise the background increases as a function of the flux of the incident

Table 1. Background on charmed particles according to topology.

Topology	fraction	signal	background	R
		(on 10^3 interactions)		
D^+D^-	10%	.50	2.30	.18
$D^\pm D^o$	43%	2.15	.32	.87
$D^o\bar{D}^o$	47%	2.35	.05	.98

particles, due to upstream interactions.

SEARCH FOR BEAUTY PARTICLES

Assuming B particles are produced with $\beta\gamma \sim 10$ and that a reasonable maximum value for their estimated mean life-time is 10^{-13} s, scanning for their decay vertices has to be confined within ~ 1.5 mm (again, different scanning schemes can be envisaged). In these conditions, the background due to interactions or to strange particle decays has about the same rate as the signal (from 9×10^{-6} to 2×10^{-6}/interaction, according to topology), but charmed particle pairs decaying within the fiducial volume are about two orders of magnitude more abundant. However, as the dominant decay mode of beauty particles is predicted to be into charmed particles, the requirement of topologies containing 3 or 4 secondary vertices is expected to reduce the background by a large amount while keeping a large fraction of the signal.

If the decaying charmed particles, main source of background, produce on average
- 2 charged hadrons
- 2 γ's
- 0.5 K^o ($\beta\gamma \sim 10$)
the probability of a tertiary interaction or decay of strange particle is now

$$P_1^{ch} \simeq 5\times10^{-4}\times L; \qquad P_1^o \simeq 8\times10^{-4}\times L.$$

The scanning paths of these last vertices must be about one half those considered previously due to the expected lower energy of the C-particles originating from B decays.

Indeed, neutral interactions cannot be assigned as originating from a given vertex, and all the contributions (both charms plus primary interaction) must be added, ending up with

$$P_1^o \simeq 2.5\times10^{-3}\times L.$$

Table 2. Background on beauty particles according to topology.

Topology	fraction	Signal (on 10^7 interactions)	background	R
B^+B^- { D^+D^- / $D^\pm D^0$ / $D^0\overline{D}^0$ }	11%	5.3	.02 } .06 } .12 / .04 }	.98
$B^\pm B^0$ { D^+D^- / $D^\pm D^0$ / $D^0\overline{D}^0$ }	21%	10.6	.16 } .53 } 1.07 / .38 }	.91
$B^0\overline{B}^0$ { D^+D^- / $D^\pm D^0$ / $D^0\overline{D}^0$ }	11%	5.3	.34 } 1.06 } 2.22 / .82 }	.67

Assuming in B decay $B \rightarrow C$ 65%, four vertices topologies would have the rates shown in Table 2. It is readily seen that, despite the very low predicted rate, background does not represent a big problem when four decays are required. In addition, if charged B's are assumed to decay only into neutral D's, certain topologies are forbidden and the corresponding background contribution may be lowered.

DETECTION EFFICIENCY

The detection of decays close to the primary interaction ($\lesssim 100 \mu$m in emulsion work) is mainly guided by the presence of prongs escaping the forward jet with a finite transverse distance to the origin (impact parameter). Decays "far" from the primary interaction can be detected because of an appreciable deflection (kink), becuase of a jump in ionization (multiprong), etc.

The visual sensitivity on transverse distances for decays close to the primary interaction may be in emulsion $\sigma \approx 1 \mu$m (down to .05 μm with suitable measurements [1]). The detection efficiency ε then depends[2] mainly on $c\tau/\sigma$ and on the charged prongs multiplicity n, as it shown in Fig. 1. In this case ε was computed at the level of 3 standard deviations and thus implies the observation in a single event of a transverse distance in the emulsion plane of few microns.

The values of $c\tau/\sigma$ for neutral and charged D's (~ 75 and 240 μm respectively) would always imply $\varepsilon > 90\%$, but these particles in most cases decay too far in order to use such calculations. For a mean life-time $\tau \sim 5 \times 10^{-14}$ s, reasonable for B

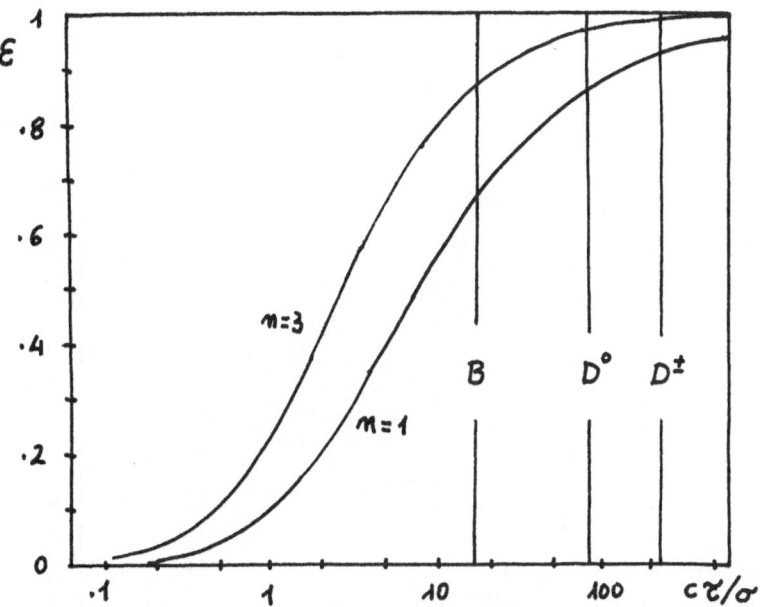

Fig. 1. Detection efficiency as a function of mean life-time
(τ), sensitivity (σ) and number of charged prongs (n).

particles, the detection efficiency would be ~65% (kinks) or ~
85% (3 prongs).

Among the decays far away from the primary star, charged
multiprong ones are detected with the highest efficiency (up to
90-95%), whereas the efficiency for kinks depends on the cut-off
value of the deflection (typically ~1°) and thus on the energy
of the decay particles. Reasonable values are in the range 50 to
80%.

The detection efficiency of neutrals decaying far from the
primary interaction strongly depends on the method used in order
to find them (volume scan, tracing back, etc) which has to be
optimized according to external factors, like the level of the
general background or the amount of volume to be scanned. Furthermore, ε depends on the number of prongs and on the opening
angles; usually it is not likely to exceed ~ 80% in a volume
scanning.

It must be remarked that the overall detection efficiency
of a multivertex event is usually higher than the product of the
single efficiencies, because once the first decay is found the
sensitivity in searching the others increases.

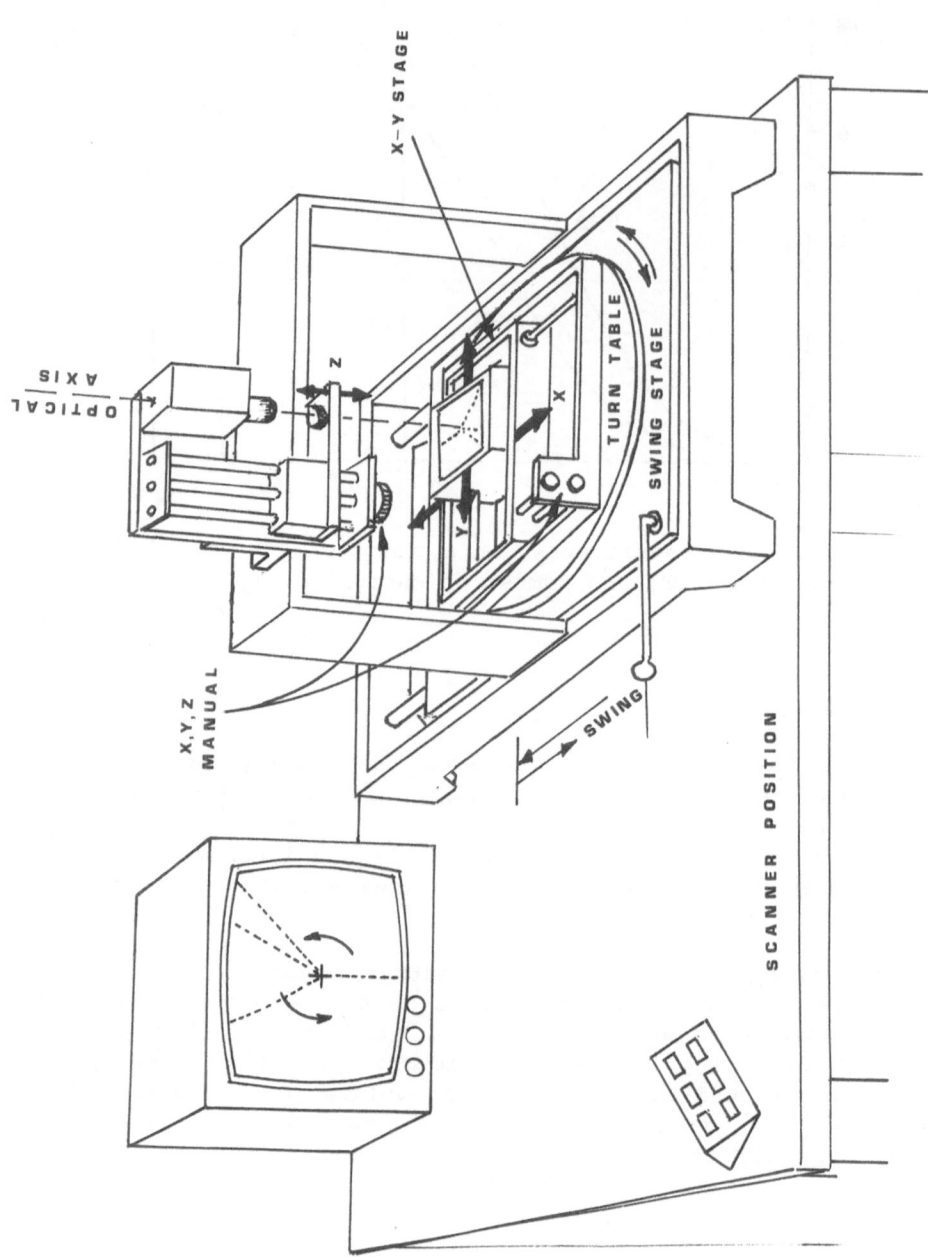

Figure 2.

EVALUATION OF THE MEAN LIFE-TIME

An estimate of the mean life-time of heavy, short lived particles can be obtained[2] from the decay path L and from the decay angles ϑ_i in the lab system even without knowing momenta and when undetected neutrals would prevent a univoque estimate of the proper decay time t. In fact,

$$\Delta = L \sin\vartheta = \beta\gamma\, ct \sin\vartheta \xrightarrow[\beta,\beta^*\to 1]{} ct \tan\frac{\vartheta^*}{2}$$

hence $\langle\Delta\rangle = A\, c\, \tau$.

Monte Carlo calculations, confirmed by the presently available data (mainly from neutrino and photon interactions), show a clear correlation between single values of Δ/c and t, suggesting a value of A around 1.

- References -
1) S.Petrera and G.Romano, Nucl.Instr. and Meth., 174:53 (1980).
2) S.Petrera and G.Romano, Nucl. Instr. and Meth. 174:61 (1980).

MEASUREMENTS OF CHARMED PARTICLE LIFETIME

G. Bellini

Istituto di Fisica dell'Università di Milano
Istituto Nazionale di Fisica Nucleare - Sezione di Milano
Milano - Italy

1. INTRODUCTION

The measurements of charmed particle lifetimes are based on experimental apparatus which combine a forward multiparticle spectrometer with a high resolution vertex detector.

The multiparticle spectrometer would provide detection and measurement of charged and neutral particles, and their identification. The vertex detector gives a strong signature to the decays and measures the lifetime.

Three categories of vertex detectors have been used until now: bubble chambers with high resolution optics, nuclear emulsions and live targets. Three experiments: CERN NA16 (LEBC-EHS), SLAC BC 72/73 (H.F.), CERN NA18 (BIBC) use bubble chambers; two experiments: FNAL E531, CERN WA58 are carried on with emulsion; one (CERN NA1 - FRAMM) is based on a Silicon detector telescope, which works as a live target.

I do not give the details of the experiments, which are explained in the papers of G. Zumerle, T. Kitagaki, N. Reay, G. Diambrini-Palazzi, E. Meroni, but I will discuss the general features of these experiments and their results.

2. EVENT SELECTION AND ANALYSIS PROCEDURE

The main features of the six experiments listed above are summarized in Tables 1 - 6: they differ in the following points:

i)   In the experiments using visual detectors the charmed particle production is studied in inclusive way: in few of them the authors search for both the partners of the associated production, but they accept also events where one charmed particle is lost.

TABLE 1

<u>LEBS - EHS NA16</u>[1]

BEAM: π^-,p   ENERGY: 360 GeV/c

PARTICLE IDENTIFICATION: $\pi/K/p$ identification: not available until
                                    now
                                  : $e^\pm$ identified in the lead glass

RESOLUTION IN EVENT MEASURING: $\sigma_M \simeq 10 - 20$ MeV

PATH DETECTOR: high resolution bubble chamber

MINIMUM DETECTABLE LENGTH: $\ell_{min} = 1$ mm for neutral D mesons

$$\ell_{min} = \ell \cdot \frac{100\,(\mu m)}{d^{max}\,(\mu m)} \quad \text{for charged D mesons}$$

(see fig. 1).

DETECTION EFFICIENCY FOR CHARMED PARTICLES: the efficiency is assu-
                              med to be a step function which switchs
                              from 0 to 1 at $\ell = \ell^{min}$

FINAL SAMPLE*: 6 neutral D's: 5 fully reconstructed,
               10 charged D's: 7 fully reconstru
               8 ambiguous with $F^\pm$ or Λ_C
               1 F^- ambiguous with Λ_C
               $\sim \frac{1}{2}$ of these decays are found in associated pro-
               ductions.

_____

(*) In this experiment the search of charmed particle is done in
    inclusive way.

TABLE II

SLAC H.F.   BC 72/73[2]

BEAM: γ ENERGY: 19.5 GeV

PARTICLE IDENTIFICATION: $\pi/K/p$ separation above $\sim$3.1 GeV/c

RESOLUTION IN EVENT MEASURING: $\sigma_M \simeq 15$ MeV

PATH DETECTOR: high resolution bubble chamber

MINIMUM DETECTABLE LENGTH: ℓ_{min} is the result of the following three requirements:
    i)   minimum decay length $\ell_{min} \gtrsim 0.5$ mm;
    ii) maximum impact parameter $d_{max} > 110$ μm;
    iii) a second track from the same decay vertex with $d_{2nd} > 40$ μm.

DETECTION EFFICIENCY FOR CHARMED PARTICLES: assumed to be a step function switching from 0 to 1 at $\ell = \ell^{min}$ where ℓ^{min} is the minimum decay length once satisfied the three requirements $\ell_{min} \gtrsim 0.5$ mm, $d_{max} > 110$ μm, $d_{2nd} > 40$ μm.

FINAL SAMPLE[*]: 10 charged decays: 7 fully reconstructed (ambiguous between D and F, even if the probability for F is smaller)[**]
        : 10 neutral decays: 3 fully reconstructed[***].

---

[*] In this experiment the charmed particle production is studied in a completely inclusive way; the mass value is not used as good signature.

[**] This ambiguity comes from the lack of particle identification.

[***] In only two events the charmed particles are produced associated.

TABLE III

E - 531$^{(3)}$

BEAM: ν ENERGY: wide band horn focused
                                        neutrino beam from 350 GeV
                                        protons

PARTICLE IDENTIFICATION: π/K separation below 3.2 GeV/c
                         K/p "   "   5.5 GeV/c
                         μ identification "   4 GeV/c
                         $e^{\pm}$ identified in the lead glass

RESOLUTION IN EVENT MEASURING: σ_M (charged particles) $\simeq$ 20 MeV
                               σ_M (neutral   "   ) $\simeq$ 16 MeV

PATH DETECTOR: emulsion

DETECTION EFFICIENCY FOR CHARMED PARTICLES: in fig. 2 the effi-
                         ciency is shown for charged and neutral
                         D's; the events are weighted taking
                         into account this efficiency.

FINAL SAMPLE[*]: 1.6 D^0's fully reconstructed + 3 with semileptonic
                 decay (obviously unconstrained);
                 2 $D^{\pm}$ fully reconstructed + 3 with semileptonic decay;
                 5 Λ_c^0 fully reconstructed + 3 with one neutral particle
                 lost;
                 3 F^+ fully reconstructed.

_____

(*) In this experiment the charmed particle production is studied
     in a completely inclusive way.

TABLE IV

<u>WA58</u>   (+ 1 event of WA45) [4]

BEAM: γ   ENERGY: 20 - 70 GeV

PARTICLE IDENTIFICATION: π/k separation above 5 GeV/c[*]

RESOLUTION IN EVENT MEASURING: $\sigma_M \simeq 20$ MeV

PATH DETECTOR: emulsion

MINIMUM DETECTABLE LENGTH: 10 μm

DETECTION EFFICIENCY FOR CHARMED PARTICLES: assumed to be constant;

no corrections are done.

FINAL SAMPLE[**]: 3 events showing an associated production with all
particles measured

5 events showing associated production with one or
two particles lost (π^0, K^0, ν)

2 events with one charmed particles lost

([*]) In the sample of the events with charmed particles, no K's are
identified.

([**]) In this experiment it is searched for associated production
events.

TABLE V

BIBC - NA18[5]

BEAM: π^- ENERGY: 340 GeV/c

PARTICLE IDENTIFICATION: no particle identification

PATH DETECTOR: bubble chamber (Freon)

MINIMUM DETECTABLE LENGTH: $\ell_{min} \simeq 3$ mm for charged D

$\ell_{min} \simeq 2$ mm for neutral D

DETECTION EFFICIENCY FOR CHARMED PARTICLES: assumed zero for $\ell < \ell_{min}$

and $\simeq 1$ for $\ell \geqslant \ell_{min}$.

FINAL SAMPLE*: 7 $D^\pm$ decays

5 D^0 decays

No events with associated production have been
reconstructed

(*) Only decays with all charged secondaries are used due to the
absence of π^0 detectors; therefore all the decays are fully
reconstructed. Only decays of particles having an invariant
mass consistent, into ±2%, with the D mass, have been accepted.

TABLE VI

FRAMM

BEAM: γ   ENERGY: 40*-150 GeV/c

PARTICLE IDENTIFICATION: π/K separation in the range 5<p<21 GeV/c

RESOLUTION IN EVENT MEASURING: $\sigma_M \simeq 40$ MeV

PATH DETECTOR: silicon live target

MINIMUM DETECTABLE LENGTH: 1.6 mm

DETECTION EFFICIENCY FOR CHARMED PARTICLES: the data are corrected
　　　　　　　　　　　　　　　　for the finite length of the target and
　　　　　　　　　　　　　　　　for the minimum detectable path

FINAL SAMPLE**: 75 associated production events identified in the
　　　　　　　spectrometer;
　　　　　　　23 events showing two decays inside the Silicon
　　　　　　　target
　　　　　　　98 $D^{\pm}$ decays in total
　　　　　　　all the decays are fully reconstructed

(*) This lower limit is introduced, in the off-line analysis using the
information from the tagging system.

(**) In this analysis the mass value is used as a constraint in the
charmed particle selection.

In the lifetime measurement with electronic technique only coherent production of one pair of charmed particles is investigated. With this limitation the multiplicity is lower, making easier the identification of the decay detector inside the Silicon telescope, and the combinatorial background is reduced because the pair of charmed particles is produced without additional pions. In addition the incident energy flows into the charmed mesons giving a high stretching γ factor and so increasing the paths of the particles before they decay.

ii) In all the experiments the selection of charm production is based mostly on the decay topology and on the presence of a second vertex near to the production point.

The topology identification is made much easier by the $\pi/K/p$ separation, which unfortunately is rather inefficient also in the experiments equipped to this purpose. In NA1 the π/K separation works a little better than in the other experiments, but in any case no more than $\sim 50\%$ of K's are identified.

A good help in fighting against the combinatorial background is the possibility to match the tracks measured in the vertex detector and reconstructed in the forward spectrometer. This track matching is possible in principle in the experiments using emulsions or bubble chamber, while cannot be carried out when the vertex detector is a live target. In this last case the number of combinations is kept within reasonable limits because the multiplicity of the coherent production of charmed particles is relatively low. Nevertheless cuts in mass are adopted in the analysis of NA1 experiment, which normally are not introduced in the data taken with visual vertex detectors (except NA18).

The search of a second vertex very near to the production point is limited by the minimum detectable length and by the finite size of the vertex detector. The lower limit, which is fixed for the live target and is a variable value for the bubble chambers, represents a starting-point from which the paths are measured. It does not introduce biases, but depresses the statistics.

The limitation due to the finite size of the vertex detector is more important for the emulsions and the live targets. Corrections are made in E531 and NA1, while the results of WA58 are not corrected at all for the inefficiency in detecting decays.

iii) Until now, only in NA1 the determination of the $D^{\pm}$ lifetime can be obtained by fitting directly the exponential $\frac{dN}{dt}$ vs t. In the experiments using visual detectors the lack of statistics does not allow to obtain $\frac{dN}{dt}$ vs t plots which give meaningful fits. In these last cases the lifetimes are obtained using maximum likelihood functions or simply the over-all arithmetic

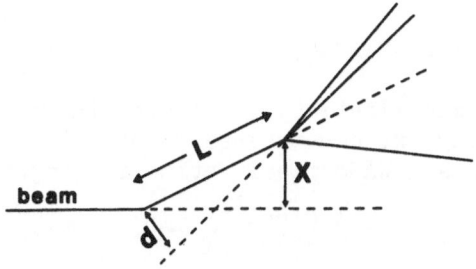

Fig. 1. Definition of the event geometrical quantities d and L.

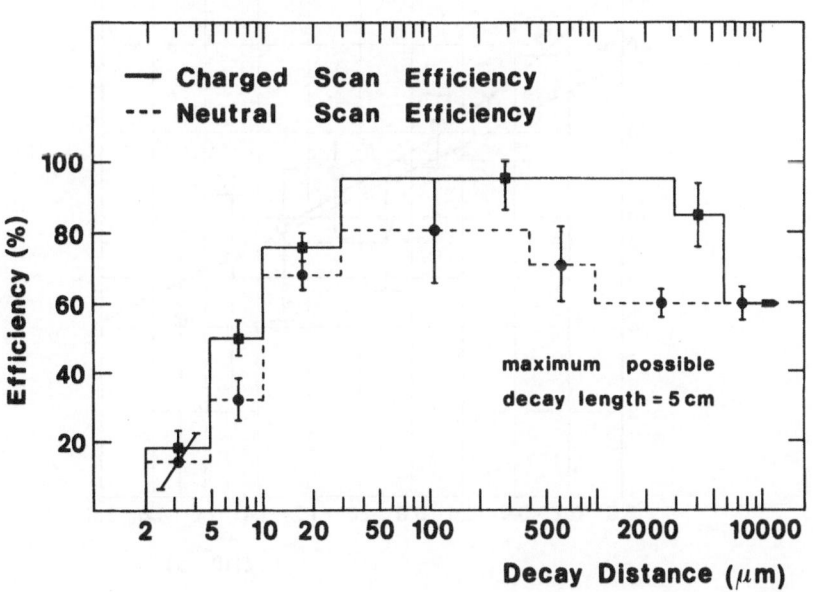

Fig. 2. Finding efficiencies of the experiment E531.

mean. In the SLAC BC 72/73 experiment the effective decay length method is employed.

3. CHARMED PARTICLE LIFETIMES

Charmed particle lifetimes are listed in Table 7 for data with statistics exceeding one constrained event.

For LEBC-NA16 and WA58 the lifetime is computed as:

$$\tau = \frac{1}{N} \frac{m}{c} \sum \frac{\ell^i - \ell^i_{min}}{p^i}$$

where ℓ_{min} is the minimum decay length which can be detected (see Table 1). In WA58 ℓ_{min} is assumed = 0. The SLAC BC 72/73 and E531 results are obtained adopting the maximum likelihood method. The BIBC data are omitted because at present they are too preliminary. The FRAMM-NA1 result is obtained by fitting the experimental plot $\frac{dN}{dt}$ vs t (fig. 3).

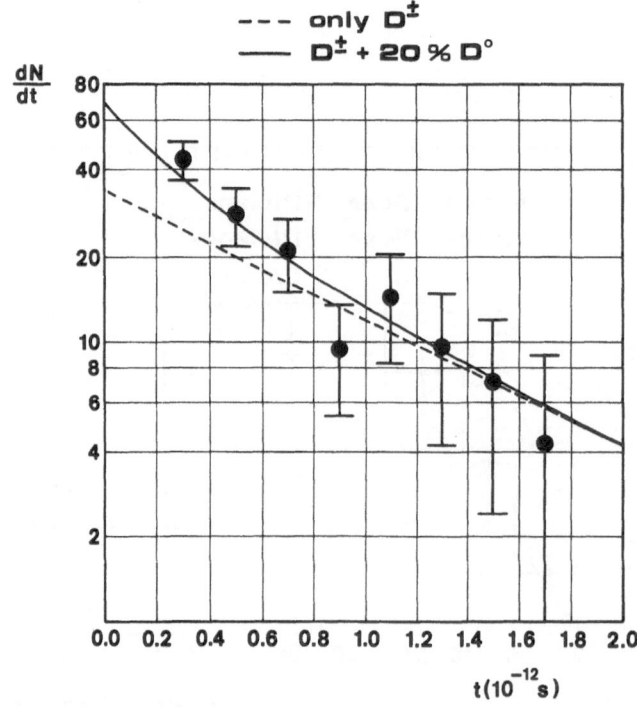

Fig. 3. Plot of the lifetimes of 98 $D^\pm$'s decaying in the target of the experiment FRAMM-NA1. The dashed line corresponds to the fit with only one exponential. The full line takes into account a further 20% contribution from a lifetime $2.5 \cdot 10^{-13}$s (D^0).

TABLE VII

	D^0			$D^{\pm}$			$F^{\pm}$			Λ_e^+		
	Lifetime $(\cdot10^{-13}s)$	Nb. of particles Tot.	fully reconstructed (f.r.)	Lifetime $(\cdot10^{-13}s)$	Nb. of part. Tot.	(f.r.)	Lifetime $(\cdot10^{-13}s)$	Nb. of part. Tot.	(f.r.)	Lifetime $(\cdot10^{-13}s)$	Nb. of part. Tot.	(f.r.)
LEBC-NA16	$2.3^{+1.7}_{-0.9}$	6	5	$6.7^{+4.9}_{-2.4}$	10	7						
SLAC-HF BC 72/73	$8.1^{+4}_{-2.0}{}^{(+)}$	10	3	$7.7^{+4.0}_{-2.0}{}^{(+)}$	10	7						
E531	$2.3^{+0.8}_{-0.5}$	16	16 $^{(++)}$	$10.3^{+10.3}_{-4.2}$	5	2	$2.0^{+1.8}_{-0.8}$	3	3	$2.3^{+1.0}_{-0.6}$	8	5
WA58	$1.34^{+0.56}_{-0.34}$	8	6									
FRAMM-NA1				$9.5^{+3.1}_{-1.9}$	98	98						

$^{(+)}$These values are obtained by means of a Montecarlo simulation using the effective decay length method. The maximum likelihood method applied only to the constrained events gives $\tau_{\pm} = 7.7^{+4.0}_{-2.0}\cdot10^{-13}$s and $\tau_0 = 8.1^{+4.0}_{-2.0}\cdot10^{-13}$s. More recently the same collaboration gives $\tau_0\approx7.7\cdot10^{-13}$ and $\tau_{\pm}\approx7.4\cdot10^{-13}$s.

$^{(++)}$Further three semileptonic D 's, obviously unconstrained, are not included. By themselves, they give $\tau_{sem}\approx7.0^{+6.7}_{-2.9}\cdot10^{-13}$s. Added to the constrained decays a lifetime $\tau\approx3.2^{+1.0}_{-0.7}\cdot10^{-13}$s is obtained.

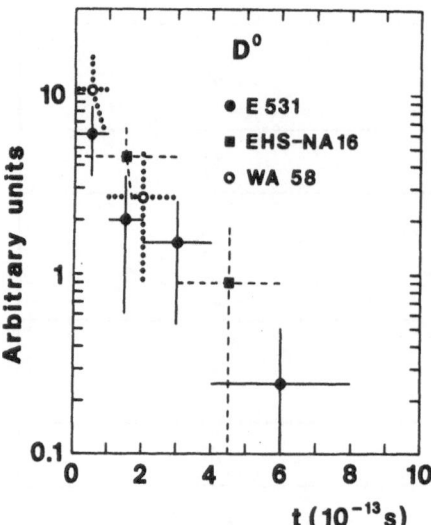

Fig. 4(a). Plot of the lifetimes for neutral D's; data of the
experiments: E531, EHS-NA16, WA58.

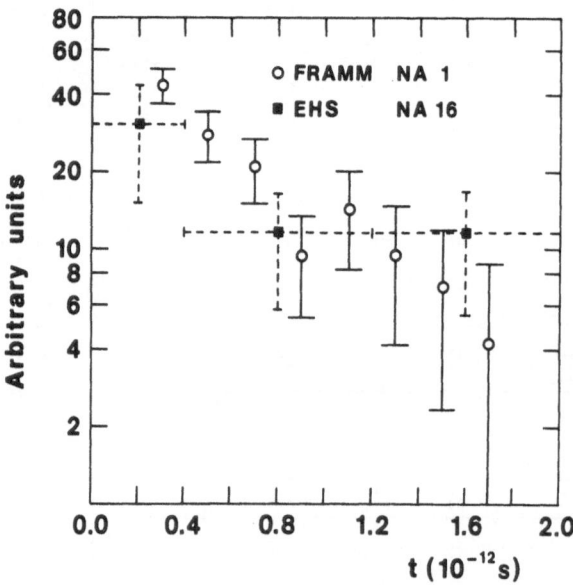

Fig. 4(b). Plot of the lifetimes for charged D's; data of the
experiments: FRAMM-NA1, EHS-NA16.

To obtain this last result one further problem had to be solved. It concerns the correspondence between the two paths per event measured in the live target and the two D particles reconstructed in the forward spectrometer. When either the combinatorial ambiguities are not solved or both the D's produced in one event decay in the same number of charged particles, the paths cannot be associated with the particles reconstructed in the spectrometer. Therefore as γ-stretching factor for both the D's, the average

$$\gamma_D = \gamma_{\overline{D}} = \frac{E_{tot}}{M_D + M_{\overline{D}}}$$

is used.

This assumption is not so wrong when only coherent interactions are taken into account, where two D particles are produced with low Q-value.

The distributions of the times of flight measured in E531, EHS-NA16 and WA58 are plotted in fig. 4a, all normalized to the same area. Similar plot for $D^{\mp}$ (FRAMM-NA1 and EHS-NA16) is shown in fig. 4b.

Two questions are now arising: do the results of the different experiments agree among them? Is there any substantial evidence for more than one neutral short-lived particle?

For the charged D's the agreement is not so bad; the "last" estimate would be very near to the FRAMM-NA1 result. For the neutral D's, the LEBC and E531 results are in good agreement; SLAC-BC 72/73 gives a lifetime value surprisingly high and WA58, a little lower.

WA58 probably lost long path decays and did not correct the data. On the other hand SLAC-BC 72/73, despite its very severe lower cuts, lost some short path decays.

The semileptonic decays observed by E531 (3) and WA58 (1) seem to show times of flight systematically longer than the other decays. In Table 8 the events of this category are summarized. Nevertheless the evidence is statistically too poor in order to adfirme that semileptonic decay lifetime is longer than the other decays.

TABLE VIII

D SEMILEPTONIC DECAYS

	Topology	Time of flight (10^{-13} sec)
E531	$D^0 \rightarrow K^- \mu^+ (\nu_\mu)$	4.2 - 7.2
	$\rightarrow \pi^- \pi^+ K^- \mu^+ (\nu_\mu)$	5.7 - 9.6
	$\rightarrow e^+ K^- (\nu_e)$	4.3 - 9.1
WA58	$D^0 \rightarrow K^0 \pi^- e^+ (\nu)$	3.4 - 7.5

In addition E531 reported a single event which is interpreted as a neutral charmed Baryon $\chi^0 \rightarrow p\pi^- K_s^0$ or $pK^- K_s^0$ with a time of life $\simeq 80 \cdot 10^{-13}$ s.

REFERENCES

(1) G. Zumerle; this Conference.
(2) T. Kitagaki; this Conference.
(3) N. Reay; this Conference and private communication.
(4) G. Diambrini-Palazzi; this Conference.
(5) E. Hugentobler; private communication.

LIFETIME OF CHARMED PARTICLES PHOTOPRODUCED

IN NUCLEAR EMULSION[*]

Presented by G. Diambrini-Palazzi

Photon-Emulsion Collaboration
[Bologna(IF and INFN)[1]-CERN[2]-Florence(IF and INFN)[3]-
Genoa(IF and INFN)[4]-Madrid(JEN)[5]-Moscow(LPI and GCPP)[6]-
Paris(LPNHE)[7]-Santander[8]-Valencia[9]]

M.I. Adamovich[6], Y.A. Alexandrov[6], J.M. Bolta[9],
L. Bravo[8], A.M. Cartacci[3], M.M. Chernyavski[6],
B. Conforto[3], A. Conti[3], M.G. Dagliana[3], M. Dameri[4],
G. Diambrini-Palazzi[4], G. Di Caporiacco[3], A. Forino[1],
R. Gessaroli[1], E. Higon[9], S.P. Kharlamov[6],
V.G. Larionova[6], R. Llosa[5], J. Lory[7], A. Mattei[3],
G.I. Orlova[6], B. Osculati[4], G. Parrini[3],
N.G. Peresadko[6], A. Quareni-Vignudelli[1], A. Ruiz[8],
K.M. Romanovskaya[6][**], M.A. Sanchis[2][***], D. Schune[7],
S. Tentindo[2], G. Tomasini[4], M.I. Tretyakova[6],
Tsai Chu[7], G. Vanderhaeghe[2], E. Villar[8] and B. Willot[7]

and

Omega-Photon Collaboration
(Bonn[10]-CERN[2]-Glasgow[11]-Lancaster[12]-Manchester[13]-
Rutherford[14]-Sheffield[15])

M. Atkinson[14], T. Brodbeck[12], G.R. Brookes[15],
P.J. Bussey[11], M. Davenport[14], J.-P. Dufey[2],
R.J. Ellison[13], W. Galbraith[15], K. Heinloth[10],
J.S. Hutton[14], R.E. Hughes-Jones[13], M. Ibbotson[13],
B.R. Kumar[14], G.D. Lafferty[2], J.B. Lane[13],
J.-C. Lassalle[2], R. McClatchey[15], D. Mercer[13],
J.V. Morris[14], D. Newton[12], G.N. Patrick[11], C. Raine[15],
A. Schlösser[10], K.M. Storr[2] and A.P. Waite[13]

---

 *) Data from the WA58 experiment.
 **) Gosniichimphotoproject, Moscow, USSR.
***) On leave of absence from the University of Valencia, Spain.

This report presents the data and the results obtained so far
by experiment WA58, performed at the CERN Super Proton Synchrotron
(SPS). It concerns mainly the charm lifetime estimates based on
the statistics available at the time of this Conference.

The data analysis performed by the WA58 as well as by other
experiments is now rapidly progressing. Therefore when the proceedings of this Conference will be published, it is likely that larger
statistics and an improved accuracy on charm lifetime data will have
been reached.

In the WA58 experiment, 5000 single emulsion pellicles,
$200 \times 50 \times 0.6$ mm^3, were exposed one after the other at an angle of
$5°$ with respect to the direction of the incident tagged photon beam
of the Omega Prime Spectrometer. The dose was such as to have 10^6
photons with energy between 20 and 70 GeV. The effective emulsion
thickness seen by the photon beam was 6.9 mm, and the photohadronic
interactions with the emulsion nuclei were selected by a trigger
logic accepting charged particle multiplicities larger than or
equal to three, whilst rejecting most of the electromagnetic interactions.

Identification of particles is performed, within a certain
range of momentum and angular acceptance, by using a Čerenkov gas
counter for hadrons, and a lead-glass wall and shower detectors for
electrons and photons.

The geometry computer program TRIDENT reconstructs the recorded
events, gives the track momentum and angles (in most cases with high
accuracy), and finally the vertex coordinates. Then the emulsion is
scanned under the microscope, over an area of 11×3 mm^2 centred on
the predicted position through all the 600 μm thickness.

When an event is found in the prescribed area, track angles are
measured in emulsion and compared with the corresponding values predicted by TRIDENT. In 50% of the cases agreement is found for the
total or a substantial number of the tracks and the event is accepted as "matched".

Then a procedure of track-following is applied to the event in
the emulsion in order to discover possible charged or neutral short-
lived particle decays. (At the beginning of the scanning, this
procedure was applied also to unmatched events.) Minimum ionization
tracks are followed from the main vertex up to their exit point from
the emulsion or up to at least 3 mm, corresponding to a proper lifetime of 10^{-12} s, if we assume a mean value p/m = 10.

Concerning neutral decays, it would be far too time-consuming
to perform a complete volume scanning in the forward direction for
all events found in the prescribed area. Nevertheless it is essential

to have a reasonably high and constant efficiency for the detection
of neutral decays -- at least in a finite range of decay length from
the primary vertex -- so as to obtain an unbiased mean lifetime evaluation. However, to make the proper cuts to the decay sample, we
need to wait for the higher statistics expected from this experiment.
In order to maximize the efficiency, in most cases the Photon-
Emulsion Collaboration adopted the following procedure.

a)   Since charmed particles are always photoproduced in pairs, when
a neutral or charged charmed particle decay has been found the
other is searched for in all the useful forward emulsion volume.
This is feasible because of the relatively low number of events involved. Therefore, even if the second charmed particle is a neutral,
its decay will be detected with reasonably high efficiency.

b)   If the TRIDENT program reconstructs some tracks not seen in the
emulsion, this could be because of some distant decay, either charged
or neutral. Then a scanning is performed along the predicted direction of these tracks.

     For instance, the V^0 decay in events 4 and 5 (Table 1a) were
found at a distance from the main vertex of 623 and 498 μm, respectively, by adopting just this procedure.

     The emulsion used for the WA58 experiment had a sensitivity of
about 30 grains per 100 μm, a grain size of 0.2-0.5 μm, and good
quality in most of the pellicles.

     The scanning started in April 1980. Work for improving the
reconstruction accuracy of the TRIDENT program was carried out and
is still continuing. Selection criteria for the events have evolved
with time according to the improvement of our understanding of the
charm features. In the following we present a list of the results
obtained up to now (November 1981), after scanning about 15% of the
emulsion pellicles, and we are confident that, from now on, the
charm-finding rate could be higher:

     8000 triggers scanned (out of 100,000);
     4000 matched events followed to search for charms;
     2000 unmatched events followed to search for charms;
     12 charm pairs corresponding to 24 decays were found.

If we add the $\overline{D^0}$ found in the previous WA45 exposure, we get 25 decays, of which 15, i.e. 7 pairs + the one single, have been reconstructed kinematically, sometimes with the addition of an unseen
neutral particle, whilst 2 more pairs are still under study and the
other 6 decays have only topological evidence.

Table 1a

Matched events

Event No.	E_γ (GeV)	Decay mode *)	Path length (μm)	τ (10^{-13} s)
1	25	$\overline{D^0} \to K^+\pi^+\pi^-\pi^-$	124	0.86
		$\Lambda_c^+ \to \Lambda^0\pi^+$	50	0.57
2	60	$D^- \to \pi^+\pi^-\pi^-(K^0)$	94	0.57-0.88
		$D^0 \to K^-\pi^+(\pi^0)$	267	0.45-0.85
3	65	$F^+ \to \pi^+\pi^-e^+(\nu)$	680	2.2-3.3
		$D^- \to \pi^+\pi^-\pi^-(K^0)$	980	7.1-14.7
		or		
		$F^- \to K^+K^-e^-(\nu)$		
4	49	$\overline{D^0} \to K^+\pi^+\pi^-\pi^-$	45	0.14
		$D^0 \to \overline{K^0}\pi^-e^+(\nu)$	623	3.4-7.5
5	32.5	$\overline{D^0} \to K^+\pi^-\pi^0$	498	2.0
		$\Lambda_c^+ \to \Lambda^0\pi^+(\pi^0)$	156	2.1-4
6	44	$F^+ \to \pi^+\pi^-\pi^+\pi^0$	570	6.0
		$\overline{D^0} \to K^+\pi^-\pi^0$	< 10	< 0.04
7	57.8	$D^0 \to K^-\pi^+\pi^0$	252	1.0
		$\overline{D^0} \to K^0\pi^+\pi^-\pi^+\pi^-$	123	0.40
WA45	64	$\overline{D^0} \to K^+\pi^-\pi^+\pi^-$	123	0.23
		X (unseen)	–	–
8	70	V^0 (under study)	23	–
		$\Sigma_c^{++} \to \Lambda_c^+\pi^+$		
		$\phantom{\Sigma_c^{++} \to}\hookrightarrow p + K^0(\pi^0)$	315	3.7-4.3
9	71.5	$D^+ \to K^-\pi^+\pi^+\pi^0\pi^0$ (under study)		
		$\overline{D^0} \to K^+\pi^-\pi^0$ (presumed decay out of emulsion)		

Table 1b

Unmatched events

Event No.	Topology	Path length (μm)
10	V^0	44
	1p	260
11	3p	32
	1p	1900
12	3p	341
	1p	3236

*) Particles within brackets were not detected and are added for momentum balance.

Table 2a

Selected D^0 and $\overline{D^0}$ decays

Decay mode	Path length (μm)	τ $(10^{-13}$ s)
$\overline{D^0} \rightarrow K^+\pi^+\pi^-\pi^-$	124	0.86
$D^0 \rightarrow K^-\pi^+\pi^0$	267	0.45-0.85
$\overline{D^0} \rightarrow K^+\pi^+\pi^-\pi^-$	45	0.14
$D^0 \rightarrow \overline{K^0}\pi^-e^+(\nu)$	623	3.4-7.5
$\overline{D^0} \rightarrow K^+\pi^-\pi^0$	498	2.0
$D^0 \rightarrow K^-\pi^+\pi^0$	252	1.0
$\overline{D^0} \rightarrow K^0\pi^+\pi^-\pi^+\pi^-$	123	0.40
$\overline{D^0} \rightarrow K^+\pi^-\pi^+\pi^-$	123	0.23
$\langle\tau(D^0)\rangle = 1.34^{+0.56}_{-0.34} \times 10^{-13}$ s		

Table 2b

Λ_c^+ decays

Decay mode	Path length (μm)	τ $(10^{-13}$ s)
$\Lambda_c^+ \rightarrow \Lambda^0\pi^+$	50	0.57
$\Lambda_c^+ \rightarrow \Lambda^0\pi^+(\pi^0)$	156	2.1-4.0
$\Lambda_c^+ \rightarrow pK^0(\pi^0)$	315	3.7-4.3
$\langle\tau(\Lambda_c^+)\rangle = 2.5^{+2.0}_{-1.0} \times 10^{-13}$ s		

Table 2c

Charged D decays

Decay mode	Path length (μm)	τ $(10^{-13}$ s)
$D^- \rightarrow \pi^+\pi^-\pi^-(K^0)$	94	0.57-0.88
$D^- \rightarrow \pi^+\pi^-\pi^-(K^0)$	980	7.1-14.7
$\langle\tau(D^-)\rangle = 5.8^{+7.2}_{-2.8} \times 10^{-13}$ s		

Table 2d

F decays

Decay mode	Path length (μm)	τ $(10^{-13}$ s)
$F^+ \rightarrow \pi^+\pi^-e^+(\nu)$	680	2.2-3.3
$F^+ \rightarrow \pi^+\pi^-\pi^+\pi^0$	570	6.0
$\langle\tau(F^+)\rangle = 3.8^{+4.0}_{-2.1} \times 10^{-13}$ s		

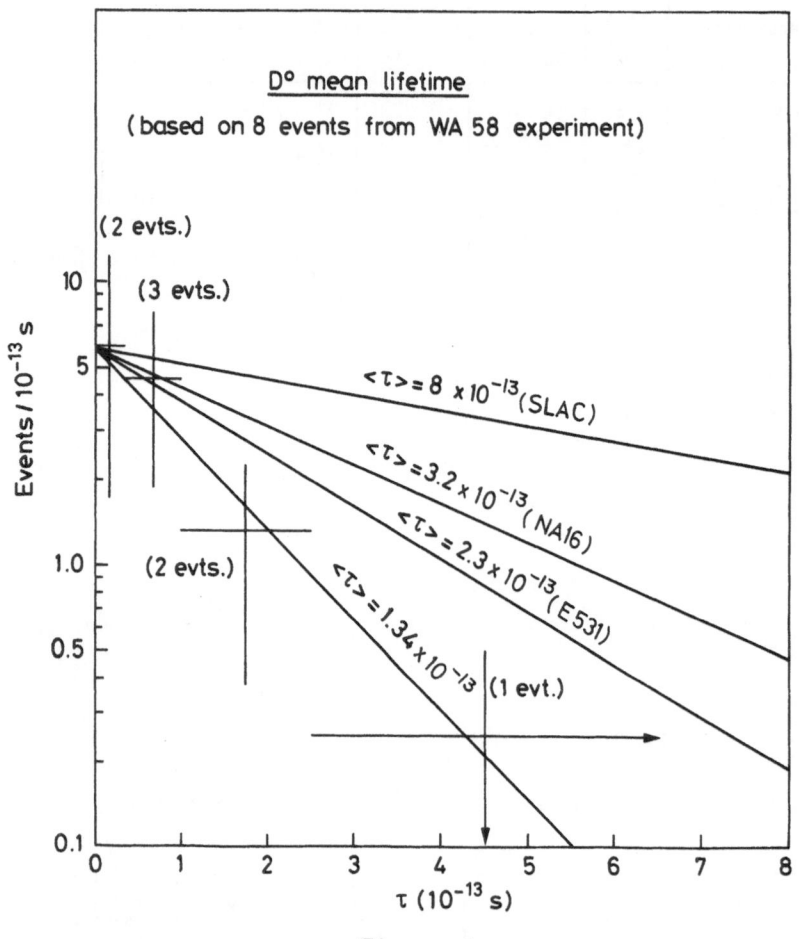

Figure 1.

The seven reconstructed pairs are:

$$2(D^0, \overline{D^0}) \ , \quad 2(\Lambda_c^+, \overline{D^0}) \ , \quad 1(F^+, D^- \text{ or } F^-) \ , \quad 1(F^+, \overline{D^0}) \ , \quad 1(D^-, \overline{D^0}) \ .$$

In Tables 1a and 1b the full list is given. Event No. 6 at $E_\gamma = 70$ GeV is of special interest. In fact the Λ_c^+ seems to come from a Σ_c^{++} strong decay. The neutral decay is still under study. The mass value obtained for the Σ_c^{++} is 2.47 ± 0.08 GeV/c^2.

Mean lifetime of the D^0 meson

For the mean lifetime evaluation we chose a sample of eight out of the twelve available D^0 and $\overline{D^0}$ decays, as listed in Table 2a. They satisfy the criteria of having a path length $\ell > 10$ μm and of having been reconstructed kinematically. When two proper lifetime values are given for the same decay, it means that an extra unseen neutral particle has been added for momentum balance, and therefore two physical solutions are obtained for the primary momentum value. The arithmetic mean of the two values is taken for the present purpose. The maximum likelihood equivalent to the over-all arithmetic mean gives the value $\langle \tau \rangle = 1/n \ \Sigma_{i=1}^{n} \ \tau_i = 1.34_{-0.34}^{+0.56} \times 10^{-13}$ s. It should be emphasized that no efficiency corrections have yet been applied, mainly because they are probably not larger than the quoted statistical error, and we await larger statistics which we expect will be obtained in this experiment.

Figure 1 shows the differential semi-logarithmic plot of the eight events from which the mean lifetime has been computed. Values from other experiments are shown for comparison.

Mean lifetimes of Λ_c^+, $D^\pm$, $F^\pm$

The values quoted in Table 2b give the mean lifetime $\langle \tau(\Lambda_c^+) \rangle = 2.5_{-1.6}^{+2.0} \times 10^{-13}$ s. From Table 2c there is only one decay assigned unambiguously to a D^- meson. If we take both decays as D, then we get $\langle \tau(D^-) \rangle = 6_{-2.5}^{+6.0} \times 10^{-13}$ s. For the only two F^+ available, we get $\langle \tau(F^+) \rangle = 3.8_{-2.1}^{+4.0} \times 10^{-13}$ s.

Note added when submitting the manuscript

At the present time (March 1982), seven new double-charm candidate events (i.e. 14 charm decay candidates) have been observed in emulsion and are now under study.

INTRODUCTION TO THE PANEL ON: LIVE TARGETS FOR LIFETIME MEASUREMENTS

N. Reay

Ohio State University

Columbus Ohio   (U.S.A.)

In a few short years, identifying and measuring lifetimes for charmed particles has grown from nonexistence into a thriving cottage industry. If this conference is any indication, the trend will continue well into the future.

My assignment is to review briefly various approaches in vertex detectors, then discuss the approach of Fermilab Tevatron Experiment 653, of yhich I am a member.

Active target charm searches may be divided into two classes, those with triggerable and those with non-triggerable targets. The former may be separated into experiments which are fully electronic and those which require optical techniques. A fully electronic scheme is shown in Figure 1a, an active target of closely spaced solid state detectors is followed downstream by a powerful charged and neutral spectrometer. In addition to mass fitting, evidence for charmed particles is given by the jump in ionization on decay, as shown in Figure 1b. This technique is best suited for diffractive processes in which the decays are separated by more than a millimeter from the primary vertex. Detection of centrally produced events is hampered both by the general increase in track multiplicity and by the heavily ionizing tracks from nuclear breakup at the production vertex. Such experiments are capable of delivering high statistics, but must contend with high backgrounds and difficulties in uniquely identifying charm decays. Professor Bellini, Meroni and others will give you more detailed accounts later on in the conference.

The advantage of incorporating optical techniques into your favorite active target is that spatial resolution is considerably enhanced (usually at the expense of data collection rate). Future experiments are being planned both with large and small high resolu-

tion bubble chambers. Professor Baltay will discuss possibilities
for charm detection arising from neutrino interactions in the Fermilab 15 foot chamber, modified to accommodate holographic techniques.
Spatial resolution smaller than 100 microns will be achieved, sufficient to detect charmed hadrons and tau leptons. Detecting taus
arising from neutrino interactions in a beam dump geometry will be
the final link in establishing the existence of tau neutrinos.

At CERN, small bubble chambers with conventional optics have
achieved 30 micron resolution in one transverse coordinate, sufficient to see charm decays with somewhat lowered efficiency.

Holographic techniques will sharpen the resolution to perhaps
6-8 microns, sufficient not only for charm but perhaps even for B
Decays. Further, this high resolution exists in both transverse
coordinates. Though the flash may be triggered, ultimately bubble
chambers are pretty much limited to recording total interactions in
a few million pictures over a several year time span. Charm yields
in hadron beams appear limited to perhaps 1000 decays. However, the
possibility of higher percentage yields in photon beams is exciting.
Such experiments will be discussed later on by Drs. Lecoq, Zumerle
and others.

High pressure streamer chambers have the significant advantage
of being highly triggerable, but have 60 micron resolution only in
one transverse coordinate. Such devices are being run by a Yale-
-Fermilab Collaboration at Fermilab. Unfortunately, they have no
representatives at this conference.

Other triggerable optical targets include thin NaI crystals and
microchannel plates. These devices typically have 40 micron resolution in one transverse coordinate, but sample only at several
hundred micron intervals along charged tracks. They will be discussed
in talks by Professors Potter and Ruchti.

Finally, there are a class of experiments which use emulsion
as a non-triggerable optical target. They have spatial resolution
smaller than one micron in both transverse coordinates and sample
every 3-5 microns along charged tracks. Such resolution is sufficient for charm and for B Decays, even for lifetimes as short as
5×10^{-15} seconds. However, proposed experiments are limited by
track densities to perhaps $1 - 5 \times 10^8$ total interactions. Emulsion
is difficult to scan, thus event selection procedures must highly
enrich the percentage of charm and B events without throwing away
too many decays. Such experiments will be discussed by Professors
Diambrini-Palazzi, Musset, Sacton and myself.

Now, in a dazzling display of egocentricity, the rest of the talk
will be devoted to Fermilab Tevatron Experiment 653. The first
step in any experimental design must clearly delineate experimental
priorities. Our goals are:
1) Get a low background sample of D^0 decays in which D^0 is separated
   from $\overline{D}^0$ by the sign of charged kaons or leptons. This sample will

be used to investigate possible anomalies in the lifetime, and
to measure the mass difference δm between neutral D CP eigenstates.
2) Obtain measurements of charged and neutral B lifetimes.
3) Look for new particles such as stable charmed baryons, new excited charmed states F^*, Λ_c^*, etc.), and exotics with extra quarks.
4) See $F \rightarrow \tau$.
5) Get precise charm lifetimes.
6) Get production dynamics, etc.

Before showing how the first two of these motivated the experimental design, permit a brief aside to discuss the ramifications
of measuring δm.

Nominally, a produced D^0 $(\overline{D}^0)$ meson will decay into K^0, $\overline{K}$
and/or positive leptons ($\overline{K}^0$, K^+ and/or negative leptons).

However, if the mass difference δm is sufficiently large,
occasionally by mass mixing D^0 mesons may decay through nominal
$\overline{D}^0$ channels, and vice versa. Should such wrong sign decays occur,
they will peak in absolute rate approximately at two lifetimes,
and the percentage of wrong sign decays will grow quadratically as
a function of proper decay time. One can then beat the effects of
systematic mis-identification without losing greatly in statistical
precision simply by looking for wrong sign decays in samples with
proper times longer than several lifetimes.

Conventional W-S theory predicts δm will be unmeasurably small,
but many higher symmetry models permit $(\delta m/\lambda) - 1$, where λ is the
inverse of the D^0 lifetime, present experiments which measure δm
directly establish only $(\delta m/\lambda)$ 0.6, 90% c.1..

Now, how do measuring δm for D^0 mesons and crudely measuring
B lifetimes focus experimental design?

Most importantly, independent of any trigger design room must
be left for kaon and lepton identification. Further, mass fitting
must be performed at the $\pm$ 20 MeV level (including π^0 fitting) so
the kaon charge may be determined uniquely even when not directly
identified. Fortunately, as parent directions are measured typically
to a milliradian in emulsion, the combinatorics of π^0 identification
are reduced greatly by demanding that transverse momenta about the
parent direction of all decay products sum to zero.

However, precision momentum determination, kaon identification
and π^0 decay opening angle measurements all act to stretch out the
experimental design, increasing the difficulty of constructing a
good experimental trigger.

Measuring B lifetimes requires overcoming two formidable experimental difficulties. As B lifetimes are unknown and may be quite
short, we must access lifetimes down to a few x 10^{-15} seconds.
Fortunately, this is easy with emulsions. The second problem is
more severe, only a few B particles will be produced per million
total interactions. The ratio of signal to noise is vanishing. We
have faced this problem by not developing risky triggers (such as

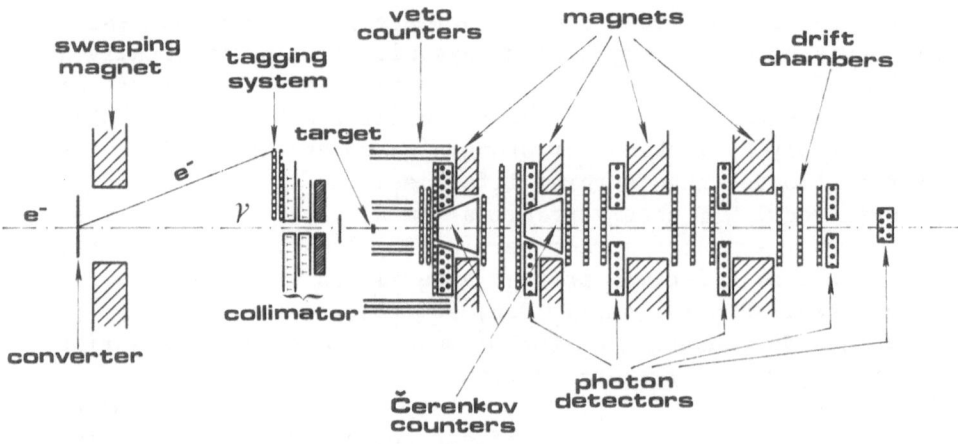

Figure 1a

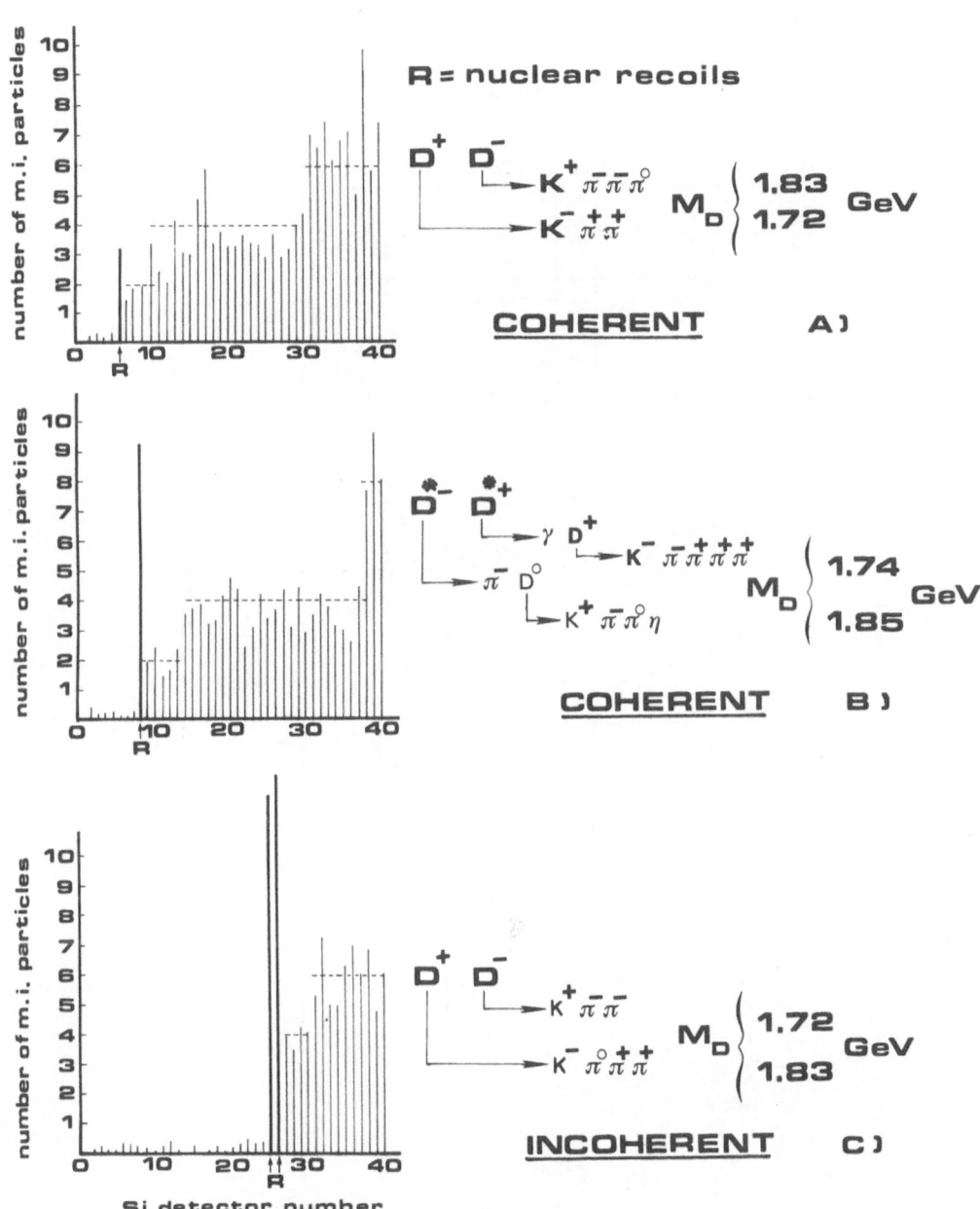

Figure 1b

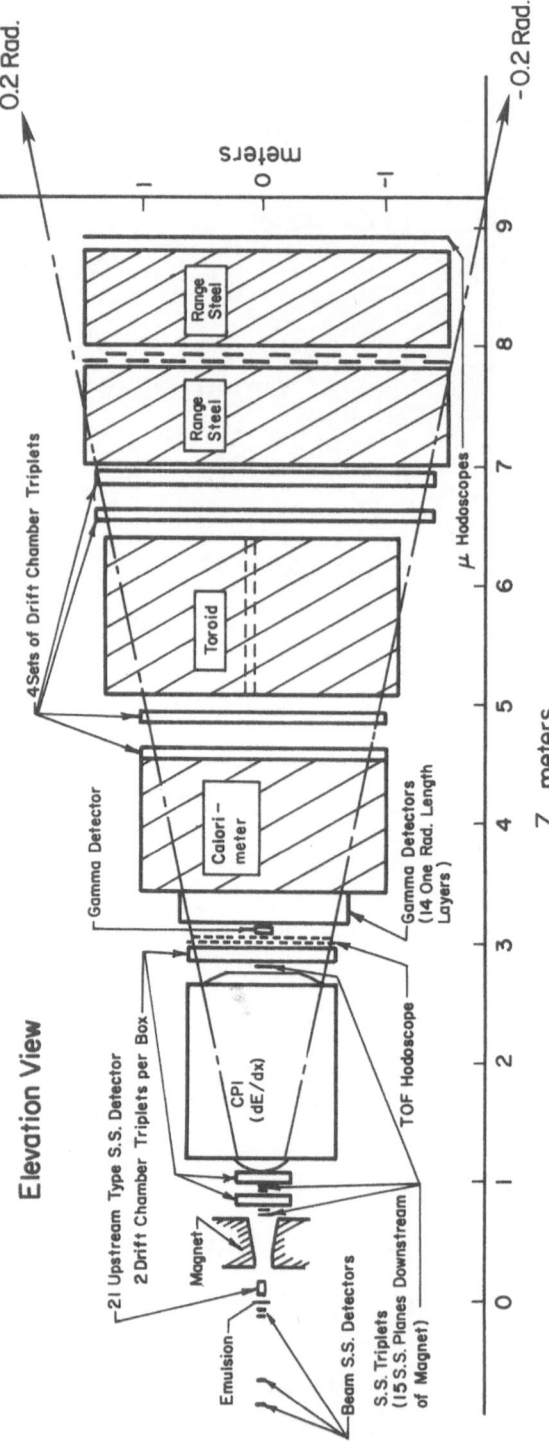

Figure 2. Experiment 653.

high $P_\perp$ muons) which take us directly to the B, but rather by setting up an event selection procedure leading to identification of a large and highly pure charm sample. Since b → c appears to occur with high probability, we can minimize the risk of obtaining nothing by scanning near the vertex for B decays in every event containing a found charm decay.

The layout proposed as a result of all this heavy thinking is shown in Figure 2. It is designed to accommodate on-line a single muon trigger, then permit searches off-line for multiprong vertices containing the muon.

The front end consists of a series of emulsion modules moved through a diffracted proton beam to guarantee uniformity of exposure. The module will be thin (∿2 cm) to permit observation of decays without undue contamination from secondary interactions.

Within centimeters of the emulsion module will be a series of 13 micron rms resolution microstrip solid state detectors, mounted with seven detectors in each of three 120^0-rotated views. Two-track separation in the forward jet will be 40 microns, sufficient to resolve more than 90% of all tracks in at least two views. The short magnet will be followed downstream by high spatial resolution detectors, dE/dx charged particle identifiers, high spatial resolution gamma detectors, a hadron calorimeter, and a toroid plus range steel muon identifier. We believe it essential to have a full complement of particle identification even though the free drift space required compromises the single muon trigger.

The actual steps in triggering and event selections will be as follows, the table indicates signal versus background surviving each cut.

1) Record any event containing a muon with greater than 5 GeV momentum by range.

2) Off-line, bound P_L > 8 GeV/c, $P_\perp$ > 0.2 GeV/c for the muon, eliminate most muons coming from kaon decay by looking for "kinks" in the track and by comparing spectrometer magnet and toroid determinations of the muon momentum.

3) Use solid state microstrip detectors with 13 micron rms spatial precision to search for tracks intersecting with the muon track at a point least four standard deviations removed from the production vertex. Demand that the net visible decay momentum be consistent with a parent particle originating near the production vertex.

4) Find surviving decay candidates in the emulsion. (Editorial comment added in writeup: Note that this step implies a powerful emulsion capability: at this stage background still exceeds signal. However, we are presently scanning the second run of E-531, a neutrino exposure, at a rate of 200 found events/ week with a finding efficiency of 90%. without improvements, this extrapolates to 20,000/year for E-653, less than a factor of two away from design.)

TABLE I

CUT	BACKGROUND SURVIVING	CHARM SURVIVING	B PAIRS SURVIVING
0) No cut	2.8×10^8	5.8×10^5	1.2×10^3
1) On-line muon cut	5×10^6	6.3×10^4	145
2) Off-line muon cut	1.5×10^6	4.4×10^4	101
These events must be analyzed in more detail.			
3) Fraction with secondary vertices passing > 2 prong cuts and which are > 4 σ from production vertex.	33,000	18,000	51
Now, must find muonic decay in emulsions			
4) 90% effic for finding decay, demand no evidence of nuclear breakup and the correct multiplicity.	740	16,300	46
Now, must scan for B's and/or associated charm decay.			
5) 0.92 scan efficiency for each decay.	Less than two associated decays with correct multiplicity and no evidence of nuclear breakup.	15,000	45

TABLE: Shows background versus charm and B decay signals expected to survive each stage of cuts. B signals are based on a 50 nanobarn cross section at an incident proton energy of at least 600 GeV. A possible factor of two increase in B yield by incorporating muon selection with $P_\perp >$ 1.3 GeV/x is not included.

After calibrating emulsion module positions using interactions in low density regions of exposure, the muonic decay candidate location will be predicted to within 1-2 pellicles and one microscope field of view (100 microns x 100 microns). Background events may be eliminated by demanding no visible evidence of nuclear breakup, and a correct decay multiplicity. We also will connect all emulsion decay tracks with electronically found tracks, demand missing transverse momentum and visible mass cuts precluding K^0, Λ and other decays.

5) Examine the event to find associated charm and/or $\overline{BB}$ decays.

As such a method is relatively independent both of particle type and decay branching ratios, it is well suited for new particle searches. The risks involved in obtaining B's are more easily eliminated and smaller than for methods requiring fancy B triggers.

However, the density of emulsion exposure has been pushed to its maximum (10^3 tracks/mm^2), emulsion scanning has been pushed far beyond previously achieved levels (70,000 - 100,000 events scanned), and data recording and analysis difficulty exceed by factors of 2-4 anything our group previously has achieved. Finally, the experiment hinges on development of large silicon microstrip detectors. Succinctly put, E-653 is not a conservative design.

Only time will tell if we've guessed correctly, but for sure the next few years will be very exciting.

LIFETIME MEASUREMENT OF D$^{\pm}$ - MESONS WITH A SILICON LIVE TARGET

E. Meroni

Istituto di Fisica dell'Università and Sezione INFN

Milano

In the following I describe some results obtained by NA1 experiment performed by FRAMM Collaboration[1] on the North Area H4/E4 beam, working in the electron mode at 150 GeV.

The purpose of this experiment is to study heavy mass states coherently produced by a photon beam off a silicon target. In particular D-meson pairs are selected in order to measure their decay path and hence their lifetime. The production mechanism is illustrated in fig. 1.

The requirement of coherence on nucleus leads to the following advantages: the coherent production implies low momentum transfer to the nucleus, and selects final states with the same quantum numbers of the incident photon. Therefore almost exclusive channels

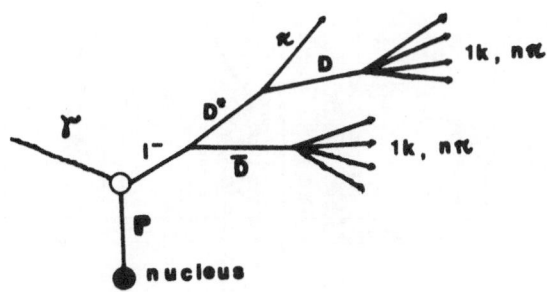

Fig. 1. Coherent photoproduction of D$\bar{\text{D}}$ pairs.

are produced and the D-pair has all the available energy. A further
reason for choosing this mechanism lies in the possibility of measuring the meson path in the silicon target. In a coherent production
the interaction point can be identified by the energy deposit of the
recoiling nucleus, which is strictly confined inside a single layer
of the target.

The apparatus

    The experimental apparatus, as shown in fig.2, consists of a
tagging system, a vertex detector and a forward spectrometer[2].
    The photon beam is produced by bremsstrahlung of 150 GeV
electrons on a lead converter, 0.1 radiation length thick. A tagging
hodoscope measures the momentum of the swept electrons providing
the photon energy with an accuracy of ± 5.%. The photon beam crosses
a set of collimators and reaches the target.
    The target is surrounded by a set of counters to veto the
majority of the incoherent reactions and events with particles
produced at angles larger than 30°.
    The forward spectrometer consists of five stacks of MWPC and
drift chambers[3],[4]. After each stack a bending magnet measures the
momentum of all charged particles produced within a cone of 90 mrad
half-aperture. This system provides a rather uniform resolution in
momentum ranging between 1. and 150 GeV/c (.5<$\Delta P/P$<1.5 %).
    In front of each magnet a shower detector detects all particles,
which would not reach the successive stack. The first two shower
detectors are sandwiches of lead and scintillation hodoscopes,
while the other three are matrices of lead glass counters. They
measure all photons produced within ± 30°, providing standard energy
resolution and a very good (± 1.5 mm) space resolution. In addition

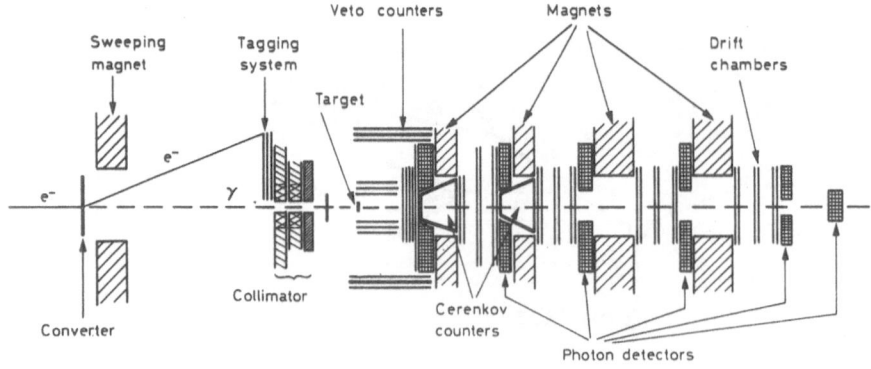

Fig. 2. Layout of the apparatus.

they identify hadrons and electrons either at the trigger level or,
coupled to the magnetic spectrometer, in the offline analysis.

   Two multicell Cerenkov counters are installed inside the first
two magnets to discriminate kaons from pions in the momentum range
4. - 20. GeV/c. The efficiency of each cell is of the order of 99.%.

The target

   The target is structured in 40 silicon detectors 300 μm thick,
spaced by 100 μm air gaps[5,6]. This telescope has been expressely
designed to detect the decay of charmed mesons with lifetimes in the
range 10^{-13}-10^{-12} sec for energy of incident photons ranging from
70 to 150 GeV/c. Altogether the target is 1.6 cm thick, corresponding to 15% of radiation length.

   The signal from each layer is proportional to the number of
minimum ionizing particles crossing the detector. The pulse height
pattern of the telescope allows the identification of the coherent
production and of the decay points of the D-mesons. Fig. 3 shows
an example of such an event: the interaction point corresponds to
the layer with a spike due to the recoil of the nucleus, while the
decay points are identified by an increase of the charged multiplicity.

   Fig. 4b shows an example of target pattern for an incoherent
event. For this kind of events the interaction point is not identi-
fiable because the nuclear fragments cross more than one layer.

   The choice of 300 μm for the layer thickness results from a
compromise between position and multiplicity resolution. Namely to
have a good signal-to-noise ratio it is preferable to work with

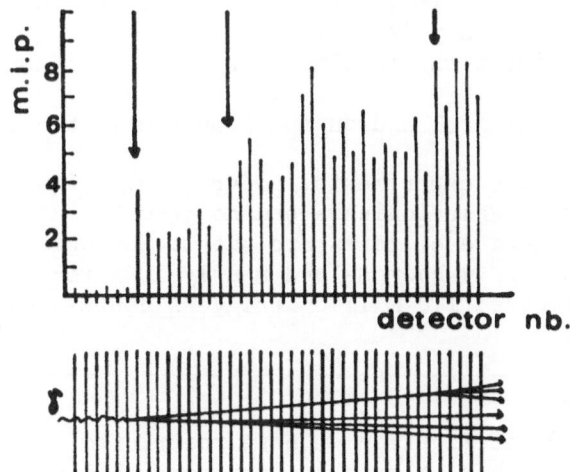

Fig. 3. Pulse height pattern in the silicon target.

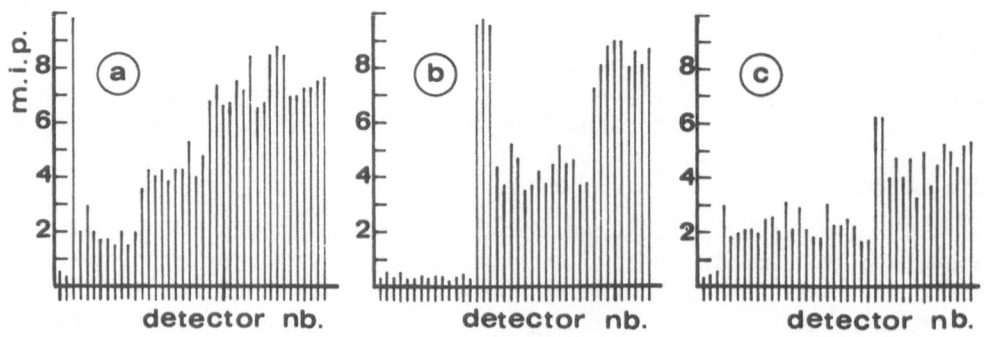

Fig. 4. a) target pattern for a coherent event
        b) target pattern for an incoherent event
        c) target pattern for an event with secondary interaction.

thick detector but this is in contrast with the requirement of measuring short decay-paths.

    Due to the fact that a photon beam is used, the counting rate
is low in the first layers of the target, becoming very high (about
10^6 particles per second) in the last ones, because of the photon
conversion into $e^+ e^-$ pairs. This means that at least for the first
layers, slower electronics with lower noise can be used. The signal
processing is described in P.F. Manfredi's talk. In fig. 5 are shown
the resulting pulse height distributions for 1-2-4-6 minimum ionizing particles obtained in the first and in the last detectors.

Trigger

    During the data taking, the shower detectors information were
used to reject electron pair background.
    In the experiment the following trigger conditions were re-
quested:
- at least 2 hadrons or 1 hadron and 1 photon in the forward spectrometer.
- nothing outside the experimental acceptance.
    Due to this trigger logic we had no geometrical cut in the ho-
rizontal plane, where $e^+ e^-$ pairs are swept. The trigger efficiency
was more than 70% for 4 hadrons and nearly 100% for photons.
    We have collected 1.8 Million triggers containing about 10^6
hadronic events.

Offline analysis

    For D-pair identification we have restricted the sample by
selecting events with 4 or more charged particles and with energy
larger than 70 GeV, which is the threshold for coherent production

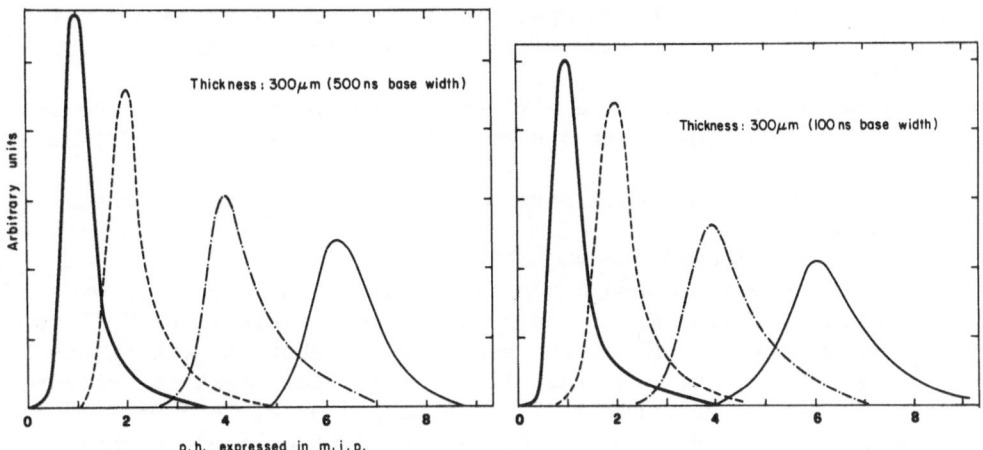

Fig. 5. Pulse height distributions for 1-2-4-6 m.i.p. crossing
a silicon detector 300 µm thick.

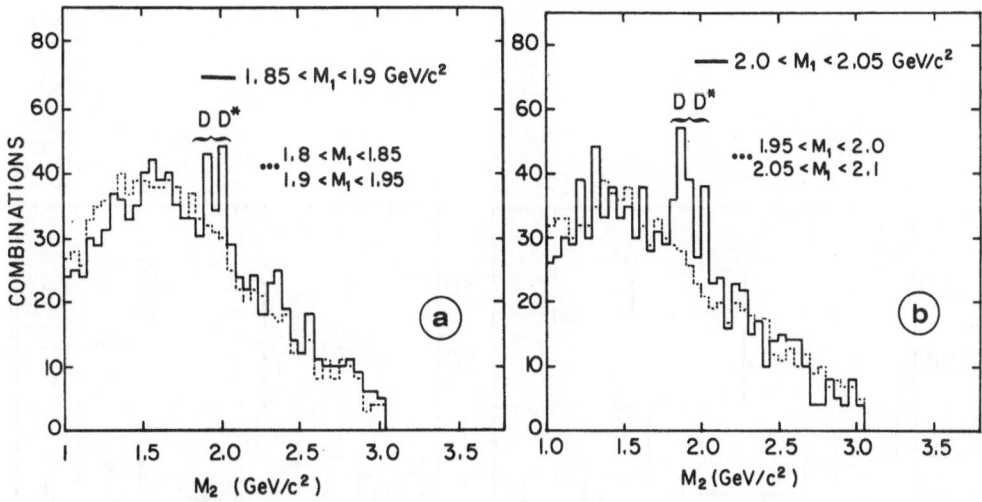

Fig. 6. Inclusive mass distribution (full line), background (dotted
line).

on silicon nuclei. These cuts reduce our sample to $\sim$ 23.000 events

Since the silicon target does not give association between the decaying D meson and the charged tracks in the spectrometer, this is done in all possible ways. The combinatorial analysis is done as following:
- the produced particles are divided in two groups so that one could be a D^+ (D^0) and the other one could be identified as D^- ($\bar{D}^0$) candidate;
- at least one group should have a K identified by the Cerenkov counters;
- only Cabibbo allowed decays are built.

These selected events should contain a pair of D mesons. If we require that the mass M_1 of one group is compatible with the D or the D^* mass, the mass M_2 of the second group shows a peak in the D or D^* region (fig. 6a - 6b full line). This clear peak is present only if M_1 is in the range 1.85 - 1.9 GeV/c^2 or 2.0 - 2.05 GeV/c^2. The background distribution (fig. 6a - 6b dotted line) is obtained when M_1 falls in the nearby mass ranges (1.8 - 1.85, 1.9 - 1.95 and 1.95 - 2.0, 2.05 - 2.1 GeV/c^2).

However, without any further information this method is not sufficient to select charmed events for a lifetime measure.

This selection can be improved by searching for specific decay channels. Fig. 7a - 7b show the mass distribution for M_2 decaying into $K^-\pi^+\pi^+$ ($+n\pi^0$'s) selecting the M_1 decay channel $D^- \to K^+\pi^-\pi^-\pi^0$ (fig. 7a, events with 6 charged particles in the spectrometer) or $\bar{D}^0 \to K^+\pi^-\pi^0$ (fig. 7b, events with 5 charged particles). These plots show that these charmed pairs are well reconstructed with a good signal-to-noise ratio.

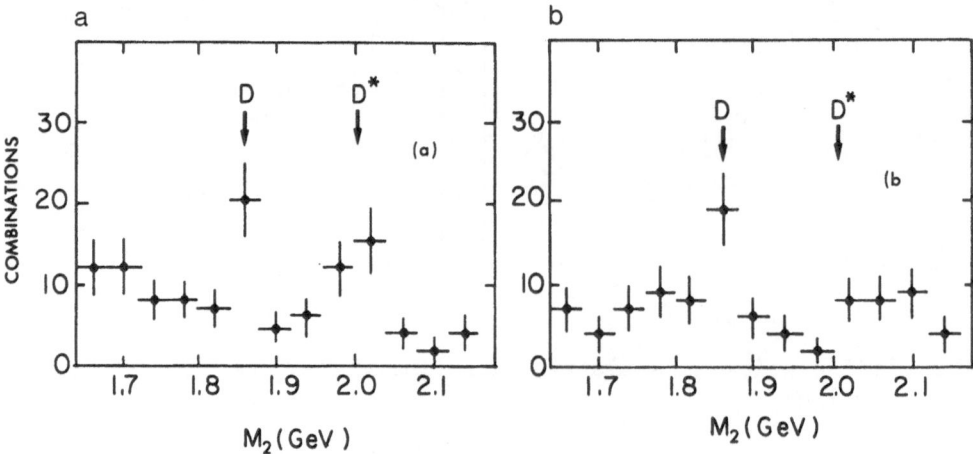

Fig. 7. Exclusive mass spectrum in the $K^+\pi^+\pi^+$($n\pi^0$) decay channels.

The resulting sample is too small to allow the measurement of the lifetime of the D-mesons.

To increase the D$^\pm$ sample, we release the above cuts by requiring only one charged group with a mass falling between 1.75 and 2.10 GeV/c^2. The selection of charmed events is made instead by requiring 6 or more charged particles in the spectrometer, together with a target pulse height pattern characteristic of charged D decays. The criteria are:
- the first 2 layers should be empty to select events produced in the target;
- a step is identified only if it extends over at least 4 layers;
- the first level after the production point should correspond to at least two minimum ionizing particles in order to avoid D$^\circ$(D$^\circ$*) -$\overline{D}^\circ$($\overline{D}^\circ$*) final states.

By applying the above requestes, we end with 86 events showing 98 decays. One of them is shown in fig. 4a.

For the determination of the proper decay time we need to know the energy of the D meson. Since the target does not provide the correspondence between the path and the reconstructed D, half of the energy of the event has been given to both mesons. This is justified by the production mechanism: the D-pair is produced in an exclusive way with a low Q-value. The resulting time distribution is shown in fig. 8, where the data have been corrected for the target acceptance. The single slope fit gives a value $\tau = (8.9^{+2.9}_{-1.7})$ 10^{-13} sec.

Two sources of background must be taken into account before quoting a value for the lifetime: the presence of faked steps due to hadron interactions or due to γ-ray conversions and the contamination of D$^\circ$ in D$^\pm$ sample.

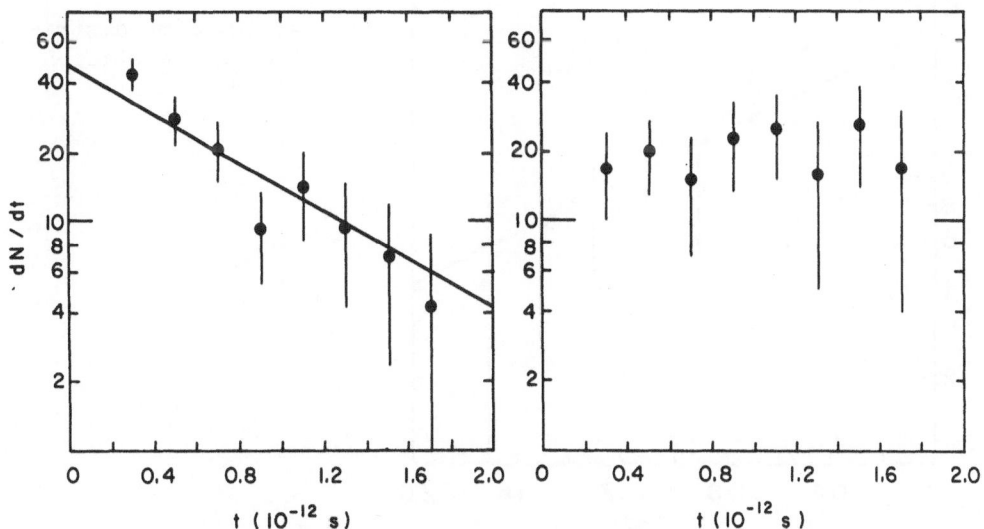

Fig. 8. Time distribution of iden-   Fig. 9. Time distribution of
tified D-mesons decays.   background events.

The majority of events with a secondary interaction are rejected by requiring that no step shows an overshoot due to nucleus recoil or breaking (fig. 4c).

The events with γ-ray conversion are identified by the presence of one or two electrons in the spectrometer.

To evaluate the remaining contamination, a sample of events with the same topology of the D-candidates but with mass lower than the selected mass range has been scanned looking for steps in the target. The time distribution of these steps, obtained by using the average γ-factors of the D-decays, is completely flat (fig. 9). We deduce from this procedure that the contamination of the D-sample is about 10%.

The second source of background in the time distribution is due to the presence of D^0 contamination. This contamination is greatly reduced by the selection criteria applied to the pulse height pattern in the target. In particular a minimum length of 4 layers for a multiplicity step (corresponding in the average to $\sim 3 \cdot 10^{-13}$ sec) suppresses D^0 decays in the hypothesis that the D^0 lifetime is around $3 \cdot 10^{-13}$ sec.

Using the known branching ratio, we have evaluated this contamination as a function of τ_{D^0}. For $\tau_{D^0} = 2.5 \cdot 10^{-13}$ sec, the contamination amounts to $\sim 20\%$.

Fig. 10 (dashed line) shows a fit which takes into account the D^0 contamination. The resulting value (fig. 10 - full line) for the $D^\pm$ lifetime is $\tau_{D^+} = (9.5^{+3.1}_{-1.9}) \, 10^{-13}$ sec.

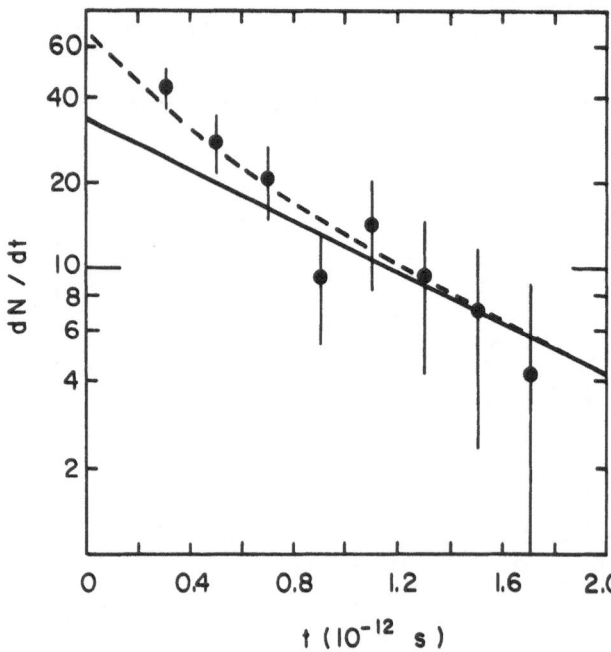

Fig. 10. Time distribution of identified D-mesons.

References

1. FRAMM Collaboration: Frascati, Milano, Pisa, Roma, Torino, Trieste
2. Status Report on NA1 experiment, 5 September 1980, CERN/SPSC/ 80-85, SPSC/M 247.
3. G. Bologna et al., Nucl. Instr. and Methods 178:587 (1980).
4. S.R. Amendolia et al., Nucl. Instr. and Methods 176:461 (1980).
5. G. Bellini et al., CERN-EP/81-145, submitted to Nucl. Instr. and Methods.
6. S.R. Amendolia et al., Nucl. Instr. and Methods, 176:449 (1980).

DISCUSSION

N. Reay: In how many of your charm events do you have a large volta ge spike in the solid state wafer in which the interaction occurs?

Answer: All the events used for lifetime measurements are coherently produced. Nearly half of them shows a signal amplitude in the interaction layer greater than the amplitude of the following detectors.

SOLID STATE TELESCOPES FOR LIFETIME MEASUREMENTS

P.F. Manfredi

Institute of Electronics and Communications
University of Pavia
Strada Nuova 106/c
27100 Pavia (Italy)
I.N.F.N. - Sezione di Milano
Via Celoria 16
20133 Milano (Italy)

I- INTRODUCTION

A telescope made of thin solid state detectors separated by narrow air gaps provides a method of measuring the lifetime of particles that after being produced in one of the detectors decay within the telescope.

A telescope of forty detectors was employed in FRAMM experiment[1,2] to measure the lifetime of charmed D^+, D^- mesons. The D pairs in FRAMM experiment are produced by 150 GeV photons. The detector where the production takes place is identified by the presence of a large signal determined by the nucleus recoil. The positions at which the produced particles decay are reconstructed from the ionization pattern along the telescope. This pattern shows where changes in multiplicity take place. An example of a pattern of pulse amplitudes appearing at the various detectors of FRAMM telescope is shown in figure 1 along with the reconstruction of the event to which it was attributed.

The vertical axis of figure 1 is calibrated in units of energy released by a single minimum ionizing particle. As shown by figure 1, there is a first group of six detectors where the number of particles detected is less than one. This means that no production has occurred yet and that the processors associated with each detector sense nothing but the electrical noise. The large signal in the seventh detector shows that production has taken place and, as a matter of fact, two minimum ionizing particles are detected by the next four silicon layers. There are, furthermore, two steps

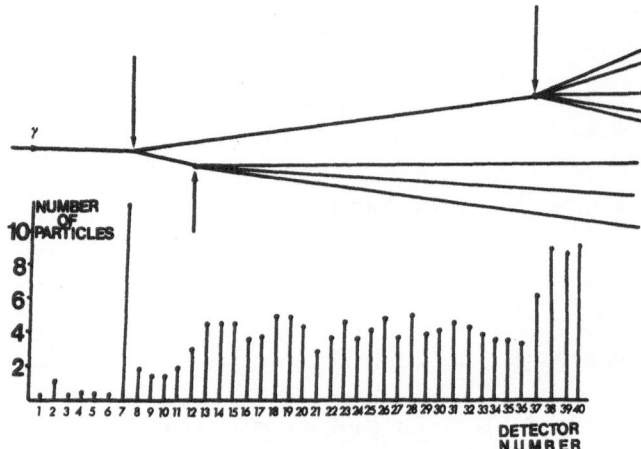

Fig. 1. Distribution of pulse amplitudes from the forty detectors
of FRAMM telescope and reconstruction of the event which
generated that distribution.

in multiplicity, one occuring between 12th and 13th detector and
one between 36th and 37th detector. They show where the first and
the second created particles decay.

According to the described principle, in order to make measurements of short lifetimes feasible the spatial resolution of the
telescope must be adequately high. The spatial resolution is determined by the geometric period of the telescope, sum of the thickness
of the silicon layer and of the air gap between contiguous layers.
The period of FRAMM telescope is 450 μm, arising from 300 μm of
detector thickness and 150 μm air gap.

The trend is now oriented to reduce the spatial period, to
improve the accuracy and to allow the measurement of shorter lifetimes.

The question therefore arises, about the smallest detector
thickness which can be employed in a meaningful way. Reducing the
detector thickness below certain limits makes the amplitudes deli-
vered by the detectors affected by the noise to such an extent
that the results supplied by the telescope would be of little use.

II- LIMITATIONS IN THE DETECTION PROCESS

The smallest thickness of a solid state layer which can be
employed in a telescope and therefore the highest geometric resolution achievable are determined by limitations in the detection
process. These limitations can be understood with reference to the
spectra of figure 2.

These spectra were obtained with a silicon detector of 200 μm
thickness and 300 mm^2 area. The line of higher energy is the Landau
distribution of 10 GeV/c negative π's, obtained by using as data
acquisition command a simple beam trigger. The shaded line is the

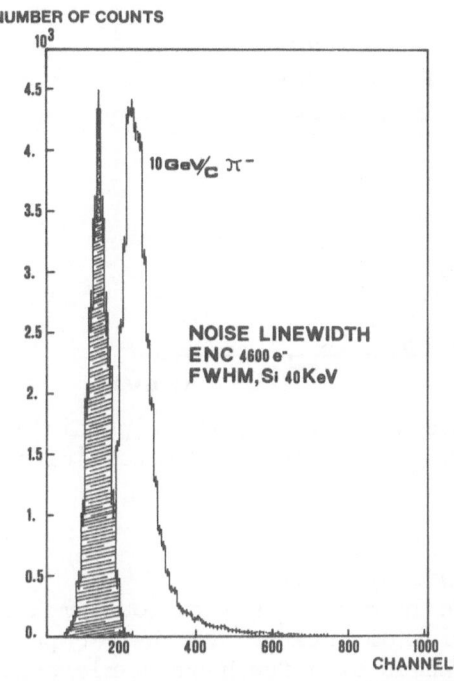

NUMBER OF COUNTS

10 GeV/c π⁻

NOISE LINEWIDTH
ENC 4600 e⁻
FWHM, Si 40 KeV

CHANNELS

Fig. 2. Landau distribution of
10 GeV/c π⁻ in a 200 μm
thick silicon detector.
The shaded curve is the
gaussian distribution
of the purely stochastic
electronic noise.

gaussian distribution of the purely stochastic noise in the electronic processor. This was obtained turning the beam off and using an
artificial trigger as data acquisition command. The difference
between the most probable energy of the π⁻'s Landau spectrum and
the centroid of the noise line, which represents the zero energy
of the system, evaluated from the detector thickness according to
the data of Esbensen et al.[3] is about 59 keV. The Landau spectrum
of figure 2 is the convolution of the physical Landau distribution
and the noise gaussian curve. As clarified by figure 2, the contribution to the width of the real Landau spectrum due to the electronic noise is big.

If the detector is simultaneously crossed by two or more
minimum ionizing particle, the energy released by such a multiple
event is the sum of as many random variables as the particles.

The random variables are independent and equally distributed,
therefore the resulting physical energy distribution is the convolution of two or more ideal Landau spectra. The final energy distribution results from a further convolution with the gaussian noise
line.

The spectra corresponding to multiplicities 1, 2, 3 in a 200
μm thick silicon detector are shown in figure 3. The spectral lines
broaden as the multiplicity increases, according to their nature
of iterated convolutions.

Landau fluctuations and electronic noise are the main sources
of inaccuracy in the measurement of the energy released. They also
limit the capability of the detector in distinguishing between

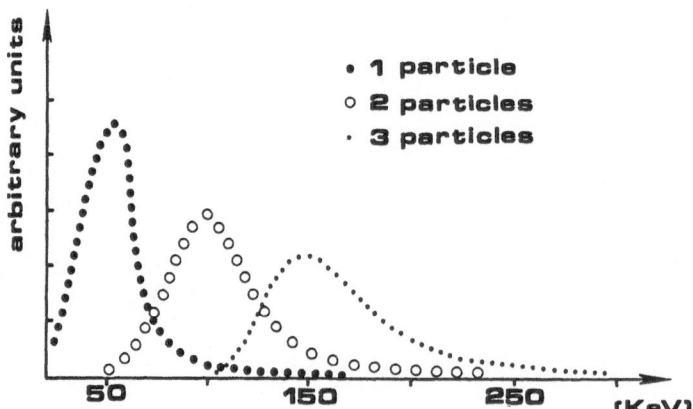

Fig. 3. Reconstructed spectral lines of one, two, three particles
        crossing a 200 μm thick silicon detectors. The particles
        are 70 GeV/c negative π's.

close multiplicities of minimum ionizing particles.

Both theory and experimental evidence, this last confirmed by
recent data show that at the energies of interest for telescope
applications and for detector thickness of a few hundred microns,
the width of the Landau distribution is almost linearly dependent
on the detector thickness[4,5]. Therefore, if the only source of uncertainty in the energy measurement were the Landau fluctuations,
the signal-to-noise ratio would be almost independent of the detector
thickness.

The presence of the electronic noise, instead, introduces a
dependence of the signal-to-noise ratio on the detector thickness.
This point will be analyzed with reference to figure 4, where the
detector is represented as a capacitive signal source connected to
the input port of a voltage-sensitive head amplifier.

The detector signal is thought of as a delta-impulse of charge
Q. The charge Q is determined by the following relationship:

$$Q = q \cdot \nu \cdot \frac{\Delta E}{\Delta x} \cdot \frac{d}{\varepsilon} \qquad\qquad (1)$$

where: q is the elementary charge
       ν is the radiation multiplicity
       $\frac{\Delta E}{\Delta x}$ is the specific energy loss, 29.5 keV/100 μm in silicon
       d is the detector thickness
       ε is the energy required to create an electron-hole pair,
         3.67 eV in silicon

Let $e_{N,TOT}$ be the total root mean square noise of the preamplifier referred to the input. Evaluation of $e_{N,TOT}$ is made by integrating the various noise contributions over the bandpass of the
filter which follows the preamplifier and then dividing the result
by the voltage gain between preamplifier input and filter output.

Let finally A be the detector surface.

The signal-to-noise ratio in the energy measurement is given by:

$$\eta = 12 \, \frac{d^2}{A} \cdot \frac{1}{e_{N,TOT}} \cdot \nu \qquad\qquad (2)$$

Equation (2) is valid for silicon, under the hypothesis, respected in most of telescope applications, that C_D is much larger than C_i. The following units have been assumed in equation (2): 100 μm for d, cm^2 for A, μV for $e_{N,TOT}$. Although for the sake of simplicity equation (2) has been derived for a voltage-sensitive head amplifier, the expression of η would be the same for a change-sensitive configuration. The contribution to the total inaccuracy in the energy measurement due to the electrical noise can be evaluated from (2) once the characteristics of the detector are known and $e_{N,TOT}$ is evaluated.

Conversely, by stating the desired signal-to-noise ratio and knowing A and $e_{N,TOT}$, it is possible to determine from equation (2) the minimum detector thickness, which can be used in the telescope.

The value of $e_{N,TOT}$ depends on the noise characteristics of the head amplifier and on the filter which follows it. It will be assumed, for reference purposed that this filter transforms the voltage step appearing at the input of the head amplifier into a triangular pulse of base width T. The value of T has to be fixed according to the counting rates the system must cope with. If λ is the counting rate, it is advisable to keep T below $0.1/\lambda$, otherwise pile-up effects would lead to too many rejected events. In high energy experiments values of λ between $5 \cdot 10^5$ and 10^6 pulses per second are quite usual and this restricts T to $100 \div 200$ ns.

At such short processing times, the dominant contribution to $e_{N,TOT}$ comes from the white term in the series noise generator of

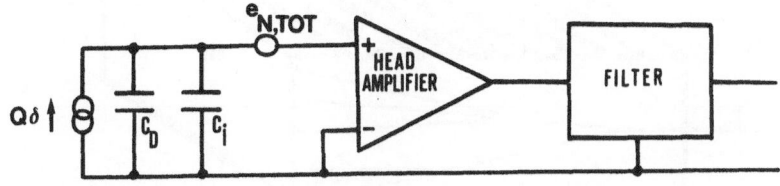

Fig. 4. Solid state detector connected to the input of a linear processor, consisting of a low noise preamplifier and a filter.

the input amplifying device. If $\frac{de^2}{df} = W^2$ is the power density of this
generator, W being a constant, the total noise $e_{N,TOT}$ is given by:

$$e_{N,TOT} = W \left[\frac{2}{T} \right]^{1/2} \qquad\qquad (3)$$

To obtain $e_{N,TOT}$ expressed in μV, W has to be specified in $nV/\sqrt{Hz}$
and T in μs. Combining (2) and (3) and assuming T = 0.1 λ, the
lowest limit for the detector thickness can be determined from the
experiment parameters according to the following relationship:

$$d^2_{MIN} = \eta \cdot A \cdot W \cdot \frac{0.37}{\nu} \sqrt{\lambda} \qquad\qquad (4)$$

where the counting rate λ is expressed in Megapulses per second.
Using, for instance, an input amplifying device with 1 $nV/\sqrt{Hz}$, a
detector with 3 cm^2 area and assuming a value of η equal to 4, the
minimum value of d in an experiment with a counting rate of 10^6 pps
would be 200 μm. In stating the desired signal-to-noise ratio it
has been considered that the value of η is defined with reference
to the root mean square noise.

The situation would be more favourable in germanium, on account
of the fact that the specific energy loss $\frac{\Delta E}{\Delta x}$ is about a factor two
larger than in silicon[3]. The telescope approach, however would
fail with germanium, as detectors with thickness in the 100 μm range
cannot be realized. As an alternative solution to the telescope
structure for lifetime measurements, the germanium structure of
figure 5 has been proposed[6,7]. The device of figure 5 is a bulk
detector with closely spaced strips on the upper face. Production
and decay points are identified, like in the case of the telescope,

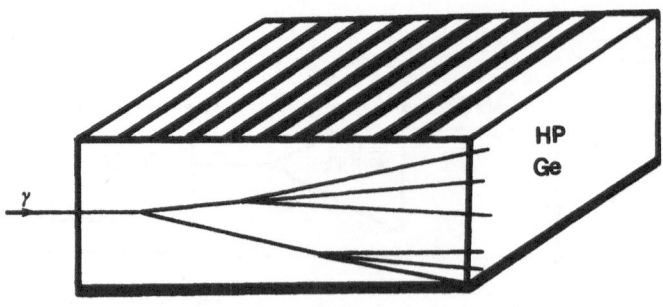

Fig. 5. Bulk germanium detector for lifetime measurements.

by the patterns of pulse amplitudes available at the strips. This device is less affected by the electronic noise than a telescope, because the capacitance presented by a strip is considerably smaller than the capacitance of a detector.

Another direction which is now being explored consists in an endeavour to reduce d_{MIN} in a silicon telescope by using for each layer in the telescope, instead of a single detector of area A, n adjacent sections of total area A. In this way the area of each section would be accordingly reduced, thus enabling a decrease in d_{MIN}.

III – LOW NOISE ELECTRONICS FOR SOLID STATE TELESCOPES

Turning the attention again to silicon telescopes, it must be pointed out that equation (2) has been deduced under the assumption that the detector capacitance is much larger than the input capacitance of the head amplifier. In this situation, if all the other parameters are left unchanged, the only way of improving η consists in reducing the noise power in the series generator of the input device. A recent effort in this sense led to the conclusion that some microwave bipolar transistor have smaller series noise than conventional junction field-effect transistors[8]. The tested bipolar transistors with an f_T of some GHz and a base spreading resistance $r_{BB'}$ between 10 and 15 Ω feature at 1 mA collector current a total equivalent series noise resistance between 22 and 27 Ω, which corresponds to values of W around 0.6 nV/$\sqrt{Hz}$. Another way of improving the signal-to-noise ratio is represented by the capacitive matching between detector and head amplifier. Capacitive matching might for instance be implemented by a transformer[9]. Detailed investigation about the limitations and the internal noise sources of a real transformer employed in capacitive matching between detector and head amplifier has been carried out[10]. The analysis has taken into account the effects of finite magnetizing inductance and of increased capacitance at the input of the head amplifier and the noise contributions due to magnetic core losses and dielectric losses in the windings. The combined action of all these undesired effects makes the transformer coupling useful in improving the signal-to-noise ratio only at comparatively large detector capacitances, near the upper limit encountered in telescope applications.

If the input device in the head amplifier is a junction field-effect transistor, capacitive matching without transformer requires the parallel connection of n field-effect transistors, n being the ratio C_D/C_i[11,12]. If instead the input device is a bipolar transistor, transformerless capacitive matching can be achieved in principle using a single device[13]. Such a possibility comes from the fact that in a bipolar transistor both transconductance and diffusion capacitance are proportional to the collector current. The collector current, therefore, can be adjusted in order to bring the value of the diffusion capacitance near to the detector capacitance.

IV – FURTHER LIMITATIONS IN ENERGY MEASUREMENTS

The discussion about the limitations in the measurement of the energy released in the detectors of a telescope has so far neglected the problems related with high counting rates. A shaping solution based upon the triangular signal considered in the previous sections as well as any other unipolar shape would lead in an ac coupled system, as most of nuclear processors are, to large baseline fluctuations at high counting rates. These fluctuations might seriously degrade the energy measurement.

Two techniques are currently employed to reduce these effects. One consists in giving to the pulses a bipolar shape rather than a unipolar one. An example of bipolar pulse with triangular lobes is shown in figure 6. Such a shaping would increase the value of $e_{N,TOT}$ of a factor $\sqrt{2}$.

The other technique makes use of a baseline restorer to reduce fluctuations in an ac coupled system working with unipolar pulses. A baseline restorer suitable for the very high counting rates encountered in telescope applications is usually of asymmetric type. Besides reducing the signal-to-noise ratio of the same factor $\sqrt{2}$, as in the previous case, it has also the disadvantage in situations where the signal-to-noise ratio is low, of impairing the linearity at small signal amplitudes.

Both solutions, bipolar shaping or baseline restoration, to conclude, introduce a further reduction on the already poor signal--to-noise ratio, and this reduction is hardly tolerable. This reason explains the effort put in the search for a new solution capable of guaranteeing the desired counting rate behaviour without sacrificing the signal-to-noise ratio. The principle followed to achieve this is based on an entirely dc coupled head amplifier-amplifier system working on strictly unipolar pulses, completed by a periodical

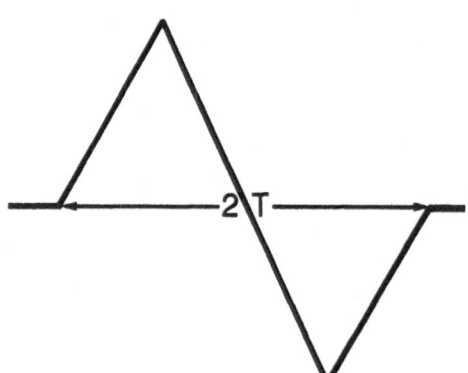

Fig. 6. Symmetric bipolar pulse with triangular lobes.

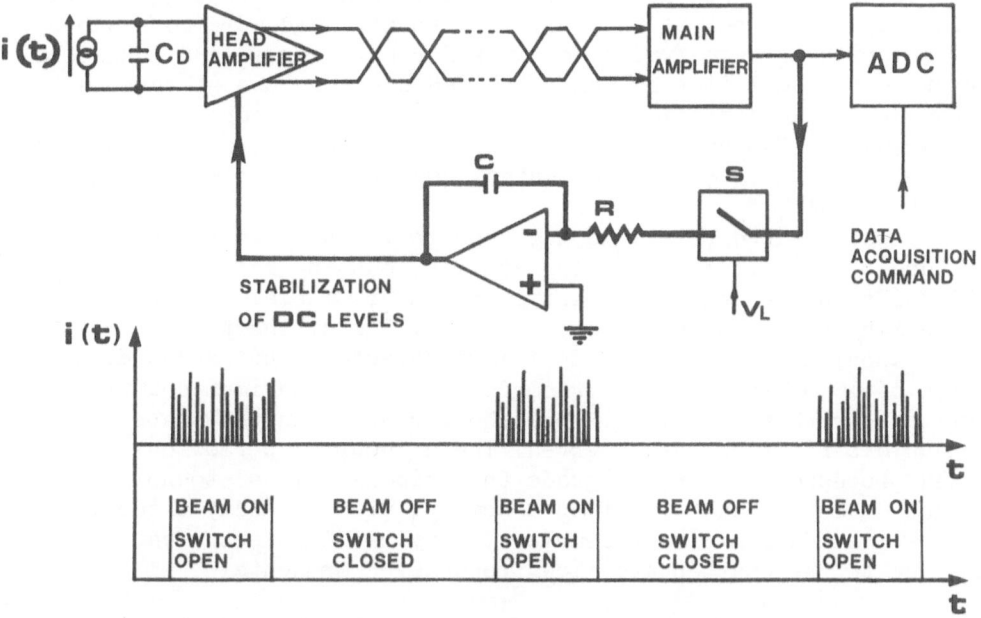

Fig. 7. Block diagram of the periodically stabilized dc coupled
        processor.

reset of the quiescent conditions[14]. The block diagram of the
adopted solution is shown in figure 7.

The upper part in the block diagram of figure 7 is the signal
path, which consists of a head amplifier connected through a twisted
pair to the remote amplifier and of a charge-sensing analog-to-digi-
tal converter. The signal channel is included in a feedback loop,
whose return path, shown by heavy lines in figure 7 consists of a
switch S and a high accuracy long term integrator. When the accelerator beam is ON, the switch S is open and the signal channel works
in a real dc coupled mode. A δ-like detector pulse, passing through
the natural time-invariant prefilter determined by the distributed
integrations reaches the output of the amplifier with an almost
gaussian shape, of about 20 ns full width at half maximum. Filtering
is completed by the time-variant function of the gated integrator
at the input of the charge-sensing ADC. The gated integrator accepts
only those pulses for which the requirements of the experiment logic
are met.

The quiescent conditions in the dc coupled head amplifier-amplifier system are periodically stabilized by taking advantage of
the dead periods of the accelerator, where no radiation falls on
the detector. Stabilization is achieved by closing the switch S at

the beginning of each dead period. The feedback loop becomes active,
restoring the output of the amplifier to zero and as a consequence,
a correcting voltage, which accounts for any thermal and long term
drift in the amplifier system establishes across the storage capacitor C. At the beginning of the next accelerator burst, S is opened
again and the feedback loop is interrupted, but the correcting voltage stored on C keeps the quiescent conditions stable in the
signal channel.

The described principle has been tested and was proven to
feature an excellent behaviour at high counting rates. Less than
.3% shift in the centroid position of an artificial line was found
increasing the counting rate up to $2.5 \cdot 10^6$ pps. The processor was
then associated with a 200 μm silicon detectors and Landau distributions of 10 GeV/c π^-'s were recorded up to the maximum counting rate
available with the beam, 10^6 pps. No variation in the shape of the
Landau distribution was observed. The dc coupled periodically stabilized processor does not degrade the original signal-to-noise ratio
of the detector-head amplifier assembly, with the only exception
of a slight decrease, noticeable at very small detector capacitances
and due to the unattenuated low frequency components.

REFERENCES

1. E. Albini et al., A live target for measuring the lifetime
    of charmed particles, XX International Conference on High
    Energy Physics - Madison - Wisconsin - July 7-23, 1980.
2. G. Bellini et al., Live target as a tool to study short
    range phenomena in elementary particle physics, Meeting
    on Miniaturization of High Energy Physics Detectors,
    Pisa, Sept. 1980.
3. H. Esbensen et al., Random and channelled energy loss in
    thin germanium and silicon crystals for positive and
    negative 2÷15 GeV/c pions, kaons and protons, Phys.
    Rev. B 18, 3:1040, Aug.1978.
4. H.D. Maccabee, D.G. Papworth, Correction to Landau's energy
    loss formula, Phys. Letters 30 A:241 (1969).
5. CERN, Genoa, Glasgow, Lancaster, Manchester, Milan, Moscow,
    Rutherford, Sheffield Collaboration, Status report on
    beauty experiment, CERN/SPSC/81-17, 17 February 1981.
6. M. Giorgi, private communication.
7. A. Menzione, Last results on detector development in Pisa.
    Silicon Detector Workshop at Fermilab, October 15-16
    1981.
8. P. D'Angelo, A. Hrisoho, P. Jarron, P.F. Manfredi, J. Poin-
    signon, Analysis of low noise, bipolar transistor head
    amplifiers for high energy applications of silicon
    detectors, to be published in Nucl. Instr. and Methods.
9. V. Radeka, Summary of noise relations for liquid argon
    ion chamber. Calorimeter Notes, Brookhaven National
    Laboratory 9-73.

10. E. Gatti, P.F. Manfredi, D. Marioli, Limitations in trans-
     former coupling between radiation detector and head
     amplifier, to be published in Nucl. Instr. and Methods.

11. C. Bussolati, S. Cova, E. Gatti, On the possibility of
     improving the signal-to-noise ratio for nuclear pulse
     amplifiers introducing decoupling networks between
     detector and amplifier input, Zeitsch. für Angew. Math.
     und Phys. 22,1:15 (1971).

12. V. Radeka, Comments on signal, noise and readout problems.
     Silicon Detector Workshop at Fermilab, October 15-16
     1981.

13. P. Rehak, Detection and signal processing in high energy
     physics. International School of Physics E. Fermi, 28
     July - 7 August 1981.

14. R.L. Chase, P. D'Angelo, A. Hrisoho, P.F. Manfredi, J.
     Poinsignon, A periodically stabilized amplifier system
     for silicon telescopes on pulsed accelerators. 1981
     Nuclear Science Symposium, San Francisco 21-23 October
     1981.

DISCUSSION

J. Appel: How does pulsed baseline restoration help during the
spill? What baseline restoration problems remain during the spill?

Answer: The principle of the periodical stabilization consists in
restoring the quiescent working conditions in the processor during
the dead times of the radiation source. If this restoration is properly realized, and this requires that when the correcting loop is
active no radiation falls on the detector, the correcting voltage
stored on the capacitor accounts for all the events of thermal and
long term drift in the processor. During the next accelerator
burst the correcting loop is open. The signal path is completely
de-coupled and therefore no baseline fluctuation occurs. The
processor working points are maintained by the voltage stored in
the capacitor during the previous correcting phase. It is true that
even in an entirely de-coupled processor there may be baseline
fluctuations during the spill. These are due to second order effects
related to improperly adjusted pole-zero cancellations, if any.

H. Newman: Did you test the radiation resistance of the silicon
detectors? What is the resistance of germanium to radiation damage
relative to silicon?

Answer: We have never followed any systematic approach to the investigation of radiation damage effects on silicon detectors. Information about radiation resistance of silicon detectors will be

provided later on by Mr. Heijne. Our observations are of indirect nature, as they are related to a possible degradation in energy resolution due to increased leakage current. As the resolutions are limited by effects of much higher importance than the variations in leakage currents, no reliable data could be collected.

A comment only: The Ga As FET is as good as the microwave bipolar transistors for short processing time, but it is very expensive ($\sim$100 $). It works at very high current (10 to 20mA).

PERFORMANCE AND APPLICATION IN HIGH ENERGY PHYSICS OF

PASSIVATED ION-IMPLANTED SILICON DETECTORS

J. Kemmer

Fakultät für Physik der Technischen Universität München

8046 GARCHING - Germany

P. Burger[+] and R. Henck

ENERTEC-SCHLUMBERGER
1, Parc des Tanneries

67380 STRASBOURG LINGOLSHEIM - France

E. Heijne

CERN - European Organisation for Nuclear Research

1211 GENEVA 23 - Switzerland

SUMMARY

The manufacturing of Ion-implanted Detectors (IP detector) is made by applying the well known techniques of the planar process : oxide passivation, photo engraving and ion implantation. Any desired detector shape can be obtained with small tolerances in geometrical and electrical properties.

Extremely good performances are obtained :

— low reverse currents less than 1 n A cm^{-2}/100 μm at room temperature,

---

(+) Presented by P. Burger.

— excellent energy resolutions : 10,6 keV for 5,486 MeV al-
phas, 1,5 keV for 14,4 keV gamma-rays with 25 mm2 area detectors, 300 μm thick,

— capability of backing out at 200°C under vacuum with no change of detector characteristics.

New possibilities open up in high energy physics :

— telescopes used as active targets or trigger units,

— microstrip detectors with pitches of 200, 100, 50 and 20 μm.

INTRODUCTION

Solid state detectors based on high purity silicon crystals
are widely used for the spectroscopy of charged particles. In the
past various types have been realized [1] :

— diffused pn - junctions
— surface barrier devices
— ion-implanted detectors.

At room temperature, the best performance in terms of energy
resolution is generally obtained with the surface barrier detectors. But, the reverse current is still mainly due to its surface
leakage components. Other drawbacks of this technology are :

— the much time consuming fabrication cycle,
— the inability of the surface barrier detectors to withstand
high temperature,
— the difficulty to make close-mounting of several detectors
(telescopes).

A new method for manufacturing silicon detectors[2] has been de-
velopped by J. Kemmer, based on the so called "planar process" wich
combines :

— oxide passivation of the surface : this is a useful method
for the reduction of leakage current[3],

— photoengraving technique which allows easy changes of the
shape of the devices by using different masks, making possible for instance the manufacturing of microstrip detectors,

— ion-implantation : by proper choice of the parameters asymmetric nearly abrupt pn-junctions, with thin dead layers
are realized.

After a description of the fabrication process of passivated
ion-implanted silicon detectors and report on the performance, some
results, comparison with other types of detectors and applications
in high energy physics are presented.

METHOD OF FABRICATION

1. Starting Material

Standard polished (111) orientation n-type silicon wafers are
used. The charge carrier lifetime is about 1-3 m·sec. The usual
thicknesses vary from 200 to 700 microns, depending on the wafer
size.

Wafer size	Usefull diameter	Resistivity	Thickness
2"	47 mm	500 to 5 K · Ω·cm	200 to 500
		5 K 52 cm to 10 K · Ω·cm	500 to 700
3"	73 mm	500 to 5 K · Ω·cm	300 to 500
4"	90 mm	500 to 5 K · Ω·cm	500 (for future)

2. Manufacturing process

The successive steps of the manufacturing process are illustrated in Fig. 1. They were performed under clean room conditions
using laminar flux boxes.

— Oxide passivation was achieved at 1030°C with a mixture of
dry oxygen and hydrogen chloride.

— Photographic masks having the desired openings were used
for the photoengraving technique. For example, for 5 x 5 mm2
area detectors, the total size of the chips were 8 x 8 mm2
resulting in 21 chips per wafer of 2 inches diameter.

— The doping was performed at room temperature using ion implantation so that nearly abrupt junctions with thin dead
layers were obtained. In order to avoid channeling effects,
the wafers were implanted at a small angle to the (111) direction. Boron was used for the p-type junction at an energy of 10-15 keV and a dose of 5×10^{14} ions.cm^{-2}. The rear
contact was also implanted using arsenic at an energy of
30 keV and a dose of 5×10^{15} ions.cm^{-2}.

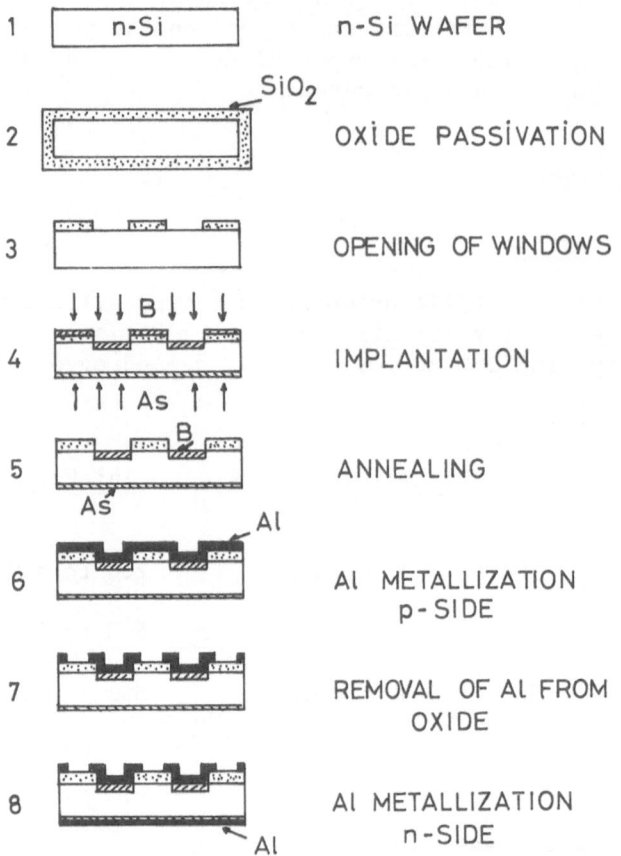

Fig. 1 : Successive steps of the manufacturing process of
          passivated ion-implanted silicon detectors.

— After implantation the wafers were thermally annealed at
  600°C during 30 min, in a dry nitrogen atmosphere.

— The wafers were then metallized on both sides with aluminium, to avoid any problem related to the sheet resistance
  of the implanted layers and to facilitate external electrical contacting.

— Aluminium between adjacent chips on the p-side was removed
  by using an appropriate mask.

— The chips were finally cut with a diamond saw.

Depending on the intended application the electrical connections were performed by pressure contacts on the metal layers, silver epoxy or ultrasonic bonding.

3. Detector properties

It appears that an estimation of the bulk lifetime of the minority charge carriers from the leakage current shows an improvement of the lifetime due probably to gettering effects during the oxydation process. Furthermore the protective action of the SiO_2 layer at the edge around the junction is very effective.

We have also tested :

— long term stability with some detectors over continuous run of about 6 months, with only small variations attributed to temperature variations.

— High current measurements (5 to 20 A/cm^2) by applying reverse bias voltages five times higher than needed for the depletion (about 600 volts for a 2 K - Ω-cm material, 300 μ thick).

We can expect that :

— Low current allows good energy resolutions.

— Overdepletion reduces pulse rise time (typical values are between 1.2 and 1.6 nsec/100 μm).

Fig. 2 shows the current-voltage characteristic for a 25 mm^2 chip, 300 μ thick : the leakage current is less than 1 nA at room temperature for a bias voltage of 200 V (depletion voltage is 130 V).

3.1. Charged-particle spectroscopy

The energy resolution capability of a 25 mm^2 area, 300 μm thick IP detector is demonstrated in Fig. 3 for an alpha-spectrum. The resolution at 5486 keV is 10.6 keV.

The dead layer thickness determined by observing the alpha particle pulse height shift with the beam respectively normal to the detector surface and at an angle of 45° is about 0.2 μm. For special applications this dead layer can be reduced below 0.1 μm by lowering the implantation energy[4].

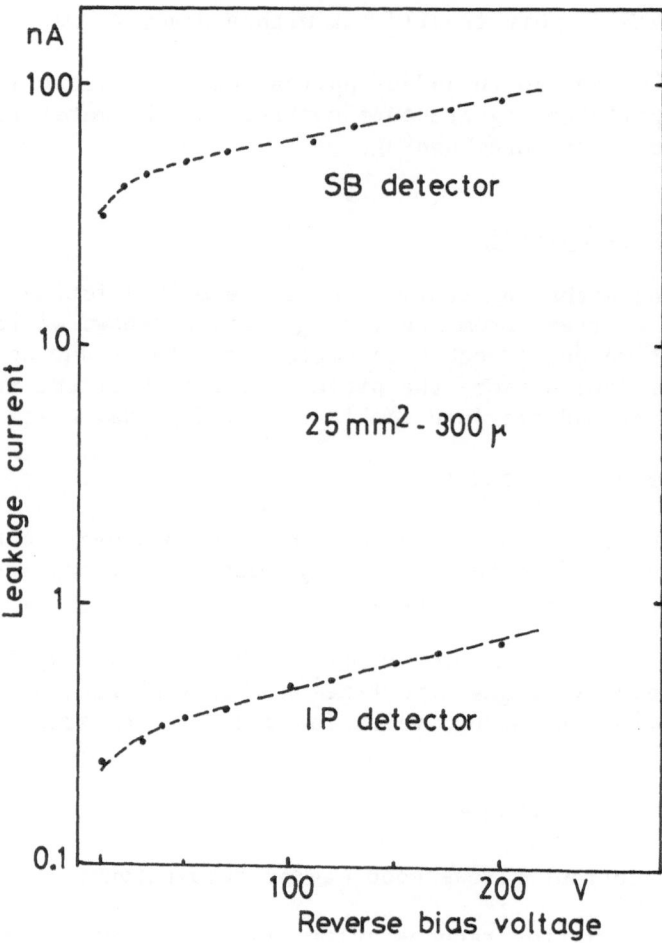

Fig. 2 : Comparison between reverse currents of a surface bar-
          rier and a passivated ion-implanted silicon detector
          of the same size (25 mm² area, 300 μm thickness).

Experiments with heavy ions[5] (70 MeV oxygen) were done using
100 mm² area IP detectors (220 to 300 μm thick). The measured resolution was about 200 keV which corresponds to the FWHM of the beam
(200 ± 20 keV).

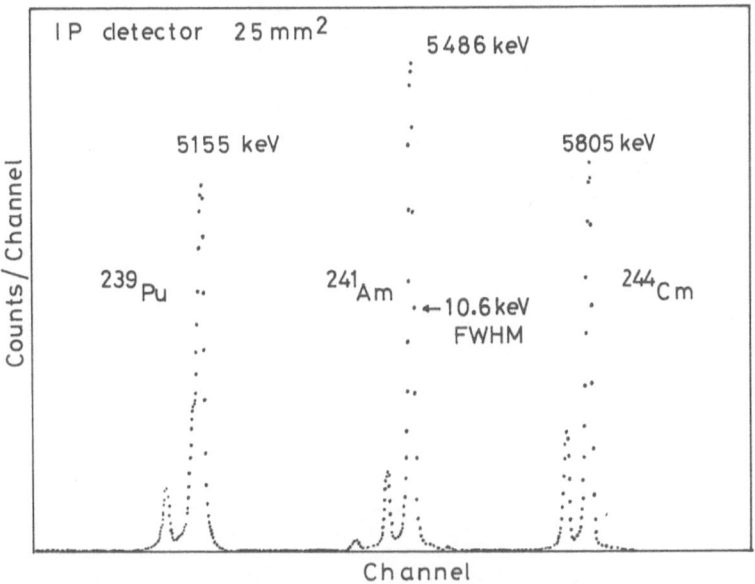

Fig. 3 : Energy spectra for ^{239}Pu, ^{241}Am and ^{244}Cm alpha-par-
ticle sources measured with an IP detector (25 mm^2
area, 300 μm thick) at room temperature.
Resolution at 5.486 MeV is 10.6 keV (FWHM).

3.2. X and 𝛾 ray spectroscopy

By selecting the chips having the lowest leakage current and
optimizing the electronic components, a noise level of 1,5 keV was
achieved at room temperature for 25 mm^2 area and 300 μ thick detectors. Fig. 4 shows an energy spectrum for ^{241}Am X and Gamma rays.
The resolution is determined mainly by the electronic noise and
therefore remains practically constant in the energy range from some
keV to 150 keV.

3.3. Baking and annealing

Ion implanted pn-junction detectors are able to withstand moder-
ately high temperature. Using pressure contacts to assure the electrical connection, a 100 mm^2 IP detector was used for tests at 200°C.
Baking was done under vacuum (10^{-6} Torr) during 15 hours at 200°C.
The results are shown in Fig. 5 : the current - voltage characteris-

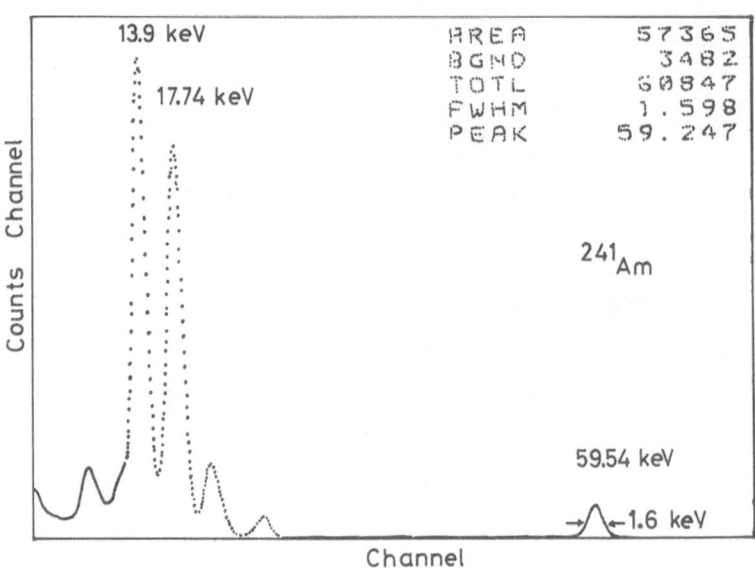

Fig. 4 :   Energy spectrum for ^{241}Am - gamma and X-radiation mea-
           sured with an IP detector (25 mm^2 area, 300 μm thick)
           at room temperature. Resolution at 59.5 keV is 1.6
           keV (FWHM).

tic remains essentially unchanged. Similar results have been reported by Hyder[6].

     The possibility of heating IP detectors is attractive for :

— use in high vacuum systems that have to be thermally cycled,

— annealing of radiation damage (experiments were done by
  Kemmer and Wagner [3]).

The actual possibilities of baking depend on the mountings :

— for a standard version, maximum temperature is 130°C,

— for a special mount with springs, maximum temperature is
  200°C,

— for a chip without mount, maximum temperature is 400°C under
  controlled N_2 atmosphere.

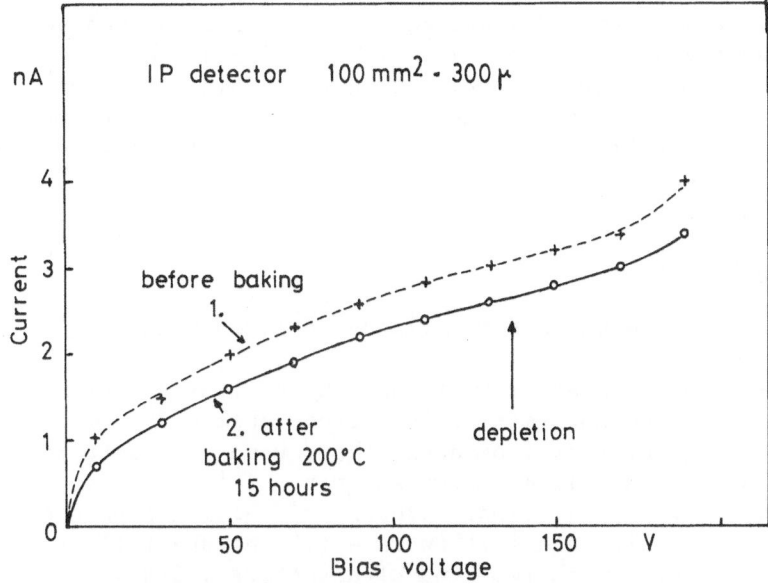

Fig. 5 : Current-voltage characteristic of an IP detector
(100 mm² area, 300 μm thick) before (curve 1) and
after baking 15 hours at 200°C (curve 2).

Fig. 6 : Telescope composed of 40 silicon detectors 160 mm²
area, 300 μm thick, 150 μm distant from each other.

HIGH ENERGY PHYSICS APPLICATION

Until recently solid state detectors have been used very little for high energy physics experiments. Two different kinds of experimental arrangements are beginning to change this situation : one is the use as live targets or trigger unit[7,8] of telescopes made up of thin silicon detectors, the other is the availability of position sensitive microstrip detectors[9].

1. Active target

Different telescopes have already been realized.

Ion-implanted detectors were used by Bellini and Diambrini. The first telescope consists of 40 silicon sheets 300 μ thick, 160 mm^2 area, 150 μm distant from each other, used for photoproduction experiments and short lifetime measurements (fig. 6). Every detector is related to an electronic processor[10] having a good energy resolution and a high acquisition rate (100 ns shaping). An other telescope was built with two kind of detectors : 5 silicon sheets 100 μm thick, 1 cm^2 area, 130 μm apart each other, and 10 silicon sheets 200 μm thick, 5 cm^2 of surface, 200 μm apart each other. This telescope was used for selecting the important events in an emulsion.

Other telescopes are being constructed :

— One for Borgeaud (Saclay) consists of 30 ion-implanted passivated silicon detectors, 300 μm thick, 200 μm distant from each other. The active areas of 50 x 40 mm^2 are divided into 24 strips with 2 mm pitch. 720 hybrid preamplifiers will be used.

— An other for Bellini and Moroni consists of 40 silicon sheets 200 μm thick, 100 μm distant from each other. The active area (20 x 20 mm^2) is divided into 4 parts (5 x 20 mm^2). 160 hybrid preamplifiers will be used.

2. Microstrip detectors

Heijne[9] et al. demonstrated the practicability of a silicon microstrip detector made by the surface barrier technique. The lower limit for the pitch was found to be 50 μm. Obviously, the passivated pn-junction technology is better adapted to such application :

— The devices, oxide passivated, are more reliable than surface barrier detectors ;

— The use of photographic masks enables more precise definition of the strips with much smaller pitches down to 20 μm.

At present planar versions of a 200 μm pitch (sensitive area
20 x 30 mm^2, 300 μm thick) and a 50 μm pitch (10 x 25 mm^2 area,
300 μm thick) have been realized, as well as detectors having concentric curved strips (radius 90 mm, length 30 mm, pitch 200 μm) used
for measuring high currents during fast beam bursts.

The silicon detector is glued with an epoxy resin into a fiber
glass printed circuit board which already contains the contacts in
the form of gold plated copper strips. The front side of the strip
detectors were contacted with an aluminium wire by ultrasonic bonding. The total current for 100 strips ranged from 130 nA to 380 nA
at 150 V (2.000 ohm-cm material) in different samples. The capacitance of a strip including stray capacitance is about 4 pF.

The electrical resistance between two adjacents strips is very
high ($>$ 1 GΩ).

The 200 μm pitch detector was tested at CERN in a beam of 10
GeV/c pions using the experimental set up and associated electronic
equipment described by Heijne et al.[9] A fast current sensitive preamplifier designed by Jarron[9] and functionning with bipolar instead
of field effect transistors was used. The electrical noise is relatively high (FWHM of 12-15 keV Si equivalent) but low enough compared to the main signal variation which originate from the fluctuations in deposited energy as given by the Landau distribution for
minimum ionizing particles.

A pulse height spectrum obtained in the beam of 10 GeV/c pions
is shown in Fig. 8. The width of the pedestal noise distribution indicates a FWHM noise of 20 keV. The energy deposition spectrum, after pedestal correction, follows a Landau distribution.

In Fig. 9, the signal height for the pions is plotted as a function of the reverse bias. Complete charge collection is obtained
above 120 V. The corresponding energy determined using a ^{57}Co source for calibration (Fig. 8) is about 80 keV for this 300 μm thick detector, in accordance with the most probable energy deposit value of
about 27 keV per 100 μm for minimum ionizing particles.

In the way the microstrip detector was positionned between wire
chambers with center of gravity read-out as sketched up in Fig. 10,
beam tracks can be selected which pass through the detector and
these should be recorded. From 10,000 track only 6 were not found.
The resulting efficiency is therefore 100 %, indicating that there
are no dead zones between the strips.

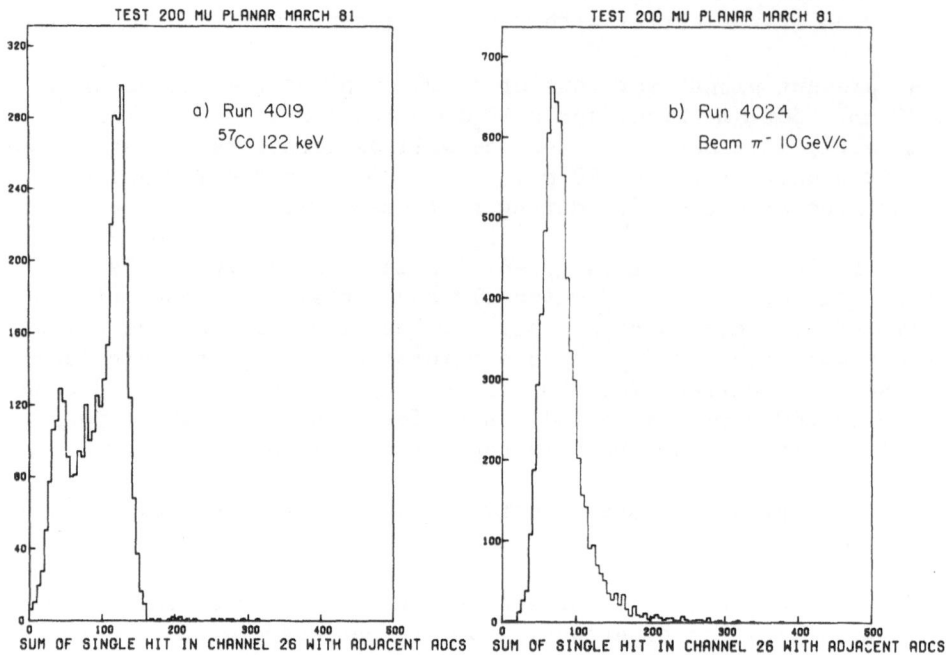

Fig. 8 :   Pulse height spectra of signals from a microstrip
          a)   for ⁵⁷Co-gamma radiation for calibration.
          b)   for 10 GeV/c pions, after pedestal correction.

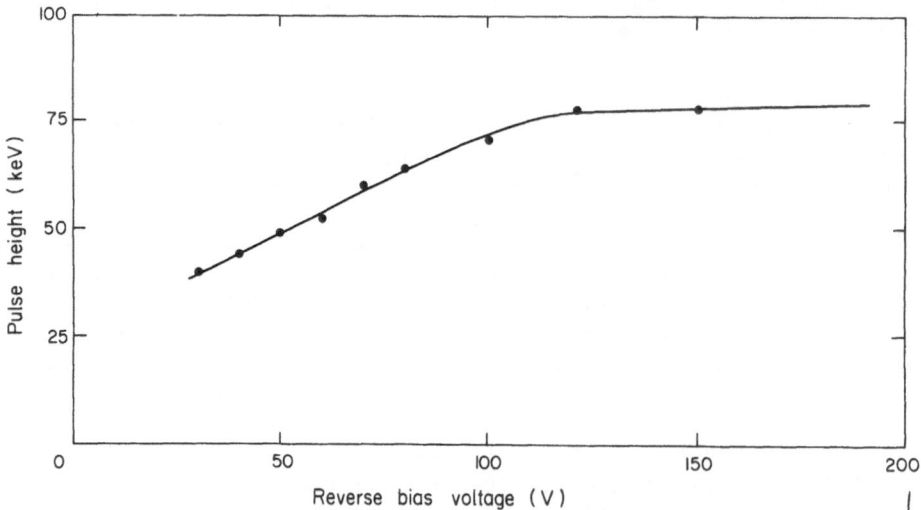

Fig. 9 :   Most probable signal height (in keV) as a function
          of the reverse bias voltage. Complete charge collec-
          tion is obtained above 120 V.

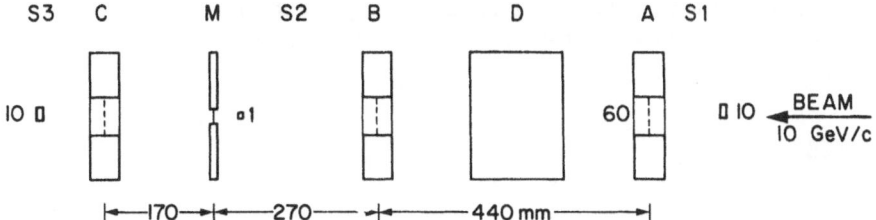

Fig. 10 : Situation of the microstrip detector (M) between
           wire chambers with centre of gravity read-out A,
           B and C. S1 and S2 are scintillators used for trigge
           ring.

Further high energy physics applications

Various applications can be expected :

— Beam profile monitor in a high intensity secondary beam.
— Fast beam hodoscope as described by Heijne and Jarron[12].
— High precision vertex detectors, by combination of several
  slices.
— Vertex reconstruction detector, possibly for nuclear emul-
  sion or mini bubble chambers.
— Specific trigger for high rate (10^8 events/s).
— Detector used inside the vacuum chambers of a storage ring.

CONCLUSION

    Using the planar process it is possible to fabricate silicon
detectors of extremely low reverse currents by the improvement of
the charge carrier lifetimes and by the elimination of surface leakage currents. In addition to their obvious application for high
performance charged particle spectroscopy, these detectors enable
the spectroscopy of X-rays with energy resolutions of 1.5 keV at
room temperature.

    In high energy physics, applications are based mainly on two
kinds of experimental arrangements : telescopes consisting of a stack
of thin silicon slices, used as active targets and position sensitive microstrip detectors. Miniaturization of the associated electronics is under way in different laboratories. The aim is to realize
a detection system having better spatial and time resolution for

multiple track events than multiwire proportional chambers and which can stand count rates in excess of 10^7 cm^{-2} s^{-1}.

REFERENCES

1. G. Bertolini and A. Coche, Semiconductor Detectors, North Holland (1968).

2. J. Kemmer, Nucl. Inst. and Meth. 169 (1980) 499.

3. G. Keil and E. Lindner, Nucl. Inst. and Meth. 101 (1972) 43.

4. H. Grahmann and S. Kalbitzer, Nucl. Inst. and Meth. 136 (1976) 145.

5. J. Kemmer and W. Wagner, Nucl. Inst. and Meth. under press.

6. A. Hyder, Nucl. Inst. and Meth. 187 (1981) 595.

7. S.R. Amendola et al., Nucl. Instr. and Meth. 176 (1980) 449.

8. M. Adamovich et al., Status report on beauty test experiment (P-137), CERN/SPSC/81-17 (1981).

9. E. Heijne et al., Nucl. Inst. and Meth. 178 (1980) 331.

10. M. Artuso et al., 10th Inter. Symp. on Nuclear Electronics, Dresden (1980).

11. FRAMM collaboration, CERN.

12. E. Heijne and P. Jarron, IEEE Trans. Nucl. Sci. NS 29 (1982) 405.

HYBRID PREAMPLIFIER FOR SOLID STATE DETECTORS

ON HIGH ENERGY EXPERIMENTS

M. Conte and A. Zambra*

LABEN - SIEL Division
Via Bassini 15
Milano - Italy

INTRODUCTION

LABEN is devoted to the research, development and production of analog and digital instrumentation for nuclear, space and industrial applications.
The wide field of experience is performed in LABEN and covers: analog signal processing and conditioning, high performance analog to digital converters, data acquisition equipment, microprocessor or minicomputer based data processing equipment, on-board scientific instrumentation and data handling, high speed technology. This gives LABEN a turn key system capability.
As a manufacturer of high level electronics instruments for physics measurement, LABEN is looking with interest at the solid state detecting devices that are coming now into a wide spread use in elementary particles experiments.

HYBRID PREAMPLIFIERS

As a first step LABEN has decided to study, develop and produce hybrid preamplifiers for low noise data acquisition from solid state detectors, this also with the collaboration of research organization as I.N.F.N.; now LABEN presents itself as a second source for this products line.
The reasons for this decision are the following:
- reduction of size and cost in experiments requiring a high number of chains

---

*Presented by A. Zambra

- improvement of performances due to the relevant technology (for example stray capacitance, matched resistor, etc.)
- fall out from space research.

Indeed LABEN has produced a complete hybrid processor for data acquisition from radiation detectors in satellite experiments and the new hybrid preamplifier belongs to a complete line where versatility as well as top performances have been assured as primary goals.

Basically four types of preamplifiers will be available in a short time and their functional diagram is shown in fig. 1

- A general purpose unit for Silicon detectors which can suit any type of application from telescope to microstrip detectors.
  The major features are:
  . bipolar input
  . short shaping, 50 + 100 ns to the base
  . resolution at $\emptyset$ input capacitance better than 15 KeV
  . slope of 100 eV/pF
  . differential output
  . driving capability for long twisted cable
  . $\pm$ 6 V power supply (120 mW)
  . 1 inch x 0,7 inch dimensions

- A low cost unit for microstrip Silicon detectors with the following features:
  . bipolar input
  . short shaping 100 ns to the base
  . resolution at 0 input capacitance 15 KeV
  . slope of 200 eV/pF
  . differential output
  . $\pm$ 24 V power supply
  . 1 inch x 0,7 inch of dimensions

- A special purpose preamplifier for very high counting rate applications with telescopes.
  The performance of this module is under finalization

- A very low noise circuit for high energy applications of Germanium detectors.
  The performance of the last module is about the same as for first type but with 2.5 KeV of resolution at $C_d = 0$

The schedule for the manufacturing and production phase is the following:
1st type in production phase
2nd type in preseries phase (hybrid prototype)
3rd type in prototype study
4th type in pre-production phase

The target of the price for the 1st and 2nd preamplifier is in order of about 16 and 12 U.S. Dollars respectively, at the middle

Functions implemented:

$$\mathcal{Z}_f = R_f \cdot C_f = 100\,K\Omega \cdot 5\,pF = 0{,}5\,\mu s$$

V_u is the single-ended output voltage with the output matched to a $50\,\Omega$ load
s is the complex frequency

1) Charge sensitive without pole-zero cancellation

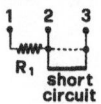

R_1 short circuit

$$V_u = -\underbrace{\frac{Q}{s \cdot C_f}}_{\substack{\text{Pure}\\\text{integr.}}} \cdot \underbrace{\frac{s \cdot \mathcal{Z}_f}{1+s \cdot \mathcal{Z}_f}}_{\substack{\text{Time}\\\text{constant}}} \cdot \underbrace{\frac{R_3}{2R_1}}_{\text{Gain}}$$

2) Charge sensitive with pole-zero cancellation

R_1 C R_2

$$V_u = -\frac{Q}{s \cdot C_f} \cdot \frac{s \cdot \dfrac{\mathcal{Z}_f}{1+R_2/R_1}}{1+s \cdot \dfrac{\mathcal{Z}_f}{1+R_2/R_1}} \cdot \frac{R_3}{2R_1}$$

Choice of the external network

$$R_1 = \frac{1}{2G}\,K\Omega$$ G is the auxiliary gain between the charge sensitive loop and the output

$$R_2 = R_1 \cdot \left[\frac{0{,}5}{\mathcal{Z}'_f} - 1\right]\,K\Omega$$ $\mathcal{Z}'_f$ (μs) is the desired value of the output time constant

$$C_2 = \frac{500}{R_2}\,pF \quad R_2 \text{ in } K\Omega$$

3) Current sensitive

1 2 C_2 3
short circuit R_2

$$V_u = -Q \cdot \frac{C_2}{C_f} \cdot \frac{R_3}{2} \quad \text{if } R_2 C_2 = 0{,}5\,\mu s$$

transresistance

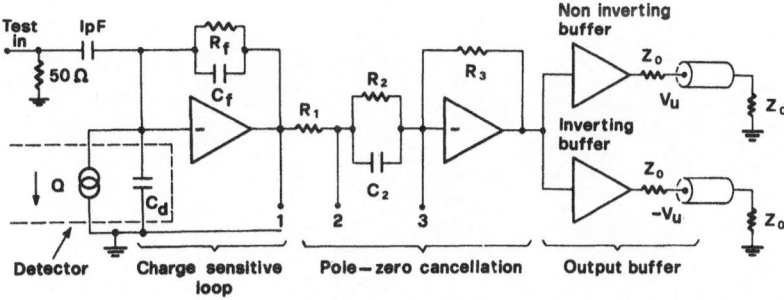

Fig. 1. Block diagram of the general purpose hybrid preamplifier for high energy physics.

of 1981 and after settling of the production.

The price is coomprensive of test and characterization on the device.

As a future step, LABEN is looking to follow the implementation of a complete low cost hybrid line, like amplifiers, stretcher and spectroscopy analog to digital converters.

LABEN experience in the field of test equipment for data handling and payload computer-based detector simulator is used for a system in testing hybrid preamplifier.

The closed loop system (controlled input stimuli-measured output) can be, at last, a good approach for on line experiment monitoring.

DISCUSSION

N. Reay: What is the power level/amplifier for your class 1 and class 2 amplifiers?

Answer: Power conjunction of type 1 is 120 mW each, for the type 2 is about 200 mW but con be reduced.

SILICON DETECTORS FOR THE NA14 PHOTOPRODUCTION EXPERIMENT AT CERN

D.M. Websdale

Imperial College, London

INTRODUCTION

The NA14 experiment[1] is installed in the North Area of the
CERN SPS and uses a high intensity tagged photon beam. The emphasis
of the experimental program is on a study of photon hard scattering
processes, for which a description in terms of the parton model and
its QCD corrections should be applicable. Examples of such processes
include deep inelastic Compton ($\gamma q \rightarrow \gamma q$), QCD Compton ($\gamma q \rightarrow gq$) and
photon-gluon fusion ($\gamma g \rightarrow q\bar{q}$). Data on these should complement the
analysis of hard photon processes; e^+e^-, Drell-Yan and deep inelastic
lepton scattering.

Most hadronic interactions of the photons are at low p_T and
described in terms of the photon as a vector meson (VMD) so
attempts to isolate the above processes need events at high p_T. A
high sensitivity is therefore required.

The 3-stage bremsstrahlung beam yields $\sim 10^7$ tagged photons of
energy greater than 70 GeV from a pulse of 3×10^{12} protons.
These photons produce $\sim$ 4000 hadronic interactions in a target
(10% radiation length) during one SPS pulse and a trigger selecting
high p_T photons in the final state allows 5 events/picobarn to be
collected in a 60 days run.

At these energies the charm photoproduction cross section is
about 1μb[2] so a few x 10^6 charmed particles will be produced. It
would be a shame to ignore these!

DETECTION OF CHARMED PARTICLE DECAYS

The NA 14 detector is shown in figure 1. It consists of a
spectrometer, electromagnetic calorimeters, threshold Cerenkov
counters and a μ-filter. Charged and neutral particles can be
reconstructed and identified over a wide solid angle.

The fraction of interactions in which charmed particles are
produced is about 10^{-2} for photoproduction. This compares favour-
ably with $\sim 10^{-3}$ for hadronic charm production at energies $\sim$ 100GeV.
In spite of this advantage the signal/background appearing in
invariant mass distributions is likely to be small unless some means
of filtering charmed events and of reducing the combinatorial background is employed.

It is planned to use silicon detectors as an active target (the
technique has been successfully used by the FRAMM Collaboration[3]),
complemented by silicon microstrip detectors downstream which serve
as a high precision chamber.

SILICON DETECTORS

The evaluation and testing of Silicon detectors for NA14 has
been directed by P.G. Rancoita at CERN and P. Borgeaud at Saclay[4].
They propose a detector layout as follows.

The active target consists of 30 Silicon wafers, each 40mm x
40mm and 300 μ thick separated by 200 μ. This assembly constitutes
$\sim$ 10% rad. length. The microstrip chamber consists of 6 planes
each 60mm x 60mm with a strip pitch of 50 μ. This will provide
3(x,y) co-ordinates with an r.m.s. precision < 20μ. Two additional
planes giving (u,v) co-ordinates are used to resolve ambiguities
in reconstruction. The silicon diodes are prepared by ion
implantation and the amplifiers used in tests to date show a noise
equivalent to < 30KeV (FWHM) compared with the signal from a
relativistic particle of 90KeV. This is satisfactory for our
application. In total 8000 channels are required.

FILTERING CHARM

The energy deposited in each layer of the active target is
measured. The known dE/dx distributions for n_i particles in layer
'i' can be used in a maximum likelihood estimate of each n_i.
Changes in multiplicity along the target can thus be detected
and these indicate the decay of a short lived state (or of a secondary interaction or $e^{\pm}$ pair conversion).

For events thus selected the secondary tracks are reconstructed
in the spectrometer. In this way a significant reduction in the
number of mass combinations per event is obtained.

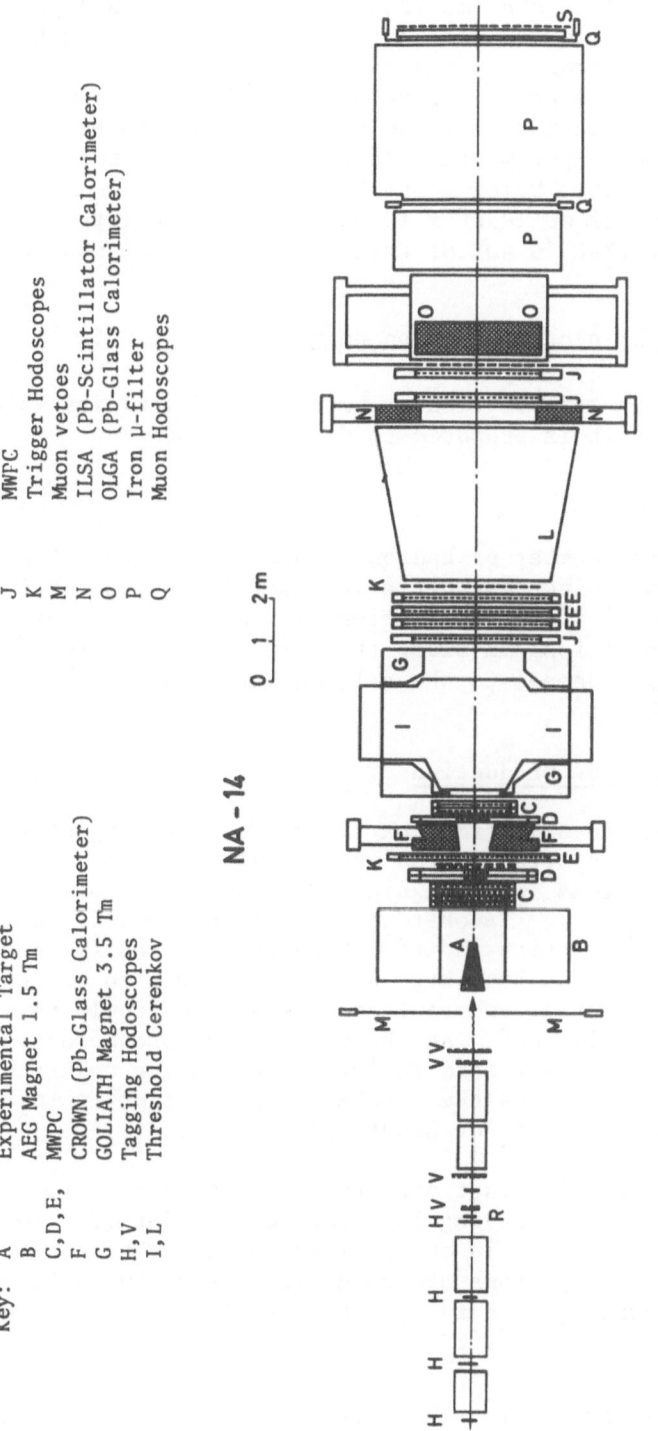

Key: A   Experimental Target   J   MWFC
       B   AEG Magnet 1.5 Tm   K   Trigger Hodoscopes
       C,D,E, MWPC   M   Muon vetoes
       F   CROWN (Pb-Glass Calorimeter)   N   ILSA (Pb-Scintillator Calorimeter)
       G   GOLIATH Magnet 3.5 Tm   O   OLGA (Pb-Glass Calorimeter)
       H,V   Tagging Hodoscopes   P   Iron μ-filter
       I,L   Threshold Cerenkov   Q   Muon Hodoscopes

NA - 14

0   1   2 m

Figure 1.   Layout of NA14 Experiment.   (Athens, CERN, Coll. de France,
            Ecole Polytechnique, LAL Orsay, Imperial College, Saclay
            Southampton, Strasbourg, Warsaw).

In evaluating the performance of detectors of charmed particle decays two parameters are important. The longitudinal precision should be less than the mean decay path $\lambda = \frac{p}{m} c\tau$ and the transverse precision better than the mean impact parameter* $b \sim \frac{1}{2}c\tau$ (indep. of momentum). For a charmed particle with 30 GeV/c momentum and $\tau \sim 5 \times 10^{-13}$ sec we have $\lambda \sim 2500\mu$ and $b \sim 75\mu$: to be compared with the corresponding experimental precision in reconstruction of $\delta\lambda \sim 300\mu$, $\delta b \sim 20\mu$. Using these figures it is estimated that a filter reduction of at least x 10 on all hadronic interactions can be achieved and that in 40% of this sample only one mass combination is required.

SOME EXPERIMENTAL PROBLEMS AND PROPOSED SOLUTIONS

I discuss in this section a number of difficulties which can be expected and how it is proposed to deal with them.

(i) High Rate Problems

Although the number of hadronic interactions is not high there will be $\sim 10^{7}$/sec electromagnetic interactions in the active target. To avoid 'pile-up' in the resolution time of 50 nsec the target will be divided into strips 2mm wide, thus limiting the rate to $<10^{6}$/sec/ strip. Radiation damage to the silicon has been tested[5] and will present no problem.

(ii) Incoherent Photoproduction

The production of charmed particles in coherent interactions on Silicon has been reported by the FRAMM Collaboration[3]. The signal for such interactions is clean since the nuclear recoil is absorbed in one silicon layer.   Incoherent production involves nuclear break-up with evaporation and recoil nucleons. The incoherent events can be rejected but there is a good reason to include them: The ratio of incoherent/coherent photoproduction could be $\sim$ 3:1 for $D\bar{D}$ production on Silicon[6]. As the mass of the produced system increases so the nuclear form factor depresses the coherent production $[t_{min} \sim (M^2/2E\gamma)^2]$. Thus the incoherent production is more effective for higher mass e.g. $\Lambda_c\bar{\Lambda}_c$ or beauty production.

The incoherent interactions are characterised by the emission of evaporation protons having low kinetic energies (T < 10 MeV). These have a short range in the silicon and saturate one or two layers. The recoil protons are more penetrating but will not have a dE/dx corresponding to a relativistic particle. These must be recognised in the microstrip chamber and the corrected dE/dx

_____

*The impact parameter b is the distance of closest approach of the extrapolated decay track to the production vertex.

applied. The interactions with recoil neutrons present no problems.
It is estimated that 50% of the incoherent events can be analysed.

(iii) $e^{\pm}$ Pairs in the Target

Pair conversions simulate an increase in multiplicity in the
target. These can be due to random pairs (not associated with the
hadronic interactions) or from π^0 decays. The randoms must occur
in a region 2mm x 2mm downstream of the interaction - these are
negligible if a 50nsec gate is used. The conversions from π^0 decay
present a problem, and will probably be the limiting factor in
the filtering efficiency. The microstrip chamber has a role to
play here since pairs separate only in the plane perpendicular to
the magnetic field. The $e^{\pm}$ tracks are thus observed as two relativistic particles in the target but as one in the horizontal chamber
strips (and this extrapolates to the production vertex). Finally
the $e^{\pm}$ can in some cases be identified in the spectrometer.

(iv) Secondary Interactions

Hadronic interactions of secondaries in the Silicon target also
give rise to an increase in multiplicity. These interactions are
accompanied by nuclear recoil or break-up which deposit a large
energy in the silicon, thus distinguishing them from a short lived
decay.

TRIGGERS FOR CHARM

This topic is to be discussed in a later session so I
summarize here possible triggers and the numbers of charmed particles
which will be produced in a 60 day run.

   (i) Interaction trigger (E_γ > 140 GeV) $\gtrsim 10^5$ charm
  (ii) γ with p_T > 500 MeV/c. to select $F \to \eta +....$ 10^4 F
(iii) electrons and/or Kaons to select $D \to Ke +....$ > 10^4 D

SUMMARY

The addition of the Silicon target and microstrip chamber to the
existing NA 14 detector will allow a study of coherent and incoherent
charm production. Lifetimes will be measured using the coherent
sample. The silicon detectors will be ready for use in 1983 and
we can expect to reconstruct $\sim 10^3$-10^4 charm decays. If the ratio
of beauty/charm photoproduction is better than 10^{-3} there is some
optimism of observing beauty.

ACKNOWLEDGEMENTS

The ideas presented here and the experimental work to test their
feasibility are mainly due to P.G. Rancoita of the NA14 Collaboration.

REFERENCES

1. Photoproduction at High Energy and High Intensity. Athens, CERN,
    London (I.C.), Orsay, Saclay, Southamton, Strasbourg, Warsaw,
    Collaboration, SPSC Proposal P109: SPSC 78-76, 78-139 (1978)
2. D. Treille. Photon and Hadron production of open Heavy Flavours.
    1981 International Symposium on Lepton and Photon Interactions
    at High Energy. Bonn, August 1981
3. G. Bellini. Contribution to this Conference.
4. P.G. Rancoita et al. Performance of Surface Barrier and Ion
    Implantation Silicon microstrip detectors. CERN EP81-06
5. P. Borgeaud et al. The Effect of Radiation on Ion-Implanted
    Silicon Detectors. Submitted to Nuclear Instr. & Methods.
6. This estimate is based on the ratio 2:1 for incoherent/coherent
    photoproduction of J/Ψ on Beryllium (B. Knapp et al. Phys Rev
    Lett 34 (1975) 1040)

DISCUSSION

J. Sacton: How will you deal with secondary interactions in the telescope?

Answer: There are 2 classes of secondary interactions
i) in the same silicon nucleus (re-interaction)
ii) in a subsequent silicon detector.
The first are much less frequent than expected in a naive optical
model. Experiments show that multiplicities grow slowly with A in
nuclear reactions and that the re-interaction probability of a
multipion system in nuclei is compatible with that of a single pion.
The second type do simulate a short lived decay but the nuclear
recoil or break up will be detected.

N. Reay: Are not secondary interactions much bigger than 1%? Won't
they cause a lot of trouble for a more inclusive trigger?

Answer: Yes, in a high multiplicity event (n > 8, say) the probability of secondary interactions is $\sim$ 20%. But the large energy deposited by nuclear recoil or break-up allows the event to be rejected.

G. Bellini: You plan to use a beam flux of the order of $\sim 10^7$ photons/
/sec, but you have no more than 10^6 photon/sec/strip. In such a way
you have no more pileup in each strip. But in each way you plan to
disentangle the pulse corresponding to the right pulse among all
the background sample forming in time from the strips of the same
plane?

Answer: We plan to use the microstrip chamber to reconstruct the
tracks geometry inside the target and to have the space position in
correspondence of all the layers.

L. Montanet: How do you get the beauty signal?

Answer: By effective mass, using charm to select the events.

H. Newman: It is very interesting to look for a long lived "beauty"
just as it is interesting to look for a long lived D^0. Any experiment which has sufficient sensitivity, such as NA14, should look
for this. Not having sufficient resolution to find B's which have
the expected lifetime should not inhibit one from looking hard
for unexpected long-lived decays.

G. Bellini: If you want to use the impact parameter to filter charm,
a multiple scattering in the target in one track in a hundred may
fake a decay if the scattering angle is bigger than some mrad.
Have you made calculations for this effect in a region of highly
non-gaussian scattering angle distribution?

Answer: Yes, the low momentum tracks have to be eliminated from
the impact parameter analysis.

R. Bizzarri: I should like to add the remark that very low momentum tracks are very likely to give a high impact parameter. To use the presence of a track missing the vertex as a signal for a charm candidate one should analyse the tracks in the spectrometer and eliminate those with a too low momentum.

Answer: In fact it is not necessary to analyse tracks in the spectro meter. Elimination of tracks whose curvature is evident in the microstrip chamber is sufficient to remove tracks of momenta less than 500 MeV/c.

J. Appel: Has anyone used or tested a Si target in a photon beam with a magnetic field? Won't the low energy electron-positron pairs curling in the target cause special rate and other problems?

P.F. Manfredi: We have used a silicon telescope in a magnetic field during the Serpukhov experiment. The incident beam was π^-. Comparison of line centroids and widths measured with and without magnetic field did not show any difference. It is true that the processing times employed in that experiment were long, about 0.5 μsec so that effects of variations in collection times may have been masked. It seems to me that the effects of magnetic fields are much more serious for a microstrip detector than for a telescope.

STATUS REPORT ON TWO NEW VERTEX DETECTORS: THE

SCINTILLATION CAMERA AND THE MICROCHANNEL PLATE

Douglas M. Potter

Department of Physics and Astronomy
Rutgers University
Piscataway, New Jersey 08854

INTRODUCTION

The recent successes in the study of charm with visual detectors have been achieved in experiments which used either bubble chambers or nuclear emulsion. Optimization of these techniques for charm has obviously been possible; further improvements to permit an investigation of beauty or of charm with large statistics are speculative. To see the difficulty is easy: the recent (and highly successful) run of LEBC in NA16 resulted in ten's of reconstructed charmed particles. LEBC was used at an interaction rate of about 10/second and about 10^6 pictures were recorded. A simple extrapolation of these experimental results leads to the conclusion that in order to accumulate a similar number of beauty particles produced with a cross-section of 10^{-3} that of charm (a) a triggerable device would be necessary and (b) an interaction rate of at least 10^4 would be required.

This report discusses two new visual detectors which have the following properties:

(a)   high interaction rates

(b)   5 - 15 µm RMS spatial resolution

(c)   triggerable and "live"

(d)   3 - 4 hits/mm of track

(e) a single 2-dimensional view

THE SCINTILLATION CAMERA

Operating Principle of the Scintillation Camera

The operating principle of the scintillation camera[1] is identical to that of the luminescent chamber[2] developed more than twenty years ago; however, substantial improvements in electrooptical technology have once again made the technique attractive. The idea behind these devices is that the light produced by charged tracks in a scintillating medium can be imaged, stored, intensified, gated, and, ultimately recorded on photographic film. If the scintillating medium is also the target in a high energy physics experiment (as is the case for the scintillation camera), an image of all charged tracks produced in an interaction can be recorded.

To implement the idea, scintillation photons produced by an interaction occurring in a NaI(Tl) crystal are focussed on the input of a four stage magnetically focussed image intensifier tube (See Figure 1). Three features of the tube are essential to the operation of the scintillation camera. First, the tube has gain sufficient to intensify and record on photographic film a single scintillation photon. Second, the phosphor screen of the first atage can be used to store the image temporarily while the decision whether to record the event is being made. Finally, by changing the potentials inside the tube, individual stages can be turned on or off; consequently, the tube can be gated to transfer the stored image to film. The image thus recorded will appear somewhat similar to that from a bubble chamber; however, the tracks consist of dots which correspond to single scintillation photons.

The scintillation camera is distinguished from bubble chambers in that it can be triggered and from spark and streamer chambers in that information from the device itself can be used as input for the decision to record the event.

Optical Considerations

A minimum ionizing particle deposits 0.484 MeV/mm in NaI(Tl). Since the scintillation process is 13% efficient about 21,000 3 eV photons are emitted per mm. The lens which images these photons onto the face of

the intensifier is an 80 mm f1.5 modified double gauss,
optimized for the spectral distribution of NaI(Tl). A
rectangular aperture was inserted into the lens to improve the spatial resolution transverse to the beam.
It reduced by a factor of two the maximum angle subtended by the lens in that direction, thereby improving
the resolution by a factor of 1.7, while reducing the
solid angle coverage by 30%. The number, N, of photoelectrons emitted from the input photocathode of the
image intensifier and reconverted to light therein per
mm. of track in the NaI(Tl) crystal can be computed as
follows:

$$N = (21,000) \times (\frac{1}{n^2}) \times (\varepsilon_1) \times (\varepsilon_2) \times (\varepsilon_3)$$

$$\times (\frac{1}{f_z} \times \frac{1}{f_y} \times \frac{1}{4\pi}) \times (\frac{1}{1+\frac{1}{M}})^2 \cong 6,$$

where

$n(\cong 1.85)$ is the index of refraction of the NaI(Tl),

$f_z(= 1.5)$ is the f-number of the lens along the
beam,

$f_y(= 3)$ is the f-number of the lens transverse
to the beam,

$M(= 2.44)$ is the magnification,

$\varepsilon_1(= 0.8)$ is the average transmissivity of the
optical system for the wavelengths of
interest,

$\varepsilon_2(\cong 0.2)$ is the average quantum efficiency of
the image intensifier photocathode,

$\varepsilon_3(\cong 0.7)$ is the probability that a photoelectron
will be reconverted to light.

(The measured yield of 4/mm agrees reasonably well with
the calculated value.)

Spatial resolution is limited by the depth of field,
which is equal to the thickness, t, of the crystal
divided by n. The RMS error per hit for all tracks
contained in a crystal of thickness, t(= 1 mm) is

$$\sigma_y = \left(\frac{t}{2\sqrt{3}}\right) \times \left(\frac{1}{n}\right) \times \left(\frac{1}{2\sqrt{3}\cdot f_y}\right) \times \left(\frac{1}{1+\frac{1}{M}}\right) = 11 \ \mu m.$$

Because $f_z = f_y/2$, the resolution, σ_z, along the beam is a factor of two worse; however, σ_z does not contribute significantly to the error in particle lifetime measurements.

Note that the resolution σ_y for a track in the plane of best focus must be determined by factors other than the depth of field, and would be expected to be considerably better. That expectation is born out by the measurements discussed later. Conversely, σ_y for a track at the edge of the 1 mm region would equal 19 μm.

Major Components of the Scintillation Camera

Target Module. The target module consists of an 18 mm diameter 1.5 mm thick disk of NaI(Tl) sandwiched between two quartz windows. Because previous tests showed that reflections in the vicinity of the target could be a severe problem, the exterior surface of both windows was multicoated to reduce reflectivity to 0.5% in the NaI(Tl) emission band.

The pulse from a PMT viewing the side of the module opposite the image intensifier was used in the trigger to signal that an interaction had occurred in the target. Discriminating between beam particles passing through the target and those interacting in it proved easy, because of the large increase in charged particle multiplicity and predominance of nuclear break-up in high energy interactions. A level of pattern recognition more detailed than that provided by a single PMT could probably be required in the trigger. Any such scheme would involve imaging and would be best implemented at a secondary phosphor of a separated function image intensifier system.

Image Intensifier. The image intensifier is an EMI type 9912, a four stage magnetically focussed tube[3] with a maximum gain of at least 10^6 and a single photoelectron spot size of about 30 μm when run in the d.c. mode. Operation in gated mode degrades both somewhat. Light amplification is accomplished by a cycle consisting of conversion of photons to photoelectrons in a photocathode, acceleration of electrons in an electric field and reconversion of electrons to photons in a phosphor.

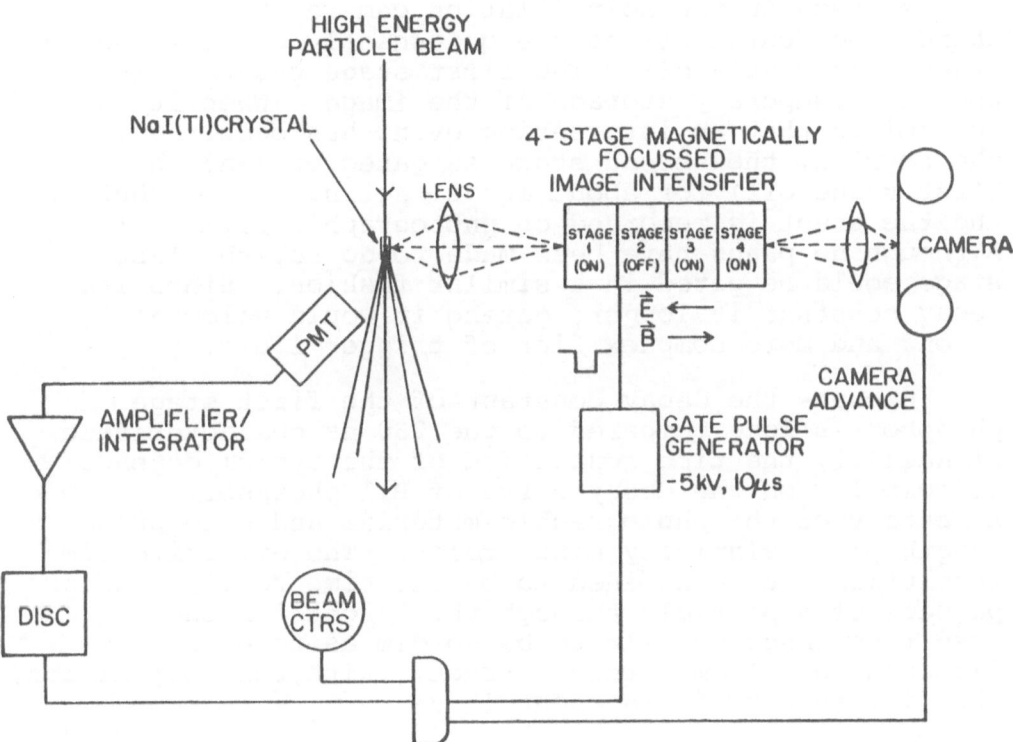

Figure 1:   Schematic of scintillation camera and test
            set up.

In a four-stage tube this cycle is repeated four times.
Focussing is achieved through the use of parallel electric and magnetic fields; electrons are required to
revolve twice about an axis parallel to that of the
tube in transit between cathode and anode. All
phosphors are P11, but due to saturation effects and
convolution of the decay distributions of successive
stages, the effective decay times vary; that for the
first stage phosphor is 15 µs, while that for the last
is about 100 µs.

     As used in the scintillation camera the first,
third, and fourth states are quiescently on, the second
stage quiescently off. The first stage phosphor is
used for temporary storage of the image. When it is
determined that an interesting event has occurred in
the crystal, the second stage is gated on (and the
first stage off) for about 10 µs - acting like a shutter -
and the event is recorded on photographic film.
Although no plans have been made to do so, the last
stage could be gated in a similar fashion. Since its
decay constant is longer, gating it would allow a
second and more complex tier of trigger logic.

     Because the decay constant of the first stage
phosphor is long compared to the 230 ns characteristic
of NaI(Tl), the time resolution of the system depends
principally on the decay curve of P11 phosphor.
Linearity of the photographic material and gate pulse
length play relatively minor roles. The effective time
resolution can be defined to be the time required after
passage of a particle through the target for the
resulting image on film to be so dim as to be easily
distinguished from a recent track. With this definition,
the time resolution is about 15 µs.

Test Results

     The scintillation camera was tested at Fermilab in
a 200 GeV/c π^- beam. Figures 2 and 3 show two interactions obtained with the trigger indicated in Figure 1.
Results of the test are as follows:

     (a) The dot size recorded on film corresponds to
about 40 µm in space.

     (b) The setting error (i.e., the RMS deviation of
the centers of the dots comprising a track from a line
least squares fit to them) is measured to be as little
as 5 µm for tracks in or very near to the plane of best

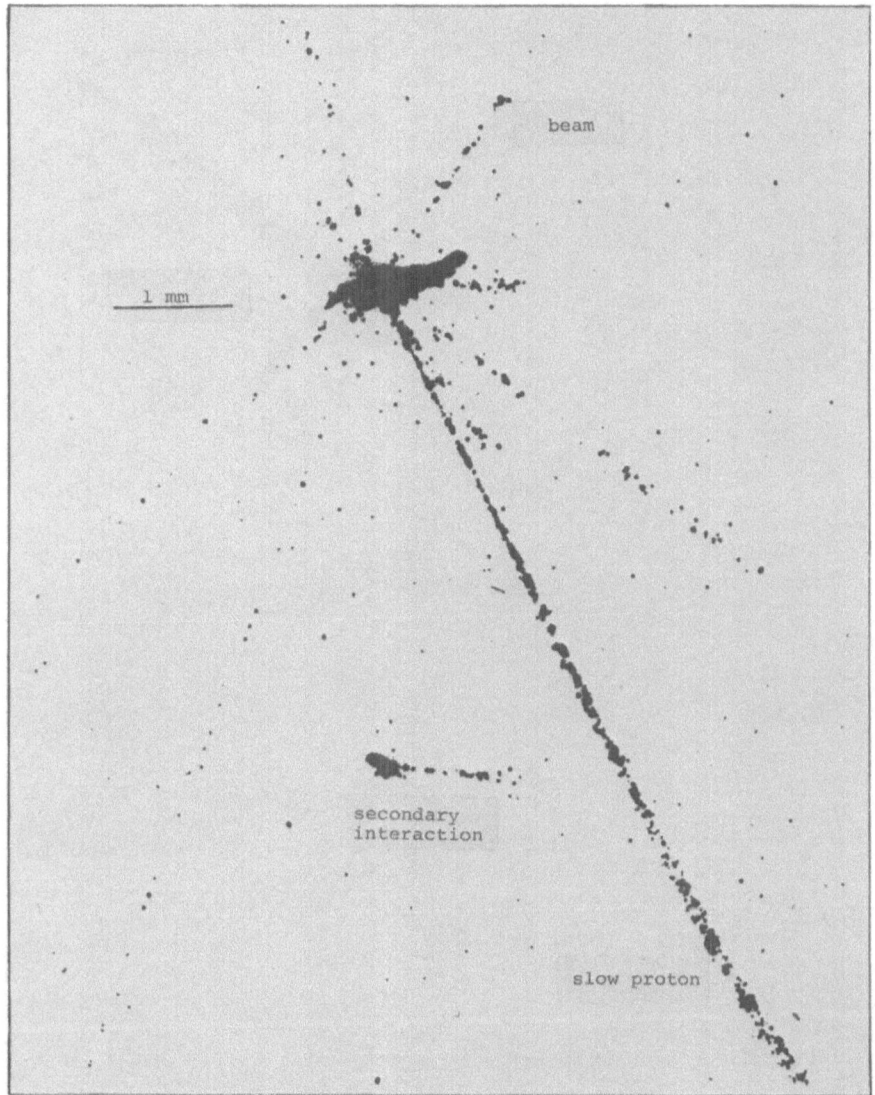

Figure 2:   A low multiplicity interaction recorded by
            the scintillation camera. A photon conver-
            sion, a secondary interaction, a slow proton
            moving out of the plane of best focus, the
            incident 200 GeV/c beam track, 2 high energy
            tracks and four other tracks can all be
            observed. (For this and the following
            figure, a better impression of the resolution
            and information content can be obtained by
            sighting in the plane of the paper.

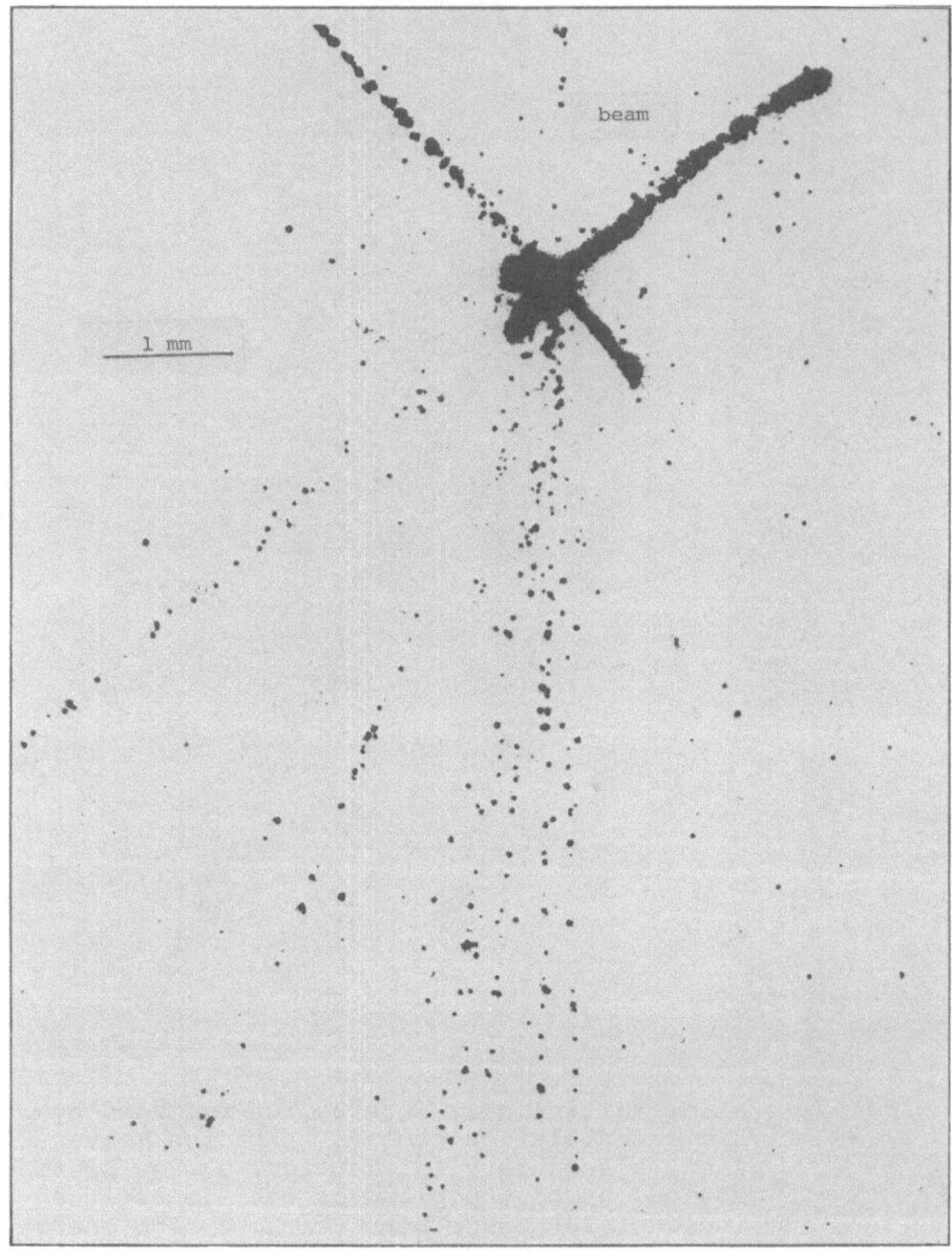

Figure 3:   A typical interaction of a 200 GeV/c
            pion recorded by the scintillation
            camera.   (See remarks in the Figure 2
            caption.)

focus and is probably limited by the device used to
digitize the film. The average setting error for tracks
outside the plane of best focus is larger than 5 μm
and is assumed to be as calculated above.

(c) The number of dots recorded on film per mm.
of minimum ionizing track in the target was found to
be about 4.

The scintillation camera was also tested for
potential use as the vertex detector in Fermilab
experiment E515. For this trial, the additional
requirement of a prompt muon signal was incorporated
in the trigger. A beam intensity of 0.35×10^6/s
(corresponding to about $1.5 \cdot 10^4$ interactions/s) was
used without causing additional difficulty in the data
analysis.

Figure 4 shows a charm candidate picked during a
scan of the test film; figure 5 is a digitized version
of figure 4 but with differentially expanded scales.

The Scintillation Camera Future

In the near future the scintillation camera will
be used as the vertex detector in Fermilab experiment
E515 (which is described in this conference by
R. Ruchti). We expect to accumulate about 10^5 pictures
with better resolution but decreased hit density.

Prospects for the more distant future are difficult
to assess. However, a variety of improvements seem
technically feasible. Increased rate capability can
certainly be achieved by decreasing the decay time constant of the storage phosphor. As mentioned previously,
a separated function image intensifier system utilizing
several stages of gating would permit use of some form
of fast pattern recognition in the trigger. Finally,
resolution could in principle be improved while retaining high dot density by employing "microbaffling" via
total internal reflection. Schemes for implementing
this last idea include fabrication of a fiber optic
plate containing a scintillating core glass[4], construction of a matrix of extruded NaI(Tl) needles[5],
and growing CsI inside a microchannel plate[6].

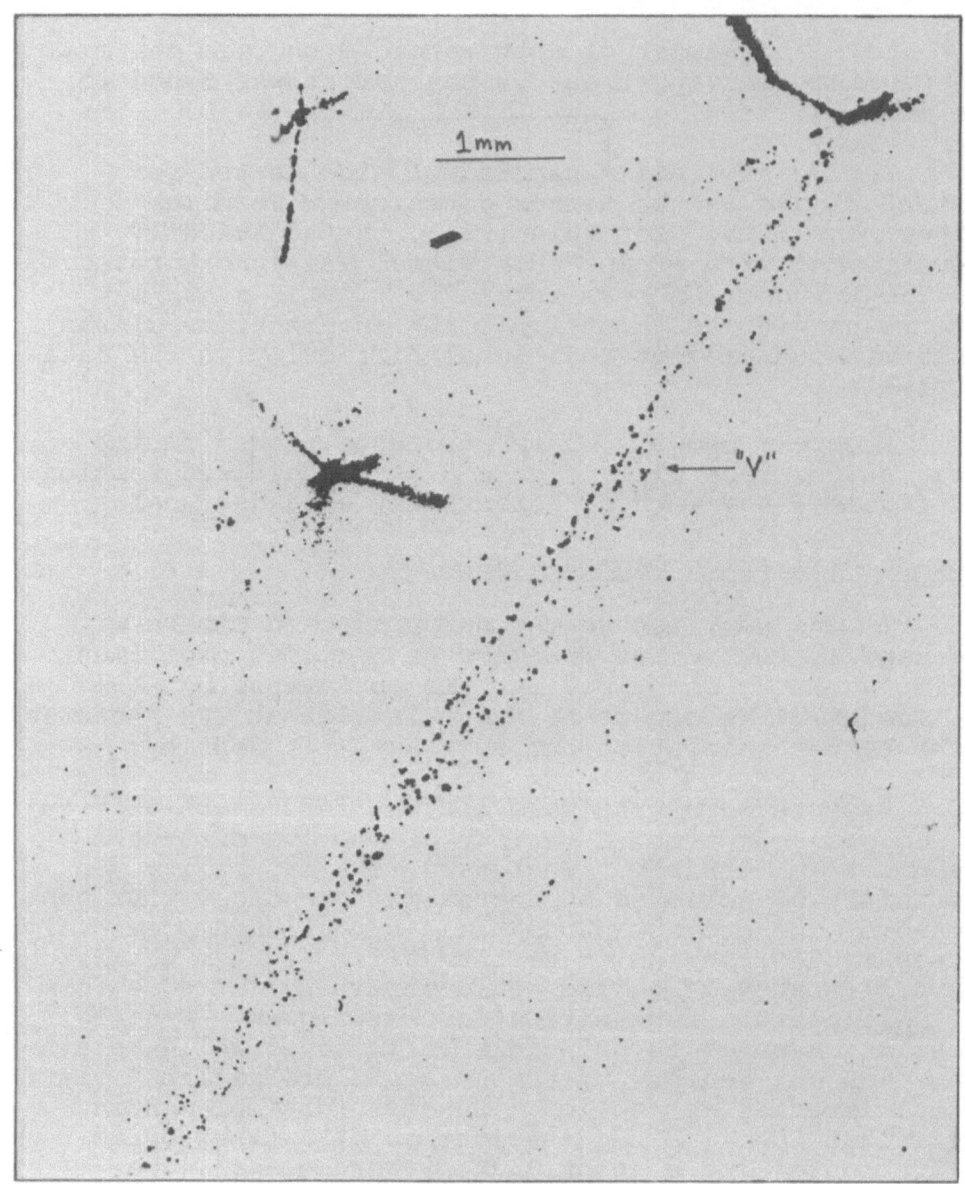

Figure 4:   Charm candidate picked during scan of test
            film.   To see "V", signt in plane of paper.

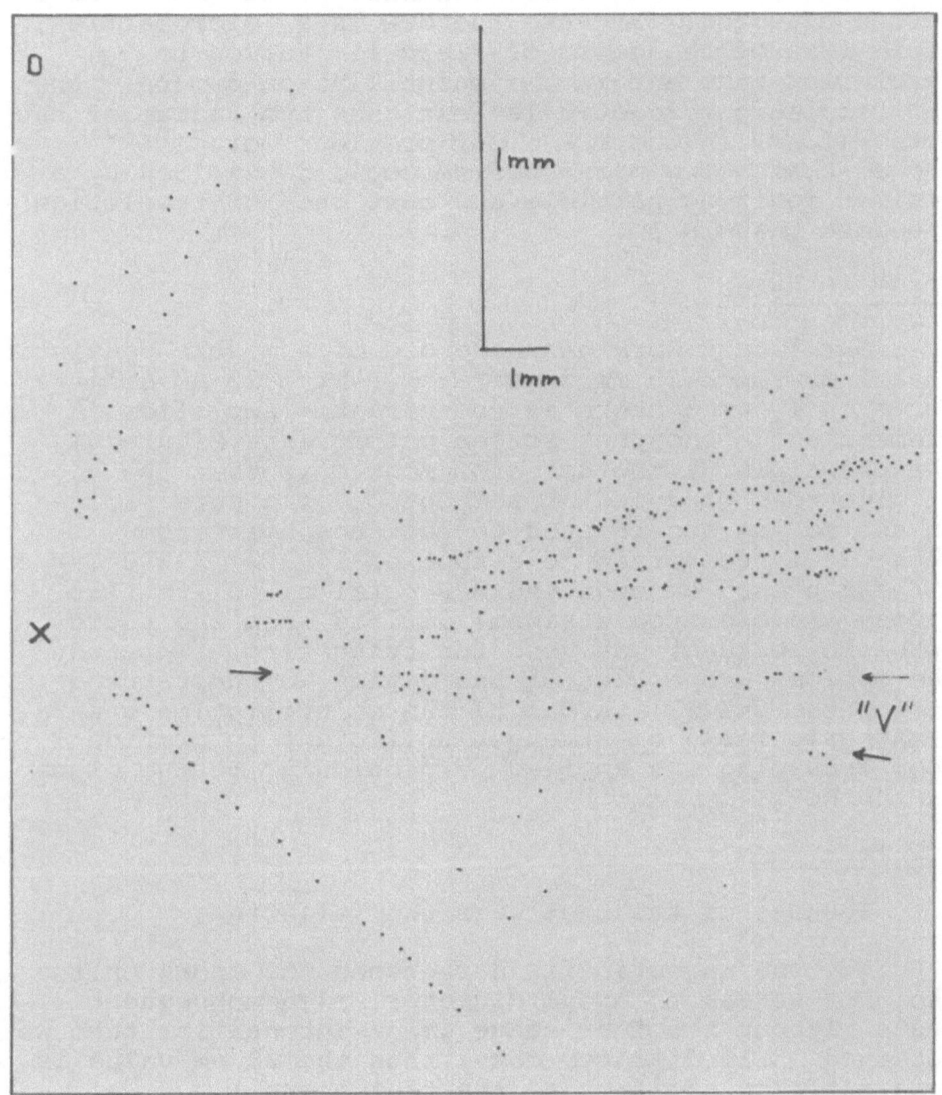

Figure 5:   Digitized version of Figure 4.

THE MICROCHANNEL PLATE VERTEX DETECTOR

The Microchannel Plate as Vertex Detector

A group consisting of K. Oba (Hamamatsu Corp.),
P. Rehak (B.N.L.) and myself has been investigating the
possibility of using a microchannel plate (MCP)[7] as a
vertex detector. Because the MCP is a "microbaffle",
it is free of the depth of field limitation to
resolution inherent in the scintillation camera; the
RMS resolution expected is less than the radius of the
pores (6 µm) in the MCP. Our previous work[8,9,10]
showed that a few pores per mm could be excited by a
minimum ionizing particle and that the RMS resolution
was less than 10 µm.

MCP Test Setup

Our latest work has made use of a module consisting
of a 1 mm thick 25 mm diameter MCP with 12 µm diameter
pores on 15 µm centers as the target - amplifier
element and a phosphor screen output (see Figure 6).
The principle of the detector module is that the cloud
of electrons produced at the output of a pore excited
in the MCP is accelerated to the phosphor screen,
where it is converted to a spot of light. A 200 GeV/c
π^- beam passed through the MCP parallel to its flat
sides. An electronic signal derived from the anode
(phosphor screen) was used for triggering. To study
the optical properties of the device we substituted
it for the NaI(Tl) target of the scintillation camera.
A magnetic field of strength sufficient to provide 1
loop focussing was applied perpendicular to the plane
of the MCP.

MCP Test Results

Results of the test were the following:

1. The magnetic field focussed the spots on the
phosphor screen to 25 µm diameter. (In subsequent
tests without the four stage image intensifier tube we
obtained 15 µm diameter dots; thus the 25 µm value is
apparently an artifact of the test setup.)

2. The magnetic field eliminated ion feedback.

3. The MCP electronic gain saturated at about
3×10^5 with a peak/valley ratio of about 10. (Almost
the entire 1 mm length of the pore was required as the
amplifier to achieve this result.)

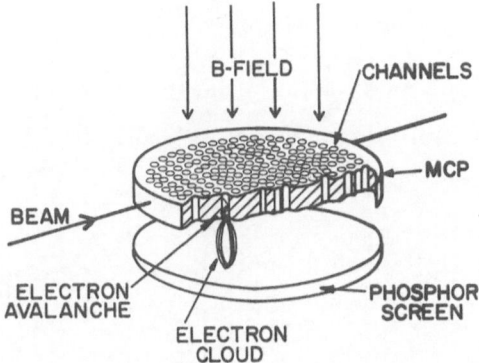

Figure 6:   Schematic of microchannel plate vertex
            detector test module.

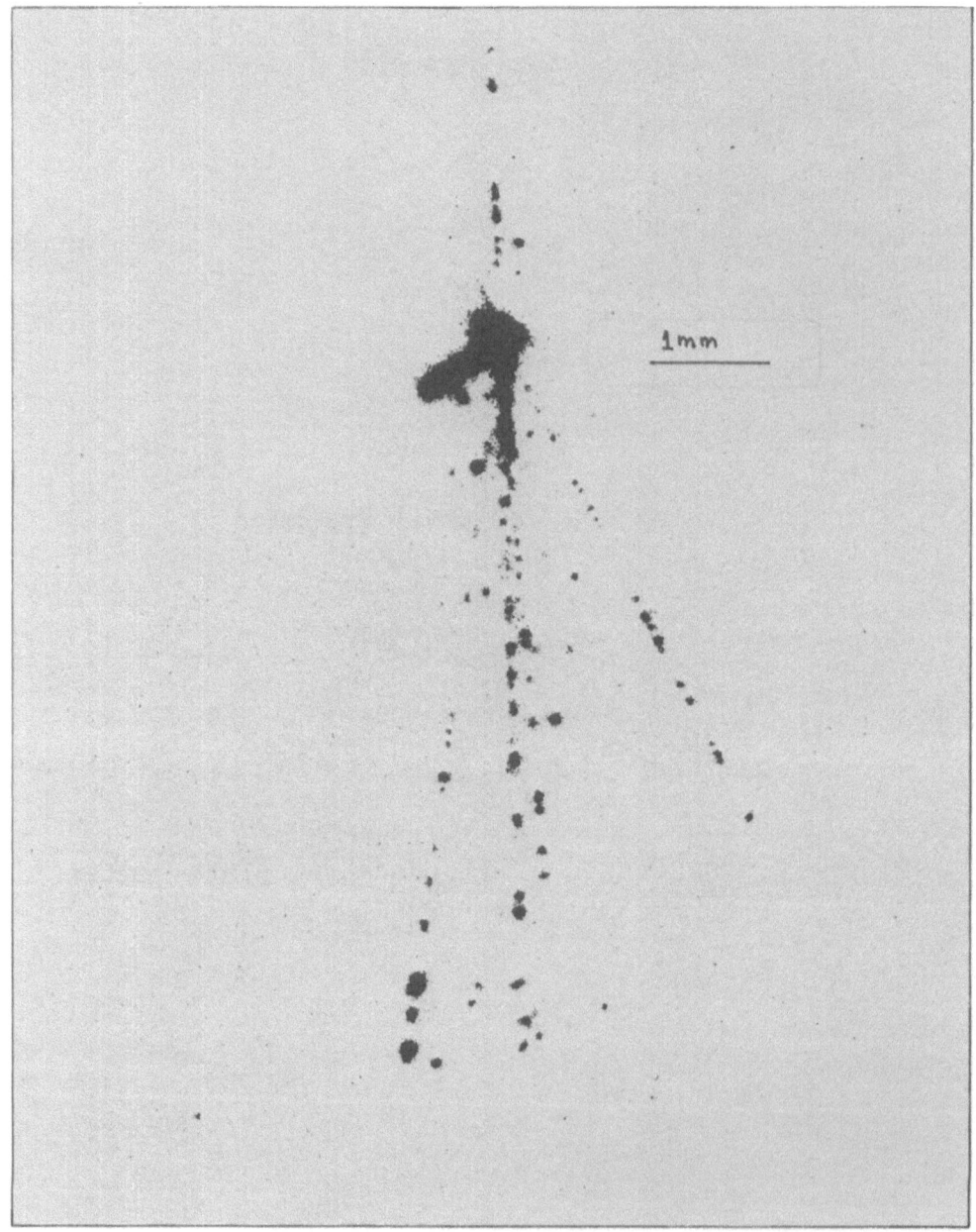

Figure 7:   An interaction of 200 GeV π⁻ in the MCP
            test module. (Both the large dot size
            and variation in dot intensity are under-
            stood to be artifacts of the test set-up.)

4. About 3 dots/mm of minimum ionizing track were recorded.

5. The RMS error per dot was about 5 μm.

The rate capability of the MCP vertex detector was not studied. However, we believe that with current technology it will probably be limited by the pore recharging time to about 10^4 - 10^5 interactions/s in a practical detector.

Figure 7 is a picture of an interaction recorded during the tests. The dot size and variation in brightness displayed here are understood to be artifacts of the test setup.

The MCP Future

We are currently evaluating the performance of several 3 mm thick MCP's; at present we know that they do not have characteristics anticipated on the basis of the tests described above, but cannot report results. (Note that a 3 mm MCP would be expected to have an active target thickness of 2 mm.) We hope to construct a viable detector in a year or so.

SUMMARY

The situation is best summarized by Table I, which lists a few features of the MCP and scintillation camera.

Table I

Detector	σ_{dot}	Dot Size	Interaction Rate	Hit Density
Scintillation Camera	5-19 μm	40 μm	$1.5 \cdot 10^4$/s	4/mm
MCP	5 μm	15 μm	10^4 - 10^5/s (expected)	3/mm

References

1. Further details can be found in D. M. Potter, "A
    Scintillation Camera for High Energy Physics"
    (to be published in IEEE Trans. Nucl. Sci.).
2. M. L. Perl et al., Proceedings of an International
    Conference on Instrumentation for High Energy
    Physics, Lawrence Radiation Laboratory, 186
    (1960).
3. F. C. Delori et al., Advances in Electronics and
    Electron Physics 33A:99 (1972).
4. A. Rogers, private communication.
5. R. Ruchti, private communication.
6. K. Oba, private communication.
7. For a review, see e.g. J. Wisa, Nucl. Instr. and
    Meth. 162:587 (1979).
8. D. M. Potter, Nucl. Instr. and Meth. 189:405 (1981).
9. K. Oba, P. Rehak and S. P. Smith, IEEE Trans. Nucl.
    Sci. NS-23:705 (1981).
10. K. Oba, P. Rehak and D. Potter, "Optimization of
    Microchannel Probe Multipliers for Tracking
    Minimum Ionizing Particles" (to be published in
    IEEE Trans. Nucl. Sci.).

PRODUCTION CROSS SECTIONS OF HEAVY FLAVOURS ON PROTONS

B. Margolis

Physics Department
McGill University
Montreal, Quebec, Canada

INTRODUCTION

The production of particles containing heavy quarks on proton targets has been studied intensively on the theoretical side and there is a considerable body of experimental data for charm production, especially for ψ and D production. There is data for ψ' production in γ-p and p-p interaction and D* photoproduction. There is also data for Λ_c production and there are reported sightings of Λ_b. We are hearing about all these measurements in this workshop.

I will concentrate here on p-p and γ-p interaction. The important mechanism in γ-p production is photon-gluon fusion and in central p-p production, away from threshold gluon-gluon fusion. There are reports that in addition to central production in p-p interaction there is copious production of heavy particles near the forward direction with a flat x distribution, a so called diffractive component. This component for charmed particles may be as large as several hundred microbarns at ISR energies and is reported seen or not seen, depending on the experiment at Fermilab and SPS energies. The theoretical description of this phenomenon may be outside the framework of straightforward perturbative QCD calculations.

I wish first to review some of the work that has been done on central production of heavy particles. This involves in simplest

form the production of free quark systems. I will then discuss and
describe a program for dressing the quarks into hadrons. An examination of the data for p-p induced production leads us to propose
a unified treatment for the production of heavy particles which is
roughly speaking flavour independent. The approach developed will
involve us with some statistical properties of a parton gas. This
program leads us to a method for estimating the diffractive production of heavy particles.

Central Production in p-p Interaction

The traditional treatment for production of heavy open flavour
is to calculate the cross section for production of a heavy quark-
antiquark pair by constituent annihilation. Then for charm or beauty
say, the contributions above DD or BB thresholds are identified with
D or B mesons[1]. The production of heavy mesons composed of quark-
antiquark pairs of the same flavour (quarkonium) has been considered
to follow either directly from constituent annihilation into the
produced state (for example gluon + gluon → η_c) or production of a
state that decays into the final quarkonium state of interest[2].

The production of ψ mesons for example is often considered to
result from production of χ states by gluon-gluon fusion followed
by photon decay into ψ. We find[3] that this mechanism yields a cross
section a few times smaller than is found experimentally. Then
again ψ' production cannot be explained this way since the branching
ratio from p-states above the ψ' is very small due to the fact that
these p-states are above threshold for DD production.

ψ production, and more generally the production of other bound
states of cc, is often estimated using a semilocal duality type
hypothesis. One takes the cross section for free cc production over
the mass range $2m_c < \sqrt{\hat{s}} < 2m_D$ and multiplies by an empirical constant F to be fitted with experiment so that[1]

$$\sigma(s,m_\psi) = F\int_{2m_c}^{2m_D} \sigma(s,\hat{s})d\sqrt{\hat{s}} \tag{1}$$

More detailed pictures of heavy flavour production developed in
the last year or two involve dressing the heavy quarks produced by
gluon-gluon fusion[3,4]. This implies higher order diagrams and since
bound states of quarks and antiquarks are involved, a departure
from perturbation theory. Along these lines we have considered DD
and BB production to be described as follows. DD production is
taken to ensue as the result of cc production followed by the capture
of two light sea quarks from the quark sea.

We calculate quarkonium production (QQ) assuming that a produced
QQ pair becomes a bound state after single hard gluon emission.

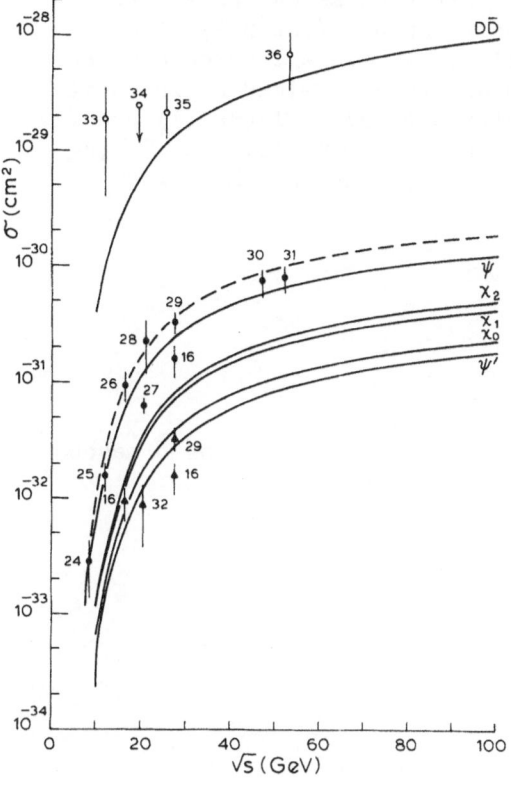

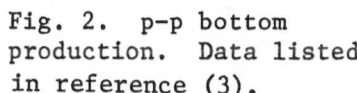

Fig. 1. p-p charm production.
Data listed in reference (3).
Dotted Curve for ψ includes
contribution from $\chi_J \rightarrow \gamma \psi$.

Fig. 2. p-p bottom
production. Data listed
in reference (3).

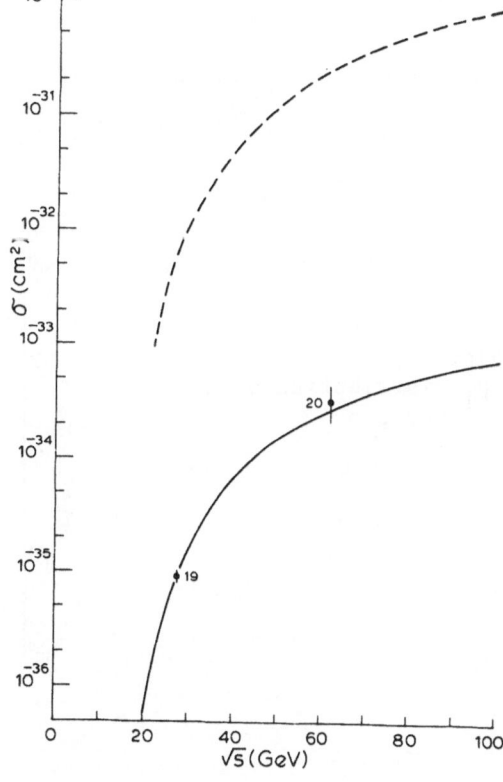

There is a high probability for soft gluon emission and absorption by the heavy quarks before light quark capture (for D,B,..) or before hard gluon emission (for $\psi,\psi',T,..$). This creates a random phase situation so that relative rates are determined by intensity ratios. Details may be found in reference (3). Figure (1) and (2) show our results for $D\bar{D}$, $B\bar{B}$ production and for ψ,ψ',χ_J and T production. Figures (3) and (4) show p_T distributions for D and B.

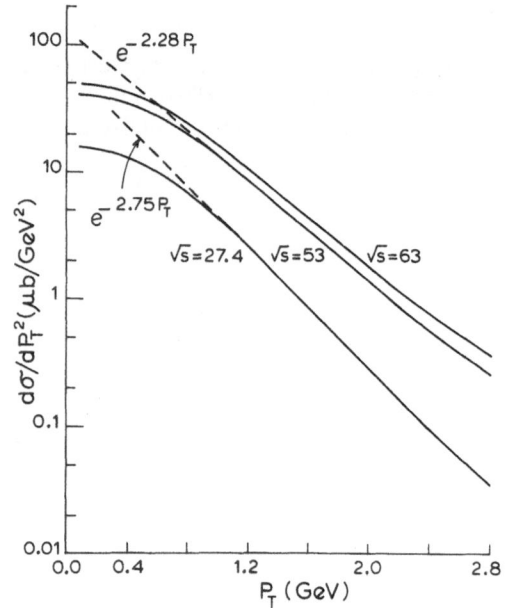

Fig. 3.
p_T distribution of D mesons for p-p.

Fig. 4.
p_T distribution of B mesons for p-p

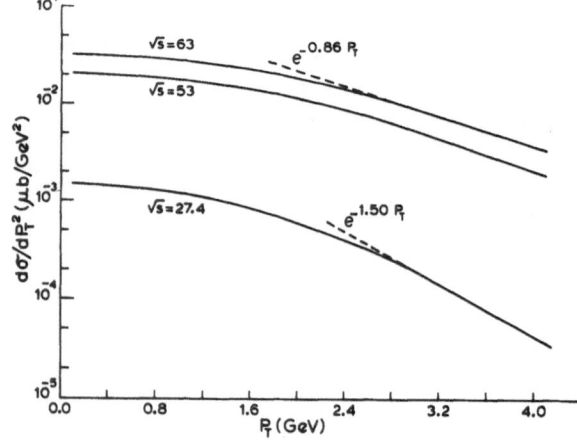

Photoproduction Reactions

Here[5] the photon fuses with a gluon from the target proton
to produce a $D\bar{D}$ on $B\bar{B}$ pair as described in the last section for p-p
production or for the quarkonium cases ψ,ψ',χ_J or T are produced again
as described above. Figure (5) shows <x> for p-p and γ-p gluons.
The small <x>γp is related to the fact that $\sigma_\psi(\gamma p)$ exclusive and
inclusive are not too different. Ref.(5) gives cross section results.

A Connection with Vector Dominance

Our calculations for photoproduction show that

$$\frac{\sigma(T)}{\sigma(B)} \ll \frac{\sigma(\psi)}{\sigma(D)} \ll \frac{\sigma(\phi)}{\sigma(K)} \tag{2}$$

at Fermilab, SPS energies. Form factors inhibit heavy quarkonium
production. We cannot calculate well using the methods outlined
above for light quarks. However we expect $\sigma_{\gamma p}(\rho)$ to be very large,
i.e. there is vector dominance for light quarks and hence for most
of $\sigma_{\gamma p}$ (total). Lowest order QCD diagrams fail for light quark production. However

$$X \equiv \sigma_{\gamma p}(\text{total}) - \sigma_{\gamma p}(D\bar{D}) - \sigma_{\gamma p}(K\bar{K}) \tag{3}$$

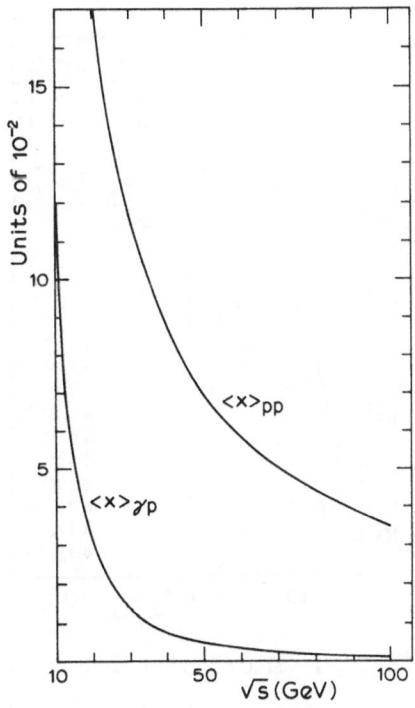

Fig. 5. Mean gluon x values.

where $\sigma(D\bar{D})$, $\sigma(K\bar{K})$ are calculated as described above, leads to the full curve of figure (6), which we compare with $\sigma_{\rho p}$(total)/216 (using $\sigma_{\pi p}$(total) and the quark model). We see that vector dominance is good for light quarks since $X \propto \sigma_{\rho p}$(total). The rising $\sigma_{\gamma p}$(total) is due in part to the rising $\sigma_{\rho p}$(total) and in part due to threshold effects in D and K production at Fermilab energies.

Central Production and Statistical Considerations

Some of the considerations of that which we have described above leads us to examine the possibility that central production of heavy resonances, in its gross aspects, can be described in a roughly flavour independent manner using statistical concepts. Figure 7 shows a plot of $\sigma m^3/\Gamma$ against $\tau^{-1}=s/m^2$ for p-p production of resonances of mass m and width Γ. The two curves are proportional to the gluon-gluon structure function

$$F_{gg}(\tau) = [\tfrac{1}{2}(n+1)^2]\int_\tau^1 \frac{dx}{x} (1-x)^n(1-\tau/x)^n \tag{4}$$

with n = 5. We see that all the boson resonances K*,f,g,K**,ϕ,ψ,ψ',T have the seme behaviour whereas Δ^{++} which is produced mainly b y fragmentation behaves in a completely different manner. It is further known that the boson resonances listed above are centrally produced.

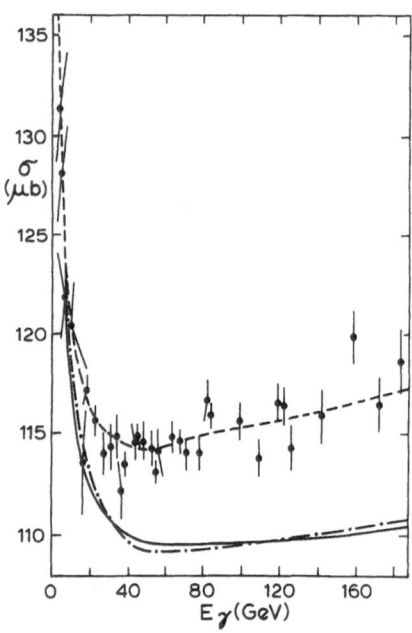

Fig. 6. Photon total cross section contributions

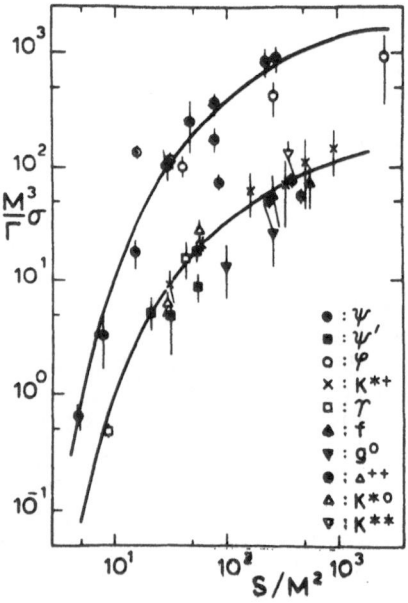

Fig. 7. $M^3\sigma/\Gamma$ plot for p-p.

We propose to describe central production in p-p interaction with a combination of parton and statistical model ideas. The production away from threshold is described as follows: (1) A fireball of mass M is produced by gluon-gluon fusion, the gluon distribution being given by formula (4). (2) The mass M fireball decays into a residual fireball of mass M' and the produced particle, mass m. The branching ratio for the decay is taken as the density of two body states of mass m and M' in a sphere of hadronic volume V divided by the total density of states of protons in a sphere of volume V. States formed from constituents of mass $m_o \ll M$ have a density

$$\rho(M) \simeq C \exp(AM^{3/4}) \tag{5}$$

We have in our statistical picture,

$$\sigma(m,s) = \int_{m_o^2}^{s} \frac{dM^2}{M^2} F_{gg}(m^2/s)\sigma_o(M)f(M,m) \tag{6}$$

where $\sigma_o(M)$ is the gluon-gluon fusion cross section and $f(M,m)$ is the branching ratio as described in part (2) above, integrated over M'. Further details can be found in reference (6). We use the two body phase space of the hadrons of mass m and M' rather than constituent phase space since the partons must transform to hadrons when the fireball M decays.

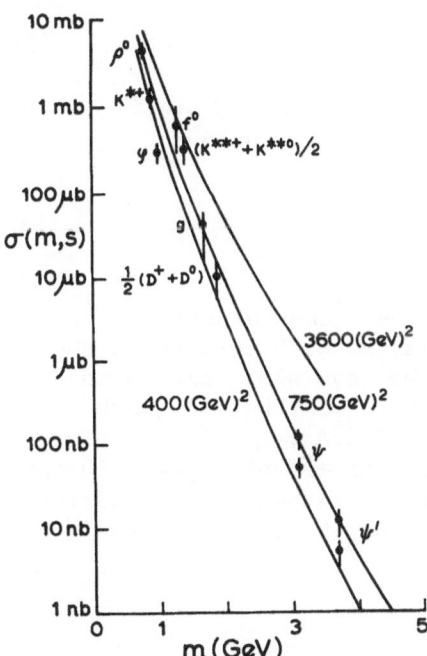

Fig. 8. p-p mass yield, $\sigma(m)/(2J+1)$.

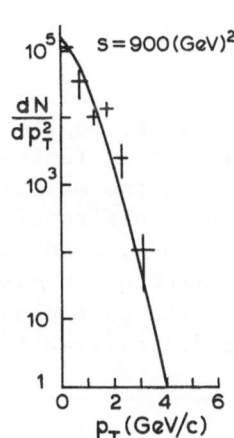

Fig. 9. p_T distribution of ψ in p-p.

The form for $\rho(M)$ given by (5) implies a temperature $T \propto M^{\frac{1}{4}}$ and because of this higher mass particles are produced much more copiously than with a limiting temperature $T \sim M_\pi$. The result of our calculation of central production $\sigma(m,s)$ are shown in figure 8. The agreement between theory and experiment is good up to masses of the order of those of the ψ family. The statistical considerations applied following gluon fusion appear to represent the effects of higher order QCD reasonably well. The weighting of fireball masses to relatively small values of M through the structure function. $F_{gg}(M^2/s)$ does not allow for substantial production of particles of the Υ family from fireball decay up to ISR energies.

The distributions in transverse momentum are obtained by suppressing the integration over momentum in the two body phase factor[6] in equation (6). Results for ψ production are shown in figure 9. Other cases show equally good agreement.

Production by Fragmentation

Fragmentation is not so easily related to QCD parton considerations. One does have the well known formula for diffractive production of mass M in p-p interaction at high energy,

$$\frac{d\sigma_D(M)}{dM^2} \approx \frac{0.54}{M^2} \text{ mb} \tag{7}$$

Assuming as in central production that these masses M are fireballs which decay according to phase space considerations then

$$\sigma(m,s) = \int_{m^2}^{m_u^2(s)} \frac{d\sigma_D(M)}{dM^2} f(M,m) dM^2 \tag{8}$$

We take $m_u^2(s) = 0.2s$ to correspond to some recent experiments[7]. For $\rho_f(M')$ we first neglect all quark masses with results as shown in figures 10 and 11 (dotted curves). We have also calculated $\sigma(m,s)$ including one massive strange, charmed or bottom quark corresponding to diffractive production of $\Lambda, \Lambda_c, \Lambda_b$. These points have been joined by a smooth curve (full curves of figures 10 and 11). The bottom quark mass reduces the cross section considerably. Generally for the particles with lighter quarks including charm we find substantial diffractive production.

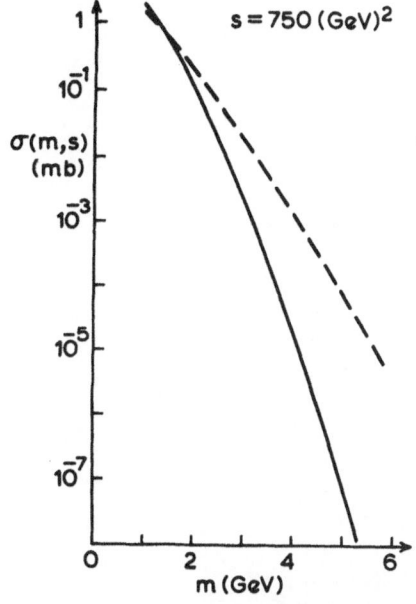
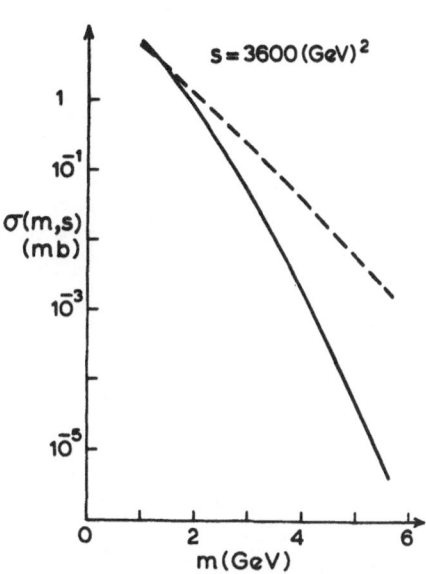

Fig. 10. Diffractive production   Fig. 11. Diffractive production
          of m in p-p.   of m in p-p.

References

1. See for example R.J.N. Phillips, in High Energy Physics-1980,
    XX International Conference Madison, Wisconsin (American
    Institute of Physics Conference Proceedings No. 68, New York
    (1980), p. 1470-1498).
2. C. E. Carlson and R. Suaya, Phys. Rev. D18, 760 (1980).
3. Y. Afek, C. Leroy and B. Margolis, Phys. Rev. D22, 86 (1980).
4. R. Baier and R. Rückl, Phys. Lett. 102B, 364 (1981).
5. Y. Afek, C. Leroy and B. Margolis, Phy. Rev. D22, 93 (1980).
6. Y. Afek, B. Margolis
        April 1, 1982.
7. See experiments of A. Zichichi et al., reported in this confer-
    ence.

PHOTON AND HADRON PRODUCTION OF OPEN HEAVY FLAVOURS

D. Treille

CERN
Geneva, Switzerland

MOTIVATIONS

The production of heavy flavours by neutrinos, photons, and
hadrons is widely studied at present, with a threefold purpose.

i) It is a source of information on the production mechanism of
heavy quarks. Our interest in such processes is obviously connected
with our prejudice that perturbative QCD, owing to the high mass
involved, should be a relevant description. However, we should
also be prepared for surprises: the discovery of abundant forward
production of charm in hadronic collisions was indeed a major one,
and it is still mysterious. Another important reason to reach some
understanding of the production mechanism, besides scientific curi-
osity, is that we need it in order to extrapolate reliably to the
future machines, have an idea of what we could expect there, and
build the relevant detectors.

ii) It is a source of information on the internal structure of
the particles involved in the collision. For instance, if one knows
how to identify and isolate the contribution of the γ-gluon fusion
mechanism in photoproduction, one will get the distribution of
gluons in the target nucleon. Another question, much debated at
this meeting, is the possible existence of an intrinsic heavy-
flavour component inside the hadrons. We have just heard about the
strange sea of the nucleon, extracted from charm neutrino production -- and so on.

iii) It is a source of information on the spectroscopy of heavy
flavours. Up to the recent past, this was considered as a monopoly
of e^+e^- rings, and this point of view is still basically correct.

However, it should be remembered that the absolute rates in photo-production, and especially in hadro-production, are much higher than at e^+e^- rings. Let us, for instance, compare the 9 nb of cross-section on the "charm factory" with the tenths or hundredths of microbarns we will be considering in charm hadronic production. Therefore if owing to the existence of a favourable production mechanism or a favourable decay mode, the usual smallness of the signal/background ratio can be circumvented, information on heavy-flavoured particles can be obtained. It may even be reasonable to think of these processes as a method of discovery.

METHODS

How do we proceed? Three methods can be used, generally in combination.

Bump Hunting

This is the classical way. We have first to ensure a reason-able signal/background ratio. This clearly depends on the projec-tile, i.e. on the proportion of relevant events we are starting with: it is well known that for neutrinos, photons and hadrons they are $\sim 10^{-1}/10^{-2}/10^{-2}$-$10^{-3}$, respectively. Then, a crucial fea-ture governing the visibility of a signal is the combinatorial level, which has to be decreased as much as possible. A good mass resolution (to fully exploit the quasi-zero width of the signal), much information on the identity of particles, the elimination of events with a too large multiplicity -- these are the usual weapons in that fight. This is well known and I will not dwell on it any further here.

I would like, instead, to discuss briefly the statistical sig-nificance of the signal that can be obtained. First, in relation to the mass calibration: Suppose you have a perfectly well cali-brated spectrometer and you are looking for a known particle. What you can do is to prepare your mass binning in advance, centring it where you expect the particle to be, and choosing the bin width in accordance with your mass resolution. This procedure will provide the maximal significance. If on the contrary you do not know your mass calibration well, and allow the signal to appear anywhere in a substantial mass interval, the freedom you take should be con-sidered in the expression of the statistical significance of your result in order to decrease it (and eventually destroy it).

Then in relation to cuts. Ignoring the physics, one is nat-urally led to try various cuts on data. It is a general tendency to consider as "good" a cut which gives one the expected signal, and to forget about the others. Even if this is indeed the right signal, the cut may have selected a favourable fluctuation and the cross-section will be overestimated.

Prompt Leptons from Semileptonic Decays

This is now widely used, and most of the results presented here have been obtained through prompt leptons. Some experiments, such as those with the beam dumps, just measure the prompt lepton yield. Others tag the event by a prompt lepton from one member of the created pair, and look for the hadronic decay of the other one. The inconvenience is that one does not know for sure which individual state has provided the lepton and therefore which semileptonic branching ratio one should use. People generally adopt reasonably weighted mean values of different branching ratios.

Seeing Short-Lived Particles

This may turn out to be the best solution. Many techniques used in connection with spectrometers have already given results. They provide the only way to measure lifetimes. But in principle they can also give unbiased information on production mechanisms and total cross-sections. Such an ideal detector has to be:
i) <u>fast</u>, because luminosity and therefore high fluxes are needed;
ii) <u>visual</u>, in the sense that it should allow each track to be attributed individually to a vertex (primary or secondary), so that inside the event you have no more combinatorial problems. A mere silicon-active target[1] does not fulfil this requirement. However, extrapolation methods from a downstream silicon microstrip detector could provide the solution. It can indeed be shown that, in order to isolate the decay vertex of a particle of lifetime τ_0, a transverse accuracy of $\sim c \tau_0$ is needed in the extrapolation. For $\tau_0 = 10^{-13}$ s, this implies ~ 30 μm accuracy, which is severe but by no means unrealistic. Figure 1 describes one such existing device[2].

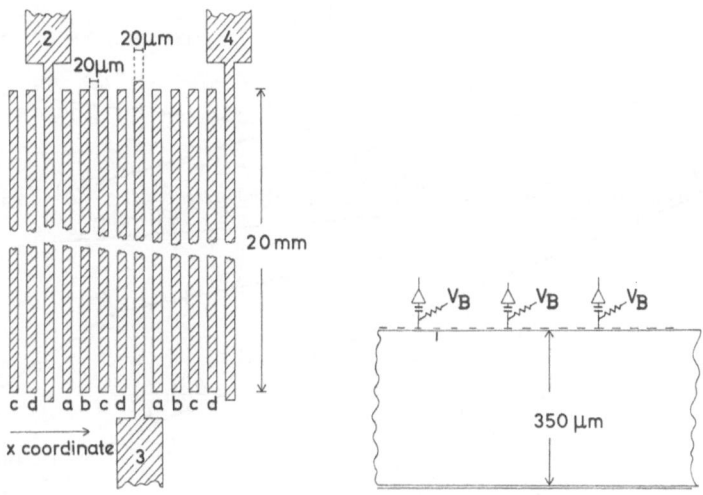

Figure 1

INPUT

When performing these studies of production mechanisms, we should know the main properties of the objects we are looking for. Some input, from theory or from other experiments, is needed. It is clear that up to now most of the information was due to the e^+e^- rings.

Again, this year, a new set of data became available, tightening the constraints. Beauty, although not seen directly, is nevertheless familiar[3]. Its semileptonic branching ratio has been measured and is around 10%. Cascading through charm seems to be its favourite decay. Decay to (ψ + anything) is not excluded at the 2% level. Its mass is well constrained, both theoretically and experimentally. All this is good news for spectroscopy. However, its lifetime is still unknown. The present limit[4] $< 5 \times 10^{-12}$ s is still far from the goal, and we do not know yet whether it will be accessible through visual techniques.

The existence of Truth has not been demonstrated but it is likely, as inferred from the properties of b decay: namely, the absence or low level of flavour-changing neutral currents[3].

About Charm, it seems that there is some hesitation about lifetimes. For instance, a question was recently raised[5] (Figure 2) about the possible admixture of long-lived baryons (csd, for instance)

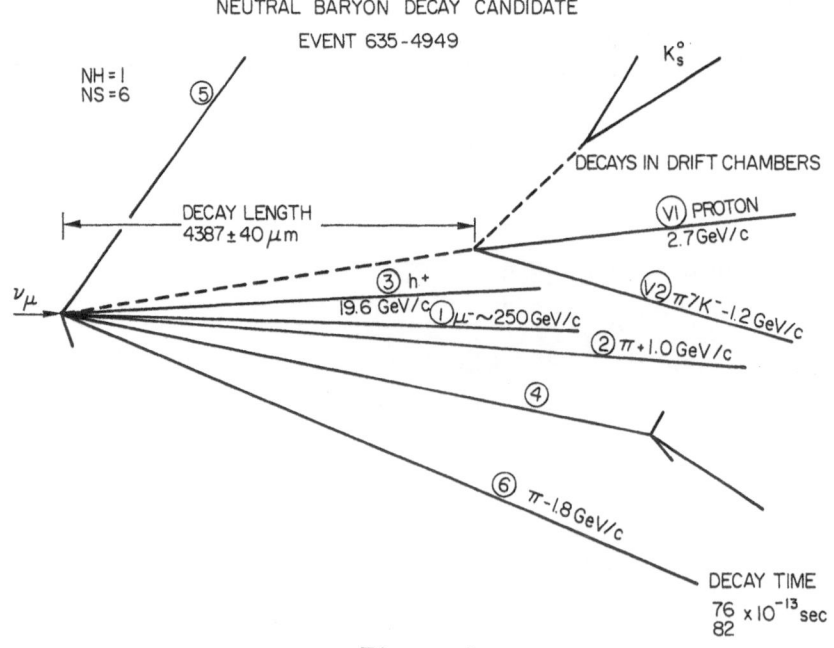

Figure 2

among the neutral decay generally attributed to D^0's. This is relevant not only for lifetime measurement but also for production mechanisms. The Λ_c properties have been determined quite well by e^+e^- experiments[6].

However, many unknowns are still present. One example is the semileptonic decay of Λ_c, and we will see that this ignorance will prevent us from drawing conclusions about some production models. Another example is the F: apart from its first observation at DASP[7], the F has been very elusive in e^+e^-, and only upper limits on its production have been obtained since[8]. From emulsions, information on F's is generally suffering from ambiguities[5].

PHOTOPRODUCTION

Survey of High-Energy Experiments

CIF[9,10]. This collaboration, using the broad-band beam at Fermilab, has obtained results at a mean photon energy of 165 GeV. They have observed $D^{*\pm}$, Λ_c and $\bar{\Lambda}_c$, D^0 and $\bar{D}^0$, and find that the production is particle-antiparticle symmetric. They extracted the cross-section, assuming a "diffractive" mechanism. Here and in the following, "diffractive" simply means that the $c\bar{c}$ quark pair has taken most of the photon energy; it does not imply a genuine diffractive mechanism. For instance, γ-gluon fusion would belong to that category. This collaboration obtained cross-sections for the channels listed above (Fig. 3), all of which have been published[10].

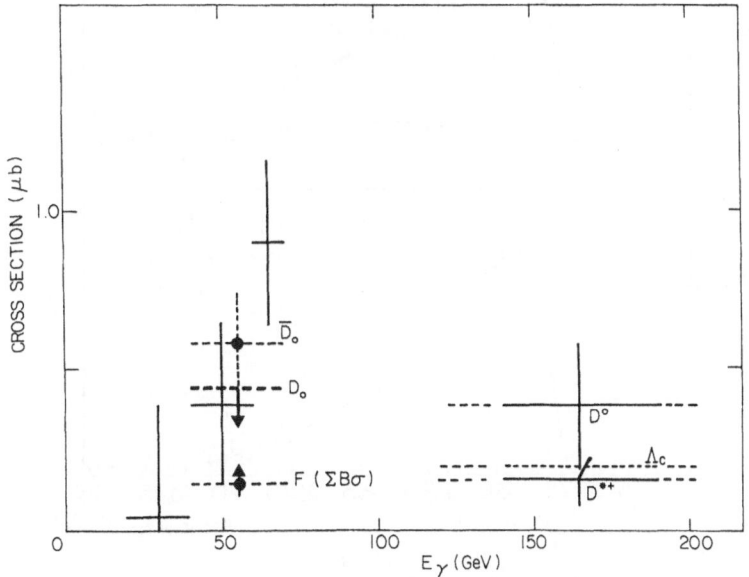

Figure 3

Adding them, with some reasonable hypothesis on the relative popu-
lations of unobserved channels, they come to a cross-section of
∿ 1 μb. This is not supposed to be the total charm production
cross-section, since, owing to the acceptance of their set-up and
of their trigger, they would not be sensitive to mechanisms such as
central production. Let us call it the "diffractive" charm cross-
section.

WA4[11,12]. This collaboration used the tagged γ beam and the
Omega spectrometer. The photon energy was 40-70 GeV. The observed
production is particle-antiparticle asymmetric: the $\bar{D}^0$ is seen,
not the D^0 (Fig. 3). This is likely to be due to $C\bar{D}^0$ associated
production, since tagging by a proton supposed to come from the C,
is beneficial to the extraction of the $\bar{D}^0$ signal.

The F is observed[13], faintly, but in four independent channels
at the same mass (Figure 4 and Table 1). The sum of observed (Bσ)'s
already reaches ∿ 100 nb. This is not in conflict with e^+e^- data or
limits. However, this would imply that the annihilation channels
leading to π's, and not observable in this experiment, do not represent the majority of F decays -- otherwise the F photoproduction
cross-section would reach an unreasonable value.

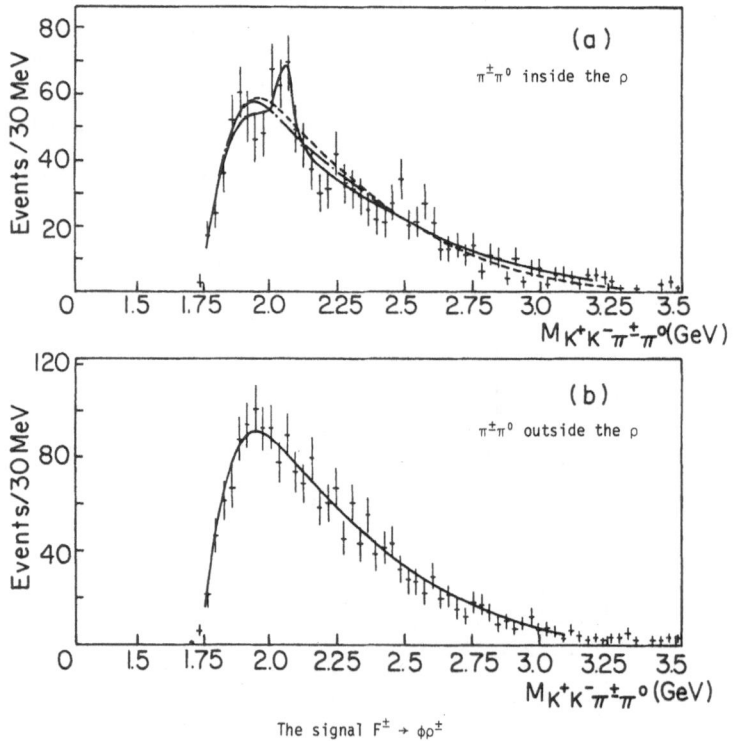

The signal $F^\pm \rightarrow \phi\rho^\pm$

Figure 4

Table 1

Mode	M (GeV)	B·σ (nb)
η3π [3]	2.021 ± 0.013	60 ± 15
η'3π	2.008 ± 0.020	20 ± 8
ηπ	2.047 ± 0.023	27 ± 7
φπ		< 4.0
φρ±	2.049 ± 0.015	33 ± 10
φ(ππ)±		< 15

This experiment has a broad x^* acceptance, and it is not likely that it would miss any production mode because of a lack of coverage. It is supposed, therefore, to measure the total cross-section. Adding up channels is a delicate operation owing to the scantiness of the numerical information available. Anyway, it is clear that a value of at least 1 μb will be obtained (Figure 5)

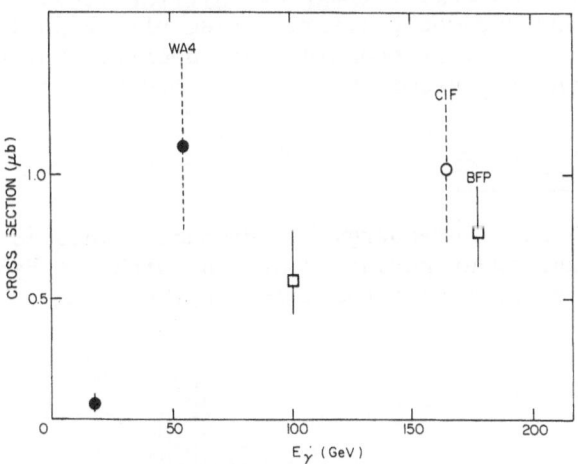

Figure 5

Virtual Photoproduction. This has been covered by Strovink[14] at this Conference. Both BFP[15,16] and EMC[17,18] measure, through dimuons (and trimuons for EMC), the "diffractive" charm photoproduction. They extract their cross-section in the framework of a γ-gluon fusion mechanism which seems to reproduce the data satisfactorily. Only the BFP Collaboration has performed the extrapolation to $Q^2 = 0$, but the two experiments are certainly in agreement

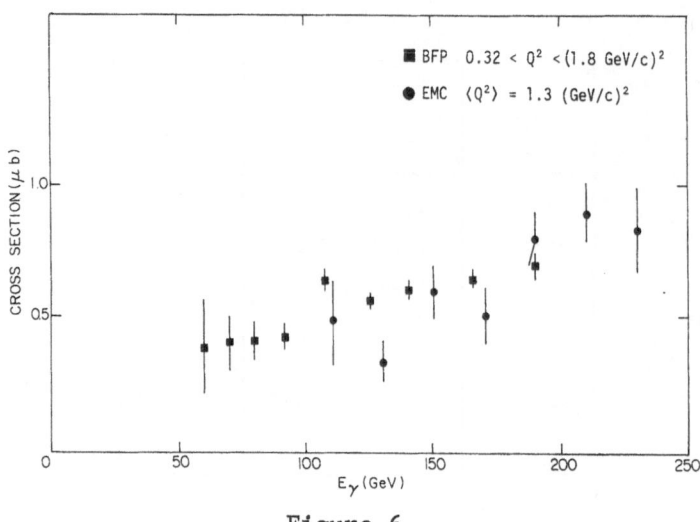

Figure 6

(Fig. 6). The BFP cross-sections are shown in Fig. 5. Again these two experiments would be insensitive to mechanisms such as central production. A qualitative argument given elsewhere[19] seems even to suggest that beyond this "diffractive" cross-section an extra amount of charm production is required. All available data are shown in Fig. 3 (channels) and Fig. 5 (σ_{tot} or $\sigma_{diffractive}$). These data are rather meagre, and one would clearly dream of a high-luminosity, wide-acceptance experiment spanning that domain. One piece of good news is that γ physics seems to have obtained high priority at the Tevatron.

The SLAC Experiment[20,21]

Figure 5 shows a low-energy measurement. This is new and is due to a nice experiment performed at SLAC with the Hybrid System (Fig. 7) fed by a back-scattered laser beam peaking at 19.5 GeV

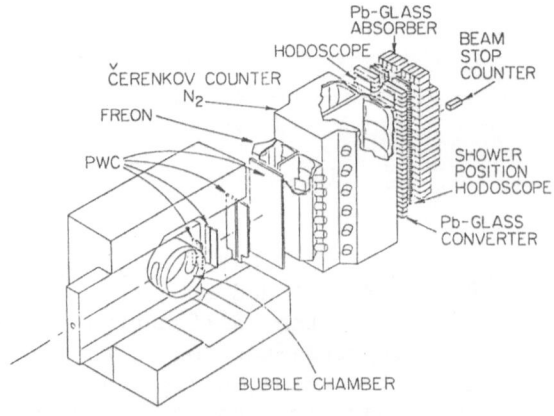

Figure 7

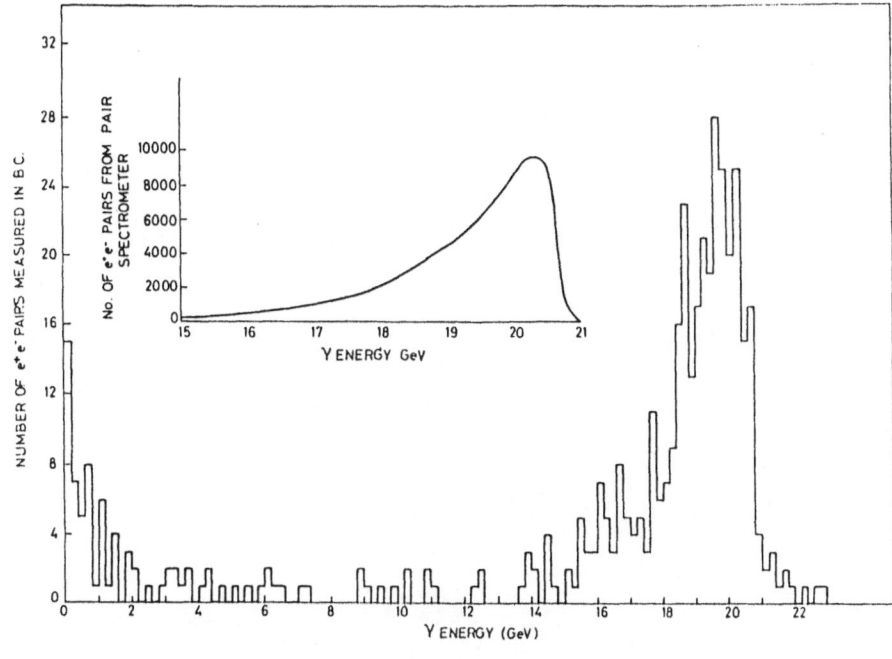

Figure 8

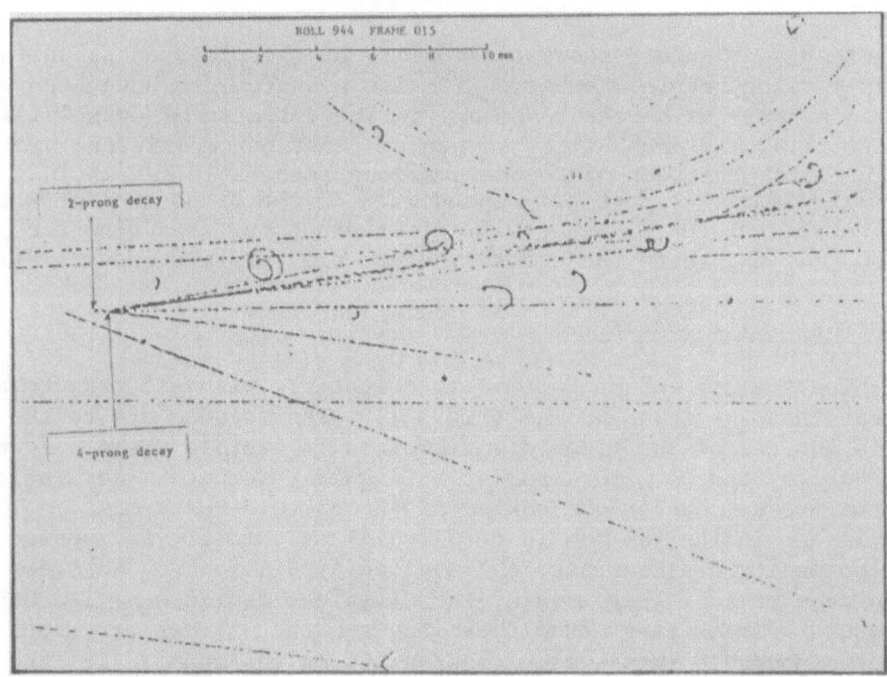

Figure 9

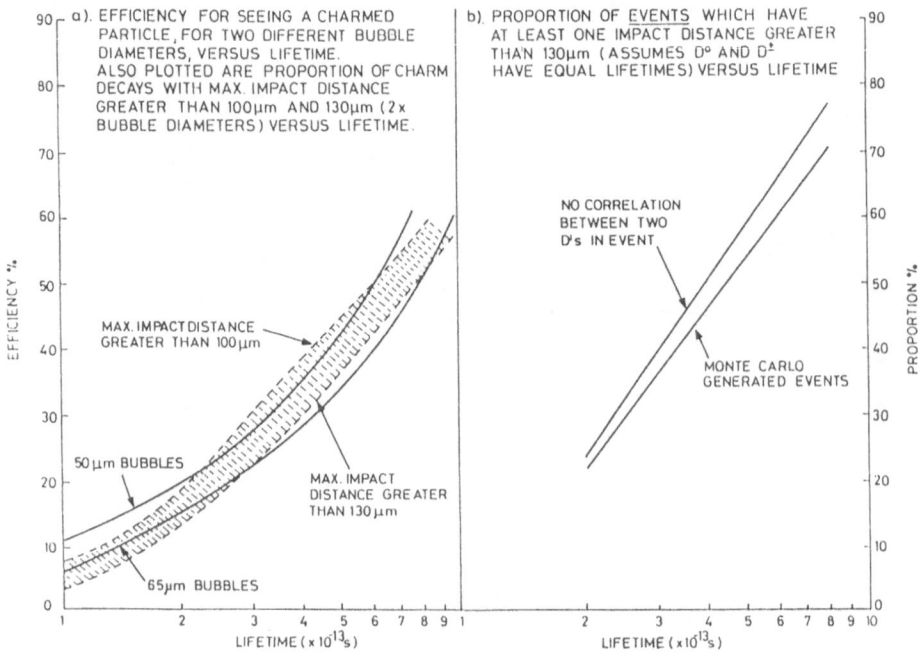

Figure 10

(Figure 8). Figure 9 shows the quality of the pictures taken in the
rapid-cycling bubble chamber. Lifetime measurements have been performed as well as a measure of σ_{tot}. The estimate of event visibility, i.e. the probability to see at least one multiprong decay
(Figure 10), has been made under various reasonable assumptions
for lifetimes. It does not depend much on the production mechanism,
and ranges between 0.25 and 0.5. This group can thus give a σ_{tot}
of $50 \; ^{+45}_{-20}$ nb, where the errors include all possible uncertainties.

Comparison with Models[22]

The γ-gluon fusion models are compatible with all measurements
except the σ_{tot} found by WA4 (Fig. 11). The freedom due to the possible choices of the gluon distributions is visible there. However
the "naive" choice with exponent 5 is quite adequate. Not indicated
is the freedom due to the choice of the charmed quark mass, m_c:
varying it inside the domain (1.5 ± 0.4) GeV, mostly frequented by
the community of theorists, changes the predictions by an order of
magnitude or so. There again, the classical choice $m_c = 1.5$ GeV is
adequate. It is clear from these figures that i) data are not accurate enough to impose many constraints on the model; ii) the WA4
point cannot be reproduced "naturally" by such a model. However,
one may ask whether it should. The γg fusion model, by construction,

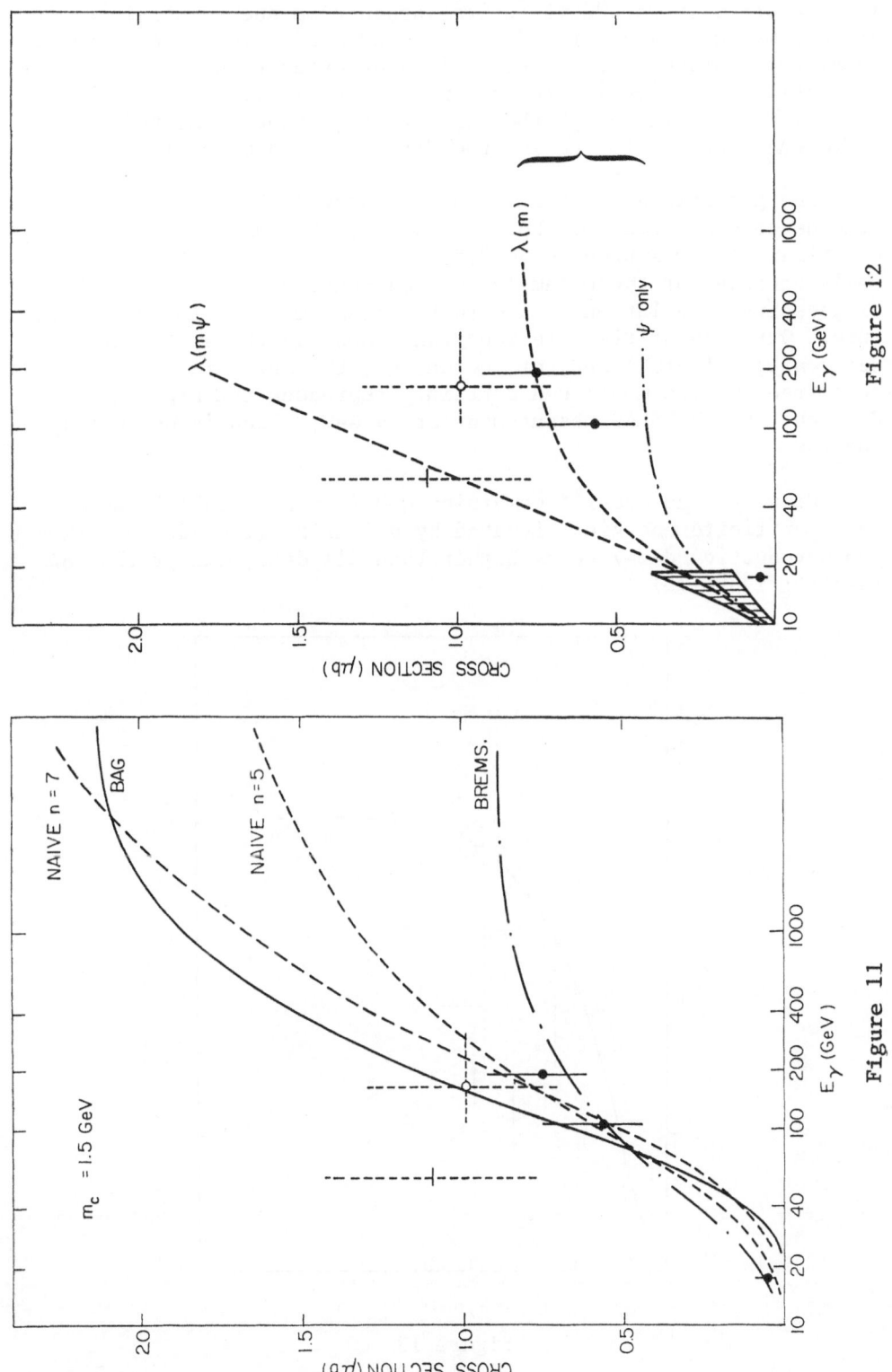

Figure 12

Figure 11

is relevant only for "diffractive" production and is unrelated to mechanisms such as central or associated production. Here one may recall what happens for ψ inelastic photoproduction: a substantial inelastic production is present, which may be related to higher-order QCD terms (although the absolute magnitude seems to be very different) but not to the lower-order γg fusion mechanism.

The generalized vector meson dominance[23] (Fig. 12) has also some degrees of freedom. For instance λ, the VDM correction term, experimentally measured to be 2.5 ± 0.3 at the ψ mass[24], is basically unknown for the ψ family and the continuum. Choosing $\lambda(m\psi)$, or guessing a variation of λ with the mass, makes a lot of difference. Other quantities are important, such as the exact variation with energy of $\sigma(\psi N)$ and, at low energy, the Re/Im ratio. With all that freedom, the model can certainly reproduce the high-energy data but not the SLAC measurement at 19 GeV, which is definitely too low.

Finally, the model[25] combining QCD (asymptotically) and a shape at finite energies dictated by unitarity (i.e. deduced from ψ photoproduction data) seems higher than all data, except the WA4 result (Figure 13)

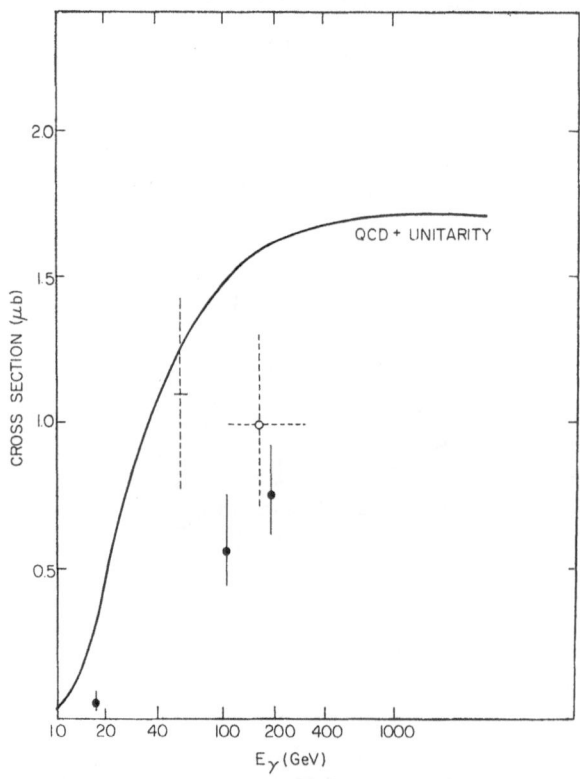

Figure 13

Production Mechanism

It seems that the type of asymmetric production observed by WA4 is very reminiscent of what happens in strangeness production. Probably mechanisms involving fusion processes and diquarks could explain it[26].

Experimentally, it would be interesting to check it against other data. Both the SLAC experiment, at lower energy, and the experiment WA58[27],[28] using emulsions (Figure 14) and the Ω' spectrometer, at the same energy, could do it. However, inspection of their reconstructed events shows that it is too early to reach any conclusion: limited statistics and frequent ambiguities make it impossible, for the moment, to confirm or disprove a substantial contribution of associated production or asymmetric behaviour of D's and $\bar{D}$'s.

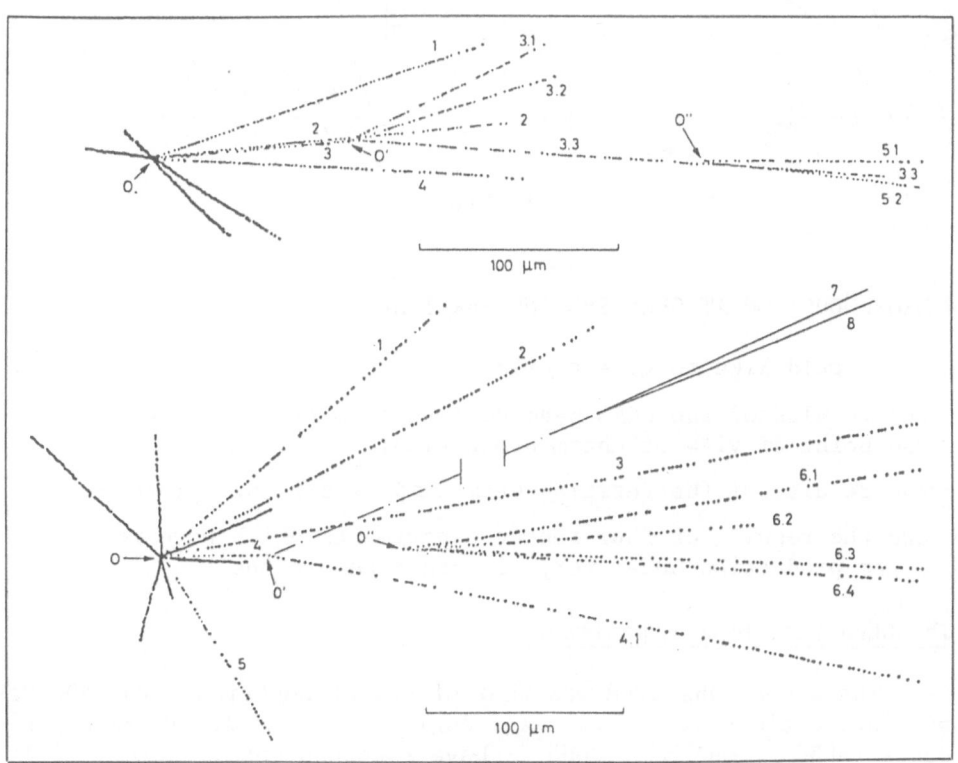

Figure 14

Y and Beauty Photoproduction

A new upper limit on Y production by muons has been given at Bonn conference[29]. It is shown in Fig. 15, where it has been re-expressed for real γ of 200 GeV. It can be seen that it falls right on the prediction of the already mentioned (QCD+unitarity) model. The QCD prediction is still much below.

For B production, the upper limit obtained from the three exotic trimuons of the EMC[30] is already below the (QCD+unitarity) model. However, large systematic errors could be present.

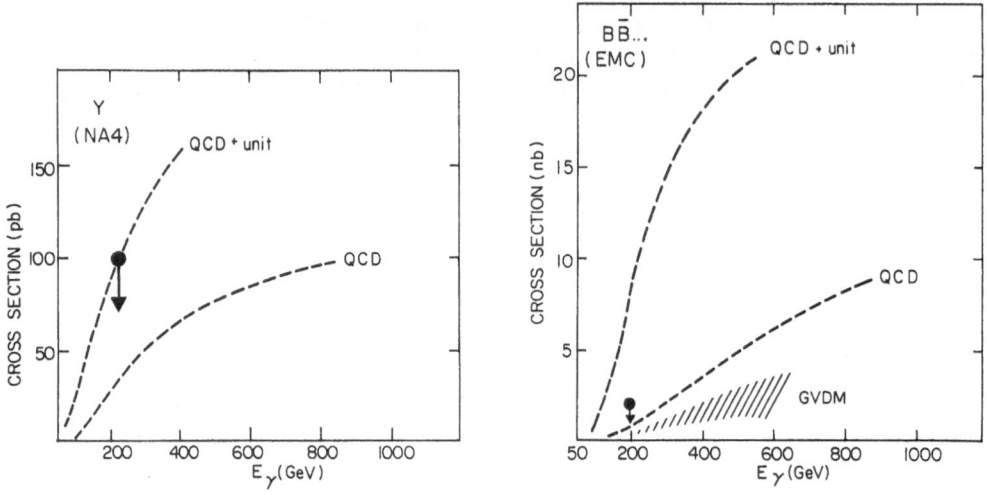

Figure 15

HADROPRODUCTION AT CERN SPS AND FERMILAB ENERGIES

I would like to give a brief description of:

- the results of the CERN beam dump experiments[31] (prompt ν) from the point of view of charm production;

- the results of the Fermilab beam dump experiment[32] (prompt μ);

- and the results of four recent experiments which have measured charm production under very different conditions.

The CERN Beam Dump Experiments

The signal observed was that of prompt neutrinos from 400 GeV pFe interactions at a very small angle, < 1.8 mrad. Three experiments (BEBC[33], CHARM[34], CDHS[35]) have recorded data. Here I will just mention what the data imply for charm hadroproduction and refer to previous reviews[31]:

i) If, as advocated by these experiments (or at least by two of them), the ν_e flux is smaller than the ν_μ flux, then charm cannot account for that effect and another explanation should be added.

ii) If, as advocated by CDHS, the ν_μ flux is larger than the $\bar{\nu}_\mu$ flux (the ratio quoted is 2.3, but the very existence of a prompt $\bar{\nu}_\mu$ signal is questioned by this experiment), then Λ_c or charmed baryon production could be a relevant explanation. If the production spectrum of the Λ_c is rather flat ($\sim$ 1-x, see later) and assuming that the totality of the prompt ν_μ signal is due to that source, it would correspond to $\sigma(\Lambda_c)B(\Lambda_c) \times$ $\times (\rightarrow \nu \dots) \leq 1$ µb. This implies either a small Λ_c production or a small semileptonic branching ratio. Furthermore, the associated charmed particle, say a $\bar{D}$, should not be produced in the same way. However, the CDHS data do not like a $(1 - |x_F|)$ distribution. With a $(1 - |x_F|)^3$ distribution, the limit on $\sigma(\Lambda_c)B(\Lambda_c)$ is $\sim$ 3.5 µb, which is much less stringent. (I am indebted to H. Wachsmuth and F. Dydak for discussions on these matters.)

iii) If the results are classically interpreted as being due to associated central $D\bar{D}$ production with a linear A dependence, one gets the cross-sections indicated in Table 2. Note that in the extraction, different assumptions on the ν_μ/ν_e ratio are made by the experiments.

Table 2

$$\sigma(pp \rightarrow D\bar{D}X) \text{ for } E\frac{d\sigma}{dp^3} \sim (1-x)^4 \, e^{-2p}T,$$

an 8% branching ratio and a linear A depence

CHARM	BEBC	CDHS
18 ± 6 µb	17 ± 4 µb	~ 10 µb
with $\nu_e/\bar{\nu}_e = 1$		with $\frac{\nu_\mu}{\bar{\nu}_\mu} = \frac{\nu_e}{\nu_e} = 2.3$

Fermilab Beam Dump Experiment[32]

This experiment was carried out by sending 350 GeV protons into a moun spectrometer (Figure 16). Using a variable density technique, this group obtained a signal for prompt µ's of either sign.

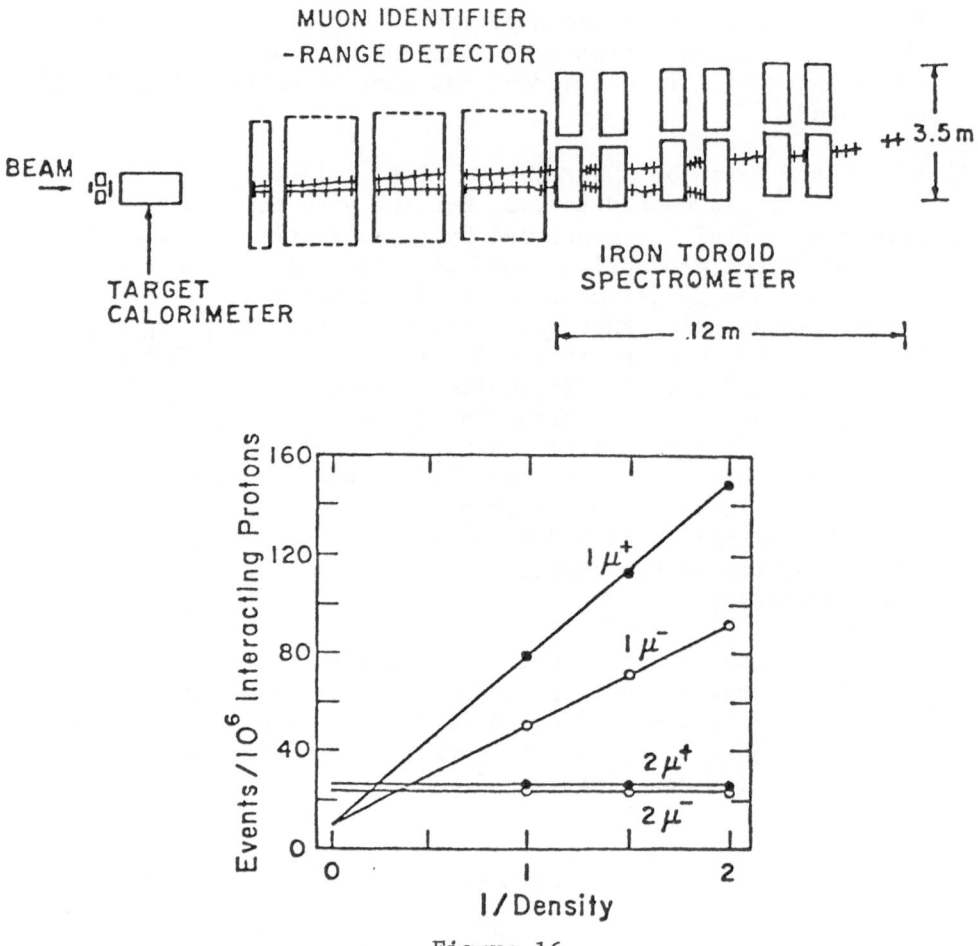

Figure 16

The angular and x domains covered are quite different from those of
the CERN beam dump experiments. Part of the Fermilab experiment
was done with a cut-off below 20 GeV for the muon; they have also
a preliminary measurement with a cut-off reduced to 8 GeV/c. From
the integral curve (Fig. 17a) giving the yield of prompt μ's above
a given momentum, they can extrapolate to the origin and get the
total μ flux. They find

$$\mu^+ = (12.2 \pm 3.8) \times 10^{-6}$$
$$\mu^- = (10.1 \pm 2.6) \times 10^{-6}$$

i.e. the equality of μ⁺ and μ⁻ yields. Therefore they do not support the particle/antiparticle asymmetry found by CDHS (note, however, the difference in the domains covered).

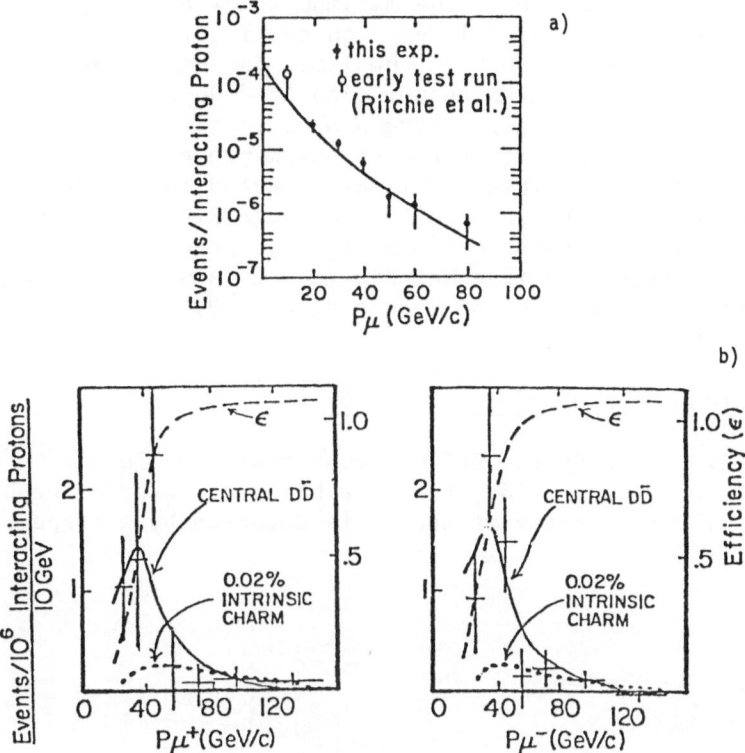

Figure 17

Both the integral curve (Fig. 17a) and the differential spectra (Fig. 17b) are well fitted by assuming central $D\bar{D}$ production,

$$E \frac{d^3\sigma}{dp^3} \sim (1-x)^{4\cdot7} \exp(-2.5 \ p_T) \ ,$$

with a total amount of 16 μb.

Also shown in Fig. 17b are curves indicating the maximum yield that can be tolerated from charm because of an intrinsic component in the projectile[36], i.e. symmetric large-x production of $\bar{D}\Lambda_c$. The

Λ_c would contribute to μ^+: the maximum tolerable $\sigma(\Lambda_c)B(\Lambda_c)$ is ~ 0.1 μb. Again, we can fall back on our ignorance of B. However, $\bar{D}$ would contribute to the μ^-; the limit is again $\sigma_{\bar{D}}B_{\bar{D}} \sim 0.1$ μb, and here B is known. Even assuming an $A^{2/3}$ dependence for the production, since this is not a hard process, this experiment shows that only 3 μb can be attributed to charm production owing to the intrinsic component. This model was invented to explain Λ_c forward production at the CERN Intersecting Storage Rings (ISR) with a cross-section ~ 100 hundred times bigger; it would therefore have to exhibit a fantastic rise between $\sqrt{s} \simeq 27$ and $\sqrt{s} = 63$ GeV. This is in contradiction with the modest rise (~ 3) it predicts between these two values.

Central Production of D^* [37] (Fig. 18)

This was done by the FPS Collaboration[38] at Fermilab, where 200 GeV π^- were sent on Be, the idea being to observe the $D^{\pm *} \to D^0 \pi^{\pm}$ cascade. The $K^{\pm}\pi^{\mp}$ decay of the D^0 is detected by a bispectrometer,

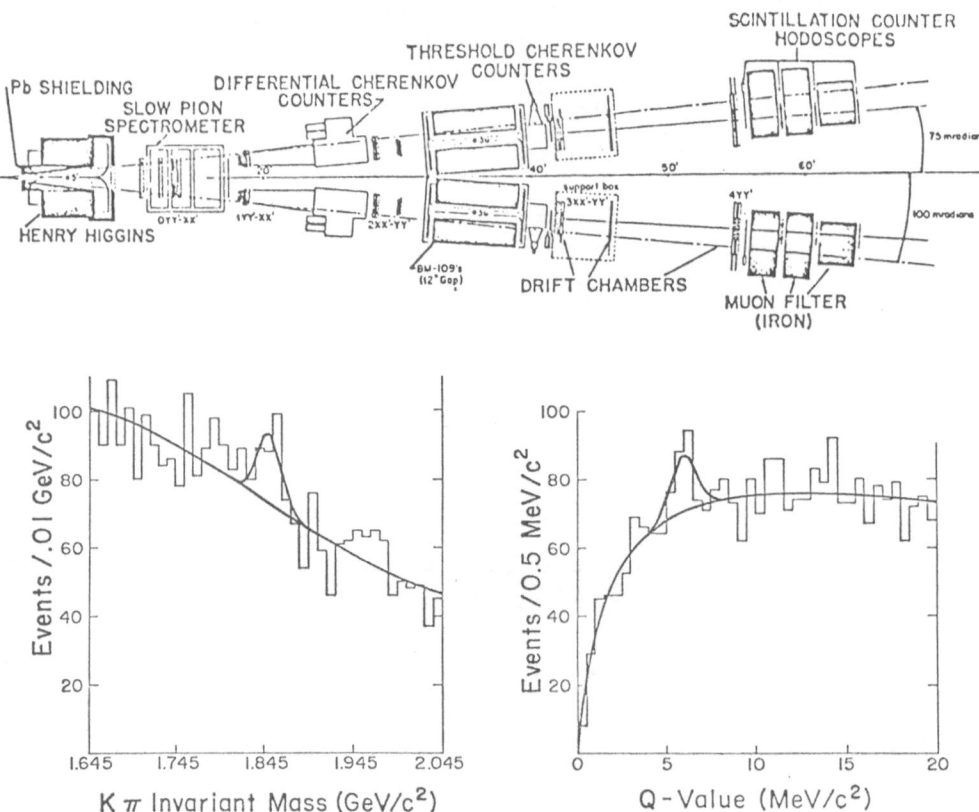

Figure 18

and the π from the cascade in a soft-pion spectrometer. Peaks are observed both in $m_{K\pi}$ and in the Q-value spectrum. A model-indepen-dent value for

$$\left. \frac{d\sigma}{dy} \right|_{y=0} = (1.6 \pm 0.5) \ \mu b$$

is found. If interpreted as central production,

$$E \frac{d^3\sigma}{dp^3} = A(1 - |x|)^3 \exp (-1.1 \ p_T^2) \ ,$$

this would correspond to a total cross-section for D^* production:

$$\sigma(D^*) = \left[\sigma(D^{*+}) + \sigma(D^{*-}) \right]/2 = (4.2 \pm 1.4) \ \mu b \ .$$

Diffractive Production of $D\bar{D}$ [39]

This experiment (using the Chicago Cyclotron set-up) (Fig. 19) has measured the production of $D\bar{D}$ systems produced in the diffrac-tive excitation of 217 GeV incident π^- on protons. The idea is to tag by a prompt μ, to select the mass range of the forward system through the recoiling proton (TOF, angle), and to observe the other

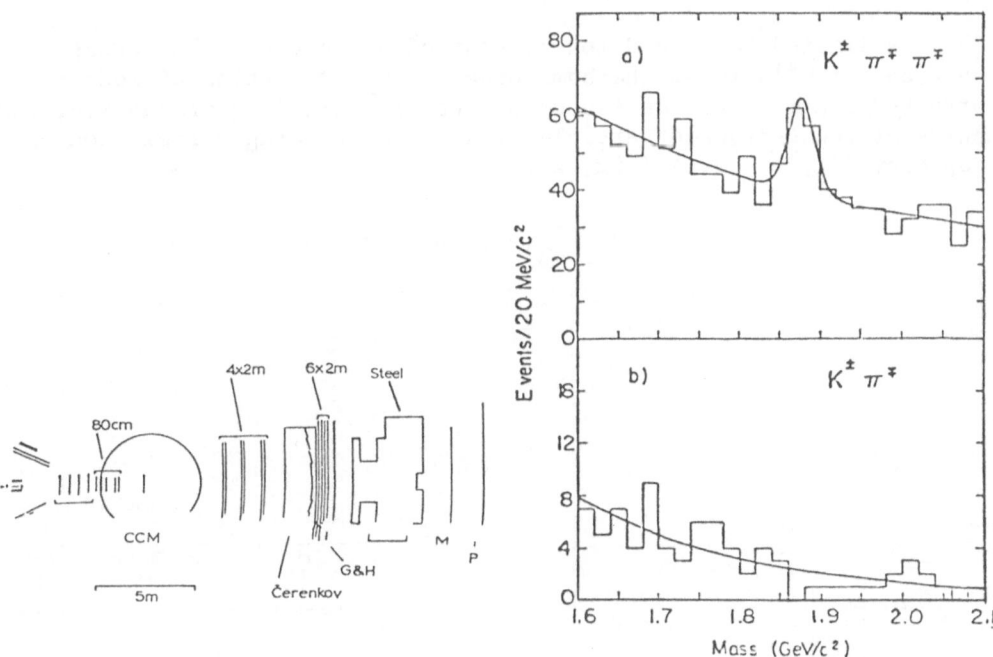

Figure 19

charmed particle in the exotic mode $K^{\mp}\pi^{\pm}\pi^{\pm}$. Because of the semileptonic branching ratios, this experiment selects mostly D^+D^- pairs.
The signals are quite clear and the yields are particle-antiparticle
symmetric. The x distribution favours the idea that the charm particles indeed come from the decay of a diffractively excited π.
Assuming such a diffractive mechanism and under various hypotheses,
the authors give a cross-section $\sigma(D^+D^-) = (7-10) \pm 4$ µb. This
should represent only part of the diffractive cross-section, since
channels such as $(D^0\bar{D}^0 \ldots)$ or $(D^{\pm}D^0)$ are missed.

Charm at the EHS[40]

Data are now available from the mini bubble chamber LEBC associated with the European Hybrid Spectrometer (EHS). Exposures to
π (360 GeV) and p (360 GeV) were performed. The trigger was an interaction one. Only a modest amount of information for track identification was available. The sensitivity corresponding to data
processed up to now is still modest: 5 evts/µb for π, $\sim$ 2 evts/µb
for p.

From their reconstructed events, the EHS group can, for a given
assumption on the lifetime, extract a cross-section for charged D
(through the three-prong decays which are supposed to represent 45%
of all decays):

$$\sigma(D^{\pm})\big|_{x_F>0} = 12.5 \pm 5 \text{ µb} .$$

This is limited to $x_F > 0$ for reasons of acceptance. The cross-
section for D^0's is of the same order. The properties of reconstructed charm particles favour a central $\left[(1-x)^{3.2}\right]$ production, and
pairs of reconstructed particles exhibit a striking correlation in
rapidity ($\langle\Delta y^*\rangle \sim 0.4$) (Fig. 20).

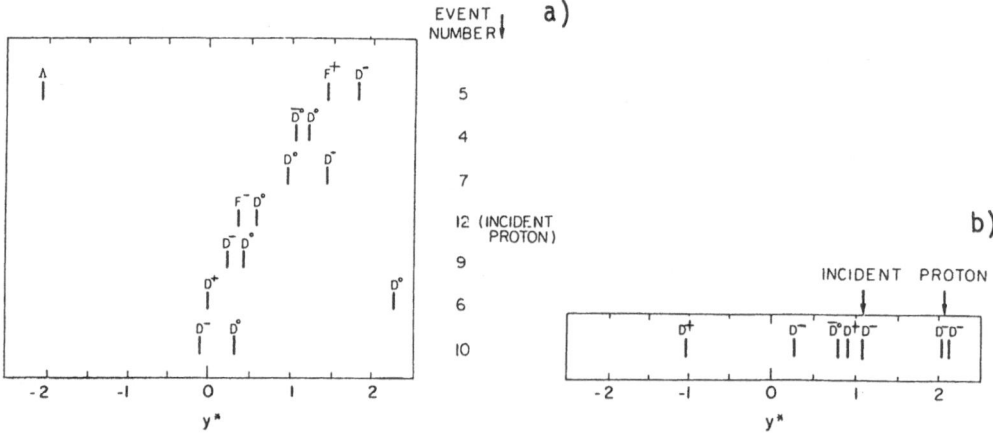

Figure 20

The results on lifetimes will be described elsewhere[41].

Note that the non-observation of Λ_c in the proton exposure is not meaningful, because of the low sensitivity they get from the only currently accessible modes, $pK\pi(n\pi^0)$.

D^* and Λ_c from the ABCCMR Collaboration[42-44]

The ABCCMR spectrometer (Fig. 21) was exposed to π^- (175-200 GeV) and p (150 GeV). The trigger was on an $e^{\pm}$. On-line (Čerenkov + total absorption counters + μ processors) and off-line filtering provided a spectacular signal/background improvement (170, finally) which, however, corresponded to a substantial reduction in the absolute yield of charm. Both a faint inclusive D signal and a clear $D^* \rightarrow D\pi$ cascade (Fig. 22) were observed with incident π's. The cross-section given under the assumption of linear A dependence and correlated production (through the decay of a 5 GeV object produced according to 1-x) is

$$\sigma(D\bar{D} + X) = 5.4 \pm 2 \ \mu b \ .$$

It would be 70% higher if no correlation and a (1-x) distribution for D^* were assumed.

In the 150 GeV proton exposure a clear Λ_c (Fig. 23) signal is seen, reinforced by some evidence for the $\Sigma_c^{++}(2440) \rightarrow \Lambda_c\pi$ cascade (Fig. 24). The cross-section, extracted assuming a rather flat (1-x) dependence for the Λ_c and a central one $\left[(1-x)^{4.5}\right]$ for the meson providing the tagging e, is pretty large

$$\sigma(\Lambda_c) = 75 \pm 50 \ \mu b \ .$$

The large error bars reflect the preliminary nature of these data. If confirmed, this is an important result. First the strong limit set by the beam dumps on $\sigma(\Lambda_c)B(\Lambda_c)$, under the assumption of a similar (1-x) distribution for the Λ_c, will force us to conclude that the $B(\Lambda_c)$ is very small (less than 1% if we were to take the numbers at their face value). Secondly, this would give a more satisfactory rise between SPS and ISR results: the Λ_c production would only rise by a factor of 3 or so, which looks reasonable in several models. However, ABCCMR should also prove that forward $\bar{D}$'s are *not* produced with substantial rates; otherwise a conflict with the beam dump results would arise. This has not yet been looked for.

The available data are plotted in Fig. 25. Except for the outstanding Λ_c, one can say that a slowly rising cross-section in the 10-20 μb region would accommodate all results.

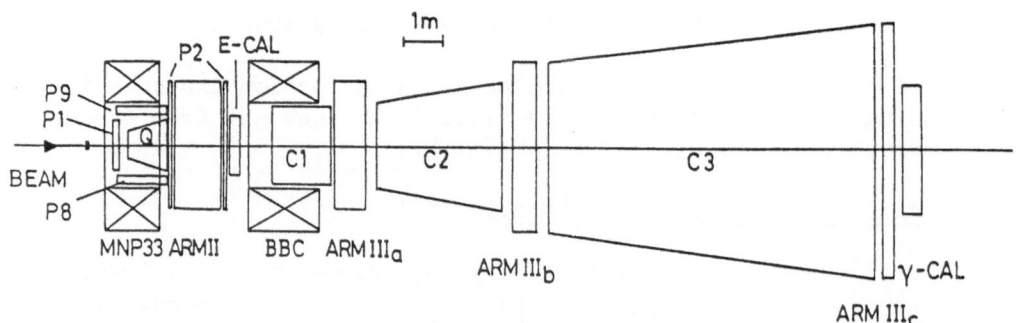

MNP33/BBC   SPECTROMETER MAGNETS

ARM II,III   DRIFT CHAMBERS (48 PLANES)

C1,C2,C3   MULTICELL THRESHOLD CERENKOV COUNTERS

P1,8,9,2   MULTIWIRE PROPORTIONAL CHAMBERS

Figure 21

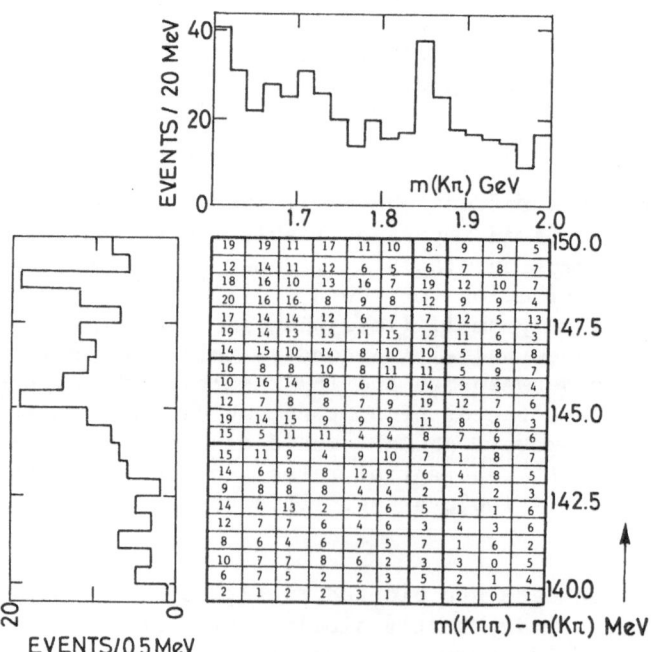

Figure 22

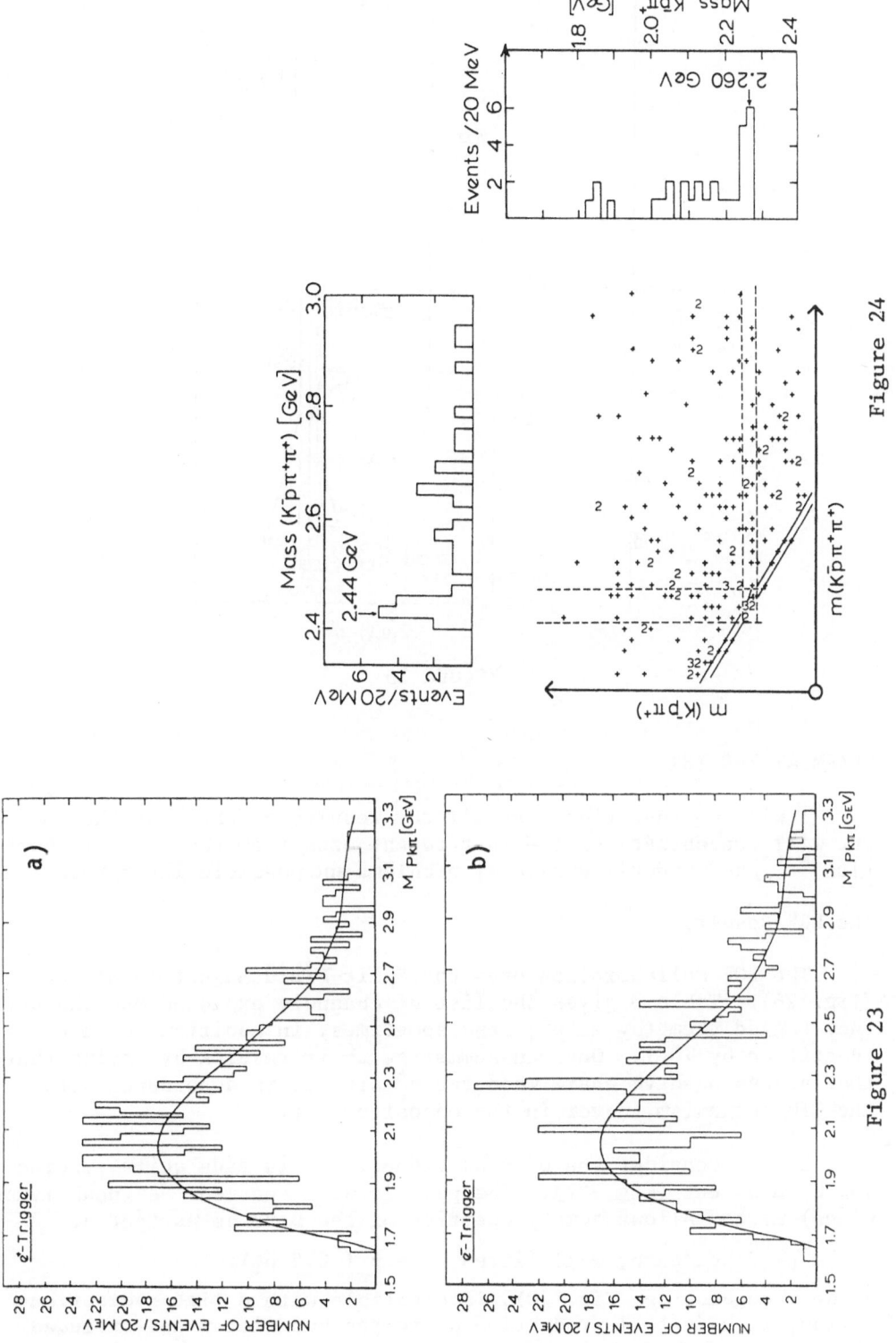

Figure 23

Figure 24

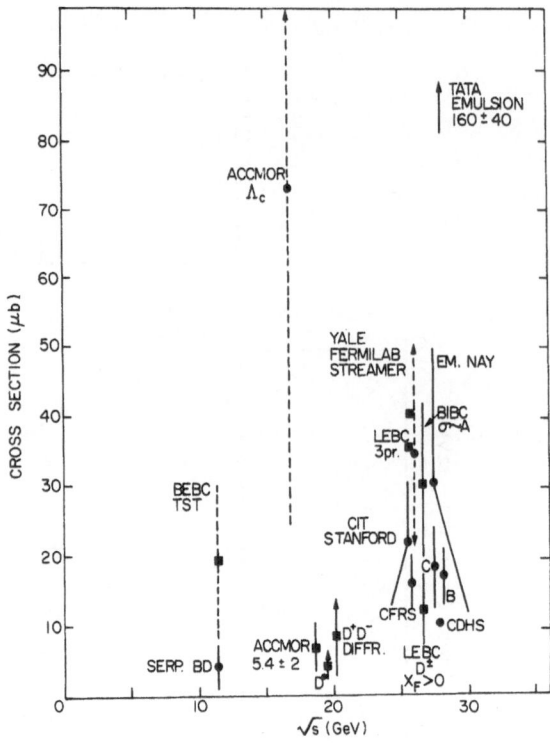

Figure 25

CHARM AT THE ISR

I will not describe here all the results on charm at the ISR[45], but will concentrate on the most recent ones from the BCF Collaboration[46,47], and underline several problems and possible interpretations.

The BCF Results

The BCF Collaboration used the Split-Field Magnet (SFM) set-up (Fig. 26). Table 3 gives the list of channels explored and the way they tagged them (by an $e^{\pm}$; and sometimes, in addition, by a K identified by TOF). One can summarize their results by saying that they always observe charm when the $e^{\pm}$ sign is in accordance with the GIM mechanism, never in the opposite case.

Let us consider the $D^0 \rightarrow K\pi$ signal. It is made quite convincing by a p_T cut (Fig. 27). The peak is well centred. A (peak minus wings) method allows some properties of the D to be extracted:

- the p_T dependence, well fitted by $\exp(-2.5\ p_T)$;

- the x dependence (Fig. 28), incompatible with a flat x distribution, with central production preferred but flat y not excluded.

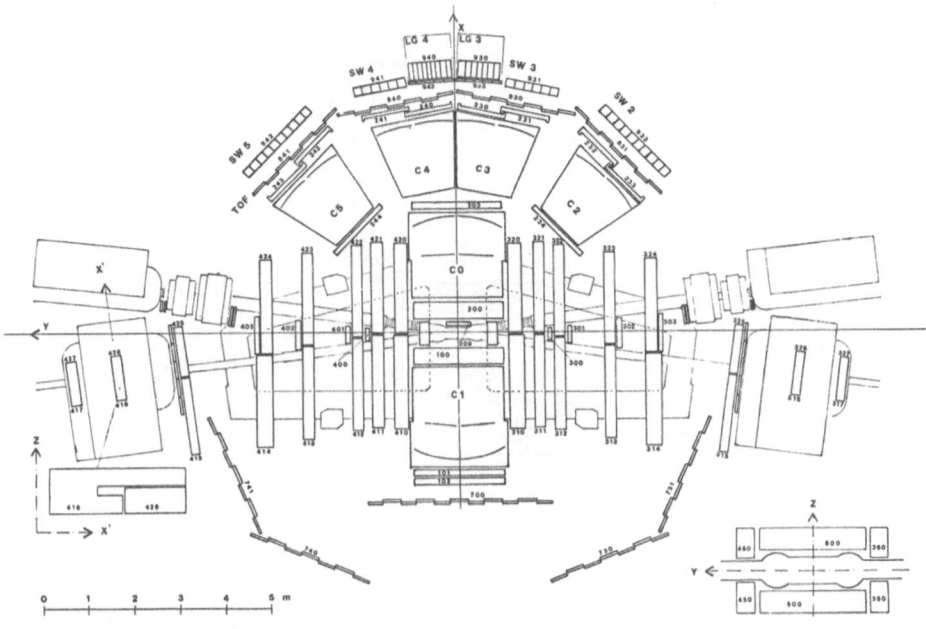

Figure 26

Table 3

Particle	Trigger	Signal
$\Lambda_c^+ \rightarrow pK^-\pi^+$	e^-, $x_L(p) \geq 0.3$	YES
	e^+, $x_L(p) \geq 0.3$	NO
$D^+ \rightarrow K_{TOF}^-\pi^+\pi^+$	e^-	YES
	e^+	NO
$D^0 \rightarrow K^-\pi^+$	e^-, K_{TOF}^+	YES
	e^+, K_{TOF}^+	NO
$D^+ \rightarrow K^-\pi^+\pi^+$	e^-, K_{TOF}^+	YES
	e^+, K_{TOF}^+	NO
$D^0 \rightarrow K_{TOF}^-\pi^+\pi^+\pi^-$	e^-	YES
	e^+	NO
$D^- \rightarrow K_{TOF}^+\pi^-\pi^-$	e^+	YES
	e^-	NO
$\overline{D^0} \rightarrow K_{TOF}^+\pi^-\pi^-\pi^+$	e^+	YES
	e^-	NO
$\Lambda_c^+ \rightarrow p_{TOF}K^-\pi^+$	e^-	YES
	e^+	NO
$\Lambda_c^- \rightarrow \bar{p}_{TOF}K^+\pi^-$	e^+	YES
	e^-	NO

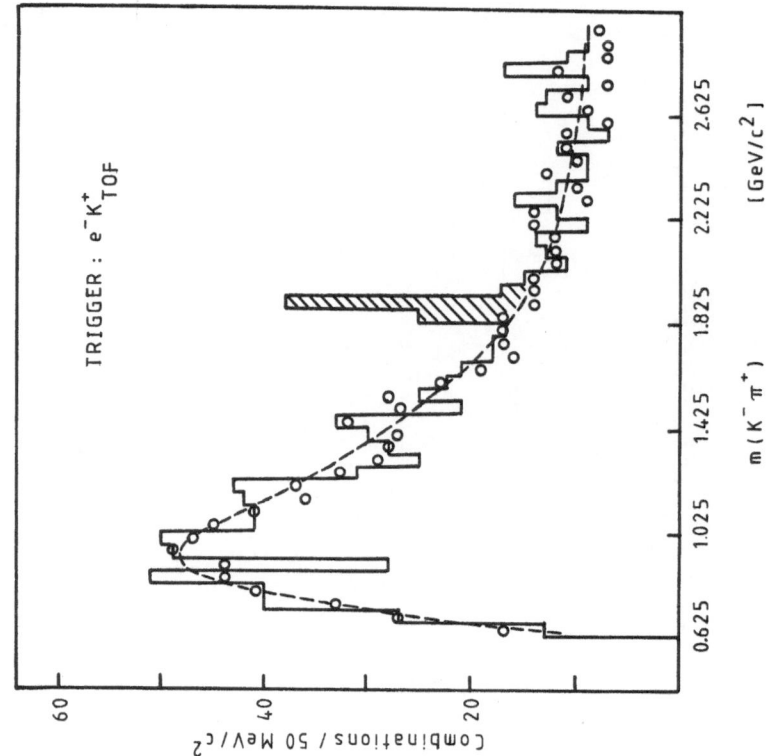

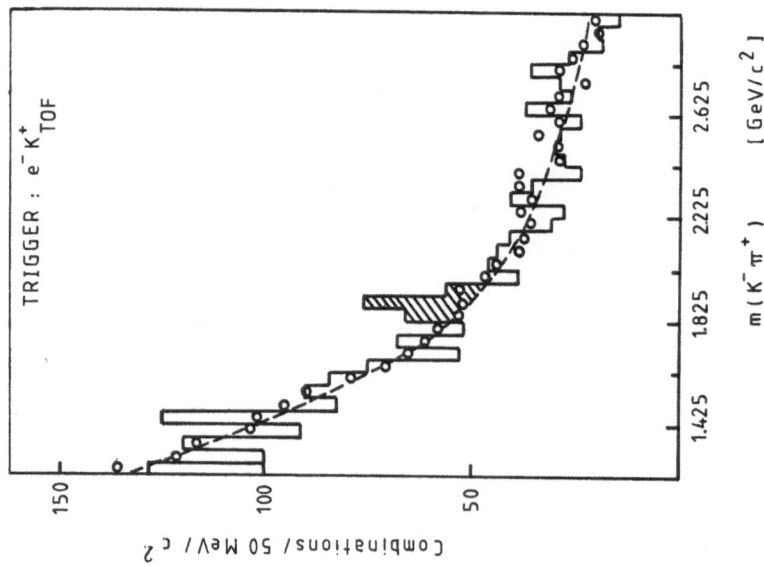

Figure 27

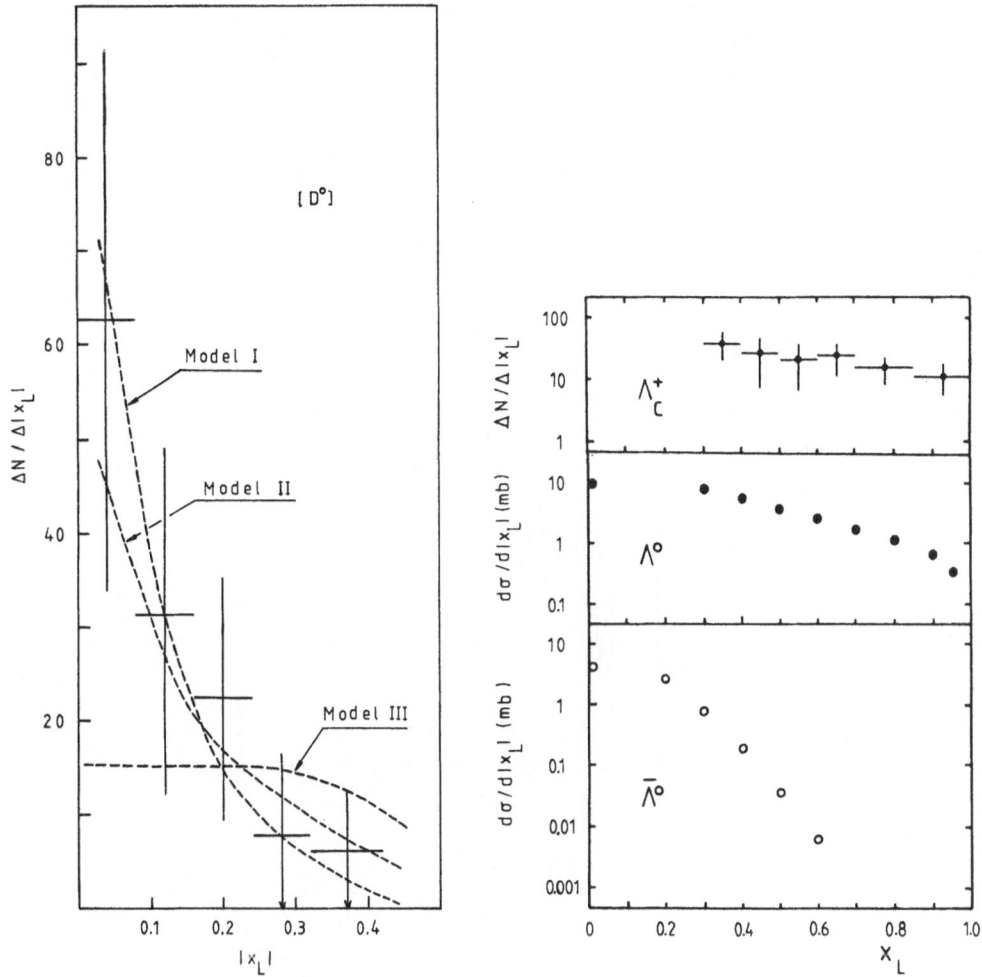

Figure 28

Figure 28 also shows the x distribution obtained in a similar way for the Λ_c: here a rather flat distribution is observed, similar to the one of the Λ, and totally different from the $\bar{\Lambda}$. This supports the general idea of the presence of leading baryons at the ISR, which the BCF group will exploit for beauty.

Figure 29 shows the results for charm production[48]. Depending on the model used to extract it, the cross-sections can vary by an order of magnitude. The lowest values -- which we are tempted to consider as the most reasonable ones -- are obtained assuming a flat x distribution for Λ_c and a central one for D's. We have seen that this is also what a direct determination favours.

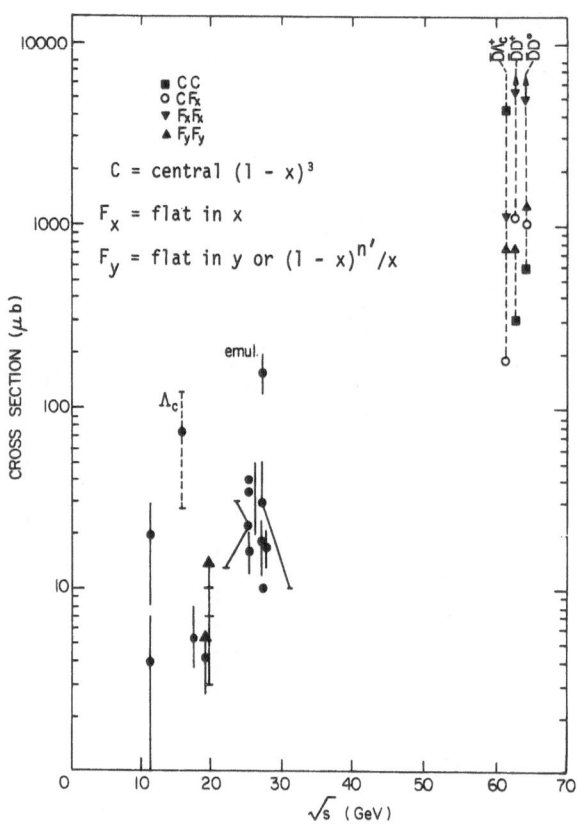

Figure 29

Some Problems

One can compare this set of results with other ISR results
(Fig. 30). First the "old" CSZ[49] measurement[50] of central charm
production through dilepton pairs. They extracted the cross-section
assuming a central process parametrized à la Bourquin-Gaillard[51],
and found values which are far below the BCF ones. It seems impossible, by a simple updating of branching ratios and other parameters,
to reconcile these two measurements. Considering now the cross-
section obtained by previous experiments at the SFM from their D^0
(e tagged)[52] and D^+ (K triggered)[53] signals, one could at first
sight conclude that there is agreement. However, the value from D^+
quoted in Fig. 30 has been obtained assuming a flat y distribution.
Since the D^+ observed is at quite a large x, if one considers it as
representing the tail of a central production, the resulting cross-
section explodes to several millibarns! These are two examples of
discrepancies existing within the various data. One could also mention a problem with Λ_c production: the LAS[54] measurement[55] seems

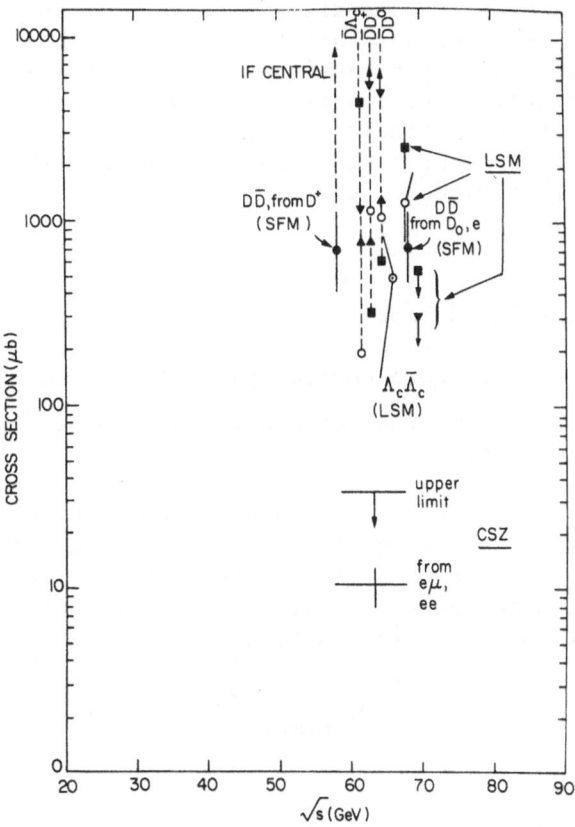

Figure 30

too high compared with others. In such discrepancies it is difficult to tell which is due to experimental distortions and which part
is due to misunderstanding, i.e. extraction of the cross-sections
under different hypotheses on production mechanism, branching ratios,
etc. Clearly a systematic and collective job of reprocessing, along
the lines previously indicated[56], would be welcome.

Another still more basic problem that such a study could help
to elucidate is the incompatibility between the e/π ratio measured
at the ISR[57] and the charm cross-section reported at present. Even
taking the lowest values, 1 mb is reached by adding up channels,
with a large fraction of charm centrally produced. However it looks
impossible, within a central production mechanism, to accommodate
more than 150-200 μb within the measured e/π (at 90°). A flat y
production would allow for more charm, as previously demonstrated
(using e/π 30° data)[58], but the charm cross-section would also be
found to be higher and the contradiction is still there. It amounts
to a factor of ∿ 5 or so.

Whatever the problems and disagreement among the data, it should not mask the important fact that charm is observed at the ISR with quite substantial cross-sections. It would simply be desirable to go from semiquantitative results to a more accurate situation. For the moment, it is difficult, for instance, to discard the possibility that ISR cross-sections, extracted from bump-hunting, could be generally overestimated.

Possible Interpretation

It is reasonable to look for:

i)   mechanisms leading to a central production, and

ii) processes explaining the forward production of baryons and eventually mesons.

In the first category, the mechanisms considered[59] (for instance gluon-gluon scattering) (see Fig. 31), generally turn out to be insufficient to explain the ISR cross-sections (remember, however, some preceding remarks on the experimental situation). Furthermore, at low p_T, their relevance is doubtful.

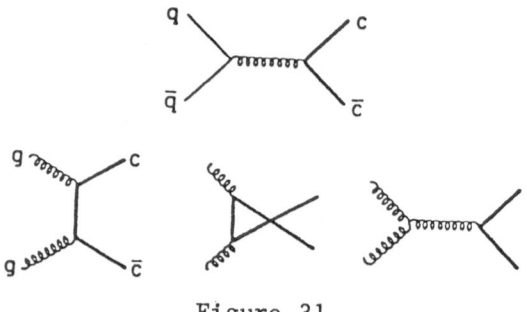

Figure 31

For the second type of production, let me simply list three attempts to explain it.

1) Diffraction excitation: This is a genuine prediction[60]. The mechanisms of Fig. 32 were considered, leading to a logarithmically

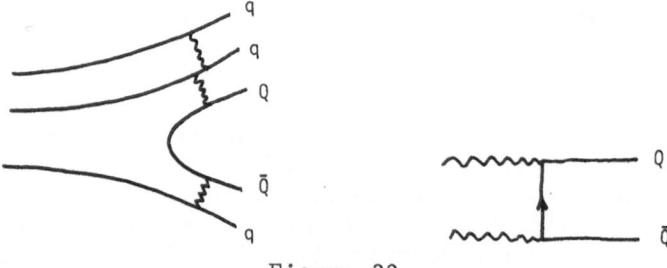

Figure 32

rising cross-section which could reach $\sim$ 150 μb at the ISR, under likely assumptions on the probability of the creation of a $c\bar{c}$ pair relative to other flavours. As with any diffractive mechanism, it predicts also $\bar{D}$ production forward and eventually D production as a competitor of the Λ_c. The $\bar{D}$ has not been observed, but one can use as an excuse the difficulty of isolating a pure sample of K^+ in a large background of protons. Another problem is that between the SPS and the ISR this model predicts only a rise of 2-3; this would lead us to expect Λ_c production at the SPS, at the level observed by ABCCMR. However, as already mentioned, the presence of forward D's, unavoidable within this model, would contradict beam-dump data.

2) Intrinsic charm[36]: This was introduced precisely to explain this forward production. We have seen in this meeting that at least two experimental results from the Fermilab beam dump on the rate of prompt μ, and from the EMC Collaboration on the charm structure function[37] (Fig. 33), do not seem to support it. This mechanism would lead to the same problems as those of the preceding one.

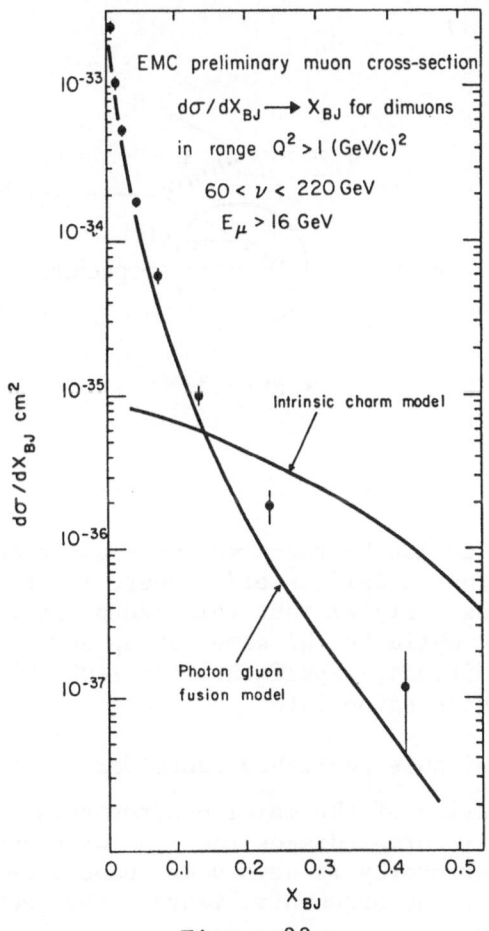

Figure 33

3) Extrinsic charm[61,62]: Some recent papers support the idea that the charm content needed is not intrinsic but extrinsic, i.e. generated by the QCD evolution itself. Resurrecting flavour excitation diagrams (Fig. 34) these authors feel they could explain the forward production of Λ_c, maintaining a more central production for the $\bar{D}$ since the flavour excitation and the subsequent rearrangement (the $\bar{c}$ expelled, the spectator c fusing with a diquark) have no reason to lead to a symmetric final state. This model raises two questions: first, it is not yet fully formulated, and the role and meaning of some quantities -- the t_{min} cut-off in Fig. 34, for instance -- has to be elucidated; secondly, since rearrangement is invoked, one may ask whether the intrinsic charm assumption could not be saved in the same way.

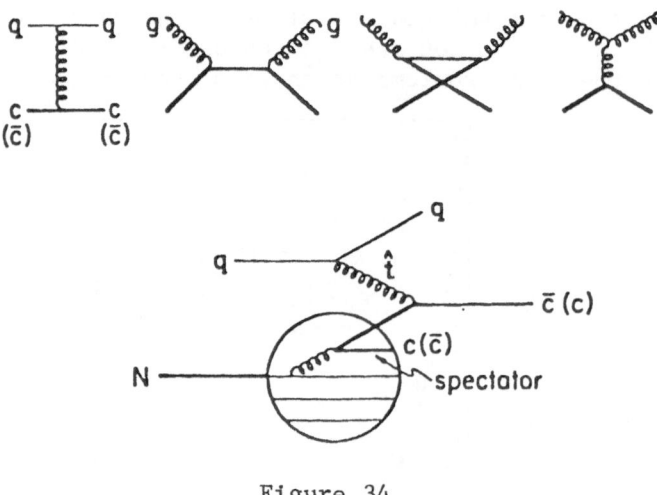

Figure 34

BEAUTY AT THE ISR

In the last few months there was much excitement about the following results: in the Split Field Magnet, experiment R415 (BCF Collaboration) sees a signal that this group interprets as a beautiful baryon Λ_b [63]; while in the same set-up and under similar but not identical conditions, experiment R416 (ACCDHW[64]) does not[65]. Let us split the discussion into:

- an exposé of available published facts for the two experiments;

- a brief presentation of the major controversial points as both teams, in still informal discussions and write-ups, analyse them. Solving this controversy is beyond my competence; I will therefore just present the arguments, leaving the debate in its contradictory state.

The Signal

The basic *a priori* idea of R415 is to exploit the leading baryon property[66] to look for beauty on a sample of events tagged by e^+. They have chosen the simplest relevant channel

$$pp \to e^+ + \Lambda_b^0 + X$$
$$\hookrightarrow pD^0\pi^-$$
$$\hookrightarrow K^-\pi^+$$

Selection of prompt e^+, supposed to come from the antibeauty cascade to charm, is performed on line (Čerenkov, total absorption counters) and off-line (dE/dx chamber and various refinements). Then a p_T cut (≥ 0.8 GeV) is performed -- mostly to get rid of a large fraction of e from charm -- as well as a cut on the momentum measurement accuracy. The final selected sample of e^+ contains $\sim$ 50% of real prompt e^+ ones and a negligible amount of contamination due to hadrons.

The leading baryon condition is imposed on a set of four particles compatible with Λ_b decay: a positive one, called a proton, required to have $x_F > 0.32$, and three others called K^-, π^+ and π^-, if the TOF information does not contradict this assessment. This set of particles has finally to satisfy the condition:

$$|y_{pK\pi\pi}| > 1.4 .$$

For the other particles (the "X" system), only charged multiplicity (≥ 4) and some topological requirement (something has to balance the e^+ p_T) are imposed.

From the events satisfying all these criteria ($\sim$ 1600), the BCF Collaboration extract the Λ_b signal by the procedure shown in Fig. 35, which basically consists in exhibiting first a D signal in the $K\pi$ spectrum and then retaining only the $K\pi$ mass domain corresponding to that D. An $\sim$ 6 s.d. signal is obtained. The number of events in the peak is $\sim$ 25 (30 combinations).

If interpreted as the Λ_b, it would correspond to a value of:

$$\sigma(\Lambda_b)B(\Lambda_b \to pD^0\pi) \sim 3\text{--}30 \ \mu b$$

depending on the production mode.

As proof of the reality of the signal, R415 refers to the difference in the e^+ p_T spectra on the peak and outside (Fig. 36).

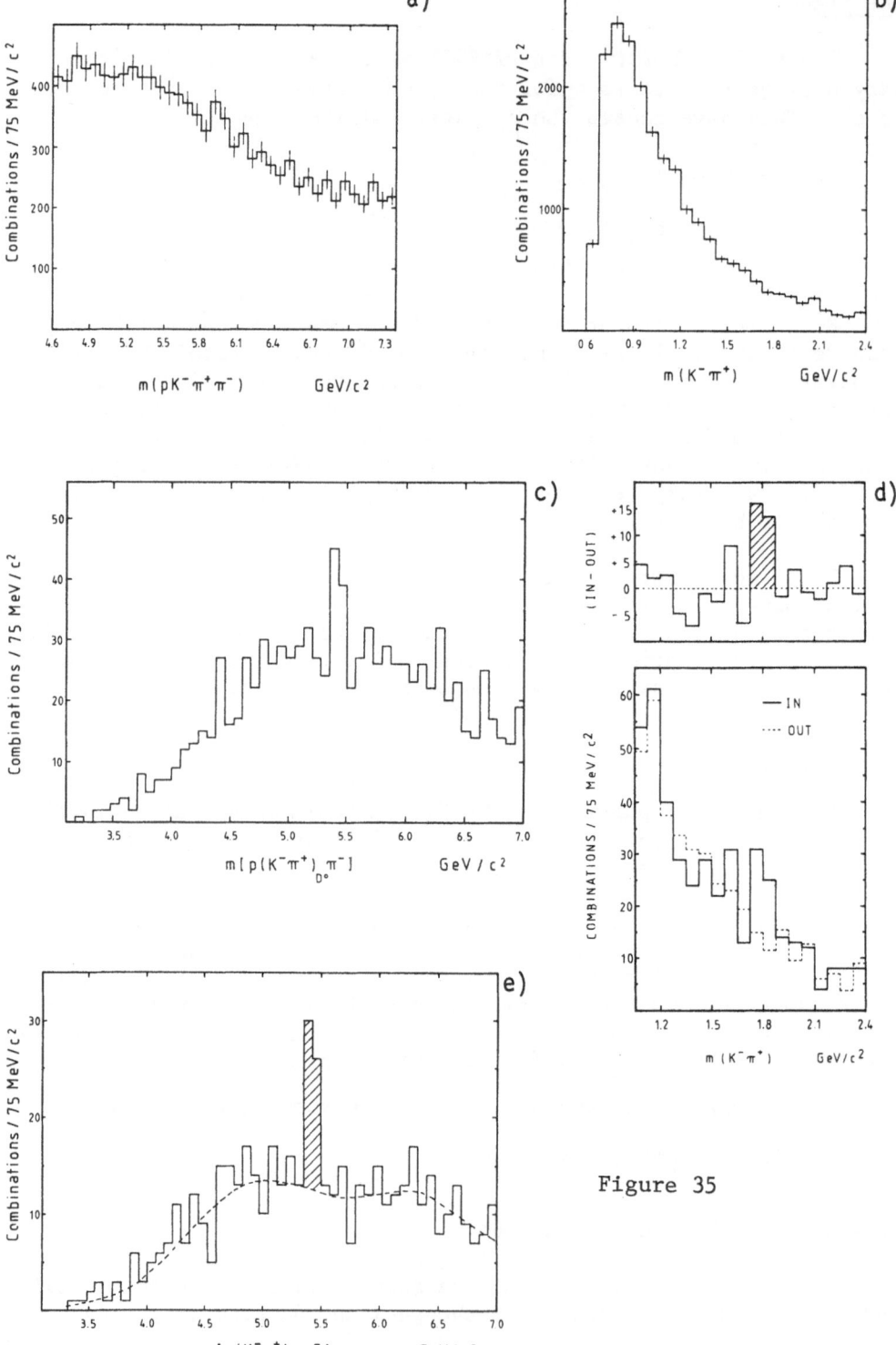

Figure 35

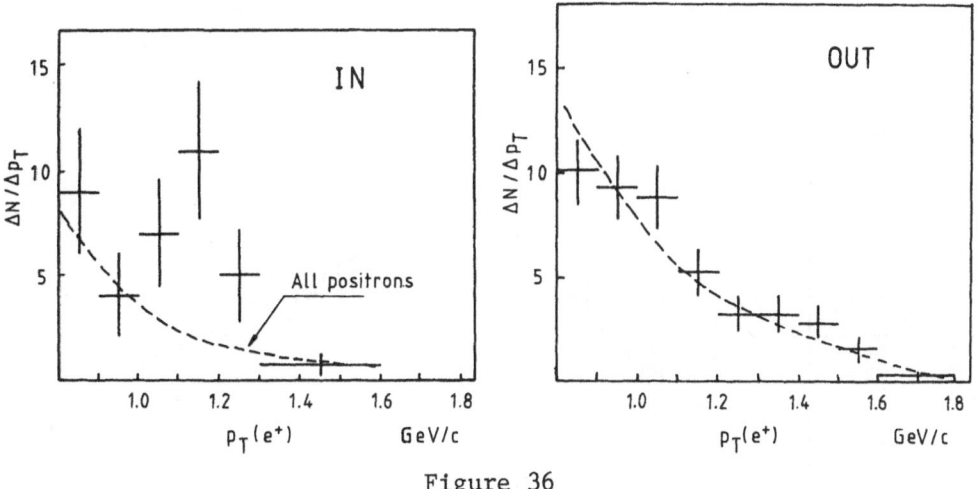

Figure 36

Remarks

Two remarks made at the very beginning are relevant here. The D, because of unknown mass calibration, is allowed to appear in a 300 MeV mass range. It is actually found 60 MeV below the nominal mass. The freedom taken does not appear in the expression of the statistical significance. Then about cuts; who can tell their exact effect on the significance when the region to retain in a mass spectrum is selected *a posteriori*, with a width and a location chosen simply to get the most favourable answer?

R416 Result[65]

With a similar integrated luminosity, this experiment has tried to check this result on previously recorded data. However, when the data were taken (with another motivation) the total absorption counters which play an important role in the selection of e^+ were not available. *A posteriori* measurements indicate that the fraction of prompt e^+ in the selected sample is $\sim$ 32%; 35% to 40% of the sample is simulated by hadrons. Experiment R416 imposes the same requirements on hadrons as does the previous experiment. The number of events retained turns out to be larger for R416, and this is attributed to a set-up that gives a better performance, and to a more efficient processing. Finally, when plotting the equivalent of the spectrum of Fig. 35, they obtain (Fig. 37), at the position of the hypothetical peak, a background level 10 times higher than the R415 one, and consider that this ratio also corresponds to the ratio expected for the magnitude of the signals. If the signal was true, they would therefore expect

$$\sim 25 \times 10 \times \frac{32}{50} = 160 \text{ events} ,$$

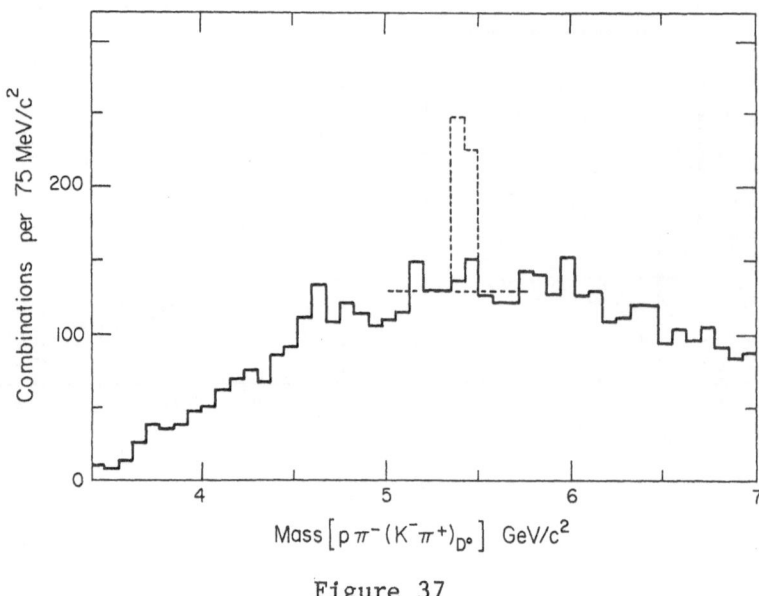

Figure 37

which they clearly do not observe (Fig. 37), and they conclude that
the two experiments are incompatible.

Controversial Points

We have seen up to now how each experiment presents its own
data. However, they also exchange several criticisms, which should
be mentioned here.

1) R416 about R415: the "numerology"

Experiment R416 points out that there is an internal inconsis-
tency in R415 results; namely, that given the R415 number of selected events N (1600), it is impossible for them to get such a signal
S (25 events). The N and S are related by

$$S = N \times \varepsilon \times \varepsilon_{SFM} \times B_D \times B_1 \times B_2 \times \rho \, ,$$

where

ε is the percentage of prompt e in the sample;
ε_{SFM} is the probability of reconstructing the K, π, π belonging
to the hadronic system;
B_D is the $(D_0 \to K\pi/D_0 \to all)$ branching ratio;
B_1 is the ratio: No. of Λ_b's/No. of beauty hadrons produced;
B_2 is the ratio: $\Lambda_b \to pD^0\pi^-/\Lambda_b \to p + \ldots$;
ρ is the ratio: No. of prompt e from B/No. of prompt e,

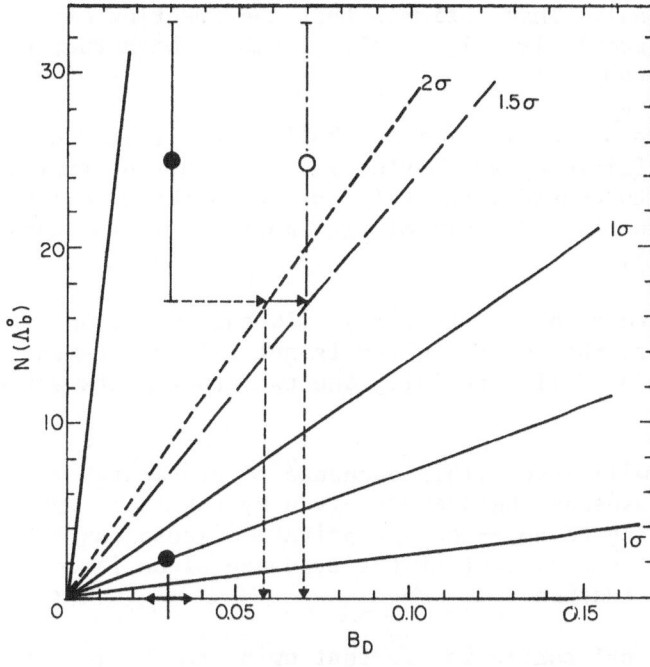

Figure 38

and ε, ε_{SFM}, B_D are known: 0.5, 0.5 and $(3 \pm 0.6)\%$, respectively
(ε_{SFM} for R415 < 0.5 according to R416). Even setting $B_1 = B_2 =$
$= \rho = 1$, which is unrealistic (although the B's which are here con-
ditional probabilities, once a fast proton has been selected, may
be quite different from their "free" values), one would expect only
S = 12 and there is already a problem. The experiment R415 answer
is explained in Fig. 38: provided B_D is substantially larger than
the nominal value, they could consider their result as being due to
a strong upward fluctuation. However, the substantial errors at-
tached to the various quantities are not sufficient to make this
fluctuation likely, since all these quantities have an upper bound
equal to one. It is clear that this problem of internal consistency
is not avoided easily. Experiment R416's own conclusion is that
neither of the two experiments, with their sensitivity, should in
fact observe the least signal: this can be demonstrated *a priori*
by assigning measured or guesstimated values to the quantities defined above.

2) <u>R415 about R416</u>

    There the controversy rests on two points:

a) The ratio of prompt leptons in the R416 sample: the R415 experiment, from its own analysis, finds that a selection made without
total absorption counters introduces much background due to hadrons

(and knock-ons). They estimate that the fraction of real prompt e^+ in the R416 sample is only $\sim$ 16%, a number which they get from two different methods.

b) The ratio of sensitivities: R415 agrees that the processing and the set-up efficiency are better in R416, but not that much better. They refuse to consider the ratio of the background levels in the final spectrum as a measure of the improvement, and admit only a factor of 4-5.

Therefore according to R415, R416 can expect only about one fourth of what they claim, which is not in disagreement with what they find. So, following R415, the two experiments are not in disagreement.

Here I will stop. This exchange of arguments, where one experiment reassesses the numbers given by the other one, is certainly the right thing to do -- but in private discussions between the two teams, until an agreement or the evidence of an irreducible disagreement is reached.

My personal conclusion is that owing to the low statistical significance of the positive result and the internal problem linked to its extraction (and also to the lower performance in selecting electrons of the experiment which does not observe the signal), the only reasonable attitude is to ask for more data. New global, precise experiments are needed, which at the same time could attempt to solve the contradictions noted above about charm production (e/π first!). I know well that this is easier to say than to do, and I apologize to the people who will have to resurrect the instrumentation needed. But this is unfortunately one of the unavoidable shackles of our discipline.

ACKNOWLEDGEMENTS

I benefited from many fruitful discussions with A. Kernan, M. Jacob, U. Amaldi, P. Roudeau, F. Richard, C. Quinn, M. Block and R.J.N. Phillips. I enjoyed the open and informative discussions I had with several members of the two SFM groups, and I appreciated their deep objectivity which always prevailed over their understandable passion. Finally, I would like to thank those I questioned so much without their being aware of it, namely the authors of excellent review talks given in the recent past, especially S. Wojcicki[67].

REFERENCES AND FOOTNOTES

1. G. Bellini et al., Nucl. Instrum. Methods 107, 85 (1973).
    G. Bellini, Talk given at the Meeting on Miniaturization of
    HEP Detectors, Pisa, 1980.
    See also NA1 Status Report, CERN/SPSC/79-112 (1979).
2. J. B. A. England et al., A silicon strip detector with 12 μm
    resolution, preprint CERN-EP/81-24 (1981), and Nucl. Instrum.
    Methods 185, 43 (1981).
3. See A. Silverman, Proc. of the Int. Symposium on Lepton and
    Photon Interactions at High Energies, Bonn, 24-29 August
    1981 (Physikalisches Institut, University of Bonn, Fed.
    Rep. of Germany, 1981), p. 138.
4. W. Bartel et al., Z. Phys. C 6, 295 (1980).
5. See J. Prentice, Talk given at the EPS Int. Conf. on High-
    Energy Physics, Lisbon, 1981.
6. G. S. Abrams et al., Phys. Rev. Lett. 44, 10 (1980).
    J. M. Weiss, Stanford preprint SLAC-PUB 2558 (1980).
7. R. Brandelik et al., Phys. Lett. 70B, 132 (1977).
8. R. Partridge et al., Stanford preprint SLAC-PUB 2731 (1981).
9. Columbia-Illinois-Fermilab Collaboration.
10. M. S. Atiya et al., Phys. Rev. Lett. 43, 414 (1979).
    P. Avery et al., Phys. Rev. Lett. 44, 1309 (1980).
    J. J. Russell et al., Phys. Rev. Lett. 46, 799 (1981).
11. Bonn-CERN-Glasgow-Lancaster-Manchester-Orsay(LAL)-Palaiseau
    (Ec. Poly.)-Paris(LPNHE)-Rutherford-Sheffield Collaboration.
12. See F. Richard, Talk given at the Int. Symposium on Lepton and
    Photon Interactions at High Energies, Batavia, 1979, eds-
    T. B. W. Kirk and H. D. I. Abarbanel (Fermilab, Batavia,
    1980), p. 469.
    B. d'Almagne, Proc. Int. Conf. on High-Energy Physics, Madison,
    1980, eds. L. Durand and L. G. Pondrom (Amer. Inst. Phys.,
    New York, 1981), p. 221 (also Orsay preprint LAL 80/31,
    1980).
    See also P. Roudeau, Orsay thesis and report LAL 80/14 (1980).
13. D. Aston et al., Phys. Lett. 100B, 91 (1981) and preprint
    CERN-EP/81-47 (1981).
14. M. Strovink, Proc. of the Int. Symposium on Lepton and Photon
    Interactions at High Energies, Bonn, 24-29 August 1981, p. 594.
15. Berkeley-Fermilab-Princeton Collaboration.
16. A. R. Clark et al., Phys. Rev. Lett. 45, 682 (1980).
17. European Muon Collaboration.
18. See G. Coignet, Talk given at the EPS Conf., Lisbon, 1981.
    J. J. Aubert et al., preprints CERN-EP/80-61 and 62 (1980).
19. M. Strovink, Invited lecture at the SLAC Summer Institute on
    Particle Physics, July-Aug. 1980 (also Berkeley preprint
    LBL 11844, 1980).
20. Birmingham-Brown Univ.-Duke Univ.-Florida State-Imperial
    College-KEK-MIT-Nara Womens Univ.-ORNL-Rutherford-Technion-
    Tohoku-Tufts-Univ. of California-Tel Aviv-Tennessee-Weizmann
    Collaboration.

21.   SLAC Hybrid Facility Photon Collab., Contributed papers
      No. 156 and A68 to the Int. Symposium on Lepton and
      Photon Interactions at High Energies, Bonn, 24-29 August
      1981.
22.   P. Roudeau, see Ref. 12.
23.   J. J. Sakurai and D. Schildknecht, Phys. Lett. B40, 121;
      B41, 489; B42, 216 (1972).
24.   U. Camerini et al., Phys. Rev. Lett. 35, 483 (1975).
25.   H. Fritzsch and K. H. Streng, Phys. Lett. 72B, 385 (1978).
26.   A. Donnachie, Z. Phys. C 4, 161 (1980).
      See also M. Fontannaz et al., Orsay preprint LPTHE 81-16 (1981).
27.   Bologna-CERN-Florence-Geneva-Moscow(LPI)-Paris-Santander-
      Valencia Collaboration.
28.   M. I. Adamovich et al., Phys. Lett. 99B, 271 (1981).
      See also Ref. 5.
29.   D. Bollini et al., Paper contributed to the Int. Symposium on
      Lepton and Photon Interactions at High Energies, Bonn,
      24-29 August 1981.
30.   C. H. Best, Paper presented at the XVIth Rencontre de Moriond,
      Les Arcs, 1981.
31.   F. Dydak, preprint CERN-EP/80-204 (1980), and Discussion
      Meetings between Experimentalists and Theorists on ISR and
      Collider Physics, Series 2, No. 4 (1981).
32.   A. Bodek et al., Rochester preprint UR783 COO-3065-299, also
      presented by J. L. Ritchie at the XVIth Rencontre de
      Moriond, Les Arcs, 1981, and by A. Bodek in a parallel
      session of the EPS Conf., Lisbon, 1981.
33.   Big European Bubble Chamber.
34.   CERN-Hamburg-Amsterdam-Rome-Moscow Collaboration.
35.   CERN-Dortmund-Heidelberg-Saclay Collaboration.
36.   S. J. Brodsky et al., Stanford preprint SLAC-PUB 2660 (1981).
37.   V. L. Fitch et al., Princeton preprint COO-3072-124 (1981).
38.   Fermilab-Princeton-Saclay Collaboration.
39.   L. J. Koester et al., Paper No. 0618 contributed to the Madison
      Conf., 1980.
40.   LEBC-EHS Collab., Paper No. 64 contributed to the EPS Conf.,
      Lisbon, 1981.
41.   See L. Foa, Proc. of the Int. Symposium on Lepton and Photon
      Interactions at High Energies, Bonn, 24-29 August 1981,
      p. 775.
42.   Amsterdam(NIKHEF)-Bristol-CERN-Cracow-Munich(MPI)-Rutherford
      Lab. Collaboration.
43.   C. Daum et al. (ABCCMR Collab.), Observation of inclusive D*
      and D^0 production in 175 and 200 GeV π-Be interactions,
      Paper contributed to the EPS Conf., Lisbon, 1981.
44.   C. Daum et al. (ABCCMR Collab.), Observation of charmed baryon
      production in 150 GeV pBe interactions, Paper contributed
      to the EPS Conf., Lisbon, 1981.

45. See F. Muller, Lecture given at the Summer Institute on Quarks
         and Leptons, Cargèse, 1979 (Plenum Press, New York, 1980),
         p. 653.
         See also A. Kernan, in Discussion Meetings between Experimen-
         talists and Theorists on ISR and Collider Physics, Series 2,
         No. 4 (1981).
46. Bologna-CERN-Frascati Collaboration.
47. M. Basile et al., CERN EP Internal Report 81-02 (1981).
48. See W. M. Geist, preprint CERN-EP/79-129 (1979).
49. CERN-Saclay-Zurich Collaboration.
50. A. Chilingarov et al., Phys. Lett. $\underline{83B}$, 136 (1979).
51. M. Bourquin and J. M. Gaillard, Nucl. Phys. $\underline{B114}$, 334 (1976).
52. D. Drijard et al., Phys. Lett. $\underline{85B}$, 452 (1979).
53. D. Drijard et al., Phys. Lett. $\underline{81B}$, 250 (1979).
54. Los Angeles-Saclay Collaboration.
55. W. Lockman et al., Phys. Lett. $\underline{85B}$, 443 (1979) and Los Angeles
         preprint UCLA 1109 (1977).
56. W. Geist, preprint CERN-EP/79-115 (1979).
57. For a review, see L. M. Lederman, Proc. Int. Symposium on
         Lepton and Photon Interactions at High Energies, Stanford,
         1975 (University, Stanford, 1976), p. 265, and J. W. Cronin,
         Lectures given at the Ettore Majorana School of Subnuclear
         Physics, Erice, 1975 (Plenum Press, New York, 1977), p. 929.
58. M. Block, private communication.
         See also I. Hinchliffe and C. H. Llewellyn Smith, Phys. Lett.
         $\underline{61B}$, 472 (1976).
59. See the review of R. J. N. Phillips, in Proc. Int. Conf. on
         High-Energy Physics, Madison, 1980, eds. L. Durand and
         L. G. Pondrom (Amer. Inst. Phys., New York, 1981), p. 1470.
60. G. Gustafson and C. Peterson, Phys. Lett. $\underline{67B}$, 81 (1977).
61. V. Barger et al., Univ. Wisconsin preprint DOE-ER/00881-215
         (1981).
62. R. Odorico, Univ. Bologna preprint IFUB 81/13 (1981).
63. M. Basile et al., preprint CERN-EP/81-38 (1981).
64. Annecy-CERN-Collège de France-Dortmund-Heidelberg-Warsaw
         Collaboration.
65. D. Drijard et al., Contributed papers No. 63 to the EPS Conf.,
         Lisbon, 1981, and No. 144 to the Int. Symposium on Lepton
         and Photon Interactions at High Energies, Bonn, 24-29 August
         1981.
66. M. Basile et al., preprint CERN-EP/81-22 (1981).
67. S. Wojcicki, Proc. Int. Conf. on High-Energy Physics, Madison,
         1980, eds. L. Durand and L. G. Pondrom (Amer. Inst. Phys.,
         New York, 1981), p. 1430.

HEAVY QUARKS IN HADRONIC COLLISIONS

Stanley J. Brodsky*

Institute for Advanced Study, Princeton, New Jersey 08540
and
Stanford Linear Accelerator Center
Stanford University, Stanford, California 94305
and
Carsten Peterson
Stanford Linear Accelerator Center
Stanford University, Stanford, California 94305
and NORDITA, Copenhagen, Denmark

INTRODUCTION

Charmed hadron production as observed at the ISR has several remarkable features:

(1) The total cross section for open charm production in pp-collisions at $\sqrt{s}$ = 63 GeV is at the 1 mb level.[1]

(2) Charmed hadrons are abundantly produced in the forward region of phase space. The pp $\rightarrow \Lambda_c X$ distribution is roughly flat in x_L;[2]

$$\frac{d\sigma}{dx_L} (pp \rightarrow \Lambda_c X) \sim (1 - x_L)^{0.4} \quad . \tag{1}$$

Also D^0 and D^+ (which carry no valence quarks in common with the proton) are produced with a flat rapidity distribution; the pp $\rightarrow D^0 X$ x_L-distribution[3] is consistent with $\sim (1 - x_L)^3$. The corresponding strange hadron cross section[4] $d\sigma/dx_L (pp \rightarrow K^- X)$ falls much steeper, $(1 - x_L)^{6 \pm 1}$

---

*Presented by S. Brodsky

(3) The Λ_c can be produced with a diffractive trigger, pp →
pΛ_cX with a cross section of the order of 240 ± 120 μb.[5]

In contrast, a standard model for charm production based on
hard scattering, gluon fusion,[6] predicts smaller cross sections
and much more steeply falling longitudinal momentum distributions
than observed (the final charm distribution are steeper than the
incoming gluons by a factor (1 – x) in perturbative calculations).
The gluon fusion model is, however, successful in explaining hidden
heavy flavor production, pp → XX → ψγX, etc.[7,8]

There is, however, another mechanism for heavy quark production, which occurs naturally in QCD. The proton wavefunction at
equal time* can be decomposed in terms of Fock state components

$$|uud\rangle, \quad |uudg\rangle, \quad |uudq\bar{q}\rangle, \quad \ldots \quad \quad (2)$$

including a small contribution for $|uudc\bar{c}\rangle$. The Fock states con-
taining heavy flavors first appear in perturbative theory via
vacuum polarization insertions in the gluons exchanged between
valence quarks. We will refer to these preexisting Fock components
as <u>intrinsic charm</u> states,[9] since they are present in the hadron
without regard to external reactions. Since all the intrinsic
quarks of a bound state tend to have the same velocity, the charm
quarks carry most of the hadron momentum in the Fock state where
they are present; i.e., the intrinsic charm quark x-distribution
can be as hard as those of the standard valence quarks. A Bag
model calculation of the cc-probability in fact gives P($|uudc\bar{c}\rangle$) ≈
1%. Thus, qualitatively the intrinsic charm mechanism can yield
cross sections of correct shape and the magnitude observed at the
ISR (2% of σ_{tot}(pp)). Furthermore, it is natural with such states
to have large cross sections for the diffractive excitation of
preexisting hadron components of the proton (at high energies
where kinematic and t_{min} effects are negligible).

There are several reasons why it is important to understand
charm production in detail:

(1) If dσ ~ $1/M_Q2$ (as suggested by the perturbative QCD-
vacuum polarization for intrinsic charm), then one expects an
appreciable production of b-quarks at the ISR and a non-negligible
production of t-quarks at the SPS and Tevatron colliders. The use
of a diffractive trigger and the possibility of production at
large x will reduce the combinatorial background in the search for
t-quark hadrons.

---

*In practice the decomposition is made at equal τ = t + z on the
  light cone in $A_0 + A_3 = 0$ gauge.

(2) In general, heavy quark production will be useful as a probe of hadron dynamics: in particular, for understanding the basic mechanisms for large x hadron production. The distinctive role of the intrinsic (preexisting) and extrinsic (created by the collision) mechanisms highlights two complementary aspects of QCD. Each contribution has its distinguishing nuclear A-dependence ($A^{2/3}$ versus A), s-dependence and x_L-dependence.

(3) In the case of the intrinsic charm component there are fundamental questions regarding the importance of non-perturbative confining forces on the heavy quark distributions.[10] From the point of view of perturbative theory or operator product expansion the leading $1/M_c^2$-contribution can be calculated using free quark propagation with up to four interactions in the heavy quark loop.

(4) Because of the heavy mass one expects strong kinematical scale breaking effects in the measured charm distribution.

(5) The presence of intrinsic charm at large x in the nucleon with strong threshold dependence has serious implications for the scale breaking parameterization of perturbative QCD, since the onset of charm masks the effects of QCD-evolution in deep inelastic structure functions.[11]

(6) For the unexpectedly high rate of observed same sign dimuons in ν-reactions[12] the 1% intrinsic charm contributes on a level consistent with the CDHS experiment (all experiments do not agree).

This talk is organized as follows: We first (Sec. 2) review the theoretical expectations for heavy quark production starting with estimates for "soft" production mechanisms and then elaborating more on what is expected from perturbative QCD with regard to open heavy flavor production. Comparisons with experimental data on $c\bar{c}$ are found in Sec. 3. In Sec. 4 a general discussion of higher Fock state decomposition of hadronic states is given and in Sec. 5 we argue for the existence of $|uudc\bar{c}\rangle$ of the 1% level and construct a model for the c(x) distributions. Hadronic production of charm is discussed in Sec. 6. In Sec. 7 our model for c(x) is confronted with data from leptoproduction experiments.

"CONVENTIONAL" THEORETICAL EXPECTATIONS FOR HEAVY QUARK PRODUCTION

The production of heavy quarks in hadronic collisions from soft mechanisms is normally expected to be very suppressed. As an example, when considering hadronic production of particles as a tunneling phenomena one finds the probability to produce a qq-pair[13]

$$P(q\bar{q}) \sim \exp\left(-\frac{\pi}{\kappa} m_\perp^2\right) \tag{3}$$

where $m_\perp = \sqrt{p_\perp^2 + m_q^2}$ and κ is the string constant ≈ 0.2 GeV^2. Using $m_u = m_d \approx 0$ MeV, $m_s = 100$ MeV, $m_c = 1500$ MeV and $\langle p_\perp \rangle = 350$ MeV one gets from Eq. (3)

$$u:d:s:c = 1:1:\frac{1}{3}:10^{-10} \tag{4}$$

The reason for the strong suppression of c-quark production is that it is very difficult to localize the energy of a substantial part of a string. Also, in other pictures one obtains a strong suppression. For example, in the statistical model[13] approach the probability for D-meson production is given by

$$P \sim \exp(-2m_D/160 \text{ MeV}) \tag{5}$$

which gives the ratio $\pi:K:D = 1:0.13:3\cdot 10^{-5}$.

However, since large masses are involved one expects calculations based on perturbative QCD to be valid. Perturbative QCD gives contributions of order $1/M^2$ in contrast to the $\exp(-\beta M^2)$-behavior for the soft tunneling processes discussed above.

Hadronic production of hidden heavy quark pairs, e.g., ψ, are well described by the hard scattering processes (see Refs. 7,8). In the case of open $Q\bar{Q}$ production the following hard scattering processes contribute[6] (see Fig. 1a,b)

$$q\bar{q} \to Q\bar{Q} \tag{6a}$$

$$gg \to Q\bar{Q} \tag{6b}$$

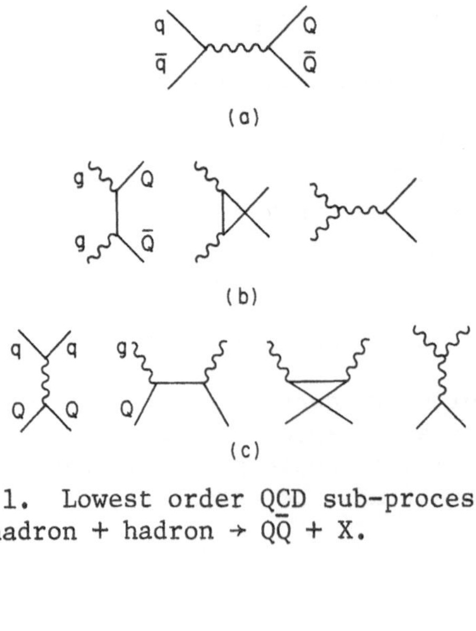

Fig. 1. Lowest order QCD sub-processes for hadron + hadron $\to Q\bar{Q} + X$.

together with the flavor excitation processes[14] (Fig. 1c)

$$qQ(\bar{Q}) \rightarrow qQ(\bar{Q}) \tag{7}$$

Predictions from the latter ones depend in detail on the understanding of the charm quark distribution in the proton.

   The gluon amalgamation process (6b) is expected to be dominant
at very high energies due to the abundance of low-x gluons. The
cross section is given by convolution of distribution functions
and the subprocess cross section $(\hat{\sigma})$

$$\sigma(h + h \rightarrow Q\bar{Q}X) = \iint_{x_1 x_2 > s_{min}/s} dx_1 dx_2 \; (Gx_1) \; G(x_2) \; \hat{\sigma}(s_1, x_2, s) \tag{8}$$

There are several theoretical uncertanties entering Eq. (8)

   i) The lower limit of Eq. (8). The true kinematical thresh-
      hold, $\sqrt{s_{min}}$, is $2m_D$ but $2m_c$ is presumably more relevant
      since the charmed hadrons are formed in a fragmentation/
      recombination process, thereby gaining energy.

   ii) The value of m_c. Most authors use $m_c = 1.6$ GeV. A
       lower value like $m_c = 1.2$ GeV, as obtained from potential
       calculations, would increase the cross section by a
       factor 4.

   iii) Higher order graphs are not yet included.

   iv) Higher twist contributions. These are unknown and could
       be important at such small masses as $m_c = 1.6$ GeV.

   v) Initial state corrections could alter the result by a
      large factor.[15]

   The cross section for $c\bar{c}$- and $b\bar{b}$-production in the FNAL/SPS-
ISR energy range from Eq. (8) is given by (see Fig. 2)

   $\sigma(c\bar{c}) = 1-50 \; \mu b$

   $\sigma(b\bar{b}) = 0.1-100 \; nb$

The energy dependence is logarithmic which is due to the 1/x-
behavior of the gluon distributions. The single particle spectrum
for the observed charmed hadrons is expected to be soft, reflecting
the incoming gluon distribution.

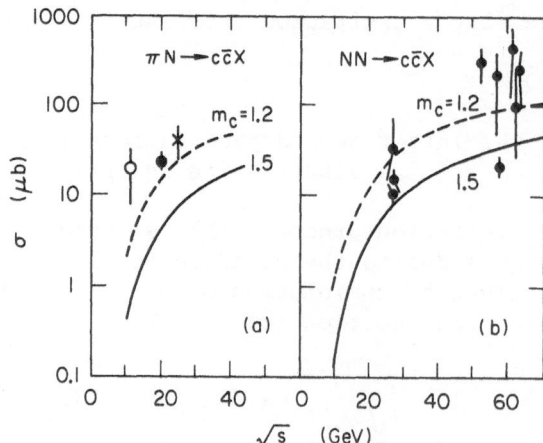

Fig. 2.   (a) $\sigma(\pi N \to c\bar{c}X)$ as a function of c.m.s.
energy, from Ref. 16.   (b) $\sigma(NN \to c\bar{c}X)$ as a function of c.m.s. energy, from Ref. 16.

COMPARISON WITH EXPERIMENTAL RESULTS ON OPEN CHARM PRODUCTION

The experimental results on charm production are reviewed in
detail in Ref. 1.   Here we only briefly mention the most important
results.   They are:

   i) At ISR one observes a large cross section (0.1-0.5 mb)
       for the reaction $pp \to \Lambda_c^+X$ (see Fig. 2b).   The cross
       sections for the other channels $pp \to D^+X$, $pp \to D^oX$ are
       in similar range.
  ii) Moreover, the Λ_c^+ seems to be produced diffractively in
       the forward region of phase space (see Fig. 3a,b).   At
       least one of the experiments has an explicit diffractive
       trigger.[5]   The x_L-distributions of $pp \to D^oX$ is $\sim(1-x)^3$
       much more forward than the corresponding strange meson
       distributions for $pp \to K^-X$ indicating that the charm
       quark carry a significant fraction of the proton
       momentum.
 iii) The situation for SPS/FNAL experiments is not so clear.
       One experiment with a diffractive trigger,[17] $\pi^-p \to D\bar{D}X$,
       observes a forwardly oriented single particle spectrum.
  iv) A signal for forwardly produced Λ_b at ISR has also been
       reported.[18]

The important discrepancy with the hard scattering approach
is the $\underline{x_L\text{-spectrum}}$ of Λ_c^+ (see Fig. 3); on general grounds one
would expect the Λ_c^+ wave function to favor configurations where
the c-quarks have the most momentum (see Fig. 4).   On the other
hand, the c-quarks produced in a hard scattering process have

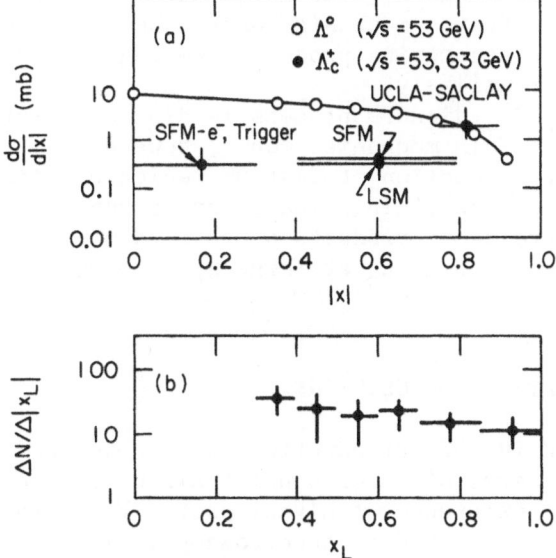

Fig. 3. (a) $d\sigma/d|x|$ for Λ_c^+ at 53 and 63 GeV.[1] The smooth curve is a fit to the Λ^0 data points. (b) Unnormalized x_L-distribution for Λ_c^+ from Ref. 2.

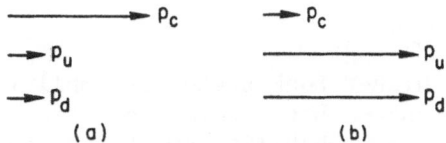

Fig. 4. (a) Typical quark momentum configuration in a Λ_c^+. (b) Typical quark momentum configuration after a hard scattering with a slow c-quark and two fast valence quarks.

small x. Hence such c-quarks would most unlikely end up in a fast Λ_c^+. The way to produce fast Λ_c^+ is to have hard c(c)-quarks initially present in the proton, i.e., $|uudc\bar{c}\rangle$ states. In fact, the similarity between $pp \to D^0(c\bar{u})X$ and $pp \to K^+(u\bar{s})X$ momentum distributions suggests that the c- and u-quark distributions are quite similar. We will discuss this <u>intrinsic charm</u> hypothesis in some detail below. Before doing so we briefly mention some recent attempts[19,20] to "improve" on the hard scattering approach.

i) In Ref. 19 it is hypothesized that a hard $c(x)$-distribution may arise from a mechanism where the c-quarks gain momentum after the scattering process. We regard such a mechanism as highly unlikely.

ii) In Ref. 20 a diquark process is discussed for producing fast Λ_c^+. This mechanism would, however, not explain the abundant production of fast D^0 (which contain no proton valence quarks). This model can be consistent only if the D^0 are decay products of charmed baryon resonances. The production of $\bar{\Lambda}_c$ at large x_L in pp-collisions would be decisive.

HADRONIC FOCK STATE DECOMPOSITIONS

As mentioned in the introduction, the proton has a general decomposition in terms of color singlet eigenstates of the free Hamiltonian. The existence of higher proton Fock states such as $|uudg\rangle$ has support from hadron spectroscopy: The p-Δ mass splitting (ΔE), which is believed to originate from the one-gluon exchange graph, is by cutting the diagram in Fig. 5 related to the probability of having extra gluon states, ($P(|uudg\rangle)$), through the relation

$$\Delta E = \sum_{\substack{\text{gluon} \\ \text{modes}}} P(|uudg\rangle) \, (E_{uud} - E_{uudg}) \quad . \tag{10}$$

The presence of higher Fock states is implicitly present in Ref. 21, where it is shown that rigorous constraints from $\pi \rightarrow \mu\nu$ and $\pi \rightarrow \gamma\gamma$ decays give a probability <0.25 for having a pion in a pure $q\bar{q}$-state at equal time on the light cone with $A_+ = 0$ gauge for a large class of wavefunctions.

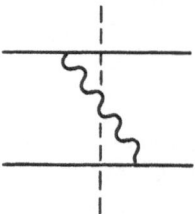

Fig. 5. One gluon exchange diagram responsible for spin-spin splitting of masses and the existence of higher Fock states containing an extra gluon.

In the next section we explore the consequences of heavy quark pairs Q$\bar{\text{Q}}$ in the Fock state decomposition of the bound state wave-function of ordinary mesons and baryons. Although proton states such as |uudc$\bar{\text{c}}$⟩ and |uudb$\bar{\text{b}}$⟩ are surely rare, the existence of hidden charm and other heavy quarks within the proton bound state will lead to a number of striking phenomenological consequences.

It is important to distinguish two types of contributions to the hadron quark and gluon distributions: Extrinsic and intrinsic. Extrinsic quarks and gluons are generated on a short time scale in association with a large transverse momentum reaction; their distributions can be derived from QCD bremsstrahlung and pair production processes and lead to standard QCD evolution. The intrinsic quarks and gluons exist over a time scale independent of any probe momentum, and are associated with the bound state hadron dynamics. In particular, we expect the presence of intrinsic heavy quarks, c$\bar{\text{c}}$, b$\bar{\text{b}}$, etc., within the proton state by virtue of gluon exchange and vacuum polarization graphs as illustrated in Fig. 6.

The "extrinsic" quarks and gluons correspond to the standard bremsstrahlung and q$\bar{\text{q}}$ pair production processes of perturbative QCD. These perturbative contributions yield wavefunctions with minimal power-law fall-off

$$|\psi(k_{\perp i}, x_i)|^2 \sim \frac{1}{k_{\perp i}^2} \tag{11}$$

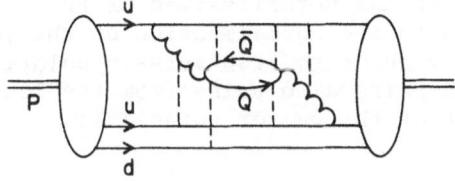

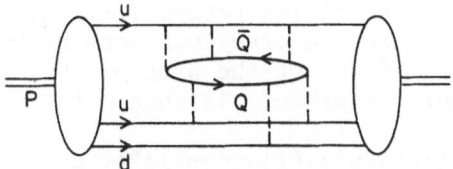

Fig. 6. Diagrams which give rise to the intrinsic heavy quarks (Q$\bar{\text{Q}}$) within the proton. Curly and dashed lines represent transverse and longitudinal-scalar (instantaneous) gluons, respectively.

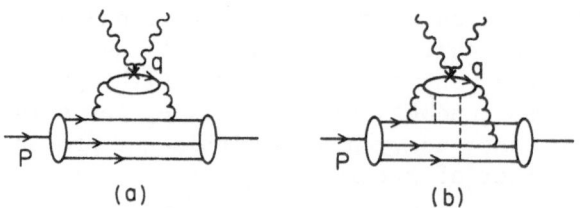

Fig. 7. (a) Example with contribution to
the deep inelastic structure functions
from an <u>extrinsic</u> quark q. (b) Example
with contributions to the deep inelastic
structure functions from an <u>intrinsic</u>
quark q.

and lead to the logarithmic evolution of the structure functions.
In contrast, the intrinsic contributions to the quark distribution
are associated with the bound state dynamics and necessarily have
a faster fall-off in $k_{\perp i}$ ($\psi \sim 1/k_i^2$ or faster).[22] The intrinsic
states thus contribute to the initial quark and gluon distributions.
A simple illustration of extrinsic and intrinsic $|uudq\bar{q}\rangle$ contribu-
tions to the deep inelastic structure functions is shown in Fig. 7a
and b. We see that the existence of gluon exchange graphs plus
vacuum polarization insertions automatically yield an intrinsic
$|uudq\bar{q}\rangle$ Fock state.

A complete calculation must take into account the binding
of the gluon and $q\bar{q}$ constituents inside the hadron (see Fig. 6) so
that the analysis is presumably non-perturbative.

We also note that the normalization of the $|uudq\bar{q}\rangle$ state is
not necessarily tied to the normalization of the $|uudg\rangle$ components
since the latter only refer to transversely polarized gluons;
Fig. 7 shows that $q\bar{q}$-pairs also arise from the longitudinal-scalar
(instantaneous) part of the vector potential.

INTRINSIC HEAVY QUARK STATES

The intrinsic heavy quark states exist on a long time scale.
Hence, an estimate of the mixing probability should be possible in
the static bag model. Such a study was done by Donoghue and
Golowich[23] in the rest frame of the proton. Summing over the
lowest states the authors of Ref. 23 obtain the result

$$P(|uudu\bar{u}\rangle):P(|uudd\bar{d}\rangle):P(|uuds\bar{s}\rangle):P(|uudc\bar{c}\rangle)$$

$$= 0.20:0.15:0.09:0.01$$

(12)

which, as far as charm is concerned, is in agreement with the
order of magnitude of the charm cross section observed at the ISR.
It should also be remarked that the results of Eq. (12) are still

consistent with previous bag calculations for the static quantities
like magnetic moments and average square radii. (For our purposes
it would be desirable to have the calculation of the intrinsic
charm content of the proton performed in the infinite momentum
frame.)

We now proceed to discuss the c-quark momentum distribution
in a $|uudc\bar{c}\rangle$ state. The general form of a Fock state wavefunction
is

$$\psi(k_{\perp i}, x_i) = \frac{\Gamma(k_{\perp i}, x_i)}{M^2 - \sum_{i=1}^{n}\left(\frac{m^2 + k_\perp^2}{x}\right)_i} \tag{13}$$

where Γ is the truncated wavefunction or vertex function. The
actual form of Γ must be obtained from the non-perturbative theory,
but following Ref. 21 it is reasonable to take Γ as a decreasing
function of the off-energy-shell variable

$$\mathscr{E} = M^2 - \sum_{i=1}^{n}\left(\frac{m^2 + k_\perp^2}{x}\right)_i . \tag{14}$$

Independent of the form $\Gamma(\mathscr{E})$, we can read off some general
features of the quark distributions:

(1) In the limit of zero binding energy ψ becomes singular
and the fractional momentum distributions peak at the values
$x_i = m_i/M$. More generally, $\mathscr{E}$ is minimal and the longitudinal
momentum distributions are maximal when the constituents with the

largest transverse mass $m_\perp = \sqrt{m^2 + k_\perp^2}$ have the largest light-cone
fraction x_i. This is equivalent to the statement that constituents
in a moving bound state tend to have the same rapidity.

(2) If one considers the proton as a state with virtual
fluctuations of $\pi^+ n$, $K^+\Lambda$, $D\Lambda_c$, etc., the most probably configurations are the closest to the energy shell, i.e., $\mathscr{E} \approx 0$. In the
case of virtual hidden charm states, the dominant configurations
thus have maximal x_c and $x_{\bar{c}}$.

(3) The intrinsic transverse momentum of each quark in a
Fock state generally increases with the quark mass. In the case
of power law wavefunction $\psi \sim (\mathscr{E})^{-\beta}$ we have $\langle k_\perp^2 \rangle \propto m_Q^2$; for an
exponential wavefunction $\psi \sim e^{-\beta \mathscr{E}^{1/2}}$, the dependence is $\langle k_\perp^2 \rangle \propto m_Q$.

In the limit of large $k_\perp$ one can use the operator product expansion near the light cone (or equivalently gluon exchange diagrams) to prove that, modulo logarithms, the Fock state wavefunctions fall off as inverse powers of $k_\perp^2$.[22] For our purpose, which is to illustrate the characteristic shape of the Fock states containing heavy quarks, we will choose a simple power-law form for the Fock state longitudinal momentum distributions

$$P_{(n)}(x_1 \cdots x_n) = N_n \frac{\delta\left(1 - \sum_{i=1}^{n} x_i\right)}{\left(M^2 - \sum_{i=1}^{n} \frac{\hat{m}_i^2}{x_i}\right)^2} \tag{15}$$

where the m_i^2 are identified now as effective transverse masses $m_i^2 = m_i^2 + \langle k_\perp^2 \rangle_i$ and the $\langle k_\perp^2 \rangle_i$ are average transverse momenta. With this choice, single-quark distributions have power law fall-offs $(1-x)^2$ and $(1-x)^3$ for mesons and baryons, respectively. (This is the most simple model for the hadronic wavefunction.)

For a $|uudc\bar{c}\rangle$ proton Fock state the momentum distribution is given by

$$P(x_1, \ldots, x_5) = N_5 \frac{\delta\left(1 - \sum_{i=1}^{5} x_i\right)}{\left(m_p^2 - \sum_{i=1}^{5} \frac{\hat{m}_i^2}{x_i}\right)^2} \; . \tag{16}$$

In the limit of heavy quarks $\hat{m}_4^2 = \hat{m}_5^2 = \hat{m}_6^2 \gg m_p^2, \hat{m}_i^2$ $(i = 1,2,3)$ we get

$$P(x_1, \ldots, x_5) = N_5 \frac{x_4^2 x_5^2}{(x_4 + x_5)^2} \, \delta\left(1 - \sum_{i=1}^{5} x_i\right) \tag{17}$$

where $N_5 = 3600$, P_5 is determined from $\int dx_1 \cdots dx_5 P(x_1, \ldots, x_5) = P_5$, where P_5 is the $|uudc\bar{c}\rangle$ Fock state probability. Integrating over the light quarks (x_1, x_2 and x_3) we get the charmed quark distributions

$$P(x_4, x_5) = \frac{1}{2} N_5 \frac{x_4^2 x_5^2}{(x_4 + x_5)^2} (1 - x_4 - x_5)^2 \quad . \tag{18}$$

By performing one more integration we obtain the charmed quark distribution

$$P(x_5) = \frac{1}{2} N_5 x_5^2 \left[\frac{1}{3} (1 - x_5)(1 + 10x_5 + x_5^2) - 2x_5(1 - x_5) \log \frac{1}{x_5} \right] \tag{19}$$

which has average $\langle x_5 \rangle = 2/7$ and is shown in Fig. 8. This is to be contrasted with the corresponding light quark distribution derived from Eq. (17) and shown in Fig. 9

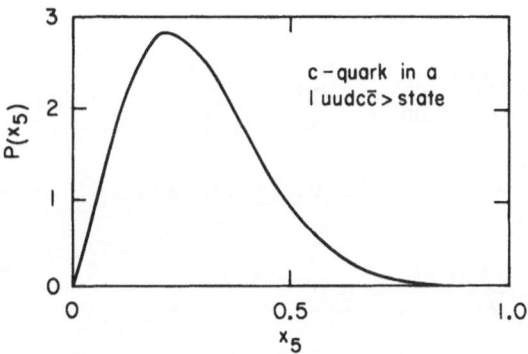

Fig. 8. The x distribution of the charmed quark in a $|uudc\bar{c}\rangle$ state.

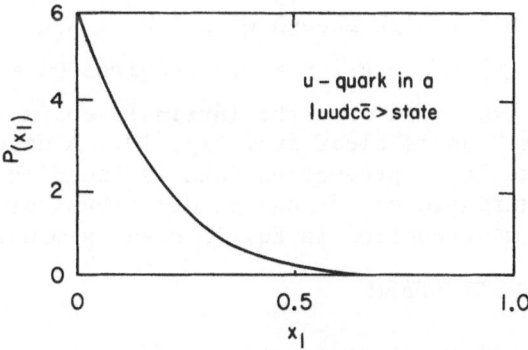

Fig. 9. The x distribution of a light quark in a $|uudcc\rangle$ state.

$$P(x_1) = 6(1 - x_1)^2 P_5 \quad .\qquad(20)$$

A more proper calculation of $c(x)$ can be done by integrating Eq. (16) over 11 variables using, e.g., exponential behavior for Γ. This was done in Ref. 24 using Monte Carlo techniques and the resulting $c(x)$ was found to be somewhat smoother than that of Eq. (19). A more detailed calculation of the perturbative diagrams in Fig. 6 yields an extra power $(1 - x_4 - x_5)$. The controlling factor in the distribution for large x is the energy denominator.

The corresponding c- and u-quark distributions in a $|udc\bar{c}\rangle$ are obtained in the same way. In order to see the contribution of the intrinsic $c\bar{c}$-pairs to the proton structure function, we use the value for $P_5 = 0.01$ from the bag model calculations discussed above. The magnitude of the charm cross section at ISR $(0.1-0.5 \text{ mb})$[1] gives for P_5:

$$P_5 = \frac{\sigma_{\Lambda_c}}{2\sigma_{inel}} \approx \frac{250 \ \mu b}{2 \cdot 30 \ mb} = 0.004 \quad .\qquad(21)$$

If the production mechanism is <u>inelastic</u> and

$$P_5 = \frac{\sigma_{\Lambda_c}}{2\sigma_{diff}} \approx \frac{250 \ \mu b}{2 \cdot 10 \ mb} = 0.01 \qquad(22)$$

if it is <u>diffractive</u>. These two possibilities will be discussed in the next section. We conclude that the charm cross section at ISR is compatible with $P_5 = 0.01$.

The charm quark distribution $c(x) = P(x_5)$ should be measurable in lepto-production for high enough Q^2 and $W^2 > W_{th}^2 = 25 \text{ GeV}^2$. Hence to measure $c(x)$ at, e.g., $x = 0.5$ requires $Q^2 = 25 \text{ GeV}^2$ ($x = Q^2/(Q^2 + W^2)$). We emphasize that the intrinsic charm sea $c(x)$ is "rare" but not "wee" as is clear from Fig. 10. A discussion on comparing $c(x)$ with lepto-production data is found in Sec. 6. In order to obtain intrinsic u-, d- and s- distributions ($|uuduu\rangle$ states, etc.) the wavefunction in Eq. 15 needs a minor modification.

HADRONIC PRODUCTION OF CHARM

Hadronic production of multiparticle final states occurs in two different ways, <u>diffractive disassociation</u> and <u>nondiffractive</u>

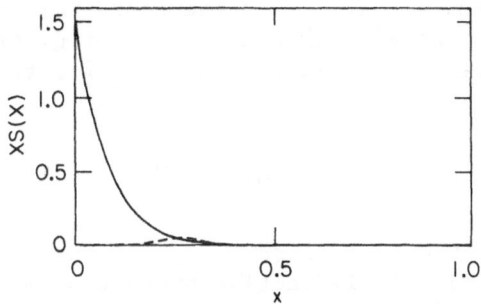

Fig. 10. Comparison of the intrinsic charm sea xc(x) (dashed line) with the total sea at Q^2 = 5 GeV^2 as parameterized by Ref. 25.

inelastic production. Although at least one experiment of Λ_c^+-production has an explicit diffractive trigger, the situation for charm production is far from settled. We will discuss the two production mechanisms below in the light of intrinsic charm.

(1) Diffractive production of high M^2-states can be interpreted as a phenomenon of short distance where perturbative QCD should be applicable to some extent. This idea was first considered in Ref. 26 in the context of charm production. Recently these questions have been studied in more detail for high mass diffraction in general in terms of so-called "transparent states."[27-30] The idea is simple and appealing: When the valence quarks of a hadron are close together the net color extension is almost zero and the hadron does not interact with other hadrons. Hence the absorptive cross section is small and the hadron scatters diffractively off the target which then appears to be transparent. This situation is, as pointed out by Ref. 28, very similar to an analogous process in QED: When e^+e^--pairs are produced in very high energy emulsion experiments, they can only be separated by distances smaller than atomic sizes. The e^+e^- has net charge zero — it is not "seen" by surrounding atoms and hence it does not ionize and give rise to visible tracks. In Ref. 30 the knowledge of the pion wavefunction in QCD at short distances is used to derive results for the pion-induced jets emerging from the "transparent" target.

As was discussed in connection with Eq. (13) one expects intrinsic heavy quark states to have large $\langle k_\perp \rangle$ and consequently

small transverse dimension. It is therefore tempting to assume
that the intrinsic heavy quark states scatter diffractively. With
this assumption one obtains in the case of 1% intrinsic charm on a
nuclear target

$$\sigma^{diff}_{charm} = 0.01 \cdot \sigma_{el} \approx 0.5 \text{ mb} \cdot A^{2/3} \quad . \tag{23}$$

This high value is encouraging as far as production of b- and t-
quarks are concerned. A diffractive production mechanism of heavy
quarks is also very favorable as far as the combinatorial background is concerned.

For the charm case the Λ_c and D-spectra can be calculated in
principle from the strong overlap between the 5-quark and the
charmed-hadron state wavefunctions, allowing for decays of excited
state, etc. For the purpose of obtaining the x_F-distribution we
use a very simple recombination mechanism for the quarks involved
in the states. Neglecting its binding energy, the Λ_c spectrum is
given by combining the u-, d- and c-quark in $|uudc\bar{c}\rangle$ to obtain

$$P(x_{\Lambda_c}) = N_5 \int_0^1 \prod_{i=1}^5 dx_i \ \delta(x_{\Lambda_c} - x_2 - x_3 - x_5) \left(\frac{x_4 x_5}{x_4 + x_5}\right)^2$$

$$\delta\left(1 - \sum_{i=1}^5 x_i\right) \tag{24}$$

(see Fig. 11) with $\langle x_{\Lambda_c}\rangle = 1/7 + 1/7 + 2/7 = 4/7$. The ISR data for
$d\sigma/dx$ (pp $\rightarrow \Lambda_c X$) is consistent with the prediction that $\langle x_{\Lambda_c}\rangle \approx 0.5$
although the data is even flatter than predicted by Eq. (24). We
expect that the low x region for charm production will be filled
in by both perturbative and higher Fock state intrinsic contributions. Assume that a hadron interacts strongly only when one of
its constituents is very peripheral, $k^2 - m^2 \approx 0$. Since $k^2 - m^2 = x\mathscr{E}$ this implies that the important interacting Fock states have
one constituent with $x \approx 0$.[31] Consequently, the spectator system
carry more momentum than in Eq. (24). This effect improves our
agreement with the data. The corresponding distribution for
$D^-(\bar{c}d)$ is given by

$$P(x_{D^-}) = N_5 \int_0^1 \prod_{i=1}^5 dx_i \ \delta(x_D - x_3 - x_5) \left(\frac{x_4 x_5}{x_4 + x_5}\right)^2$$

$$\delta\left(1 - \sum_{i=1}^5 x_i\right) \tag{25}$$

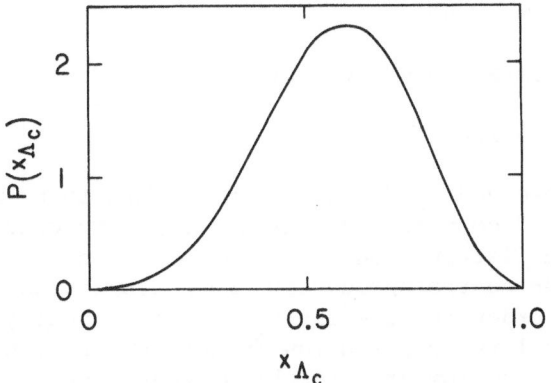

Fig. 11. The x distribution of the Λ_c^+ from the intrinsic charm component of the proton.

with $\langle x_{D^-} \rangle = 1/7 + 2/7 = 3/7$, and is shown in Fig. 12. The $D^+(c\bar{d})$ distribution would, in principle, be obtained from the $|uudc\bar{c}d\bar{d}\rangle$ Fock state of the proton, where the $d\bar{d}$ could be extrinsic or intrinsic. Assuming that the $\bar{d}$ momentum is small, the D^+ distribution should be close to that of the c-quark shown in Fig. 8. These predictions apply for forward production ($x_F \gtrsim 0.1$), where perturbative contributions and higher Fock state contributions can be neglected. Spectra for pion induced reactions are obtained in the same way.

In addition to charmed mesons and baryons, the J/ψ may also be produced diffractively from the intrinsic charm component of the proton. Compared to the charm production cross section at FNAL energies

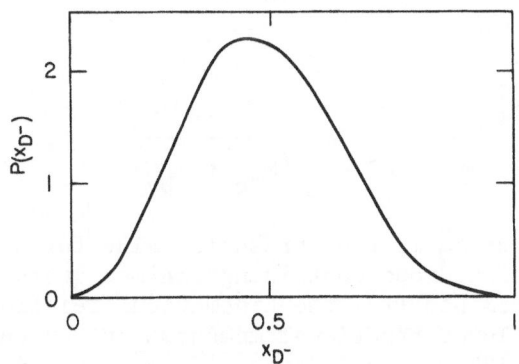

Fig. 12. The x distribution of the D^- from the intrinsic charm component of the proton.

$$\sigma(\pi N \to DX) \simeq 20 \ \mu b \quad , \tag{26}$$

J/ψ production data around 200 GeV give[32]

$$\sigma(\pi N \to \psi X) \simeq 100 \ nb \quad .$$

Further, the observed x_F-distribution appears to be more strongly peaked near $x \simeq 0$ compared to what would be expected from the intrinsic charm distribution. Evidently most of the ψ production comes from other central production mechanisms such as gluon and $q\bar{q}$ fusion. In order for the intrinsic charm model to be consistent, there must be a large suppression factor for the ψ production from the intrinsic charm compared to the D production.

$$\left. \frac{\sigma(\pi N \to \psi X)}{\sigma(\pi N \to DX)} \right|_{\text{intrinsic charm}} \lesssim 5 \times 10^{-5} \quad . \tag{27}$$

As was shown in Ref. 9, such a suppression factor is obtained using the intrinsic charm wavefunction and taking flavor and color suppression into account.

(2) A simple mechanism for <u>inelastic</u> charm <u>production</u> is gluon excitation of preexisting c-quarks (see Fig. 1c). This process is discussed in Ref. 33.

To study the energy dependence of the "diffraction" mechanism with "intrinsic" heavy quarks we will use the empirical formula for high mass diffraction[34]

$$\frac{d\sigma}{dM^2} = \sigma_0 \ \frac{1}{M^2} \tag{28}$$

valid for $M^2 \gtrsim 2$ GeV^2. The integrated charm cross section is given by

$$\sigma = \sigma_0^c \int_{M_0^2}^{M_1^2} \frac{dM^2}{M^2} = \sigma_0^c \ \log \frac{(1-x_1)s}{\left(M_{\Lambda_c} + M_D\right)^2} \quad , \tag{29}$$

where in this case M_0^2 is the threshold value for associated production of a pair of hadrons containing charmed quarks. The upper limit M_1^2 is determined from the kinematical relation $M_1^2 = s(1-x_1)$ where x_1 is the lower fractional momentum cut on the recoiling proton. In the ISR pp $\to$ p $\Lambda_c X$ experiment[5] one triggers on events with $x_p \geq 0.8$. If we assume that essentially all the charm cross section $\sigma_c \sim 300 \ \mu b$ is due to diffractive production, then we can determine $\sigma_0 = 77 \ \mu b$. From this we predict that at SPS and FNAL energies ($s \cong 400$–600 GeV^2), the total pp → charm cross section should be of the order of 150 μb. Clearly this prediction is larger than present experimental data at SPS/FNAL with both pion and proton beams.[35] The energy dependence thus seems to be stronger than what is implied by Eq. (29).

Concerning production of heavy quarks on nuclear targets one expects an $A^{2/3}$-dependence from the intrinsic charm model. This is in contrast to the perturbative hard scattering cross section, which should be proportional to A.

As far as the production of b- and t-quarks are concerned, one can argue on general grounds that the probability of a hadron to contain an intrinsic heavy quark pair should fall as

$$P_{Q\bar{Q}} \propto \frac{\alpha_s^2(R^{-2})}{R^2 m_Q^2} \tag{30}$$

where R is a hadron size parameter. Assuming $\sigma_c \simeq 300$ μb, $m_c = 1.5$ GeV, $m_b = 5$ GeV, and $m_t = 30$ GeV and using Eq. (29), one obtains the cross sections for b- and t-quark production as shown in Table I.

Table 1. Cross section for b- and t-production at ISR and Tevatron energies from Eqs. (29) and (30). The numbers in parentheses are the conventional perturbative QCD-predictions.

	ISR ($\sqrt{s} = 63$ GeV)	Tevatron ($\sqrt{s} = 2000$ GeV)
b	15 μb (0.5)	70 μb (2)
t ($m_t = 20$ GeV)	0	3 μb (0.1)

THE INTRINSIC CHARM AND LEPTOPRODUCTION EXPERIMENTS

An important test of the intrinsic charm content of the proton is the direct measurement of the charm quark distribution in deep inelastic scattering:

$$F_2^c(x,Q^2) = \frac{4}{9} x \left[c(x,Q^2) + \bar{c}(x,Q^2) \right] \tag{31}$$

As is clear from Fig. 10, the intrinsic charm sea is very small compared to the total sea. However, it should be visible in experiments explicitly looking for leptoproduction of charm. This is the case in dimuon production (Fig. 13a)

$$\mu^\pm N \to \mu^\pm \mu^\pm X \tag{32}$$

where one of the final state muons originates from charm decay.

There are, however, a number of complications:

a) The model dependence of the charm fragmentation function and the associated experimental acceptance corrections.[36]

b) A strong scale-breaking effect associated with a high mass threshold:

$$W^2 = (p+q)^2 > W_{th}^2 = (m_D + m_{\Lambda_c})^2 \simeq 17 \text{ GeV}^2 \quad . \tag{33}$$

The W^2-threshold enters explicitly in the Bjorken condition. Let x_i^+ be the light cone momentum fractions $x_i^+ = (k_i^0 + k_i^3)/(P^0 + P^3)$ of the hadronic constituents with $\Sigma x_i^+ = 1$ and $\Sigma k_{\perp i} = 0$. The Bjorken condition for putting the final state on shell (p^--conservation) is then (we neglect $k_\perp \cdot q_\perp$ terms)

$$\frac{2M_P \nu}{Q^2} = \frac{1}{x_{Bj}} = \frac{1}{x_c} + \frac{1}{Q^2}\left[\frac{m_{c_\perp}^2}{x_c} + \frac{m_{\bar{c}_\perp}^2}{x_{\bar{c}}} + \sum_{i=u,u,d}\frac{m_{i_\perp}^2}{x_i} - M_P^2\right]$$

$$\tag{34}$$

$$\geq \frac{1}{x_c} + \frac{W_{th}^2 - M_P^2}{Q^2} \quad .$$

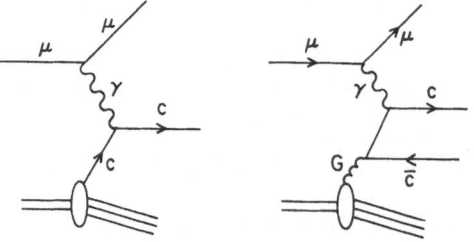

Fig. 13. Lepto-production of charm from the intrinsic charm sea and via the proton-gluon fusion model, respectively.

Thus, in general, the light cone momentum fraction of the charmed quark is larger than the Bjorken value x_{Bj} with the excess controlled by W_{th}^2/Q^2. Since $c(x,Q^2)$ falls with x, this means that $F_2^c(x_{Bj},Q^2)$ increases with Q^2 for fixed x_{Bj} unless $Q^2 \gg 17$ GeV^2. The usual rescaling variable

$$\zeta = x_{Bj} + \frac{m_c^2}{Q^2} \tag{35}$$

is incorrect since it ignores the heavy mass of the spectator system.

The EMC-[37] and BFP-data[38] which are binned at fixed x_{Bj} do show significant rise with Q^2. This kinematic effect has to be extrapolated to $Q^2 \gg W_{th}^2$ before accurate comparisons with the intrinsic charm distribution can be made. Threshold factors of the form $(1 - (W_{th}/W)^2)^n$ may be useful for the parameterization of the data.

c) The $c(x)$-distribution as measured in deep inelastic scattering at large Q^2 differs from that determined in low momentum transfer hadron-production because of standard QCD-evolution. This tends to further suppress $F_2^c(x,Q^2)$ at large x and Q^2 (see Fig. 14).

The comparison of the intrinsic charm production (see Eq. (19)) with data was done in Ref. 39. The limits on intrinsic charm is $\lesssim 0.5\%$. However, the comparison does not include the threshold suppression from Eq. (34), so that the net result is not inconsistent with the predicted form and 1% normalization of intrinsic charm. A definitive comparison requires a detailed analysis of the scale breaking effects.

A very interesting implication of intrinsic charm for νN and $\bar{\nu} N$ charge current reactions is the production of beauty quarks $(\nu \bar{c} \rightarrow \mu^+ b$ and $\nu \bar{c} \rightarrow \mu^- \bar{b})$.[9] The subsequent leptonic decay of the b and $\bar{b}$ then leads to same-sign muon pairs (see Fig. 15). The experimentally observed rate of same-sign muon pairs is unexpectedly high, although the different experiments disagree with an order of magnitude.[12] Using the intrinsic charm distribution with present limits on the left-handed c-b coupling the c-b process almost agrees with the CDHS data.

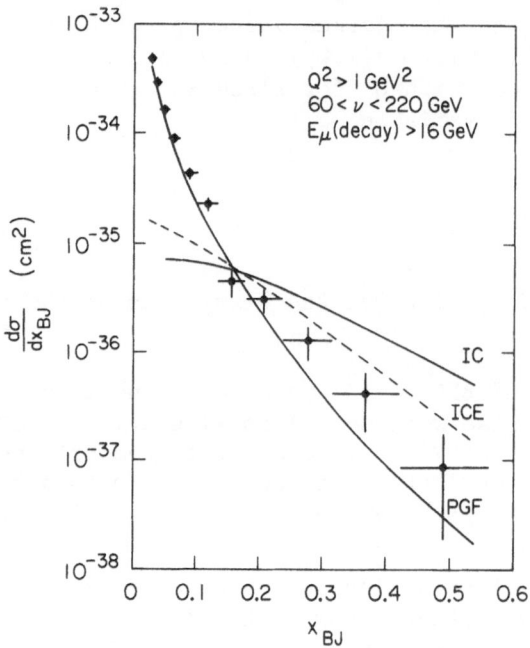

Fig. 14. Variation of $d\sigma/dx_{Bj}$ with x_{Bj}
for dimuons in the range $Q^2 > 1$ GeV^2,
$60 < \nu < 220$ GeV, decay muon energy
>16 GeV. The horizontal bars represent
the bin widths. The figure is taken
from Ref. 39. The curves are:

    PGF: photon-gluon fusion model,
     IC: intrinsic charm model,
    ICE: intrinsic charm model with
          maximum Q^2 evolution.

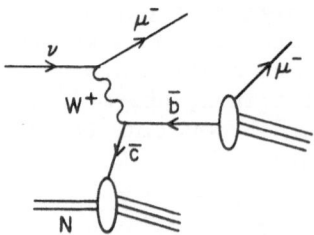

Fig. 15. Same sign dimuon
pair production from the
intrinsic charm component
of nucleons.

CONCLUSIONS

There are a number of theoretical and phenomenological issues related to intrinsic charm:

(1) Do the Fock states containing heavy quarks have a small transverse dimension? Note that the structure of the energy denominator in Eq. (15) implies that all the quarks in the $|uudc\bar{c}\rangle$ state have larger $k_{\perp}$ than in the $|uud\rangle$ state.

(2) How much of the strange sea can be attributed to the intrinsic $|uuds\bar{s}\rangle$-Fock state rather than standard evolution? A phenomenological analysis given by R. Phillips[24] gives $P(|uuds\bar{s}\rangle) \approx 0.031$.

(3) What is the correct mechanism for the high energy excitation of the charm component? Note that, in general, gluon (or Pomeron) interactions occur coherently with all the quarks of the nucleon Fock state. In time-ordered perturbation theory the charm production can occur before, during, or after the hadronic interaction.

(4) A more detailed calculation on the intrinsic charm wavefunction may be possible in Bag models, lattice calculations or more directly from the QCD equations of motion.[40] The magnitude of the p-Δ hyperfine splitting can give a bound on the intrinsic gluon and $q\bar{q}$ Fock state components.

(5) Much more experimental information is needed to unravel the role of the different QCD contribution

    a) The x_L-dependence of Σ_c, $\Lambda_{\bar{c}}$, D^0 and D^+. The $\Lambda_{\bar{c}}$ distribution is particularly important since it determines the $\bar{c}(x)$ without complications from valence quark recombination or resonance decays.

    b) The physics of intrinsic charm can depend in detail on the nature of the incoming hadron; K, π, p and γ.

    c) The threshold dependence, $(1 - s_{th}/s)^n$, of heavy quark production must reflect the nature of the production mechanism.

    d) The nuclear A-dependence separates the intrinsic and hard scattering contributions. The A^{α}-behavior is a function of x; at large x we expect $\alpha = 2/3$.

    e) Hidden charm states, X, ψ, ..., should be seen at some level at large x from the intrinsic charm.

In conclusion, a valence-like charm quark distribution c(x) in the nucleon at the 1% level accounts qualitatively for hadron induced charm production in magnitude, shape and diffractive features at ISR energies. There is no contradiction with the EMC-data on $F_2^c(x)$ provided the appropriate threshold dependence is taken into account.

In any event, the determination of the charm quark distribution is important for understanding the Fock state structure of the hadronic wavefunctions and as a probe of hadron dynamics in the nonperturbative domain.

ACKNOWLEDGEMENTS

Work supported in part by the Department of Energy contract DE-AC03-76SF00515 and by the Swedish Natural Science Research Council under contract F-PD8207-101.

This talk was also presented at the Topical Conference on Forward Collider Physics, Madison, Wisconsin, December 10-12, 1981.

REFERENCES

1. For a review see, e.g., C. Heusch, Lectures given at the 1981 SLAC Summer Institute on Particle Physics, SLAC-PUB-2876.
2. M. Basile et al., Nuono Cimento Lett. $\underline{30}$, 481 (1981); ibid $\underline{30}$, 487 (1981).
3. M. Basile et al., CERN EP/81-125.
4. S. Singh et al., Nucl. Phys. B $\underline{140}$, 189 (1978).
5. K. L. Giboni et al., Phys. Lett. $\underline{85B}$, 437 (1979) and A. Kernan, private communication.
6. H. Fritzsch, Phys. Lett. $\underline{67B}$, 217 (1977). F. Halzen, Phys. Lett. $\underline{96B}$, 105 (1977). L. M. Jones and H. W. Wyld, Phys. Rev. D $\underline{17}$, 759, 1782, 2332 (1978). M. L. Gluck and E. Reya, Phys. Lett. $\underline{79B}$, 453 (1978); $\underline{83B}$, 98 (1979). M. Gluck, J. F. Owens, and E. Reya, Phys. Rev. D $\underline{17}$, 2324 (1978). J. Babcock, D. Sivers and S. Wolfram, Phys. Rev. D $\underline{18}$, 162 (1978). C. E. Carlson and R. Suaya, Phys. Rev. D $\underline{18}$, 760 (1978); Phys. Lett. $\underline{81B}$, 329 (1979). H. Georgi et al., Ann. Phys. $\underline{114}$, 273 (1978). K. Hagiwara and T. Yoshino, Phys. Lett. 80B, 282 (1979). J. H. Kuhn, Phys. Lett. $\underline{89B}$, 385 (1980). J. H. Kuhn and R. Ruckl, MPI-PAE/pTH 7/80. V. Barger, W. Y. Keung and R.J.N. Phillips, Phys. Lett. $\underline{91B}$, 253; $\underline{92B}$, 179 (1980); Z. Phys. $\underline{C6}$, 169 (1980). Y. Afek, C. Leroy and B. Margolis, Phys. Rev. D $\underline{22}$, 86, 93 (1980).
7. C. Peterson, Proc. of XII International Conference on Multiparticle Dynamics, University of Notre Dame, Indiana, June 22-26, 1981.
8. R. Raja, Proc. of XII International Conference on Multiparticle Dynamics, University of Notre Dame, Indiana, June 22-26, 1981.

9. S. J. Brodsky, P. Hoyer, C. Peterson, and N. Sakai, Phys.
    Lett. 93B, 451 (1980); P. Hoyer, in High Energy Physics —
    1980, Proceedings of the XX International Conference,
    Madison, Wisconsin, edited by L. Durant and L. G. Pondrom
    (AIP, New York), 1981; S. J. Brodsky, C. Peterson and
    N. Sakai, Phys. Rev. D 23, 2745 (1981).

10. See, e.g., the discussion of A. H. Mueller, in Proceedings of
    International Symposium on Lepton and Photon Interactions

11.   at High Energy, Bonn, West Germany, August 24-29, 1981.

11. D. P. Roy, Phys. Rev. Lett. 47, 213 (1981).

12. J. Knoblock, Proceedings of International Conference on Neu-
    trino Physics and Astrophysics, Maui, Hawaii, July 1-8, 1981.

13. For a more detailed discussion, see C. Peterson, Proceedings
    of the Topical Workshop on Forward Production at High-Mass
    Flavors at Collider Energies, College de France, Paris
    (1979).

14. B. L. Combridge, Nucl. Phys. B 151, 429 (1979).

15. G. T. Bodwin, S. J. Brodsky, G. P. Lepage, Phys. Rev. Lett. 47,
    1799 (1981); SLAC-PUB-2860.

16. R. Phillips, in High Energy Physics — 1980, Proceedings of the
    XX International Conference on High Energy Physics, Madison,
    Wisconsin, edited by L. Durand and L. G. Pondrom (AIP, New
    York, 1981).

17. L. J. Koester, in High Energy Physics - 1980, Proceedings of
    the XX International Conference on High Energy Physics,
    Madison, Wisconsin, edited by L. Durand and L. G. Pondrom
    (AIP, New York, 1981); D. E. Bender, Ph.D. thesis, 2980,
    University of Illinois (unpublished); J. Cooper, Proceedings
    of the XV Recontre de Moriond, 1981 (unpublished).

18. M. Basile et al., Nuovo Cimento Lett. 31, 97 (1981).

19. V. Barger and F. Halzen, Phys. Rev. D 4, 1428 (1981).

20. R. Horgan and M. Jacob, Phys. Lett. 107B, 395 (1981).

21. S. J. Brodsky, T. Huang and G. P. Lepage, SLAC-PUB-2540; T.
    Huang, in High Energy Physics — 1980, Proceedings of the XX
    International Conference on High Energy Physics, Madison,
    Wisconsin, edited by L. Durand and L. G. Pondrom (AIP, New
    York, 1981).

22. S. J. Brodsky and G. P. Lepage, Phys. Rev. D 22, 2157 (1980)
    and S. J. Brodsky, Y. Frishman, G. P. Lepage, and C.
    Sachrajda, Phys. Lett. 91B, 239 (1980), and references
    therein.

23. J. F. Donoghue and E. Golowich, Phys. Rev. D 15, 3421 (1977).

24. R.J.N. Phillips, Rutherford Laboratory Preprint RL-82-004.

25. A. J. Buras and K.J.F. Gaemers, Nucl. Phys. B 132, 249 (1978).

26. G. Gustafson and C. Peterson, Phys. Lett. 67B, 81 (1977).

27. J. F. Gunion and D. E. Soper, Phys. Rev. D 15, 2617 (1977).

28. J. Pumplin and E. Lehman, Zeitschrift für Physik C9, 25 (1981).

29. G. Gustafson, LUTP 81-1, talk given at the "IX International
    Winter Meeting on Fundamental Physics," Siguenza, Spain,
    February 1981.

30. G. Bertsch, S. J. Brodsky, A. S. Goldhaber and J. F. Gunion, Phys. Rev. Lett. <u>47</u>, 297 (1981).
31. C. Peterson, work in progress.
32. J. Badier et al., Proc. Lepton-Photon Conf. at Fermilab, 1979, p. 161; CERN/EP 79-61.
33. R. Odorico, Phys. Lett. <u>107B</u>, 231 (1981).
34. M. G. Albrow et al., Nucl. Phys. B. <u>108</u>, 1 (1976).
35. R. C. Ruchti, Proc. of XII International Conference on Multi-particle Dynamics, University of Notre Dame, Indiana, June 22-26, 1981.
36. R. V. Gavai and D. P. Roy, Ziet. Phys. <u>C10</u>, 333 (1981).
37. H. Best, Proc. of XVI Recontre de Moriond, 1981.
38. M. Strovink, Proc. of 10th Int. Symp. on Lepton and Photon Interactions at High Energy, Bonn, West Germany, August 24-29, 1981.
39. J. J. Aubert et al., CERN-EP/81-161 (1981).
40. See, e.g., S. J. Brodsky, T. Huang and G. P. Lepage, SLAC-PUB-2868.

DISCUSSION

W. Reay

Comment 1: Everyone defines a different variable z, but in
terms of our definition, for our 40 charm events we find a peaking
at z = 0.5.

Comment 2: We see no B production and HPWF measures a ratio
$\mu^-\mu^-/\mu^+\mu^-$ which is flat below B threshold. This is weak evidence
against intrinsic charm.

Comment 3: A Rochester-Caltech group presented results at the
Gordon conference corresponding to a limit of 0.05% on intrinsic
charm from muons measured in a beam dump experiment. Can you or
someone from this experiment comment on this?

R. Coleman

New results from the Caltech-Fermilab-Rochester-Stanford prompt
muon experiment indicate diffractive cross sections from 280 GeV
pion and proton beams are small ($\sim$5 μb/nucleon).

Answer

It is clearly very difficult to reconcile the FNAL and ISR
results unless there are very strong threshold effects suppressing
charm production below 400 GeV and/or the nuclear number dependence
is much weaker than A^2. Measurements at the lowest possible ISR
energy would clearly be valuable. Fit to threshold factors of the
form $(1 - W_{th}^2/W^2)^n$ and analyses of the A-dependence of the heavy
flavor cross sections are clearly important for sorting out the
production mechanism.

J. Sacton

The experimental data on like-sign dimuon production that you
have shown only represents part of the picture. There exists new
data from CDHS, which on quite good statistics, are more than a
factor of 10 down as compared to CFNRR and HPWFOR. Also CHARM has
provided new data which, if correctly scaled for the different cuts
used on the second muon momentum (4 GeV vs. 9 GeV), agree quite well
with CDHS. Admittedly both European collaborations are working at
lower energies than the American experiment.

H. Newman

Can one simply relate the fragmentation functions of e and b quarks as seen in hadroproduction to what is expected in e^+e^-? Is the relationship between the two types of experiments clouded by the complicated states you expect to exist inside the nucleon?

Answer

Independent of the origin of the charm quarks there are final state interactions of a scattered quark in the nucleon which smear its transverse momentum and degrade its energy due to induced particle production in the central rapidity region. (See G. T. Bodwin, S. J. Brodsky, and G. P. Lepage SLAC-PUB-2860.) At very high energies where these finite energy effects can be neglected one expects factorization of the fragmentation functions. The correlations between charm particles in the current quark fragmentation region and the associated charm particles produced from the spectator quark depend in detail on the charm-pair production model for leptoproduction.

D. Korbel

If all intrinsic quark probabilities for $u\bar{u}$, $d\bar{d}$, $s\bar{s}$, and $c\bar{c}$ as given in the MIT Bag model are added together they sum up to 50%. How much of the nucleon's momentum do they carry and how much is the gluon's momentum distribution affected? Can this be measured using deep inelastic sum rules?

Answer

Since the intrinsic $u\bar{u}$, $d\bar{d}$, and $s\bar{s}$ sea quarks in the nucleon are dominantly at low x, I believe there is no conflict between the measured gluon momentum fraction and the bag model estimates. It appears natural from the QCD equation of state for hadronic wavefunctions at equal time on the light cone that the sea quarks and gluons are substantially due to hadronic binding effects rather than evolution alone.

B. Roe

What experiments distinguish between these different models?

Answer

In principle, the models for heavy quark hadroproduction are distinguished by their predictions for the beam, target, s, and x_L dependencies, the relative importance of diffractive production and flavor correlations. Clearly direct measurements of the charm quark distribution in leptoproduction (preferably by direct observation of charmed hadrons) and observation of flavor correlations in the central and target rapidity regions are crucial.

W. Reay

Another good place to look for intrinsic charm effects is to look for associated charm production in neutral current neutrino interactions.

G. Diambrini-Palazzi

It was predicted some time ago that perturbative QCD would allow for a sizeable asymmetry ratio in photoproduction of charm particles by polarized photons. Could this be used for testing the different production models?

Answer

This may be possible; however, the polarization predictions based on gluon-induced subprocesses are severely modified when longitudinal gluon polarization is included in the next-to-leading log calculations. This has been recently analyzed by S. Mahmood.

P. Musset

Was intrinsic charm included in the calculations reported by R. Ordorico?

R. Odorico

The calculations involved only the use of perturbative QCD. Charm production results only from the conversion of gluons into charmed quark-antiquark pairs according to perturbative QCD.

Answer

One part of the Odorico calculation which I do not understand is the origin of the strongly asymmetric momentum distribution of the struck and spectator quarks, since the standard splitting function for g → c$\bar{c}$ in QCD is symmetric, independent of the momentum transfer of the probe. In addition, in the two-photon QED ee → eeμμ process, the analogous distribution is symmetric, I believe, even at low q^2.

R. Odorico

The kinematical situation is different from that present in hadronic charm production.

R. Odorico

Parton concepts are meant to be used for interacting hadrons. In the interaction partons are likely to go off-shell. Don't you think then, that one should be very careful about making a too critical use of quark masses?

Answer

It seems likely that the running quark mass for heavy flavors probably does not vary rapidly in typical hadroproduction sub-processes, but I agree that one should be cautious.

L. Montanet

Could you give a quantitative estimate for F meson production from a K^+ beam at 200–300 GeV/c?

Answer

In the intrinsic charm model where the F would be produced from a rearrangement of the intrinsic strange and charm quarks in the kaon Fock state, one would expect that K → F is comparable to π → D. Such measurements would be very valuable for distinguishing models.

H. Newman

Is the reason there is such a fast side in heavy flavor production with increasing s simply higher order QCD corrections? One encounters a similar effect in e^+e^- thresholds for new flavors where QCD corrections give a very strong enhancement, and a fast rise at threshold. Could there also be structure in the threshold?

Answer

The effects you suggest are certainly possible. In addition, compared to direct photo-induced reactions, hadronic production is inefficient in utilizing beam energy since the gluons which initiate the production subprocesses typically carry only a small fraction of the hadron momentum. One can estimate using counting rules that a hadronic gluon-induced reaction can be expressed near threshold by a factor $(1 - W_{th}^2/W^2)^n$ (with $n \sim 6$ for protons, $n \sim 4$ for mesons) relative to direct photoproduction.

B. Foster

What would you expect for the baryon spectrum of baryons in e^+e^- if your ideas were correct?

Answer

The charm and bottom baryons would be very fast, but the decay down to normal baryons would smear the resulting distributions to low momentum. The prediction of fast heavy hadrons fragmenting from heavy quarks follows from the models of Bjorken and Suzuki.

B. Foster

So the TASSO data on Λ production in e^+e^- annihilation would not rule out your ideas?

Answer

Detailed calculations are needed, but I don't believe there is a problem. Measurement of the charmed baryon fragmentation function would be very useful.

PRODUCTION OF HEAVY FLAVOURS IN MUON IRON INTERACTION

Volker Korbel

DESY
Notkestraße 85
D-2000 Hamburg 52

I. The Experiments

We discuss some new experimental results on virtual high energy photoproduction of heavy quarks on nuclei. Virtual photons are produced by the scattering of high energy muons, the momentum and energy spectrum of the photons is given by QED. The photoproduction cross section for heavy quarks is found to be small. Thus dense and long targets of several meter length as well as high intensity muon beams are used.

Results are now available essentially from the following experiments:

1) Michigan State University - Fermilab Collaboration, MSUF
2) Berkeley - Fermilab - Princeton Collaboration, BFP
3) European Muon Collaboration at the CERN SPS, EMC[+]

The heavy quarks can be identified by their fragmentation into heavy mesons and weak semileptonic decay into secondary muons, e.g. $D^{\pm} \rightarrow \mu^{\pm} \nu K^{o(*)}$, $D^{o(\bar{o})} \rightarrow \mu^{\pm} \nu K(F)$.

All 3 experiments use calorimeter targets, thus being able to measure the "missing" energy carried away by neutrinos. An other way to detect the production of heavy quarks is the search for $\mu^{+}\mu^{+}\mu^{-}$ events where the generated $\mu^{+}\mu^{-}$ mass spectrum peaks at the mass of a hidden heavy quark state like J/ψ, ψ', T.

The MSUF experiment is finished and dimuon results for beam en-

---

[+] The EMC is a Collaboration of the following institutions or universities: CERN-DESY(Hamburg)-Freiburg-Kiel-Lancaster-LAPP(Annecy)-Liverpool-Oxford-Rutherford-Sheffield-Torino and Wuppertal

ergies of 150 and 270 GeV are published [1]. BFP has also results published for dimuon and J/ψ production at 209 GeV beam
energy [2]. The EMC has made measurements of multimuon final
states at muon beam energies of 120, 200, 250 and 280 GeV. The
results of 280 GeV energy are published [3], the analysis of
the 250 GeV data is partly finished [4] and other preliminary
results are available.

II. Models

The production of heavy quarks is interesting as the large
masses involved imply a short distance scale. Such hard processes are of special interest for tests of QCD. The quark
production process can be described using different models:

The photon gluon fusion model (PGF) [5] describes the interaction as the fusion of a photon with a gluon producing a
quark antiquark pair (Fig. 1a).
An other picture of the process is used in quark-antiquark
fusion models [6]. Here the photon is thought to be a virtual
quark antiquark state (Fig. 1b). The charm part of the photon
structure function is thought to be well known [7].
The classical model for photoproduction of vector mesons is
the Vector Mason Dominance model (VDM, [8]). The photon couples
directly to a heavy vector meson which than again scatters as
a hadron off the nucleon (Fig. 1c).

III. Results

a) Open charm production

In the following I will concentrate mainly on preliminary
EMC results which are compared with results of BFP.
The branching ratio for decays of charmed mesons into muons
can be measured by comparing the ratio of the produced 3
muon and 2 muon final states. EMC finds a ratio:
$B_{Charm} \to \mu X = 7 \pm 3$ %. The charmed quark fragmentation
function is badly known up to now. It can be measured,
using the assumptions that

1) the charmed quark momentum spectrum is given by a
   model as e.g. by the first order PGF model, [5c]
2) all charmed quarks fragment into D-mesons and
3) the D-mesons decay with equal probability in
   $\mu\nu K$ and $\mu\nu K^*$

The comparison between the given quark distribution and the
measured longitudinal momentum distribution of the decay
muons yields the fragmentation distribution $F_{C \to D}$ (z) =
$0.4 \exp^{(1.6 \pm 1.6)z}$ with $z = (\frac{P_D}{P_C})_{Lab}$

The cross section for 2μ events measured within kinematical
cuts is shown in Fig. 2 as a function of Q^2, [3a]. The background from muons produced by pion or kaon decay within the

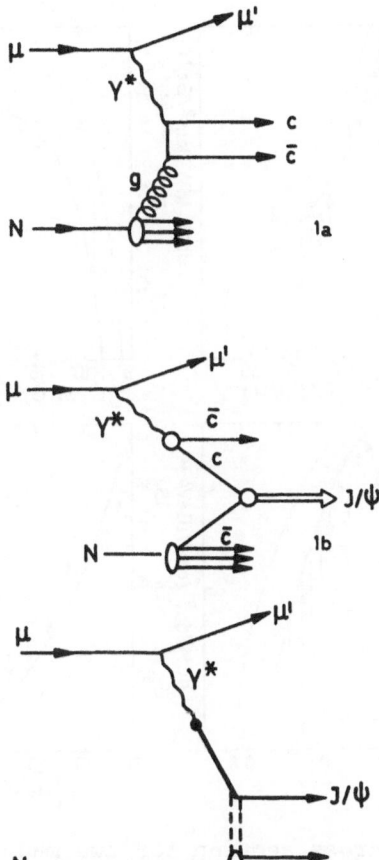

Fig. 1 a) The photon gluon fusion process
        b) J/ψ production via c̄c fusion
        c) J/ψ production in the Vector Meson Dominance model.

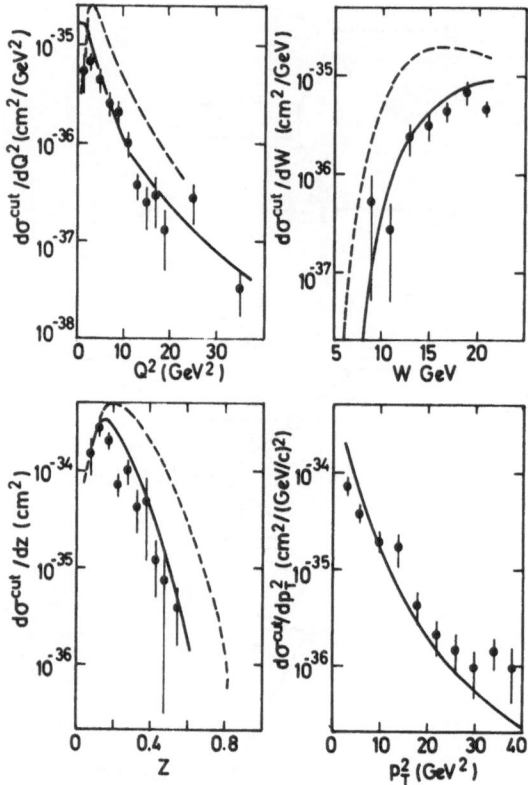

Fig. 2 Differential cross section for two muon final states in the
        kinematic region $y < 0.91$, $Q^2 > 1$ GeV^2, $\Theta_{\mu_1} > 7$ mrad and
        $E_{\mu_2} > 16$ GeV as measured by EMC. The π,K decay back-
        ground contribution is subtracted.

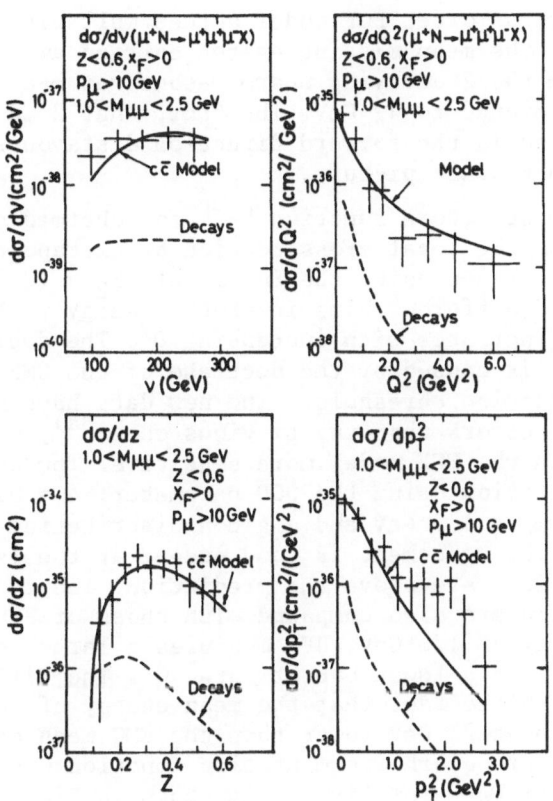

Fig. 3 Differential cross section for trimuons inside the given kinematic cuts and for $E_{Calorimeter} > 20$ GeV. The solid curves are the calculated absolute values from the photon gluon fusion model [5c]. The dashed curves show the calculated background from π,K decays being already subtracted from the data. z and p_t^2 refer to the $\mu^+\mu^-$ pair.

hadronic shower in the calorimeter is 10 ± 5 % and is subtracted. The cross section predicted by the first order PGF model
is in good agreement with data in contrast to a calculation
using a Buras Gaemers charm quark sea [9] from where the quarks
are knocked out carrying the whole energy of the virtual photon.

Fig. 3 shows cut cross section for $\mu^+\mu^+\mu^-$ events [3b] where the
mass of the produced $\mu^+\mu^-$ pair was between 1 and 2.5 GeV.
Again the first order PGF model agrees well with the data.
Furthermore the mean missing energy carried away by the 2 neutrinos from the 2 decaying charm mesons was measured to 44 ± 7
GeV and predicted to 42 GeV. The fact, that 2 additional muons
were observed in the forward direction disfavours again any
single struck quark picture.

The nucleon structure function $F_2^{c\bar{c}}$ for charmed quarks can be
derived from the total cross section by extrapolating over the
unmeasured z range using the PGF model. $F_2^{c\bar{c}}(Q^2)$ is shown in
Fig. 4 for 3 different bins in photon energy ν. F_2 is zero at
$Q^2 = 0$ and increases with increasing Q^2. The decrease observed
at large Q^2 is caused by the decrease of the CMS energy towards
the $D\bar{D}$ production threshold. The new data have much smaller
statistical errors than the previous ones [3a], making a comparison with the PGF model more sensitive. The curve is a PGF
model calculation using the QCD parameter: $\Lambda = 0.15$ GeV, the
quark mass $m_c = 1.5$ GeV and a gluon distribution $G(\eta) =$
$3(1-\eta)^5/\eta$. The agreement is good except at the largest Q^2 where
the data points are above the prediction. The results in the
highest ν bin are also compared with those of BFP [2b] at a mean
photon energy of 178 GeV. BFP measures a larger cross section
at small Q^2 and a lower one at large Q^2. The difference is
mainly due to the fact that the mean energy of the BFP data
points is about 22 GeV lower than the EMC mean energy. Also a
different charm quark fragmentation function was used. The Q^2
dependence of $F_2^{c\bar{c}}$ for fixed x is shown in Fig. 5. Here the
CMS energy increases with Q^2 leading to an increase of the
charm structure function above threshold. Also plotted is the
total nucleon structure function F_2 of hydrogen for x = 0.03
and of iron for x = 0.25 and x = 0.45. These data are from EMC
single arm measurements [10]. For the lower x bin a clear increase of the structure function by about 30 % over the measured range in Q^2 is observed. A comparison with the lower curves shows that at that x the open charm production contributes
to the observed scaling violation of F_2 only by about 30 %.

b) Intrinsic charm in the nucleon?
In 1979 several experiments at the CERN ISR found a relatively
large diffractive production of D^+ and Λ_c^+ [11]. This led to the
suggestion [12] that some intrinsic charm of the order of 1 %
could exist in the nucleon at low Q^2. Such heavy quarks carry

a larger fraction of the nucleon momentum than the lighter ones. Thus one expects to find them at larger x.

Fig. 6 shows the cross section for charm production within kinematical cuts and integrated over Q^2. The curve labelled PGF is a photon gluon fusion model prediction with standard parameters as quoted above. For x > 0.2 the data points are above the PGF model. Nevertheless the prediction of the intrinsic charm model is clearly above the data (IC). The charmed quark distribution is assumed to evolve with Q^2 according to the Altarelli-Parisi equations. The dashed curve represents such a evolved distribution where the QCD coupling α_s is fixed to 0.4 and the charmed quark mass is neglected [13]. The data points are still below the curve and give a limit on the intrinsic charm contribution of 0.59 % with 90 % confidence level. Nevertheless the center of mass energy for the ISR results of 53 GeV is much larger than the EMC energy of 13 - 15 GeV in the large x range. It may be possible to find a reasonable threshold development in order to get agreement between the intrinsic charm model and the data [14].

c) J/ψ production

EMC has also measured J/ψ production by triggering on 3μ events in 280 and 250 GeV muon iron interactions. The mass spectrum of the produced $\mu^+\mu^-$ pair is shown in Fig. 7. The width of the observed J/ψ peak is mainly due to the multiple scattering of the produced muons in the iron target.

Events with z > 0.95 were called elastic, where $z = E_\psi/E_\gamma$. The virtual photoproduction cross section for elastic and inelastic J/ψ's is shown in Fig. 8 for 2 ν bins. There is also observed an elastic nuclear coherent contribution which is subtracted from the data points in the following. The Q^2 dependence has been fitted to the mass propagator $(1 + \frac{Q^2}{M^2})^{-2}$ as used in

Table I

VDM models. The results for M are listed in table I. M is in the range of the J/ψ mass as expected from VDM models but for elastic and inelastic production a slight increase with ν is

<ν>	M (GeV) z < 0.95	M (GeV) z > 0.95
90 GeV	2.3 ± 0.2	2.9 ± 0.3
150 GeV	2.7 ± 0.3	3.6 ± 0.4

Results of a mass propagator fit to the Q^2 distribution of the J/ψ

observed. The EMC results can be compared with the results of BFP. Unfortunately both groups use slightly different definitions for the 'elastic' and 'inelastic' cross sections. EMC

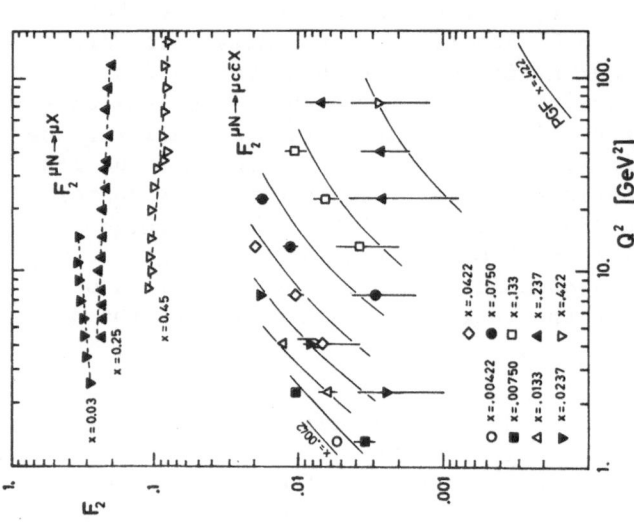

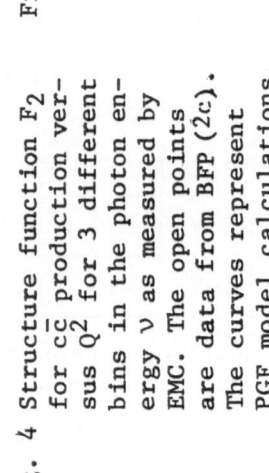

Fig. 5 Structure function F_2 for $c\bar{c}$ production versus x for 9 different bins in x between x = 0.0042 and x = 0.42 as measur-
ed by EMC. The results are compared with PGF model calculations. Also shown is the total structure function F_2 also
from EMC for x = 0.03, (measured on H_2) 10a) and for x =
0.25 and x = 0.45 (measured on Fe) 10b). The symbols used
are the same as in the corresponding bins of $F_2^{c\bar{c}}$.

Fig. 4 Structure function F_2
for $c\bar{c}$ production versus Q^2 for 3 different
bins in the photon en-
ergy ν as measured by
EMC. The open points
are data from BFP (2c).
The curves represent
PGF model calculations.

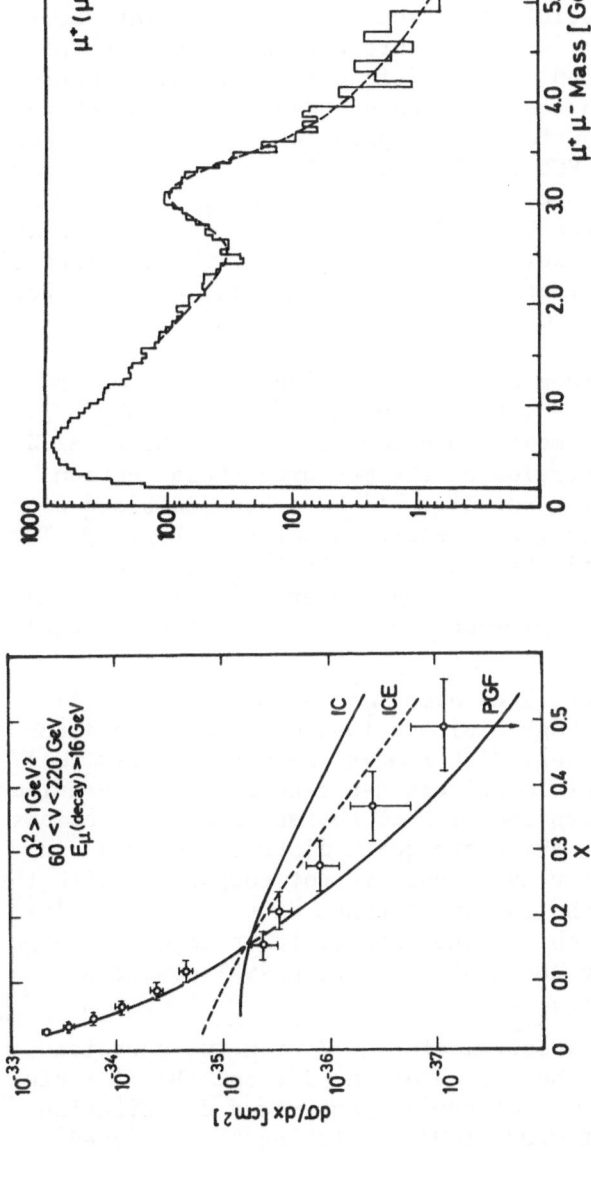

Fig. 7   Mass distribution of the produced $\mu^+\mu^-$ pair events of the type $\mu^+N \rightarrow \mu^+\mu^+\mu^-$ X measured by EMC. Only events with $P_\mu > 10$ GeV and $60 < \nu < 200$ GeV are used.
The curve is a fit to a Weibel distribution background and a gaussian distribution centered of the J/ψ mass. The results of the fit are $m_{J/\psi} = 3.082 \pm 0.007$; $\sigma = 0.206$ GeV.

Fig. 6   Variation of $d\sigma/dx$ with x for dimuon events in the range $Q^2 > 1$ GeV^2, $60 < \nu < 220$ GeV and a decay muon energy > 16 GeV. The horizontal bins represent the bin widths. The curves are:
PGF : photon gluon fusion model,
IC : intrinsic charm model,
ICE : intrinsic charm model with maximum Q^2 evolution.

separates both data sets at z = 0.95 whereas BFP uses a cut at
z = 0.9.
The propagator masses measured by BFP [15] are M = 2.7 ± 0.5 GeV
for the 'elastic' and M = 3.0 ± 0.2 for the 'inelastic' sample.
These values are averaged over all measured photon energies.
Fitting PGF model results with such an ansatz gives a mass parameter around 2.7 GeV.

Fig. 9 shows the cross section for elastic and inelastic J/ψ
photoproduction extrapolated to Q^2 = 0 as a function of the photon energy ν. BFP measures due to the different elasticity cut
a slightly larger 'elastic' cross section and a smaller 'inelastic' one. But both experiments measure within the errors the
same total J/ψ production cross sections for all photon energies. At the largest photon energies observed the total photoproduction cross section for J/ψ's approaches 50 nb. The results
are also compared with PGF model calculations. The first order
calculation [5] describes the elastic cross section fairly good,
the higher order calculation of Baier and Rückl [16] using a quark
mass of 1.25 GeV agrees well with the shape and the absolute size
of the inelastic cross section. This calculations are made for
inelastic J/ψ's with z < 0.9.

In the PGF model the gluon momentum distribution G(η) may be
extracted from the data [5]. For fixed photon energy only a limited range in the gluon momentum η contributes to the total J/ψ
cross section. Thus inversion of the measured cross section
yield G(η). Data at lower energies [17] have been combined with
this data to extract G(η). The result is shown in Fig. 10. The
curve with the parametrization G(η) = $3(1-\eta)^5/\eta$ gives a good
representation of the data. The range covered in Q^2 is to small
to observe within the experimental errors a variation of G(η)
with Q^2.

The z-distribution, integrated over p_t^2 is shown in Fig. 11. A
strong peak is seen at high z with a long decreasing tail towards large inelasticities. Higher mass charmonia decays to J/ψ
can not contribute significantly as has been discussed previously [3d]. The data are compared with c$\bar{c}$ fusion model predictions [18]
using various forms for the charm quark distribution c(x) in
the nucleon [12, 9]. All predictions are not compatible with the
data. The curve using the intrinsic charm distribution of [12]
is for 100 % intrinsic charm. The data at low z impose an upper
limit on such intrinsic charm of 2.5 % with 90 % confidence.

d) <u>ψ' production</u>
A fit on the $\mu^+\mu^-$ mass spectrum of Fig. 7 with two gaussian
distributions fixed at the masses of the J/ψ and the ψ' including an exponential and a polynomial background distribution
yielded a ψ'/ψ ratio of 0.025 ± 0.011. Taking the measured

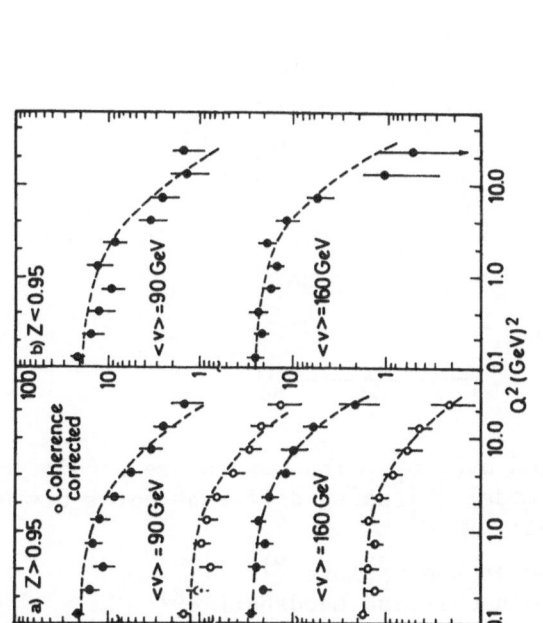

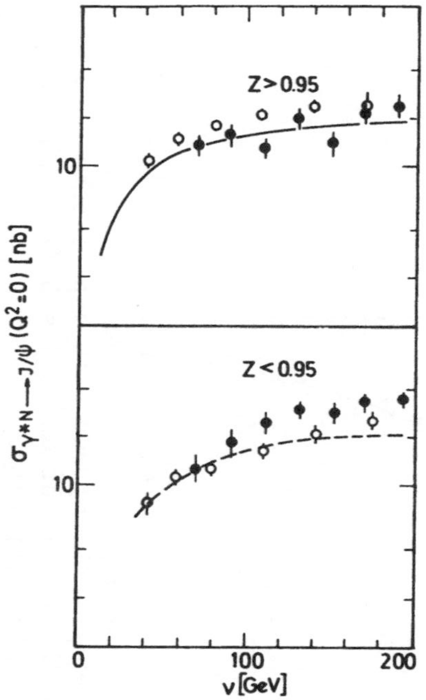

Fig. 9 Real potoproduction cross section as extrapolated to $Q^2 = 0$ versus the photon energy ν.

    a) for high z after subtraction of the nuclear coherent contribution
    b) for low z.

The full points are from EMC, the open circles are from BFP. The full curve for high z represents a first order PGF model calculation 5d), the dashed curve at low z shows the result of the color singlet model 16) for m_c=1.25 GeV.

Fig. 8 σ_{eff} ($\gamma^* N \to J/\psi X$) versus Q^2 for different samples:

    a) $\langle\nu\rangle$ = 90, high z
    b) $\langle\nu\rangle$ = 160, high z
    c) $\langle\nu\rangle$ = 90, low z
    d) $\langle\nu\rangle$ = 160, low z

The open circles show data after subtraction of the nuclear coherent corss section. The curves are fits to the VDM form $1/(1 + Q^2 / M^2)$.

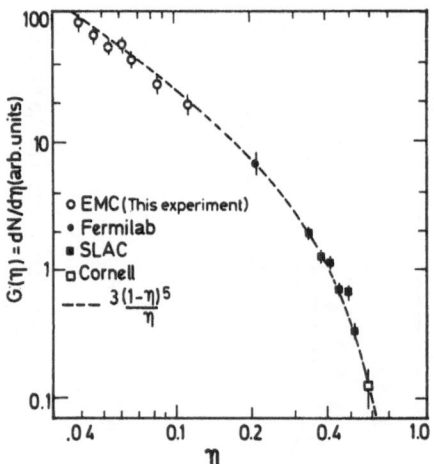

Fig. 10   Gluon momentum distribution derived using the first order
          PGF model [5]. The data points are from EMC (open circles)
          and from lower energy photoproduction experiments [17].
          The curve shows the distribution $G(\eta) = 3\,(1-\eta)^5/\eta$ where
          the normalisation is arbitrary.

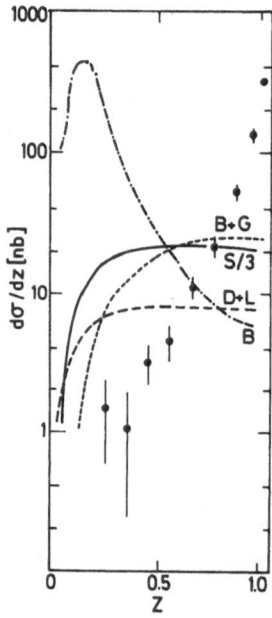

Fig. 11 $d\sigma/dz$ integrated over p_t^2 . The curves are predictions of
          the $c\bar{c}$ fusion model [18] using different forms of the nu-
          cleon charm sea:

          B + G = Buras and Gaemers [9]
          D + L = Donnachie and Landshoff [18]
          B   = Brodsky et al. intrinsic charm [12]
          S/3   = strange sea /3.

branching ratio into 2 muons one deduces a ratio of the ψ' to ψ production cross section of 0.22 ± 0.10 for a mean photon energy of 140 GeV. Preliminary EMC results on hydrogen and deuterium give a similar ratio.

e) Ypsilon production
In the $\mu^+\mu^-$ mass spectrum one event has a mass of 10 GeV. This is used to give an upper limit for ypsilon production. The extrapolated background predicts 0.14 events for the mass range between 9.1 and 10.3 GeV. From that an upper limit for the cross section times branching ratio in $\mu^+\mu^-$ can be derived

$B \cdot \sigma_{\mu N \to \mu YX} \leq 5.2 \cdot 10^{-38}$ cm^2 (90 % CL), E_μ = 250 GeV. A similar result is quoted by BFP [2b] $B \cdot \sigma_{\mu N \to \mu YX} \leq 2.2 \cdot 10^{-38}$

cm^2 (90 % CL), E_μ = 209 GeV.

f) Open beauty production
The EMC observes wrong sign 3 muon events with large missing energy and large transverse momentum p_t [4a]. Two events are of the type $\mu^+ N \to \mu^+\mu^+\mu^+ X$ and one $\mu^+ N \to \mu^+\mu^-\mu^- X$. The background expected from the decay of charm, π, K is about one event.

Such events can be produced by the decay of a pair of beauty mesons, where one muon is produced by the semileptonic decay of such a meson and the other one by the semileptonic decay of a charm meson in the decay chain. Assuming that the mean branching ratios for both charm and beauty meson decay are 10 % an upper limit for $B\bar{B}$ production is derived:

$$\sigma_{\mu N \to \mu B\bar{B}X} < 12 \cdot 10^{-36} \text{ cm}^2 \text{ (90 % CL)}$$

Wrong sign events can also be produced by charm pair production with D^0-$\bar{D^0}$ mixing. Assuming that half of the charm produced are D mesons whose different charged states are produced with equal probability leads to an upper limit for D^0-$\bar{D^0}$ mixing of 20 % (90 % CL). This is similar to previous limits [19].

REFERENCES

1) C. Chang et al, Phys. Rev. Lett. <u>39</u> (1977) 519
    D. Bauer et al, Phys. Rev. Lett. <u>43</u> (1979) 1551

2) a) Clark et al, Phys. Rev. Lett. <u>45</u> (1980) 682
    b) Clark et al, Phys. Rev. Lett. <u>45</u> (1980) 686
    c) Clark et al, Phys. Rev. Lett. <u>45</u> (1980) 1465
    d) Clark et al, Phys. Rev. Lett. <u>43</u> (1980) 187
    e) Clark et al, Phys. Rev. Lett. <u>45</u> (1980) 2092

3) a) EMC, J.J. Aubert et al, Phys. Lett. <u>94B</u> (1980) 96
    b) EMC, J.J. Aubert et al, Phys. Lett. <u>94B</u> (1980) 101
    c) EMC, J.J. Aubert et al, Phys. Lett. <u>89B</u> (1980) 267
    d) EMC, J.J. Aubert et al, CERN-EP/80-84
    e) EMC, O.C. Allkofer et al, Nucl. Instrum. & Meth. <u>179</u> (1981)
                                                                    445

4) a) EMC, J.J. Aubert et al, Phys. Lett. <u>106B</u> (1981) 419
    b) EMC, J.J. Aubert et al, Phys. Lett. <u>110B</u> (1982) 73

5) a) M. Glück and E. Reya, Phys. Lett. <u>79B</u> (1978) 453
    b) M. Glück and E. Reya, Phys. Lett. <u>83B</u> (1979) 98
    c) S. Leveille and T. Weiler, Nucl. Phys. <u>B147</u> (1979) 147
    d) T. Weiler, Phys. Rev. Lett. <u>44</u> (1980) 304
    e) V. Barger, WY Keung, R.J.N. Phillips, Phys. Rev. <u>D20</u> (1979)
                                                                    630
6) K.T. Mahanthappa and J. Randa, Phys. Rev. D23 (1981) 696
    A. Donnachie and A. Voudas, Z. Physik <u>C7</u> (1981) 257

7) R.J. DeWitt et al, Phys. Rev. <u>D19</u> (1979) 2046

8) J.J. Sakurai and D. Schildknecht, Phys. Lett. <u>40B</u> (1972) 121
    and references therein
    T. Bauer et al, Rev. Mod. Phys. <u>50</u> (1978) 261

9) A.J. Buras and K.J.F. Gaemers, Nucl. Phys. <u>B132</u> (1978) 249

10) EMC, J.J. Aubert et al, Phys. Lett. <u>105B</u> (1981) 315
    EMC, J.J. Aubert et al, Phys. Lett. <u>105B</u> (1981) 322

11) D. Drijard et al, Phys. Lett. <u>81B</u> (1979) 250
    K.L. Giboni et al, Phys. Lett. <u>85B</u> (1979) 437
    W. Lockmann et al, Phys. Lett. <u>85B</u> (1979) 443
    D. Drijard et al, Phys. Lett. <u>85B</u> (1979) 452
    Nuov. Cim. Lett. <u>33</u> (1982) 33

12) S.J. Brodsky et al, Phys. Lett. <u>93B</u> (1980) 451
    S.J. Brodsky et al, Phys. Rev. <u>D23</u> (1981) 2747

13) R.J.N. Phillips, private communication

14) D.P. Roy, Phys. Rev. Lett. 47 (1981) 213

15) M. Strovink, Proceedings of the 1981 International Symposium on Lepton and Photon Interactions at High Energies, BONN (1981) 594

16) R. Baier and R. Rückl, Universität Bielefeld, Bi-TP 81/30

17) B. Knapp et al, Phys. Rev. Lett. 34 (1975) 1040
T. Nash et al, Phys. Rev. Lett. 36 (1976) 1233
U. Camerini et al, Phys. Rev. Lett. 35 (1975) 483
B. Gittleman et al, Phys. Rev. Lett. 36 (1975) 1616

18) A. Donnachie and A. Vourdas, Z Physik C7 (1981) 257
K.T. Mahanthappa and J. Randa, Phys. Rev. D23 (1981) 696
Y. Afek, C. Leroy and B. Margolis, Phys. Rev. D22 (1980) 93

19) G. Goldhaber et al., Phys. Lett. 69B (1977) 503
G.J. Feldman et al., Phys. Rev. Lett. 38 (1977) 1966

HADRONIC PRODUCTION OF HEAVY FLAVORS

R. Odorico

Istituto di Fisica dell'Università and I.N.F.N., Bologna

ABSTRACT

Problems met by perturbative QCD models in describing hadronic production of open charm are reviewed. The possibly important role of the often neglected flavor excitation diagrams is discussed. Ways are suggested to overcome difficulties in their calculation.

Recent experimental surveys on the subject can be found in[1,2]. The theoretical understanding of the data has been summarized by Phillips[3] at the Madison Conference (1980). The main conclusions about conventional perturbative QCD models based on fusion diagrams ($GG \rightarrow c\bar{c}$, $q\bar{q} \rightarrow c\bar{c}$, see Fig. 1) are:
i) they provide an adequate description of ψ and Υ meson production;
ii) they predict a too low σ(open charm);
iii) they are not able to reproduce the observed diffractive production of Λ_c^+.

Such difficulties may prompt consideration of alternative approaches to hadronic production of open charm. A number of them have been proposed recently, like diffractive dissociation[4], intrinsic charm[5], diquark recombination[6].

But, before discarding perturbative QCD models in this context, one should better clarify the role of a poorly known and often ne-

411

glected contribution, that of flavor excitation diagrams (qc $\rightarrow$ qc, Gc $\rightarrow$ Gc, see Fig. 1). Their calculation requires knowledge of the charm structure function $c(x,Q^2)$ at the evolution scale of interest. In addition they present a $d\widehat{\sigma}/d\hat{t} \propto 1/\hat{t}^2$ divergence (gluon exchange) in the kinematically allowed region, which requires special (?) treatment.

Serious consideration to flavor excitation diagrams was first given by Combridge[7] in an extensive paper dating back in 1979. In the calculations that he presents the Buras and Gaemers[8] parametrization is assumed for $c(x,Q^2)$. In the treatment of[8] the charm sea is evoluted as normal sea, i.e. without taking into account any mass effects, except for a delay in the evolution which is assumed to start at $Q_0^2 = 1.8$ $(GeV/c)^2$ (coming from fits to the SLAC–MIT electroproduction data). This value is also assumed for the $\hat{t}_{min}$ cutoff on the divergence, without further justification. Partly because of the unconstrained evolution of the charm sea, and partly because of the identification of the evolution scale with $\hat{s}$, the energy invariant for the parton subprocess, Combridge finds a huge and fast rising charm cross section. We know at present that the Buras and Gaemers parametrization does not provide even a qualitative description of the recently appeared BFP[9] and EMC[10] data for the charm structure function $F_2(charm)$.

Interest in flavor excitation diagrams has been revived in a recent paper by Barger, Halzen and Keung[11]. They assume $x\,c(x,Q^2) =$ $x(1-x)$ at the evolution scale appropriate for the calculation, $Q^2 \simeq 4m_c^2$, m_c being the charm quark mass. As to $\hat{t}_{min}$, it is assumed $\hat{t}_{min} \approx m_c^2$ but uncertain by factors. They enphasize the role of the charm quark spectator which is assumed to fragment into charmed hadrons (especially Λ_c^+) with the latter carrying all the nucleon momentum left by the interacting charm quark. As to the problem of meeting upper bounds on $c(x,Q^2)$ at large x from deep inelastic scattering data (which is a sensitive issue for the intrinsic charm model, in a way similar to theirs), it is argued that at large Q^2 ordinary QCD degradation should change $c(x,Q^2)$ back into a normal sea shape. With these assumptions one is able to reproduce the observed level of $\sigma(charm)$ and the hard spectrum of Λ_c^+.

These calculations show that flavor excitation may represent the dominant contribution in charm hadroproduction and that it may account for the existing data. However, in order to claim that the

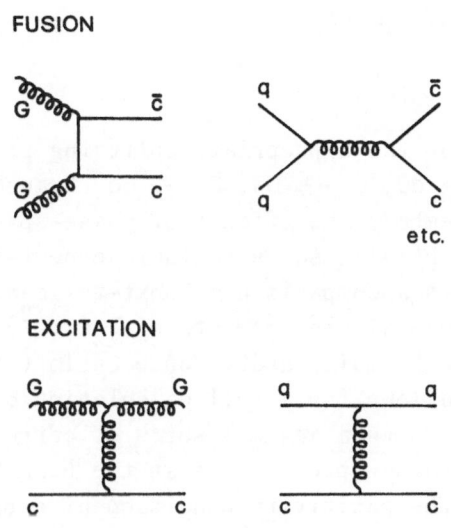

Fig. 1. Typical fusion and flavor excitation diagrams.

perturbative QCD approach is really adequate for the process at hand
one must first provide a quantitative calculation of the evoluted
$c(x,Q^2)$ and of the charm spectator spectrum, $dN/dx_{spectator}$,
according to perturbative QCD and taking duly into account mass
effects. Conventional evolution equations are insufficient in this
respect, since they are inclusive equations and therefore cannot
handle phase-space constraints correctly. One needs an exclusive
approach to the LLA evolution and the only one available at present
is the QCD Monte Carlo.

The QCD Monte Carlo is a mathematical technique for handling
the QCD LLA cascade[12]. Its only input is the elementary emission
probability:

$$dP_E = \frac{\alpha_s(K^2)}{2} \frac{dK^2}{K^2} \int_{\mathcal{E}_1(K^2)}^{1-\mathcal{E}_2(K^2)} P(z) \, dz \qquad\qquad 1)$$

where $P(z)$ is one of the appropriate splitting probability
functions (for $q \rightarrow qG$, $G \rightarrow GG$, $G \rightarrow q\bar{q}$ branchings), and
$\mathcal{E}_1(K^2)$, $\mathcal{E}_2(K^2)$ embody the effects of phase-space bounds. One
can also modify the $P(z)$'s so as to include next-to-leading
effects. Fig. 2 shows a comparison of next-to-leading corrections
to the evolution of non-singlet moments of $F_2(x,Q^2)$ calculated by
the standard analytic formulae and by Monte Carlo ($\overline{MS}$ scheme)[13,14].
The QCD Monte Carlo allows for a full calculation of the final state,
including transverse momenta and all sorts of correlations. It is
also easy to impose phase-space bounds at the branchings. The latter
can be derived from the positivity condition at a branching:

$$K_T^2 = z(1-z)(K^2 - K_1^2/z - K_2^2/(1-z)) \geqslant 0 \qquad\qquad 2)$$

where K_T is the transverse momentum generated at the branching;
K^2, K_1^2 and K_2^2 are the square (virtual) masses of the parent, and
of the two secondaries, respectively; z is the momentum fraction
of the parent taken by the first secondary (K_1^2). In a spacelike
evolution $K_1^2 \ll K^2 < 0$. If the branching involves charm pair
creation, $G \rightarrow c\bar{c}$, it must be $K_2^2 \geqslant m_c^2$. Immediately after the
branching, one has then the inequality[15]:

$$x_{spectator} \geqslant x_{active} \, m_c^2/Q^2 \qquad\qquad 3)$$

where x_{active} is the momentum fraction of the (spacelike) c
quark which (after further evolution) will be active in the interaction process, and $x_{spectator}$ represents the same quantity for

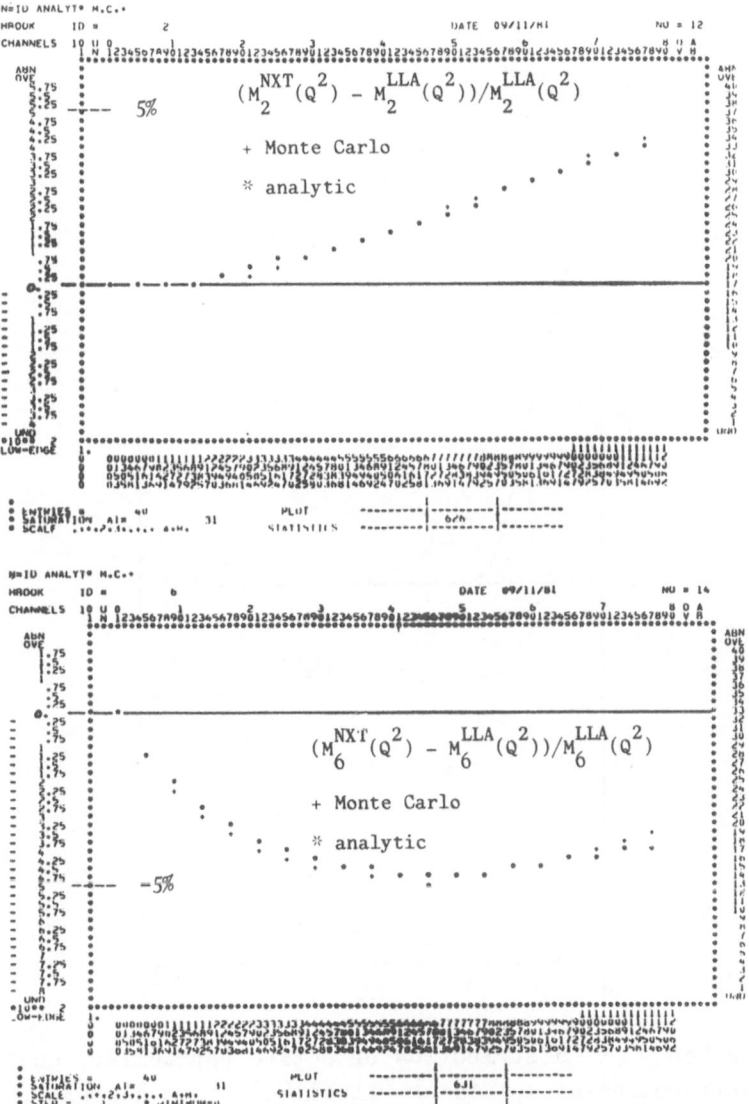

Fig. 2. $(M_n^{NXT} - M_n^{LLA})/M_n^{LLA}$ versus $\log(\log(Q^2/\Lambda^2)/\log(Q_0^2/\Lambda^2))$, where M_n^{LLA} and M_n^{NXT} are the nth moments of the non-singlet structure function calculated in the LLA and including next-to-leading corrections, respectively ($\Lambda = 0.3$ GeV , $Q_0^2 = 1$ $(\text{GeV/c})^2$, $Q_{MAX}^2 = 200$ $(\text{GeV/c})^2$). Stars: analytic calculation. Crosses: Monte Carlo results. $n = 2$ above, $n = 6$ below.[13,14]

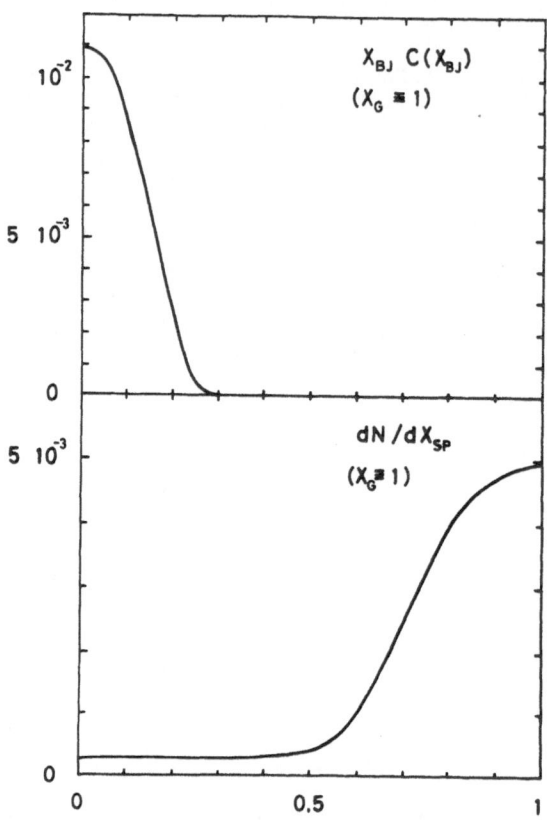

Fig. 3. Bjorken's x distribution of active (spacelike) charm quarks and x distribution of spectator (timelike) charm quarks from gluon initiated QCD cascades as calculated by the QCD Monte Carlo ($x_G \equiv 1$ for the initial gluon, $\alpha_s = \alpha_s(Q^2 + 4m_C^2)$, $Q^2 = m_C^2$, $m_C = 1.5$ GeV)[15].

the (timelike) c quark which does not participate in the interaction. For $Q^2 \gg m_c^2$ the inequality is ineffective and the spectra of the two c quarks become essentially equal (deep inelastic scattering regime), but for $Q^2 \approx m_c^2$, which applies to hadroproduction of charm, the inequality strongly favors the flow of momentum to the spectator quark. Considering LLA cascades started by gluons (which dominate the yield of charm) at $Q^2 = 4m_c^2$ ·one obtains the x distributions of Fig. 3 [15]. In an actual calculation these distributions must be convoluted with the initial gluon distributions. From Fig. 3 one learns that, according to perturbative QCD, even at low Q^2 $c(x,Q^2)$ remains peaked at x $\approx$ 0 , differently from what assumed by[11]. However, the spectator c has a substantially harder spectrum which,hopefully, can accomodate the observed diffractive production of Λ_c^+ .

A complete calculation of the charm cross section requires appropriate fixing of the $\hat{t}_{min}$ cutoff, and an hadronization model is required in order to discuss the longitudinal spectra of charmed hadrons. Work along these lines is in progress[16]. For $\hat{t}_{min}$ one can observe that the transverse momentum distribution of charmed hadrons much depends on this parameter, therefore the latter can be fixed by fitting the corresponding experimental distributions. In other words, the interaction radius which is left undefined by the $d\hat{\sigma}/d\hat{t}$ $\propto$ $1/\hat{t}^2$ divergence can be directly taken from the data. As to the hadronization model for the final state, I think that the Field and Feynman quark fragmentation model is inadequate in our case. This model was introduced to deal with electron-positron annihilation, deep inelastic scattering and large p_T processes, that is processes in which the fast fragmenting quark is far in phase-space from other fast quarks, so that recombination with them into low-mass hadrons is kinematically impossible. But in charm production we have to deal, for instance, with the hadronization of the spectator charm quark, which is quite close in phase space to the other light quark spectators. Use of the recombination model or, more in general, of recombinations ideas seems inescapable.

REFERENCES

1. S. Wojcicki, Proc. Int. Conf. on High-Energy Physics, Madison (1980), p. 1430.
2. D. Treille, Proc. Int. Symp. on Lepton and Photon Interactions at High Energies, Bonn (1981).
3. R. J. N. Phillips, Proc. Int. Conf. on High-Energy Physics,

Madison (1980), p. 1470.

4. G. Gustafson and C. Peterson, Phys. Lett. 67 B, 81 (1977).

5. S. J. Brodsky, P. Hoyer, C. Peterson and N. Sakai, Phys. Lett.
    93 B, 451 (1980);

    S. J. Brodsky, C. Peterson and N. Sakai, Phys. Rev. D 23, 2745
    (1981);

    D. P. Roy, Phys. Rev. Lett. 47, 213 (1981);

    R. V. Gavai and D. P. Roy, Tata preprint, TIFR/TH/81-7 (1981).

6. R. Horgan and M. Jacob, CERN-TH 2824 (1981).

7. B. L. Combridge, Nucl. Phys. B 151, 429 (1979).

8. A. J. Buras and K. J. F. Gaemers, Nucl. Phys. B 132, 249 (1978).

9. A. R. Clark et al. (BFP Collaboration), Phys. Rev. Lett. 45,
    1465 (1980).

10. C. H. Best (EMC Collaboration), Talk at the XVI Rencontre de
    Moriond, Les Arcs (1981), Rutherford preprint RL-81-044 (1981);

    G. Coignet (EMC Collaboration), Talk at the EPS Int. Conf.,
    Lisbon (1981).

11. V. Barger, F. Halzen and W. Y. Keung, Madison preprint DOE-ER/
    /00881-215 (1981).

12. R. Odorico, Nucl. Phys. B172 (1980) 157;

    G. C. Fox and S. Wolfram, Nucl. Phys. B 168, 285 (1980).

13. R. Odorico, Phys. Lett. 102 B, 341 (1981).

14. A. Sansoni, Thesis, University of Bologna (1981).

15. R. Odorico, Phys. Lett. 107 B, 231 (1981).

16. R. Odorico, in preparation.

DISCUSSION

P. Musset: Does the Monte Carlo QCD calculation contain the "intrinsic" charm mechanism?

Answer: No intrinsic charm is assumed. The aim of the calculation
is to show that "extrinsic" charm alone, i.e. charm generated
exclusively in the QCD evolution process, is enough to explain the
data.

SILICON MICROSTRIP DETECTORS, A NEW TOOL FOR CHARM PHYSICS

E. Heijne[+] and P. Jarron

CERN
EF Division
CH-1211 Geneva 23

PARTICLE DETECTION IN SILICON

The characteristic property of a monocrystalline semiconducting material is the small separation between the electronic conduction band and the valence band. A small amount of energy, Eg = 1.12 eV in the case of silicon, is therefore sufficient to excite an electron into the conduction band. The hole left in the valence band behaves like an independently moving positive charge carrier and has a mobility which is comparable to that of the free electron (1500 cm² V⁻¹ s⁻¹ for e⁻ and 600 cm² V⁻¹ s⁻¹ for the hole[1]. At a given temperature there is equilibrium between the generation and recombination of free electrons and holes. Impurities and crystal defects act as generation or recombination centers and reduce therefore the mean free carrier lifetime. The introduction of electrically active donor or acceptor impurity atoms ("doping") results in a n-type or p-type conductivity semiconductor with an excess of electrons respectively holes. The product of the concentrations of electrons n and holes p remains however constant

$$n \cdot p \propto e^{-\frac{Eg}{kT}}$$

Free charge can be generated in excess of this thermal equilibrium value by electromagnetic or particle irradiation. It has been found that on the average 3.62 eV is deposited in the silicon for every generated electron-hole (e-h) pair. This value is apparently independent of the type of particle. The difference

---

[+]Presented by E. Heijne

419

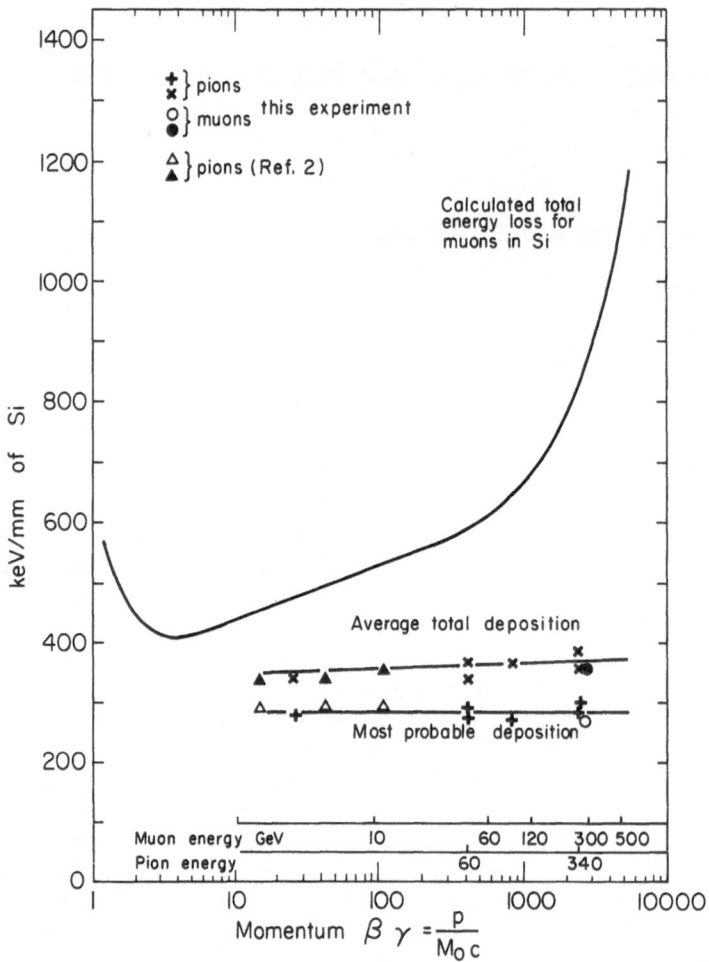

Fig. 1.   Total energy loss for muons in silicon, including
          ionization loss, pair production, Bremsstrahlung and
          nuclear loss (calculated by Claude-Serre[3]) compared
          with measured energy deposition in thin silicon
          detectors. The lower points are the most probable
          deposit (peak of the Landau distribution), the higher
          points represent the mean values. Data were obtained
          with pions, protons and muons by Esbensen et al.[2] and
          Heijne, Jarron and Viertel (unpublished).

with the bandgap energy (3.62 - 1.12 eV) is spent on other types
of excitations in the Si lattice. If the mean free carrier
lifetime is sufficiently long, the radiation-produced charge can
be integrally collected by applying an electrical field to the
semiconductor. In the case of Si at room temperature however, a
rectifying diode structure with non-injecting contacts must be
used, to avoid a large steady current, which would mask the small
radiation induced signals.

Minimum ionizing particles deposit energy in thin silicon
detectors, and the energy spectrum is similar to the one described
by Landau for energy loss. The most probable energy deposit is
about 27 keV per 100 μm of Si and this energy deposit is nearly
independent of momentum, as shown in fig. 1 (see also ref. 2).
The specific energy deposit is however a non-linear function of
the detector thickness, and for a 1 mm thick detector the value is
about 32 keV per 100 μm of Si. To collect all the charge
generated in the detector thickness, the electrical field, and
therefore the diode space charge region must extend from the
rectifying side to the back contact on the silicon wafer, and this
imposes the use of very high resistivity, very pure silicon,
containing less than 10^{13} cm^{-3} doping atoms. The thickness X_D
of the diode space charge region, or "depletion layer", can be
found by solving the Poisson equation, and depends on the applied
reverse bias voltage V_B and the doping density N_D, which appears
macroscopically as the material resistivity ρ (in k Ω cm)

$$X_D = \sqrt{\frac{2\varepsilon\varepsilon_0}{q\,N_D}}\ (V_0 + V_B) \simeq .5 \sqrt{\rho(V_0 + V_B)}$$

q is the unit electrical charge, V_0 is the built-in diode
voltage (about .8V), ε and ε_0 are the relative and absolute
dielectric constants.

The signal formation in semiconductor detectors is basically
different from the process in gaseous MWPC or drift chambers.
There is no multiplication in the semiconductor, and the signal is
induced on the contacts as soon as the electrons and holes start
separating from the initially neutral cloud of excited carriers.
The drift velocity is proportional to mobility and field, up to a
saturation velocity of 10^7 cm s^{-1} at room temperature, which can
be attained at a field of $\sim 10^4$ V cm^{-1}. The total collection
time in a 400 μm thick detector at 200 V is then about 8 ns for
electrons and 20 ns for holes, but the rise time of the current
signal is only about 6 ns. The integrated charge from a single
minimum ionizing particle in such a 400 μm thick detector is
about 5 fC or 30 000 e-h pairs. Such a signal is shown in fig. 2
at the output of the fast preamplifier (described in sect. 3).

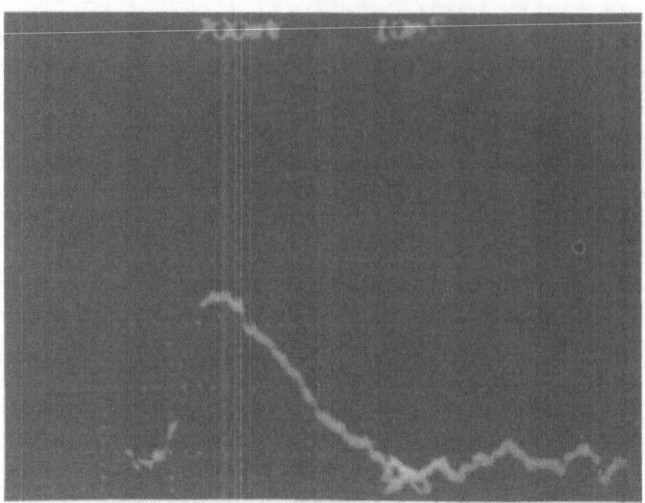

Fig. 2.   Typical signal at the output of the preamplifier
          (10 ns/div.) for a particle which deposits about 120 keV
          or 5 fC of charge.

MICROSTRIP STRUCTURES

Silicon detectors of 1 mm² to 10 cm² area and a few μm to a
few mm of thickness are available commercially since a long time;
initially mostly diffused diode structures, more generally now
with a Schottky surface barrier as the blocking contact. Also
lithium drift (for thick diodes) and ion-implantation (for
resistive surface layers) have been used. Recently, J. Kemmer[4,5]
proposed an improved ion-implantation process, which involves
planar oxidation and photo lithography techniques, now commonly
used in the electronic industry. This process leads to very low
diode reverse leakage current and low noise detectors.

Occasionally, silicon strip detectors and silicon "checker
board" detectors were used for position sensitive particle
detection, as early as 1963, but for nuclear physics the position
sensitive resistive readout was brought to perfection[6]. The
spatial resolution $\Delta x/L$ is limited by the position noise N_x,
which is > 50 keV for 100 pF detector capacity and a 2 μs
signal shaping time. For signals of several MeV this leads to a
resolution of better than 1%, but the particle rate and the time
resolution are limited.

$$\frac{\Delta x}{L} = \frac{N_x}{E}$$

Because of the very small signals of minimum ionizing
particles, around 100 keV in a 400 μm thick silicon detector,
silicon microstrip detectors are proposed for detection in high

energy physics experiments, where good spatial resolution is
required[7,8]. It is expected that the application of micro
electronics silicon technology could result in detecting elements
of 5 to 20 μm diameter.

The first prototypes for use in high energy test beams were
made with surface barrier technology[7,8] and had 1 mm respectively
.2 mm elements. Results of the tests will be discussed in
sect. 4. More recently[9] a detector was built with 100 elements
of 50 μm pitch, covering a 30 x 5 mm area. The masks and the
mounting used during the evaporation are illustrated in fig. 3.
In fig. 4 enlarged pictures of this mask and the resulting
aluminium pattern on the silicon are shown. These detectors were
manufactured in close collaboration with Enertec-Schlumberger in
Strasbourg (France).

Fig. 3. Masks and mounting used for the vacuum evaporation of a
50 μm strip pattern on silicon surface barrier
detectors. The right-hand mask is used for the
reinforcement of the contacts on the epoxy edge.

An alternative approach for building microstrip detectors
makes use of the already mentioned improved ion-implantation
planar ("IP") process. This eventually leads to a better pattern
definition, much lower leakage currents and a lower electrical
noise. In fig. 5 a cross section is shown of two such
ion-implanted strip detectors, which were also manufactured in
collaboration with Enertec. In a suitable mounting, such IP
detectors can be annealed to 200°C, or possibly higher
temperatures, which could be useful in a ultra-high vacuum
environment or for the curing of radiation damage.

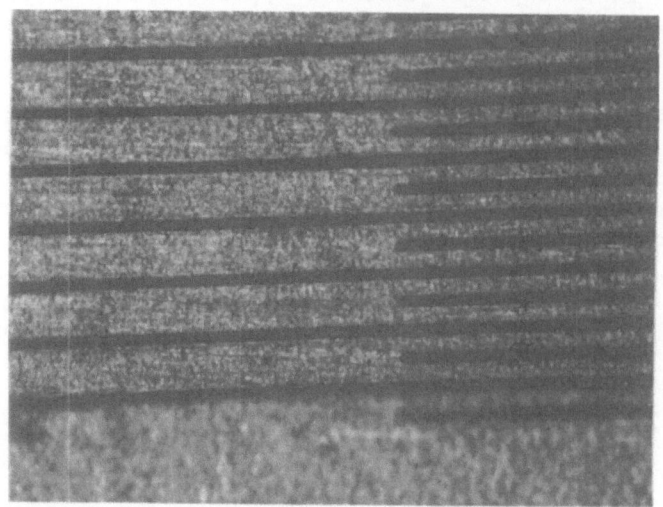

(a) Photo micrograph of the mask shown in fig. 3.
Contacts are made alternatively to the left and the
right, so that there is one contact every 100 µm.

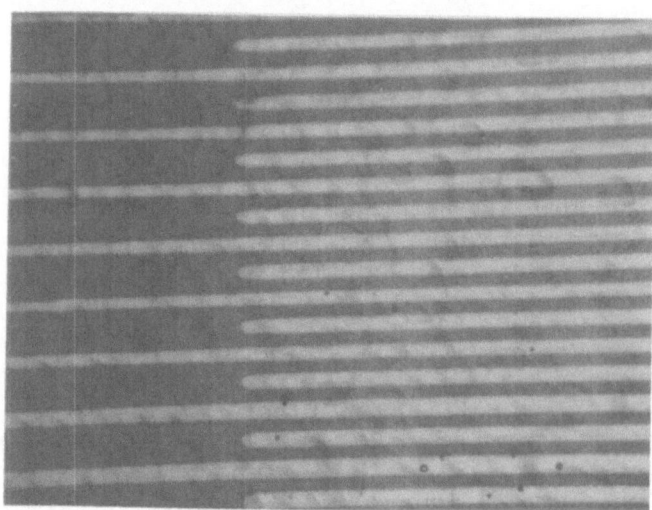

(b) Aluminium strips on the silicon surface, evaporated
through this mask.

Fig. 4.

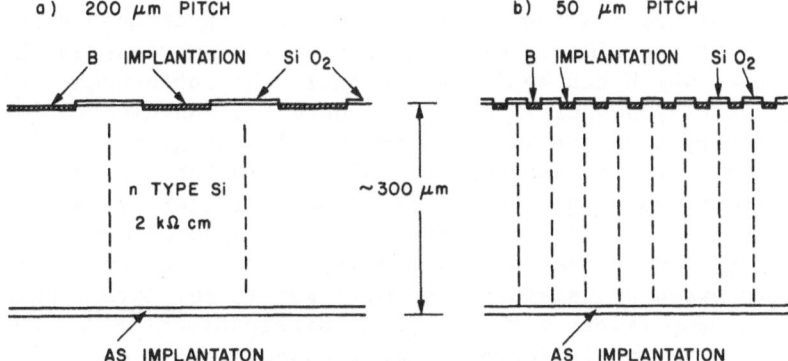

Fig. 5.   Structure of planar versions of a 200 μm pitch and a
          50 μm pitch parallel microstrip detector.

     If one compares the strip structure for particle localization
with the resistive read-out detector, the former has a number of
advantages for the use in high energy physics experiments:

(a) Fast signals can be obtained with a relatively thin detector
    (.05 g cm^{-2}, corresponding to 30 cm of Ar), which enables
    high particle rates and good timing ($\sim$ 1 ns).

(b) High multiplicity events can be accepted, although a true 2
    dimensional discrete structure would be much better than
    combined strip detectors, because of the ambiguities in strip
    combinations.

(c) If a parallel electronic treatment is used, the information
    can be used to establish fast triggers.

(d) The position resolution does not depend directly on detector
    noise, which may increase with time, due to radiation damage.

(e) Room temperature operation is possible.

Clearly, the main disadvantage is the multitude of electronic
channels. The number of channels for 10 μm resolution detectors
for any realistic system rapidly attains a few times 10^4 for a
fixed target experiment or a few times 10^6 for a collider
experiment.

ELECTRONICS FOR SILICON MICROSTRIP DETECTORS

Successful application of silicon microstrip detectors in high
energy physics experiments will be critically dependent on the
possibilities which can be found to handle the large number of low
noise electronic channels. Moreover, detector and electronics
have to be very close to minimise noise. An integrated design of
detectors, electronics and the mechanical structure is necessary
in view of the required high density. Several iterations in
miniaturization of the electronics have already been made or are
being studied. One makes use of the improved characteristics of
bipolar microwave transistors, to build small, relatively cheap,
low noise preamplifiers with low power dissipation[8,9,10]. When
dealing with the special case of small capacitance microstrip
detector elements (< 10 pF), an extremely simple current
sensitive preamplifier can be used[8]. In fig. 6 a schematic
diagram is shown, and in fig. 7 such a preamplifier is shown,
besides a microstrip detector with a 5 x 30 mm central sensitive
region (100 strips of 50 μm).

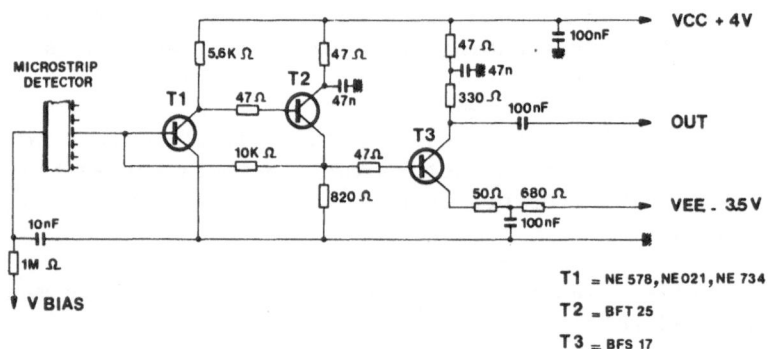

Fig. 6.   Electronic scheme of a current sensitive low noise
          preamplifier to be used with a low input capacitance
          microstrip detector.

Depending on the experimental situation, the signal output of
the preamplifier can be immediately discriminated or has to be
transmitted to an electronics barrack. In fig. 8 is shown a
complete electronic chain which provides both logic and fast
analog signals, transmitted over nearly 100 m. The implementation
of this system is illustrated in fig. 9 and signals are shown in
fig. 10. The r.m.s. electronic noise is about 1400 electrons,
which corresponds to a 15 keV FWHM in Si.

Fig. 7. Preamplifier and line-driver circuits shown along with a surface barrier microstrip detector with 100 strips of 50 μm pitch.

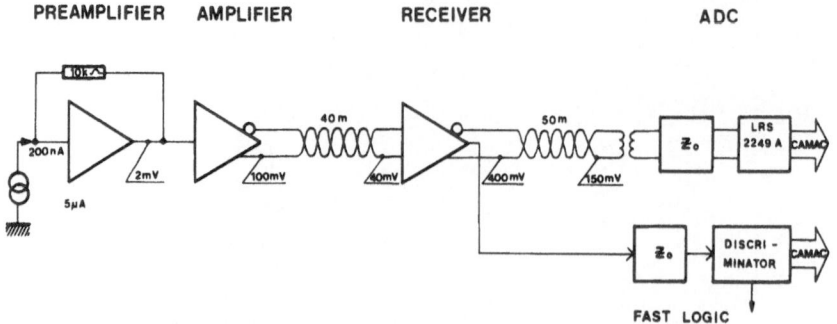

Fig. 8.   Schematic drawing of a complete read-out chain for a
          microstrip, which provides a fast analog output signal
          and a discriminated logic signal.

    The electronic noise in general depends on the detector and
the preamplifier. The diode reverse current and the cross talk
resistance determine the detector noise. In the course of time,
the detector noise may become worse due to increased reverse
current by radiation damage and incomplete charge collection. The
preamplifier noise is mainly determined by the detector
capacitance and the noise performance of the input transistor[9,10].
With the presently designed electronics, a sufficient signal/noise
ratio is obtained with a detector thickness of 250-400 μm silicon.

    A completely parallel linear signal processing, as described
here, certainly cannot be used for future large detectors. A
reduction in the number of channels can be obtained by a charge
interpolation method on the detector level. Alternatively,
parallel to serial conversion using digital techniques, or
possibly analog devices like charge transfer devices (CCD), may
diminish the number of outputs to an acceptable quantity. Also
the off-line data analysis has to remain manageable, therefore
some on-line data reduction seems needed, possibly in connection
with a high level trigger logic.

RESULTS OF BEAM TESTS

    A typical spectrum obtained for a single strip in a test beam
of 10 GeV π^- is shown in fig. 11. The detector thickness was
280 μm. The channel numbers obtained in the pulse height
analysis can be converted to keV, using a ^{57}Co calibration, and
the zero-energy corresponding to the position of the pedestal
peak. The small tail on the low energy side of this Landau type
distribution appears as a result of the "hit" definition. It is

Fig. 9. Mounting of the 48-channel preamplifier-driver modules
         next to the detector, which is shown without its
         precision support. The two NIM modules below are the
         12-channel receiver driver modules, which can be located
         40 m away from the detector.

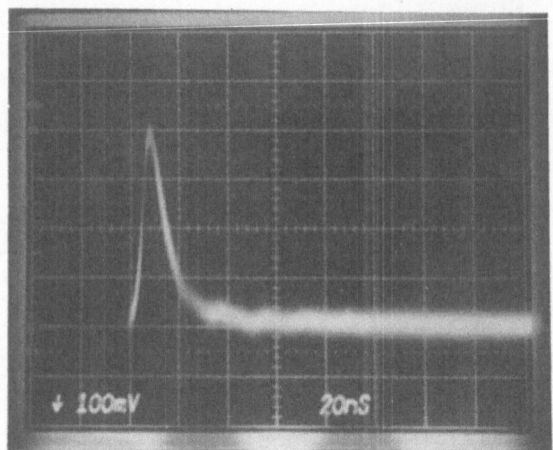

Fig. 10.
Output signal of receiver
module with pole-zero
cancellation. The input
charge of the pulser was
5 fC.

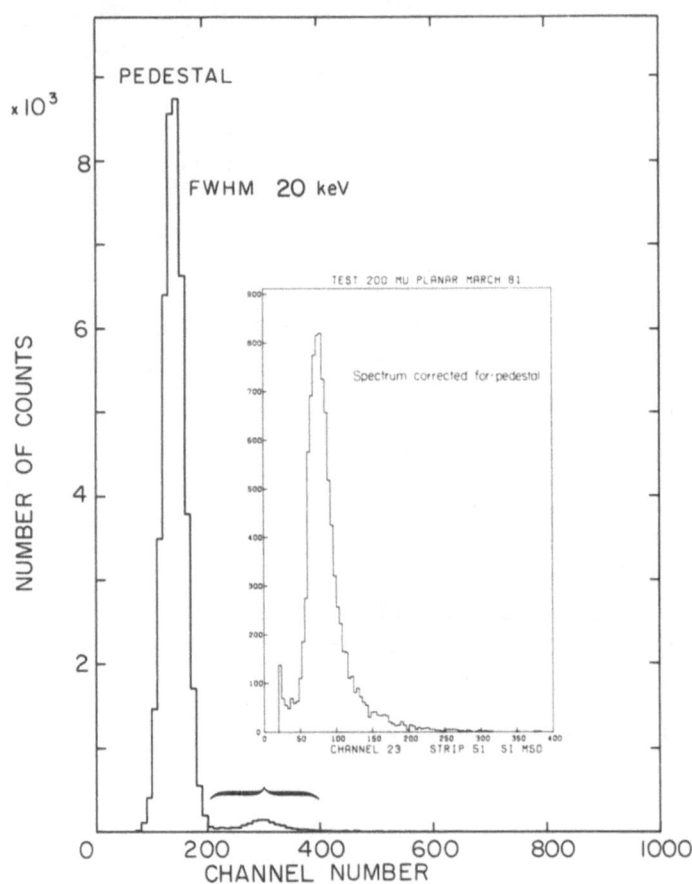

Fig. 11. Typical spectrum for a strip. The pedestal width is a
measure for the electronic noise. The Landau spectrum of
10 GeV π^- is shown enlarged in the insert.

clear that the particle signals are well separated from the noise
peak. In our definition, a strip is <u>hit</u>, if its signal exceeds
the average pedestal value by more than three standard deviations
of the pedestal noise distribution (PND).

Sometimes one finds several adjacent strips hit. This is
called a <u>cluster</u>. A <u>double hit</u> is then by definition a cluster of
only two adjacent hits. This double hit phenomenon was studied by
rotating the detector 45° in the beam. It was verified that the
sum of signals in three adjacent strips (fig. 12) was √2 times
the signal at perpendicular incidence. Therefore, no signal loss
occurs between the strips.

Obviously, also an efficiency measurement can be used to
ascertain that there are no dead regions between the strips. The
set up of fig. 13 was used to this aim. In table 1 an analysis is
given for 49181 events which were triggered by all three
scintillators.

SILICON SURFACE BARRIER MICROSTRIP DETECTOR

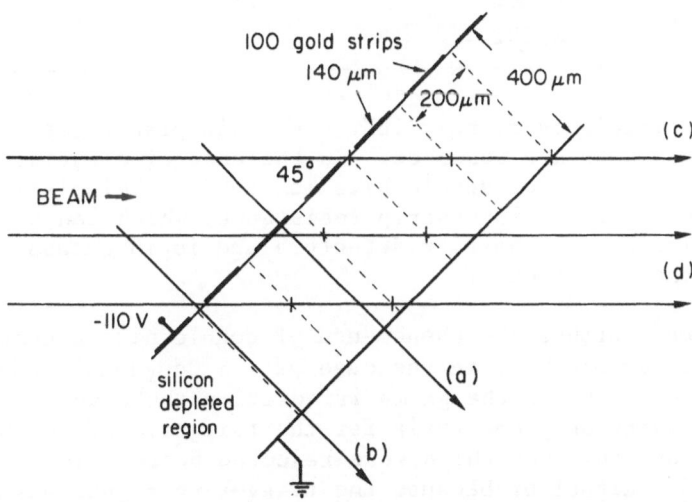

Fig. 12. By comparing signals at 45° incidence with those in the
         normal case it could be verified that within our
         measurement accuracy (5–10%) no signal loss occurs for
         particles crossing the inter-strip boundaries.

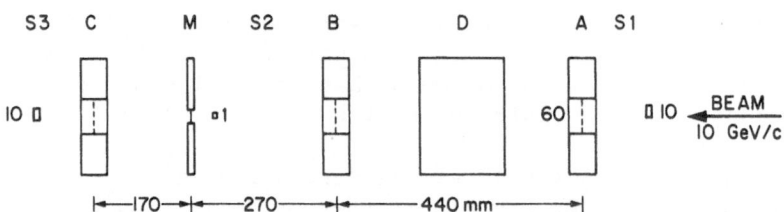

Fig. 13. Situation of the microstrip detector M between wire
         chambers with centre of gravity read-out (A, B and C).
         S1, S2, S3 are scintillators used to trigger the read-out.

Table 1. Analysis of hits on 200 μm IP detector (run 4026 beam)

All events	49181		100%		
No hits in Si	295		.6%		
Only 1 cluster	44042		89.6%		
		1 hit		41965	85.3%
		2 hits		1952	4.0%
		3 hits		84	.2%
		>		41	.08%
2 Clusters	4440		9.0%		
		2 x 1 hit		4099	8.3%
3 Clusters	355		.7%		
	49132		99.9%		97.9%

The inefficiency is only .6%. For this planar detector only
4% of double hits are observed. In the case of a surface barrier
detector the number of double hits is 15-25%. This is mainly
related to the lower interstrip resistance, which can be
∿ 100 kΩ for surface barrier detectors and is in excess of
1 GΩ for the IP detectors.

To study further the phenomenon of double hits a comparison
was made of double hits in the case of a [57]Co gamma irradiation
and the beam test. In the gamma irradiation the events were
recorded, using only one strip for the trigger. Of 100 000 events
somewhat more than one third were rejected because more than 10
strips had a signal or because the triggering signal was below the
off-line threshold of three standard deviations of the PND. The
remaining sample had 9% double hits, which is twice the amount in
the beam test. The energy spectra for single hits and double hits

in both cases are compared in fig. 14. Double hits in the case of
^{57}Co yield exactly the same total energy as single hits, but

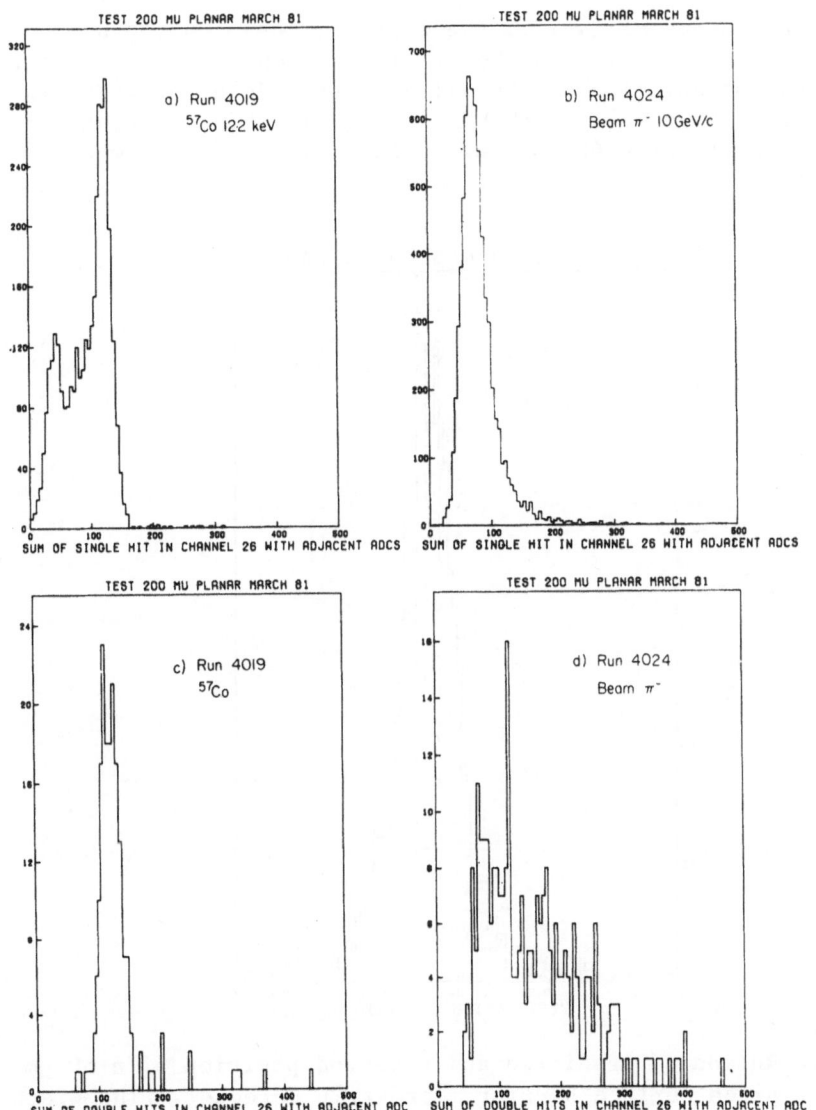

Fig. 14. Comparison of single and double hit energy spectra for
^{57}Co (left) and 10 GeV π^- (right).

double hits in the beam shown an excess of events with high energy
content. Therefore it is concluded that double hits do not all
originate from simple charge division, but that a significant

fraction of them are special events with an extended delta
electron or gamma shower. Further study of the double hits seems
however needed.

In the set-up of fig. 13 the particle position can be
predicted with the wire chambers. The residu between prediction
and actual measurement on the silicon strip detector is shown in
fig. 15 for 16 GeV π^-. The FWHM is 140 μm, which corresponds
to a resolution of $\sim$ 60 μm, nearly the limit of these wire
chambers.

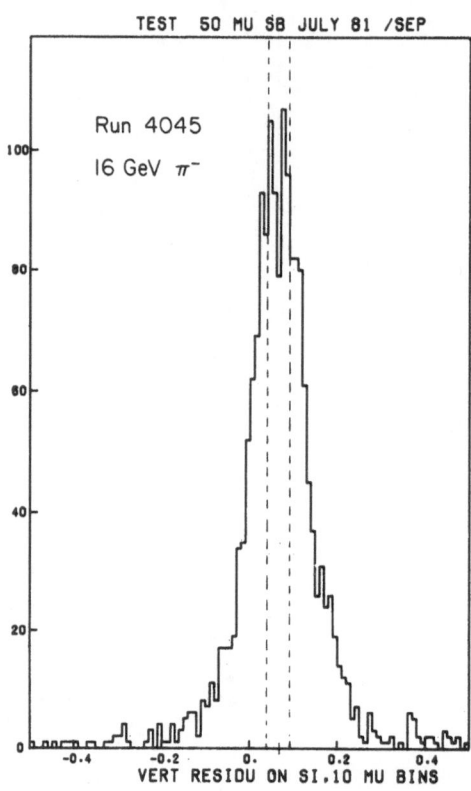

Fig. 15. Residu of predicted and observed position of a 50 μm
         pitch surface barrier microstrip detector. The width of
         one detector element is indicated.

RADIATION DAMAGE

Radiation damage in silicon is caused by the creation of
charged defects in the crystalline lattice which disturb the
electronic band structure. Single vacancies can be created by the

excitation of a silicon atom from its lattice position, to which it is bound by an energy of $\sim$ 25 eV. These single vacancies are mobile above 55° K and recombine with other vacancies or with impurity atoms, to form stable "point-defects", like vacancy-vacancy, vacancy-oxygen or vacancy-phosphorus. The predominant defects will depend on the type of silicon, the impurities present, the irradiation temperature, etc., and it is difficult to give general quantitative rules, valid for all cases of irradiation.

The study of radiation damage can be conducted at the microscopic level with a number of methods. Some information on the effect of GeV muon irradiation was obtained using the Thermally Stimulated Current (TSC) arising from certain deep electronic levels[11,12].

Macroscopically, the reverse diode ("leakage") current is for detectors the easiest accessible parameter which is influenced by the irradiation. The point-defects act as generation-recombination centers for e-h pairs and therefore increase the current and the noise in the diode. But also the space charge distribution, and therefore the capacitance can be modified by irradiation.

From experience in using silicon detectors in a high energy muon flux, the following conclusions can be drawn[12]:

(a) The damage caused by minimum ionizing particles themselves may be relatively unimportant compared to that which is caused by a "hidden" component like low energy neutrons or electromagnetic cascades.

(b) Degradation with time is not only related to irradiation, there may be other effects.

(c) Electronics often breaks down before the detector, especially when it contains the more sensitive MOS devices.

(d) If one introduces a _damage tolerant design_ both for detector and electronics, the lifetime can be extended by several orders of magnitude.

From the foregoing discussion it must be clear that it is not justified to quote numbers with general validity. But signs of radiation damage will appear beyond 10^{10}-10^{12} cm^{-2} and detectors can be functioning up to 10^{11}-10^{14} cm^{-2}. In a test of 4×10^{16} γ cm^{-2} of ^{60}CO, which corresponds to about 2×10^7 rad, some detectors degraded considerably, others were still quite unaffected.

It is possible to remove some of the radiation damage by annealing the detector at 150-350°C. If one intends to do this, the detector and its environment have to be designed very carefully, because several metals like Na and Au have a high diffusion speed even at 150°C, and they create deep levels in the silicon, which are worse than the radiation damage levels. In this context one should remember that the initial electrical characteristics of the detector grade silicon are determined by $\sim 10^{12}$ cm^{-3} doping impurities for 5×10^{22} atoms cm^{-3}. Therefore even very slight contamination has considerable influence.

FUTURE DEVELOPMENTS

Silicon particle detectors may prove to be useful for the detection of short lifetime particles, events with high multiplicity or as detectors inside a vacuum chamber. At present, the spatial resolution is already better, or at least equal to that obtained in the gaseous or liquid detectors. High particle rates can be accepted because of the small element size, and the short collection time of both positive and negative charges. Therefore they may be suitable to study low cross section phenomena in intense beams.

Whereas the application in fixed target experiments appears rather straightforward, and the more so the higher the enery, applications in colliding beam experiments pose a number of problems. The most severe is the solid angle to be covered, and therefore the most immediate use of silicon detectors in colliding beam experiments may be limited to the forward detection region. There also one may find enough place for the electronics, which has to be close to the detector elements. For the central detection region, silicon could only enter into consideration if more sophisticated methods for connections, read-out and data processing have been developed[13,14].

Experimental groups at CERN, Pisa, Saclay, Munich, Milano, Brookhaven and Berkeley are at present working on various aspects of microstrip detectors. The commercial firm Enertec/Schlumberger near Strasbourg (France) is taking an active part in the production of these devices. Other firms like Ortec/EG & G, EMI (Ltd) and Centronic are intending also an effort in this direction.

At present, a number of silicon microstrip detectors are installed in charm experiments at CERN, and this will enable a better evaluation of possibilities and problems. It is hoped that realistic solutions can be found for the electronics problems, for new contacting technologies and for true two-dimensional, or even three-dimensional detection.

It may be concluded that detector development reached the demonstration level, but that most of the work is still ahead if reliable devices are to be produced in significant quantity. The electronics development will be ultimately determining the usefulness of the multielement microdetectors. However, to the extent that both detector and electronics development consist in applying existing technologies to a new field of application, a rapid progress can be expected.

References

1.   S.M. Sze, Physics of Semiconductor Devices, New York 1969, Wiley-Interscience.
2.   H. Esbensen et al., Random and channeled energy loss in thin germanium and silicon crystals for positive and negative 2-15 GeV/c pions, kaons and protons, Phys. Rev. B18 (1978) 1039.
3.   C. Richard-Serre, Evaluation de la perte d'energie unitaire et du parcours pour des muons de 2 à 600 GeV dans un absorbant quelconque, Genève 1971, CERN Yellow Report 71-18.
4.   J. Kemmer, Fabrication of low noise silicon radiation detectors by the planar process, Nucl. Instr. Meth. 169 (1980) 449.
5.   J. Kemmer et al., Performance and applications of passivated ion-implanted silicon detectors, IEEE Trans. Nucl. Sci NS-29 (1982) 733, see also these Proceedings.
6.   E. Laegsgaard, Position sensitive semiconductor detectors, Nucl. Inst. Meth. 162 (1979) 93.
7.   S.R. Amendolia et al., A Multi electrode silicon detector for high energy physics experiments, Nucl. Instr. Meth. 176 (1980) 449.
8.   E.H.M. Heijne et al., A silicon surface barrier microstrip detector designed for high energy physics, Nucl. Instr. Meth. 178 (1980) 331.
9.   P. D'Angelo et al., Analysis of low noise, bipolar transistor head amplifiers for high energy applications of silicon detectors, Nucl. Instr. Meth. 193 (1982) 533.
10. E. Heijne and P. Jarron et al., A fast high resolution beam hodoscope using silicon microstrip detectors, IEEE Trans. Nucl. Sci. NS-29 (1982) 405.
11. H.M. Heijne et al., TSC defect level in silicon produced by irradiation with muons of GeV energy, Radiation effects 29 (1976) 25.

12.   E. Heijne et al., Radiation damage: experience with silicon
      detectors in high energy particle beams at CERN,
      proceedings of meeting on mimiaturization of high energy
      physics detectors, Pisa 1980, Plenum Press, in press,
      also as: Internal report CERN/EF/BEAM 81-6.
13.   T. Ludlam, Semiconductor devices as track detectors in high
      energy colliding beam experiments, IEEE Trans. Nucl. Sci.
      NS-28 (1981) 549.
14.   V. Radeka, Proceedings of Fermilab Workshop on semiconductor
      detectors, ed. T. Ferbel, Fermilab 1982.

RESISTIVE CHARGE PARTITION WITH MESD

Ettore Focardi

INFN - Sezione di Pisa and
Scuola Normale Superiore
I-56100 Pisa

INTRODUCTION

After the early applications of standard silicon detectors in the high energy physics field such as live targets to study nuclear coherent-incoherent interactions in silicon induced by hadron beams or beam radiation monitoring, we started in our laboratory the development of telescopes detectors for a rather different and non-conventional use: fine granularity detectors for lifetime measurement of decaying charm mesons and high precision position sensitive detectors.

In section 1 the performances of the first multi electrodes semiconductor detector made in Pisa will be discussed.

Section 2 is dedicated to explain the charge partition principle and, in particular, the resistive charge partition method that we mean to use as a reasonable compromise for high space resolution without paying with an expensive high strips density.

Preliminary results on a recent space precision measurement are described in section 3.

Last section contains the description of proposed applications in experiments.

1. <u>Mesd</u>

Following the experience acquired in the whole project and
in the construction of part of the silicon target for charm decay
used in FRAMM NA1 experiment at CERN [1], we began the study and
the development of new semiconductor detectors for high space
precision measurements.

Position sensitive detectors with high strip density have
been built in the past and found current applications in other
fields of research such as space flight physics [2], and
extensively reported in literature. We have built our first
multi electrode silicon detector (MESD) to be operated as
proportional chamber to measure the impact position of particles
traversing the detector. For simplicity, the first prototype has
been made by surface barrier junction. We have used an n-type
silicon crystal with resitivity of ∿ 20000 ohm x cm : the
junction was formed by Au while the ohmic contact was made with
Al. The formation of an oxide layer between metal and
semiconductor seems important for the rectifying property. Since
we have used high resistivity crystals, no special precaution has
been taken to reduce the global reverse current, already
reasonably low. Typical current-voltage characteristic curve for
the whole detector is shown in fig. 1.

We have tested our first prototype at CERN SPS with a
π⁻beam of 40 GeV [3]. The result in fig. 2 shows that the
pulse height signal of a minimum ionizing particle hitting a
strip is well separated from the background. This detector
(800 μm thick) was operated in the total depletion mode.
Fig. 3 shows the pulse height spectrum of a strip when particles
crossing the adjacent one are selected in trigger. By comparing
fig. 3 with fig. 2 it can be seen that cross-talk effects are
very small, accounting at most of the right hand side tail of the
noise distribution of fig. 3.

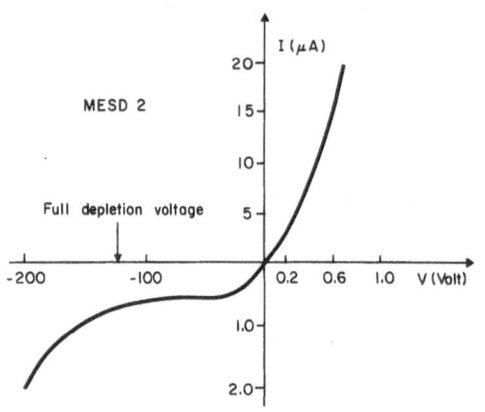

Fig. 1. - Current vs. voltage
characteristic curve.

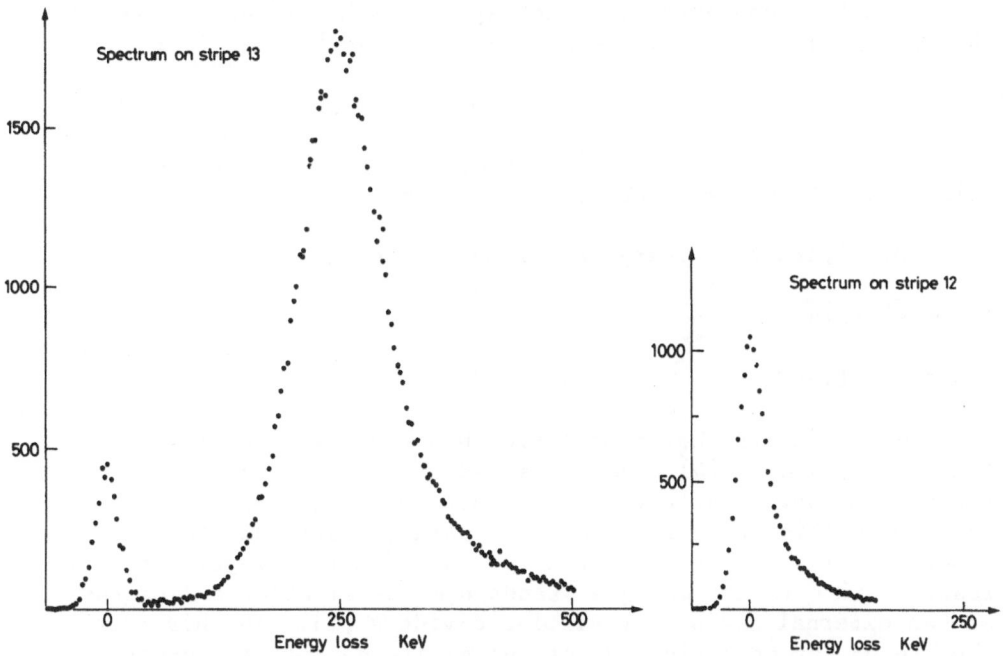

Figs 2,3. - MESD 1, 145 Volt bias, Spectrum on strips 13 and 12.

2. Charge partition method

The space resolution for detectors operated as above is
determined by the granularity of the deposited strips, the
resolution between tracks is the same as space precision.
Therefore detectors with high strip density appear to be a good
instrument to measure tracks in high multiplicity events; they
can be built and have in fact been built in the past [4] and more
recently with high electrodes density up to 1 strip/20 μm.

A rather different approach to the high precision
requirement without paying with an often intolerable enormous
number of read outs is to make use of the charge partition
principle among adjacent electrodes. In this way the number of
strips can be kept rather low by increasing the spacing in
between. A limit to the obtainable resolution is due to the
diffusion of charges produced in the crystal by the incident
particle. To obtain an extimate of the thermal diffusion of the
charges in silicon we can consider a localized charge
distribution that, in absence of other effects, diffuses by
multiple collisions according to a gaussian law.

With this approximation the expected standard deviation of the transverse spatial distribution of charge "cloud" is :

$$\sigma = \sqrt{2Dt}$$

where $D = 35$ cm^2 sec^{-1} [5] is the diffusion coeffecient for silicon and t is the carriers transit time.

Two alternative charge partition approaches can be used :

 i) resistive

ii) capacitive.

The resistive charge partition has been used for a long time. Many groups [6] have deposited thick resistive layers on the semiconductor surfaces, the results obtained were very poor in terms of linearity from point to point, mostly due to inhomogeneities in the thickness of the resistive layer. Better results are obtained using a standard multi electrodes detector and an external chain of resistive dividers [2]. In this case the use of series resistors between electrodes of the detector and preamplifier introduces an extra source of noise.

Our approach is to use proper resistivity of the crystal operated in slightly undepleted mode. A sketch of the working principle of a surface barrier n-type silicon detector used to measure only one coordinate is shown in fig. 4. Aluminium strip electrodes, as ohmic contacts are deposited on the surface opposite to the junction, obtained by gold evaporation. The bias voltage is set to such a value below the total depletion that the interstrip undepleted silicon layers act as a continous low resistance divider.

Then the impact position of particle crossing the interelectrode space is reconstructed by means of the pulse heights measured on the two adjacent strips. If x is the distance between one strip and the impact point of the particle, assuming linearity, it is proportional to the quantity :

$$x \quad \alpha \quad \frac{Ph_1 - Ph_2}{Ph_1 + Ph_2}$$

Ph_1 and Ph_2 being the pulse heights measured on the two adjacent strips.

In order to check this principle we have built a MESD, 800 μm thick with 8 Al strips (120 μm wide, 660 μm or 1330 μm spaced, 10 mm long).

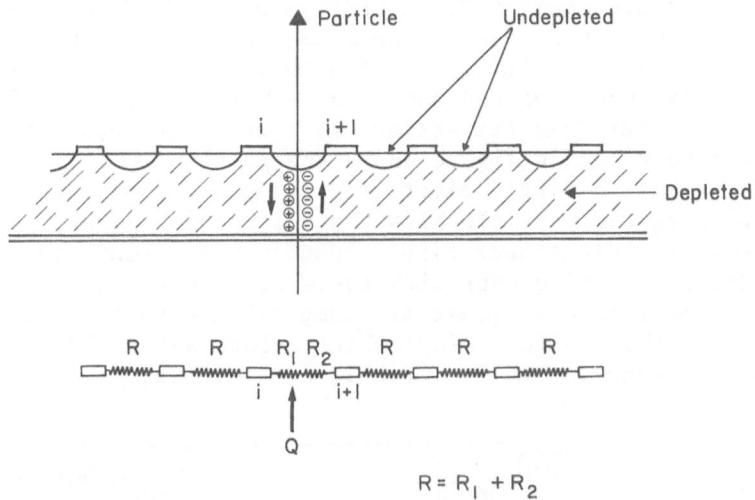

Fig. 4. - Working scheme of resistive charge partition in an MESD

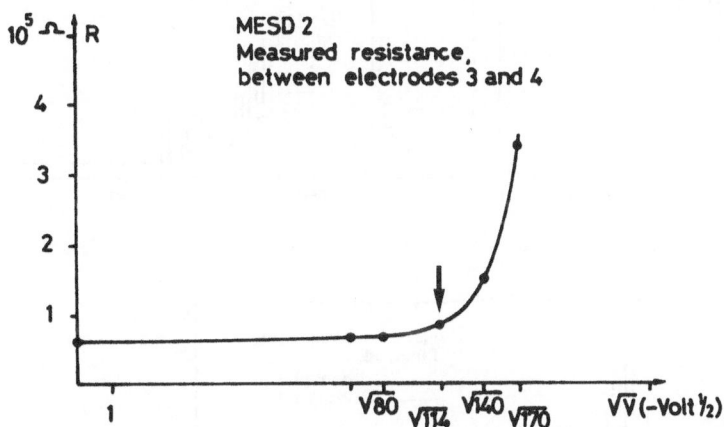

Fig. 5 - Resistance between two adjacent aluminium strips versus √V (proportional to the depletion depth).

The resistance between adjacent strips 1.3 mm apart versus the square root of the bias voltage is shown in fig. 5 ; the arrow indicates the nominal total depletion voltage. When the voltage is increased over the total depletion value, the resistance increases and the region where a visible charge partition still occurs is limited to a narrow interval in the middle of the interstrip space.

We have tested this MESD at CERN PS with 7 GeV hadrons beam
[7]. Fig. 6 shows the results corresponding to the total depletion
voltage : in this case it is expected that the interstrip region is
still slightly underdepleted and a pulse height flat distribution is
expected for a particle traversing the interstrip space if charge
partition occurs. This figure shows the pulse height histograms when
one particle hits a single electrode, 'single hit' events (solid
line), and when the particle hits the interstrip region, in this case
we have only two electrodes fired, 'double hit' events (dashed
line). The number of events with three or more electrodes fired is
< 1%. The observed histograms are compatible with those expected :
a Landau distribution for 'single hit' events and a flat distribution
folded with a Landau tail for 'double hit' events.

Fig. 6.
Pulse height dist-
ribution for sig-
nals collected on
one strip, for a
bias corresponding
to the total
depletion

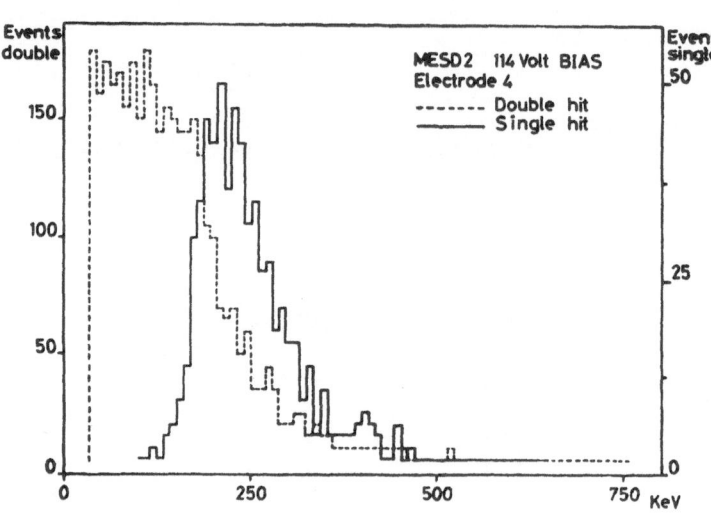

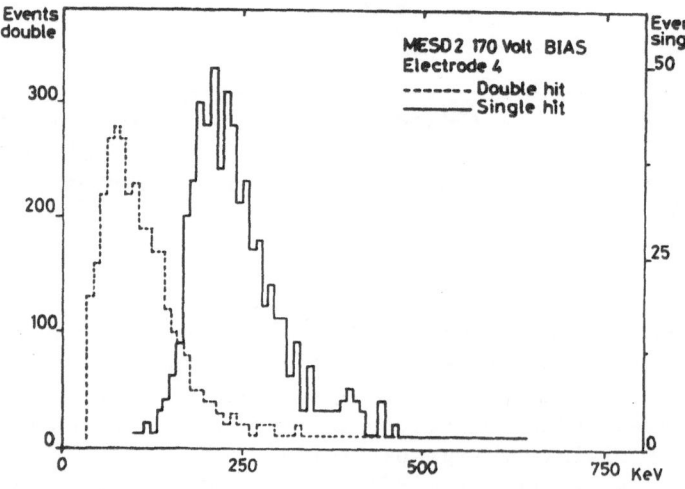

Fig. 7.
Pulse height dis-
tribution for sig-
nals collected on
one strip, for a
bias corresponding
to over-depletion

Fig. 7 shows analogous histograms for a bias set in over depletion : in this case we have a reduction of 'double hit' events, and the corresponding dashed curve approximates a Landau distribution centered at one-half the peak value of the 'single hit' Landau distribution.

The expected precision of MESD is :

$$\sigma = 1 \, \frac{\sigma_{noise}}{<E_{loss}>}$$

that is $\sigma = 85$ μm if $\sigma_{noise} = 15$ keV, $<E_{loss}> = 240$ keV and $1 = 1320$ μm. The thermal diffusion contribution expected for this detector is $\sigma = 17$ μm.

The resolution obtained is $\sigma \sim \sigma_{D.C.} < 120$ μm, where $\sigma_{D.C.}$ is the resolution of the gas drift chambers used to define a narrow beam line. Furthermore also the beam divergence is included in this value. The other way to use the charge partition, via capacitive coupling between eletrodes, has been explored recently [8] and has been described by other speakers.

3. Space resolution measurements

To better measure the space precision limit of MESD, operating with resistive charge partition, we have built 3 identical MESD using n-type silicon crystals each of them 900 μm thick, with resistivity of 10000 ohm-cm and with 22 Al strips 100 μm wide, 300 or 600 μm spaced, 20 mm long. The ohmic contact strips were obtained evaporating Aluminium through a Cu-Be mask, the junction evaporating Gold on the opposite surface. We have tested the 3 MESD at CERN SPS with a π-beam of 40 GeV.

Fig. 8.
Measured resistance
between two adjacent
electrodes.

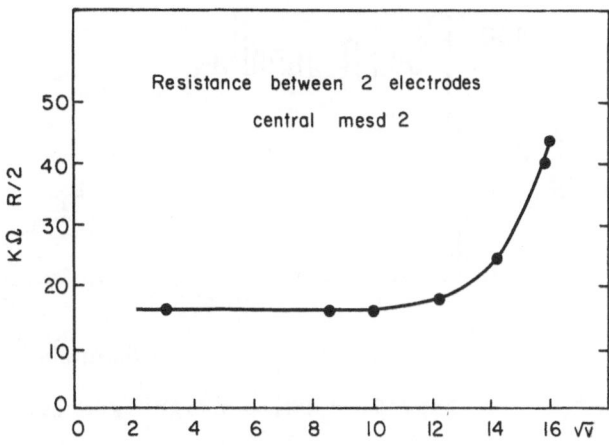

The three cambers were parallel and the central one was movable by a micrometric movement ; each chamber was set to a voltage slightly under the nominal total depletion voltage. Typical measured resistance between two adjacent electrodes of the central MESD is shown in fig. 8.

Fig. 9 shows a pulse height spectrum of one strip where it is clear the flat distribution due to the resistive charge partition.

It is possible to extract the space precision by imposing a linear relation between the impact positions of particles given by each of the 3 MESD.

Fig. 10 shows the distribution of the deviations from the linear relation and the standard deviation corresponds to the space precision $\sigma = 22$ μm, that is 7% of the interstrip distance.

4. Future applications

We have in mind some applications of multi electrode silicon detectors operated in charge division mode.

In the new colliding beam machines the high energy interactions $p\bar{p}$, e^+e^- are characterised by high multiplicity events and the identification of few tracks coming from a secondary vertex, separated from the beam crossing point, can be a signature of a heavy flavour weak decay.

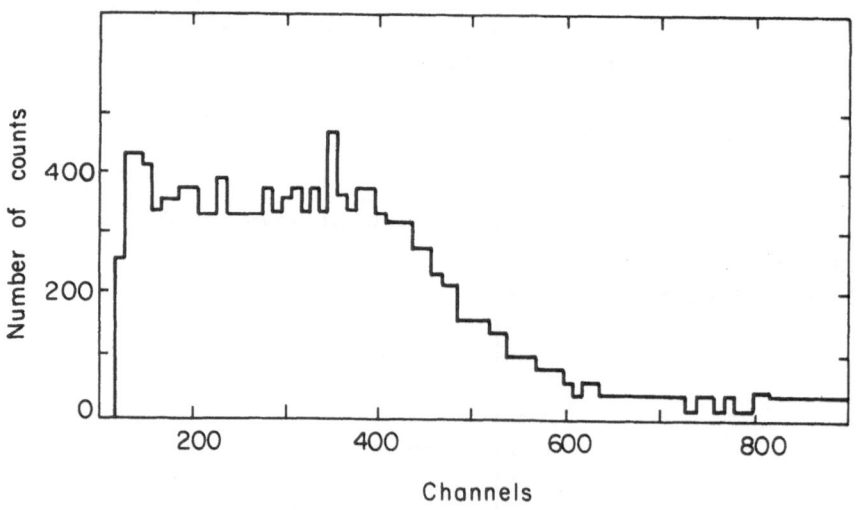

Fig. 9. - Pulse height spectrum of one strip

Fig. 10.
Deviations from the linear
relation between impact
position of particles

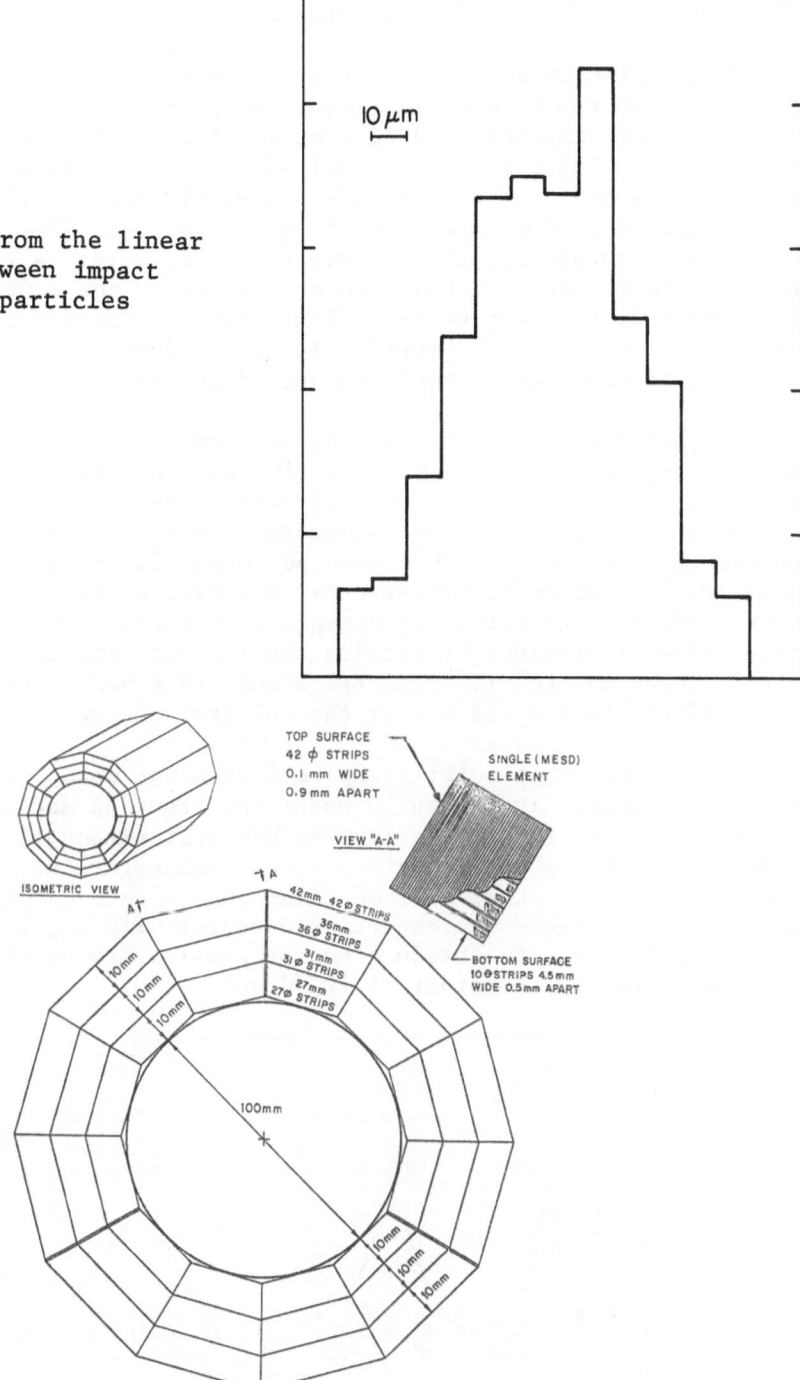

Fig. 11. - Array of MESDs to be located around the intersection
of Fermilab pp collider for secondary vertices detection.

With this goal in mind I will describe a proposed system of
MESD for Fermilab pp collider reported in fig. 11.

Inside the vacuum pipe of the collider, four concentric
barrels, each composed of 12 sectors of MESD, cover the full
azimuth. Each crystal is 5 cm long and 1 mm thick with strip
electrodes on the two faces: longitudinal electrodes .1mm wide
and .9 mm spaced on the back side, perpendicular θ strips are
4.5 mm wide and .5 mm apart on the junction side. This
granularity should satisfy the request to correlate a track in
the MESD telescope to a track reconstructed in the central drift
chambers. In this way we have 2160 read-out channels with 1/3 of
4π for the solid angle covered. It is possible to increase the
solid angle coverage by putting more identical arrays in a row.

With the achieved resolution, as shown in the previous
section, the system would be sensitive to reconstruct secondary
vertices coming from decays of particles with lifetimes in the
range of 10^{-13} sec and would allow to observe an appreciable
number of charm decays. The good performances of silicon
detectors in the dE/dx measurements can also be used in the
search for free fractionally charged particles. The expected
resolution is obtained by scaling the NA1 data for four detectors
and is shown in fig. 12. The 1/3 e and 2/3 e peaks are separated
respectively by 9 σ and 6 σ of the integral charge peak.

In the field of multi strip semiconductor detectors a new
fine grain target is presently under construction and test in our
laboratory. The idea is to improve the measurement of D mesons
lifetimes in the photoproduction with FRAMM apparatus,
substituting the actual silicon telescope, with a granularity of
400 μm with a unique semiconductor detector with a granularity
of 100 μm in order to obtain a better sensitivity on the
lifetime range of the order of 10^{-13} sec.

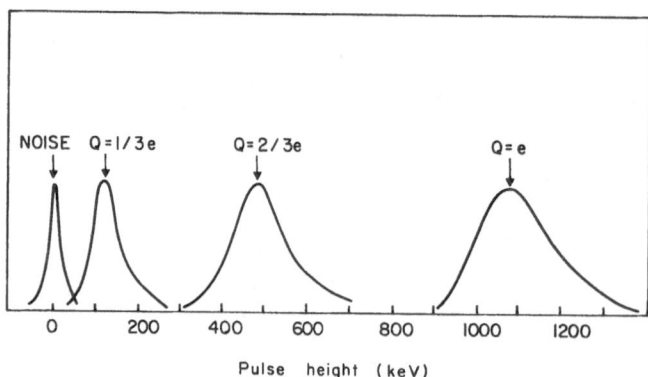

Fig. - 12. - Charge resolution expected at Fermilab pp
collider with the array telescope of NA1 experiment.

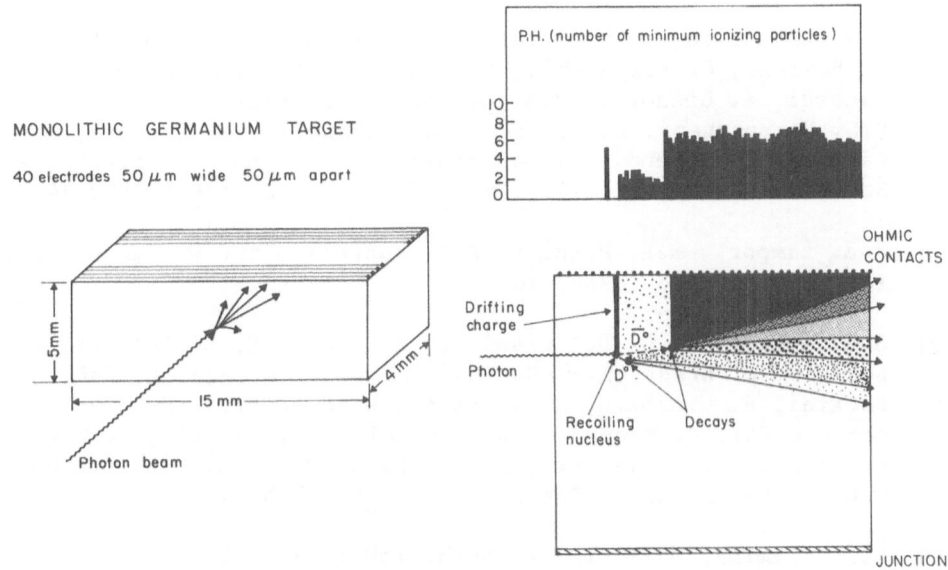

Fig. 13. - Sketch of the Germanium multi-electrode target.

The solution studied in Pisa is a monolithic Germanium detector
operated at 77 °K in high vacuum [9]. The expected energy
loss of a minimum ionizing particle in Ge is a factor of two
more than in silicon, due to the increased density and Z.
Furthermore the required energy to produce an electron-hole
pair in Germanium is 1.3 times smaller than in silicon; then
the pulse height increases by a factor of 2.6 and it is
possible to use detectors as thin as 100 μm.

Our prototype has the following characteristics:

n-type Germaniun crystal with $5 \cdot 10^{10}$ impurities/cm³,
5 mm long in the beam direction, 200 mm wide and 5 mm thick;
junction and ohmic contacts will be made respectively by
implantation of B and diffusion of Li. The electrodes are
strips deposited on one face of the crystal, orthogonal to the
beam direction; the longitudinal cells are defined by the
charge collection of the subsequent strips. To avoid the
charge partition between adjacent cells the chamber must be
operated in over-depletion mode. The thermal diffusion
is estimated less than 20 μm. Fig. 13 shows a sketch of the
detector and a picture of a simulated $D^0\bar{D}^0$ decay in the target.

REFERENCES

[1]   S.R. Amendolia, G. Batignani, F. Bedeschi, E. Bertolucci,
      L. Bosisio, C. Bradaschia, M. Budinich, F. Fidecaro, L. Foà, E.
      Focardi, A. Giazotto, M.A. Giorgi, M. Givoletti,
      P.S. Marrocchesi, A. Menzione, D. Passuello, M. Quaglia, L.
      Ristori, L. Rolandi, P. Salvadori, A. Scribano, R. Stanga, A.
      Stefanini, M.L. Vincelli, Nucl. Instr. Meth. 176 (1980) 449.

[2]   J.E. Lamport, M.A. Perkins, A.J. Tuzzolino and R. Zamow, Nucl.
      Instr. Meth. 178 (1980) 105.

[3]   S.R. Amendolia, G. Batignani, F. Bedeschi, E. Bertolucci, L.
      Bosisio, C. Bradaschia, M. Budinich, F. Fidecaro, L. Foà, E.
      Focardi, A. Giazotto, M.A. Giorgi, M. Givoletti, P.S.
      Marrocchesi, A. Menzione, D. Passuello, M. Quaglia, L. Ristori,
      L. Rolandi, P. Salvadori, A. Scribano, R. Stanga, A. Stefanini,
      M.L. Vincelli, Nucl. Instr. Meth. 176 (1980) 457.

[4]   I.S. Fleming, Nucl. Instr. Meth. 150 (1978) 417.

[5]   C. Jacoboni, C. Canali, G. Ottaviani and A. Alberigi Quaranta,
      Solid State Electronic 20 (1977) 77.

[6]   E.J. Ludwig, Rev. Science Instr. 36 (1965) 1175.

[7]   M.A. Giorgi, Silicon multiwre proportional chambers and their
      applications in high energy experiments, Proc. Meeting on
      Miniaturazization of High Energy Physics Detectors, Pisa,
      September 1980 Plenum Press, London 1981.

[8]   J.B. England, B.D. Hyams, L. Hubbeling, J.C. Vermeulen, P.
      Weilhaumer, Nucl. Instr. Meth. 185 (1981) 43.

[9]   G. Bellini, L. Foà and M.A. Giorgi, Semiconductor Detectors for
      lifetime measurements and high space resolution, to be
      published on Phys. Reports C.

A HIGH RESOLUTION TELESCOPE USING CCD PHOTODIODES ARRAYS

M. Bocciolini[2], G. Di Caporiacco[2][+], A. Forino[1],
R. Gessaroli[1], G. Parrini[2] and A. Quareni-Vignudelli[1]

(1) Istituto di Fisica-Sezione INFN, Bologna, Via Irnerio 46
(2) Istituto di Fisica-Sezione INFN, Firenze, Largo Enrico Fermi 2

A relatively new approach to the problem of high spatial resolution in vertices reconstruction is here reported.

The method is based on the use of CCD area image sensors generally used as television image scanners. These devices consist of a large number (up to 2.10^5 or more) of very small MOS photosensors ($10^2 \div 10^4$ μ m^2). We rely on detecting the charges produced by the crossing of a minimum ionizing particle in the depletion layer under the MOS gate.

Owing to the small thickness of the depletion layer (<10 μm) the expected signal is small (< 800 el.).

We expect that the precision in determining the particle position is of the order of a single pixel.

Figure 1 shows the layout of an array. The output is serially arranged and the information comes out row by row. The device is sensitive during the read-out too. The output frequency can be about 10 MH giving an output time (for 2.10^5 photodiodes) of about 20 ms. This large output time is the main disadvantage of this device, the advantages being the unique output line that simplifies the electronics, the absence of ambiguities for the crossing points and the accuracy (of the order of 10 μm).

Several arrays have been tested for their sensitivity to minimum ionizing particles. The best results have been obtained with

---

[+] Presented by G. Di Caporiacco.

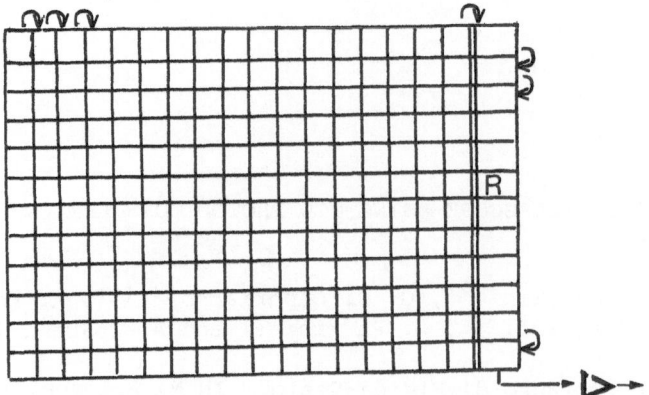

Fig. 1.   The arrows indicate how the charge is transferred from
one element to the other. The charge configuration (row
by row) is moved at the same time towards the output shift
register. The content of the shift register R is serially
read in the time between two transfers.

the GECMA357 (now the name has been changed into EEV-P8600) area
image sensor, which is also suitable for its topology. Figure   2
shows the output of a row when the array is exposed to a 22 MeV/c
LINAC electron beam of the Institute of Radiology of Florence University. A signal is clearly evident. Table 1 gives a summary of
the test conditions and of the main results. It is remarkable that
the collected charge is larger than that due to the depletion layer.
This means that some mechanism (diffusion.....) contributes in
collecting the charge produced in the substrate. This makes it
easier to use these devices as minimum ionizing particle detectors.
It has to be pointed out however that a dark fixed pattern due
to individual differences among the photosensors is present and
should possibly be corrected in order to discriminate the signal
from the background. The masking of the array is not practical
because of the very large number of elements. The lowering of the
temperature may be a practical method, the dark current in the
photodiodes being the main source of this pattern.

The time resolving power of the array is extremely poor. The
situation becomes however definitely better when the arrays are
arranged for the construction of a telescope. In fact, outputting
the planes of the telescope alternatively in opposite directions,
only the "in time" tracks maintain the alignment of the crossing

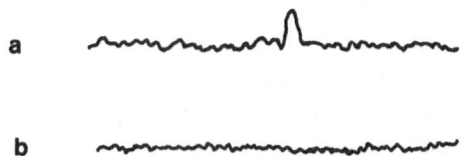

Fig. 2. Oscilloscope picture of a single row of the array. a) array exposed to 22 MeV/c electrons; b) array not exposed.

TABLE I

Array	EEV P8600 (GEC MA357)
Overall dimensions	8.5 x 12.5 mm^2
Photosensor dimensions	22 x 22 μm^2
Silicon thickness	300 μm
Depletion layer thickness	$\approx$ 10 μm
Number of photosensors	221760
Number of rows	576
Number of columns	385
Read-out frequency	10 MH
Total read-out time	20 ms
Read-out time for one row	40 μs
Incident particles	22 MeV/c electrons
Operating temperature	- 20 °C
Charge due to the dark current	$\approx$ 6000 electrons (1 photodiode)
Maximum difference for the dark current charge	$\approx$ 1500 " "
Collected charge in a crossing point	$\approx$ 4000 " "

points. The "out of time" tracks will appear no more aligned. The resolving time, i.e. the read-out time of a single row, is thus lowered to about 40 μs. Obviously all the track crossing points are recorded and that resolving time is reached only after the track reconstruction.

The maximum number of crossing tracks depends on the ability of the software to properly reconstruct the tracks in a strong background of "out of time" points. A program has been developed to test the limiting values of this background. The results are that an "in time" event of about 15 tracks is correctly reconstructed with a background of about 400 points on each array, due to the tracks of 8 "out of time" events and about 300 non interacting primaries. The events were generated by simulating a 300 GeV/c pion beam impinging on an emulsion target in front of the telescope. This consists of seven area sensors and it is inside a 2T magnetic field. The interactions were generated with the production of Beauty particles, decaying via charmed D particles.(Figure 3).

The spatial precision in vertices reconstruction (taking into account also the multiple scattering in the dense target) is about 10 μm in the plane perpendicular to the beam and about 200 μm in the beam direction. The spatial resolving power is about 30 μm in the perpendicular plane and about 700 μ m in the longitudinal direction. The reconstruction program gives a 75% of resolved vertices (main vertex from the farthest decaying D) assuming a mean lifetime of 10^{-14}s for the B and 5.10^{-13}s for the D.

This device is therefore suitable to study short lived particles.

The maximum beam intensity is limited by the maximum amount of background. From the simulation program this limit comes out to be about 300 pions on the telescope during the read-out time of 20 ms, giving a beam intensity of $1.5.10^4$ pions/sec and 3.10^4 pions/burst of SPS, for a telescope with a single array per plane (about 1 cm^2). Naturally the useful area of the telescope can be increased by using several arrays for each plane, thus increasing the maximum number of pions too.

A test run is foreseen at the beginning of the next year with high energy pions. This test will be devoted to the measurement of the efficiency of the arrays to minimum ionizing particles (that seems to be near 1), and to the determination of the best work conditions. This test will be carried out with 2 planes. A further test is foreseen at the end of the next year with 7 planes and an emulsion target to check the reconstruction program.

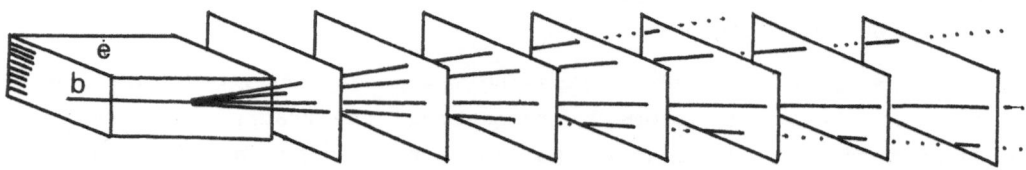

Fig. 3. Set-up of the target and of the telescope. b) Beam;
        e) Emulsion stack.

DISCUSSION

L. Foa': Maybe I have missed something or I am too far on the book,
but in the slide showing your reconstructed tracks I do not see
any correlation between points and reconstructed tracks. Could you
show it again and comment on this?

Answer: It is really not easy to see the correlations on the picture.
In fact the correlations appears when the points are tested to be
quasi aligned and have to satisfy some definite ratio on their
mutual distances. When a track is curved by the magnetic field the
curvature have to be almost costant. The reconstruction program
starts from the farthest plain going towards the nearest to the
target, working on the y-z coordinates and using the x coordinate
to check they are quasi-equidistant.

G. Moneti: What is the thickness of this device in radiation lengths?

Answer: The whole telescope of 7 planes has a thickness of about
1/10 of radiation length.

H. Heijne: The charge generated by black and grey tracks in the
bulk of the CCD may diffuse slowly to the pixels and this could
constitute an important background. Could you comment on this
problem?

Answer: I have not up to now any data on the response of this CCD
to black and grey tracks. In fact I have now only one GEC MA 357
and this has a frontal glass and a rear cover of plastic that prevents
to make such test. I think that the output signal is not very
different if compared with that of a minimum ionizing particle,
except that the signal is bigger. In fact the spread of the charge
generated by black and grey tracks is about the same of minimum
ionizing tracks, and, as I have said it does not seem that the
diffusion has to give a strong contribution to the collection of
the charge from the not depleted region. In fact it seems that the
pixel collects charge from a distance very large (or the order of
100 μm or more) without seeing a corresponding number (about 4 for
each side) of pixels interested in the collection. In other words
the signal is narrow. But I have not tested this condition and
hence this is only my feeling.

SILICON DETECTOR ELECTRONICS FOR H.E. PHYSICS EXPERIMENTS

Aleksandar Hrisoho

Laboratoire de l'Accélérateur Linéaire
Bâtiment 200
91405 ORSAY CEDEX FRANCE

INTRODUCTION

Silicon detector electronics, for H.E. physics experiments must supports high rate. Short processing time (filter time constant) is needed to avoid pile-up effects. Hence, the series noise of the amplifier will be the dominant one, and the paralled noise will be attenuated.

In this case bipolar transistors can be used, instead of FET for the head amplifier. The advantage of the bipolar transistors is that high g_m (transconductance) for low current can be obtained (compared to FET) and a low dissipation amplifier can be designed. The parallel noise due to the base current is reduced when short filter time constant is used.

For silicon telescope detectors, as live targets or as multiplicity recognizers, the detector capacity, C_D, is between 100 pF and 1000 pF. In this case, small and stable input impedance of the head amplifier is needed.

The condition $C_D.R_{in} \ll$ processing time (where R_{in} - the head amplifier input impedance) has to be satisfied. If not,

- instabilities of R_{in} will affect the final resolution,

- the "balistic deficit" of the injected charge will give a poorer signal to noise ratio.

μ-strip detectors have capacitances up to 10 pF. The counting rate/strip is smaller. In this case R_{in} can be higher.

The energy released in the silicon sensitive region is
= 30 keV/100 μm. To detect this charge, low noise amplifiers are
needed.

The design of the head amplifier is partially dependent on the
processor employed. Time-variant filtering is attractive, as it is
implemented by using gated integrators of the commercial charge-
sensing A.D.C. Hence, <u>current sensitive</u>, or more generally, head
amplifier reproducing the detector current shape, represents a suitable solution.

HEAD AMPLIFIER CONFIGURATION

Three types of head amplifiers have been analysed :

- Grounded base configuration, Fig. 1.

Transistors with very small collector-base capacitances have
been chosen . The impulse response is an exponential one with time
constant $\tau_L = R_L C_L$. High value for R_L is chosen for low noise.

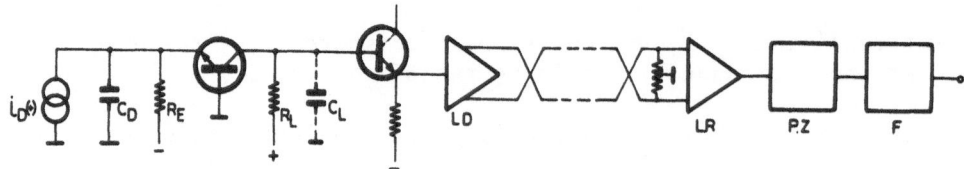

Figure 1.   The grounded base configuration:
            L.D. - Line driver; L.R. - Line reciever;
            P.Z. - Pole-zero cancellation; F - Shaping filter
            amplifier.

from R_L. The stray capacitance $C_L \doteq 2pF$. Hence the constant $R_L C_L$ has to be cancelled. This is done by the pole-zero cancellation circuit.

A filter circuit is added to obtain an optimized weighting function for low noise.

The input impedance is given by :

$$R_{in} = \frac{kT}{qI_E} = \frac{26}{I_E[mA]} \tag{1}$$

k : Boltzman constant
T : absolute temperature
q : electron charge
I_E : head transistor current

The imput impedence can be made relatively small by proper choice of the emitter current I_E. Hence, this configuration is recommended when the detector capacitance C_D is high (more than 20 pF).

- Resistive feedback current amplifier, Fig. 2.

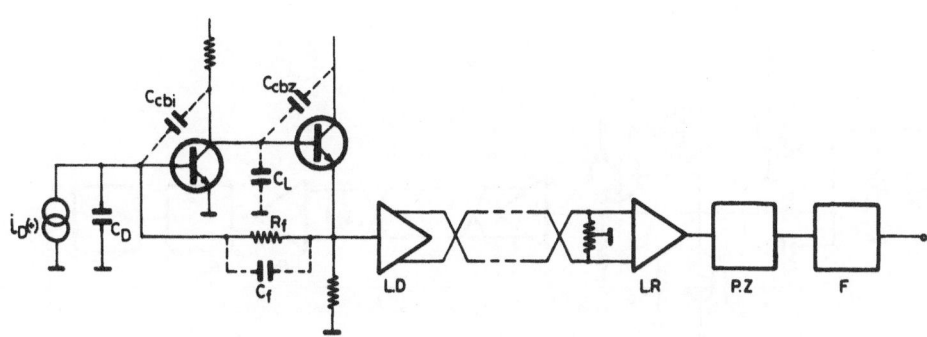

Figure 2.   The resistor feedback configuration.

The impulse response is an exponential with time constant
$\tau_f = R_f C_f$. The stray capacitance $C_f \approx 0.2$ pF. Much higher value for
R_f can be chosen. Low supply voltage is needed, which gives low
dissipation. This configuration can operate with a supply voltage
of 3.5 V and 25 mw dissipation including the driver circuit is obtained.

The input impedance, given by :

$$R_{in} = \frac{26}{I_E} \cdot \frac{C_{cb1} + C_{cb2} + C_L}{C_{cb1} + C_f} \tag{2}$$

where C_{cb1} and C_{cb2} are the collector to base capacitances for the
first and the second transistors and C_f the feedback stray capacitance, is higher than that given by (1). Hence, this configuration is more recommended for detectors with low C_D.

Pole-zero cancellation and filter circuit are necessary.

- Charge amplifier configuration, Fig. 3.

Similar to the resistive feedback amplifier, this configuration is more complicated. Well suitable for FET, but also bipolar

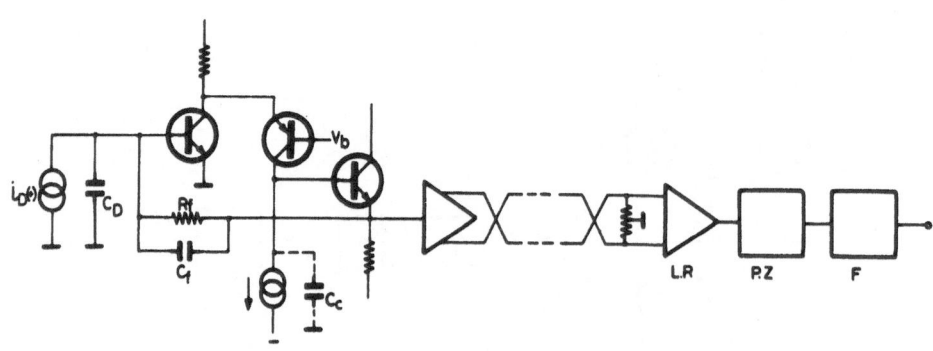

Figure 3.   Charge amplifier configuration.

transistors can be used. The input resistance is more stable and is given by :

$$R_{in} = \frac{1}{g_m} \cdot \frac{C_c}{C_f} \tag{3}$$

g_m : the transconductance.

By definition in a charge amplifier the feedback capacitance C_f is about $1 - 2$ pF, and the input resistance is low. When FET are used, high current has to be employed to obtain good g_m.

FET current in the order of 10 mA will give a $g_m \approx 10$ mA/V which can be obtained with a bipolar transistor at 0.26 mA.

NOISE ANALYSES

The study of noise is based on thermodynamics, circuit theory and statistical analysis. The noise gives a random wave shape and this random time varying function is called a random process $v(t)$. A sample of $v(t)$ at a time (t) is a random variable with a probability density function $f_t(t)$. This function is generally a gaussian distribution around a mean $E(v)$. As a measure for the noise the variance of the distribution $f_t(t)$ is used.

The noise generated by a current (composed of a series of independent events arriving at random) is called shot noise. This noise can be represented, in the time domain, by a current source of randomly distributed current impulses defined by $q_o \delta(t)$ ($\delta(t)$ - the Dirac function and q_o - the electron charge) with a mean rate $n(I)$ depending on I.

The mean square noise current per Hz of this generator is[1] :

$$\overline{i_n^2} = 2 q_o^2 \cdot n(I) \tag{4}$$

and with $n(I) = \dfrac{I}{q_o}$; the well known relation for $\overline{i_n^2}$ can be obtained :

$$\overline{i_n^2} = 2 q_o I . \tag{5}$$

Noise generated by a resistor R is determined by studying the thermal equilibrium behaviour. The mean square noise per Hz for thermal noise of a resistor R is :

$$\overline{i_n^2} = \frac{4kT}{R} . \tag{6}$$

We can introduce an equivalent current reproducing the same

noise as that one generated by the resistor R.

$$\frac{4kT}{R} = 2 \ q_o I_R = 2 \ q_o^2 \ n(R) \tag{7}$$

The equivalent noise source for a resistor R will be defined by a current source, in parallel with R, of the randomly distributed current impulse $q_o \delta(t)$ with a mean rate n(R) depending on R :

$$n(R) = \frac{2}{q_o^2} \cdot \frac{kT}{R} \tag{8}$$

By using Thevenin's theorem the resistor noise can be replaced by a voltage noise source, in series with R, defined by $Rq_o \ \delta(t)$ voltage impulses with mean n(R). The mean square voltage noise per Hz will be :

$$\bar{e}_n^2 = 2 \ (Rq_o)^2 \cdot n(R) = 4kTR \tag{9}$$

which agrees with the well known relation for the voltage noise generator of a resistor R.

The noise generated by a transistor is shot type noise. It can be entirely represented by two equivalent noise sources :

- a parallel one, the noise measured at the output with the input open circuit ;

- the series one, the noise measured at the output with the input short circuit.

The parallel noise is represented by a current generator of pulses $[(q_o \delta(t), n(I_B)]$ depending on the base current of the transistor.

The series noise is represented by a voltage generator of pulses $[q_o(r_e + r_b) \ \delta(t), n(r_e + r_b)]$ depending on r_b - the base spread resistance, $r_e = \frac{1}{2g_m}$ the equivalent noise resistance due to the collector current noise.

The mean squared noise current for the parallel noise generator is :

$$\bar{i}_n^2 = 2q_o I_B \tag{10}$$

and the mean square voltage noise for the serie noise generator is :

$$\bar{e}_n^2 = 4kT \ (r_e + r_b) \tag{11}$$

As the signal is a current source, the comparison with a current noise source is more convenient. Hence, the series noise generator $[q_o \ (r_e + r_b) \ \delta(t), n \ (r_e + r_b)]$ can be replaced by a parallel one. Using Thevenin's theorem , and supposing that the capacitance at the amplifier input is C_i, the current noise generator in parallel with C_i will be given by :

$$i_n(t) = C_i \frac{de_n(t)}{dt} = C_i (r_e + r_b) \cdot q_o \delta'(t) \tag{12}$$

with a mean $n(r_e + r_b)$ and a spectrum :

$$\overline{q}_n^2 = 2 [C_i (r_e + r_b) \cdot q_o]^2 \cdot n(r_e + r_b) = 4kT (r_e + r_b) \cdot C_i^2 \tag{13}$$

The signal to noise relation for an amplifier is conventionally expressed as an equivalent noise charge producing an output equal to the rms noise level[3]. This equivalent noise charge, ENC, depends on the location of the noise source in the circuit, on the noise spectrum and on the filter type used with the amplifier.

For a current noise source defined by delta impulses $[q_o \delta(t) \cdot n]$ the mean square equivalent noise charge $\overline{ENC}^2$ for all impulses arrived before the measuring time t_m will be[2] :

$$\overline{ENC}^2 = \frac{1}{2} \overline{i}_n^2 \int W(t)^2 \, dt \tag{14}$$

W(t) : The weighting function referred to the measuring time t_m (which is normally the time of the maximum of the impulse response) is defined for $t < t_m$ and at time t has a value W(t) expressing the output waveform amplitude at $t = t_m$, obtained from a $\delta(t)$ impulse of unit charge which occurs at time t.

For a noise source defined by $C_i (r_e + r_b) \cdot q_o \delta'(t)$, $n (r_e + r_b)$ the mean square equivalent noise charge $\overline{ENC}^2$ for all impulses arrived before the measuring time t_m will be :

$$\overline{ENC}_s^2 = \frac{1}{2} \overline{q}_n^2 \int W'(t)^2 \, dt \tag{15}$$

The contributions of independent parallel and series noise sources are expressed as an equivalent total noise charge ENC_{tot} at the amplifier input by :

$$ENC_{tot}^2 = ENC_p^2 + ENC_s^2 \tag{16}$$

ENC_p^2 is the sum of all parallel noise sources.
ENC_s^2 is the sum of all series noise sources.

For the grounded base configuration :

$$ENC_p^2 = \frac{1}{2} \left[2q_o [I_{B1} + I_{B2}] + 4kT \left(\frac{1}{R_E} + \frac{1}{R_L} \right) \right] \int w(t)^2 \, dt \tag{17}$$

The term $4kT \left(\dfrac{1}{R_E} + \dfrac{1}{R_L} \right)$ accounts for noise due to parallel resistances at the input. I_{B1} and I_{B2} are the bases currents of the first and second transistors.

There is a useful rule giving the equivalence of noise from a current I and a resistor R. Any resistance noise source can be replaced by a current noise source and vice versa using the relation :

$$RI = 50 \cdot 10^{-3} V = 50 \text{ mV} \qquad (18)$$

The series noise will be :

$$ENC_s^2 = \frac{1}{2} C_{in}^2 \, 4kT \, (r_e + r_b) \int w'(t)^2 \, dt \qquad (19)$$

For the resistor feedback and the charge amplifier $I_{B2} = 0$ and $\left(\dfrac{1}{R_E} + \dfrac{1}{R_L} \right) = \dfrac{1}{R_f}$.

It is possible to make the design such that the noise contributions from the resistors are negligible. Assuming that r_b is negligible one can find that the minimum total noise is obtained when $ENC_p = ENC_s$.

This condition can be satisfied by choosing properly I_E. The emitter current of the first transistor :

$$I_E \text{ opt [mA]} = \frac{C_i \, [pF]}{t_F \, [nS]} \qquad (20)$$

where t_F is the filter time constant, for a triangular impulse response t_F is the base width of the triangular output.

The maximum value of I_E is limited by :

$$r_b < r_e = \frac{1}{2g_m} = \frac{13}{I_E \, [mA]} \qquad I_E < \frac{13}{r_b} \qquad (21)$$

In this case the condition $r_b < r_e$ will be satisfied. An approximated formula for the noise corresponding to the optimum conditions is given by :

$$ENC_{opt} = 225 \sqrt{C_i}$$

where C_i the detector capacitance + strays and $R_{in} C_i \ll t_F / 10$
R_{in} - the input impedance.

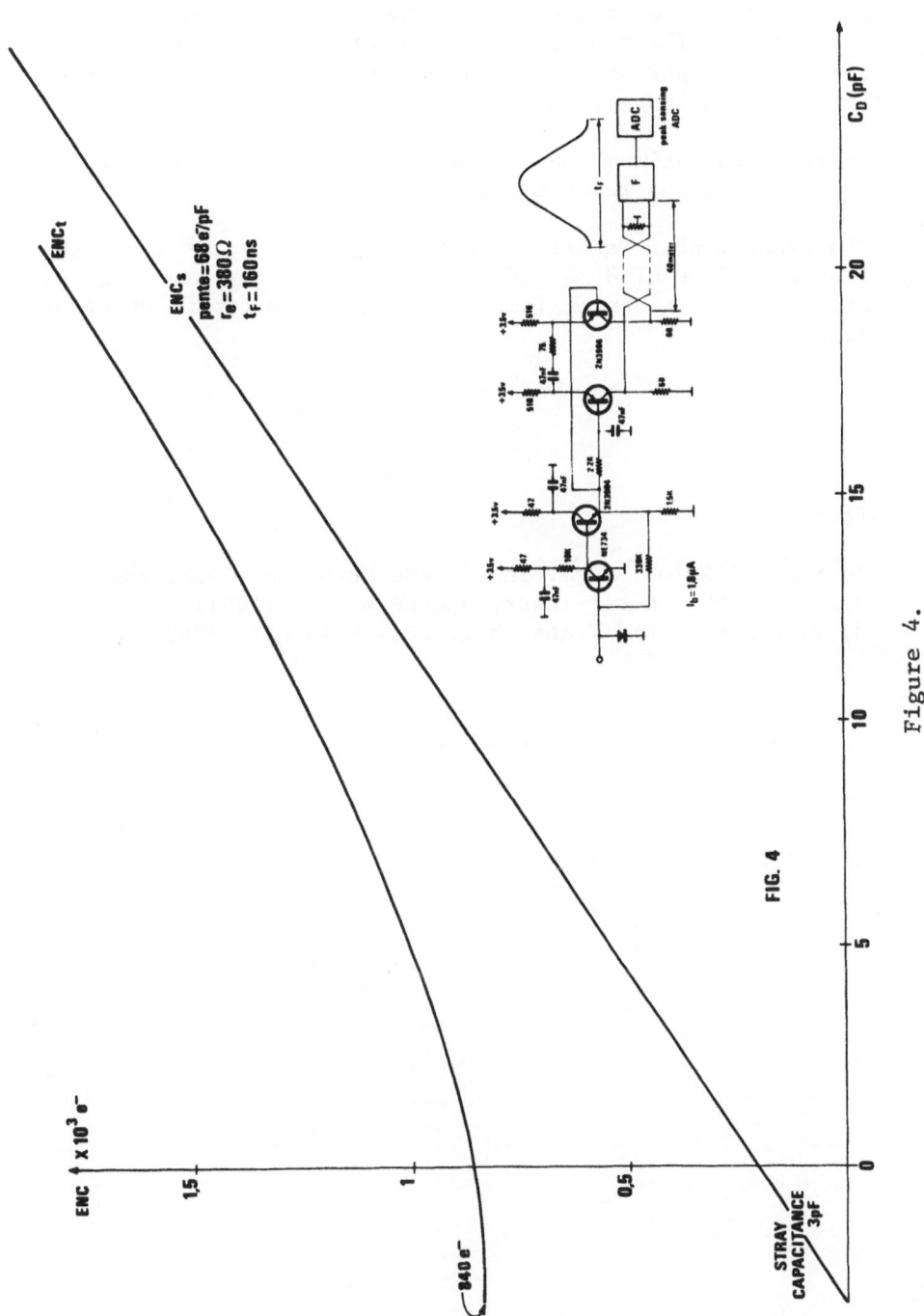

Figure 4.

CONCLUSION

As long as noise is considered, the amplifier configuration is not important. The choice of the transistors and the design áre essential. The comparison has been made using the same transistors for the three configurations.

There is an optimum when the current I_E of the first transistor is matched to the input capacitance C_i.

The experimental results obtained with a resistor feedback amplifier with I_E = 0.18 mA and t_F = 160 ns are given on Fig. 4. The total noise is given in [rmse] as a function of the detector capacitance C_D.

REFERENCES

1. M.O. DEIGHTON, Nucl. Instr. and Meth. 58 (1968) 201-212.
2. A. HRISOHO, Nucl. Instr. and Meth. 185 (1981) 207-213.
3. V. RADEKA, IEEE Trans. Nucl. Sc. NS15 n°3 (1968) 455.

HEAVY FLAVOUR PRODUCTION IN HADRON-HADRON COLLISIONS

(Presented by G. D'Ali)

M. Basile, G. Bonvicini, G. Cara Romeo, L. Cifarelli,
A. Contin, G. D'Ali, P. Di Cesare, B. Esposito, P. Giusti,
T. Massam, R. Nania, F. Palmonari, G. Sartorelli,
G. Valenti and A. Zichichi

CERN, Geneva, Switzerland
Istituto di Fisica dell'Università, Bologna, Italy
Istituto Nazionale di Fisica Nucleare, Bologna, Italy
Istituto Nazionale di Fisica Nucleare, LNF, Frascati, Italy
Istituto di Fisica dell'Università, Perugia, Italy

1. INTRODUCTION

Open charm production in hadron-hadron interactions has already been extensively studied and reported by several authors[1-5]. Recently, also the first open beauty state has been identified as the Λ_b^0 baryon, in the experiment R415 performed at the CERN Intersecting Storage Rings (ISR)[6], where charmed-particle production distribution has also been widely studied[7-11]. The purpose of this talk is to report on this experiment, putting particular emphasis on the common features of charm and beauty production in hadronic interactions:

i) Charm production cross-sections at the ISR are very high compared to lower energy results, suggesting a very strong energy dependence (Figure 1 shows a summary of the charm cross-section measurements as a function of $\sqrt{s}$). The Λ_b^0 cross-section estimate too, seems to be rather high compared to expectations.

ii) In proton-proton interactions, Λ_b^0 and Λ_c^+ production distributions show strong similarities, suggesting the same "heavy-flavour production mechanism". Moreover, charm- and beauty-production rates are compatible with following a mass law:

$$\sigma(m_f) \propto 1/m_f^2 \; ,$$

where m_f is the mass of the produced flavour.

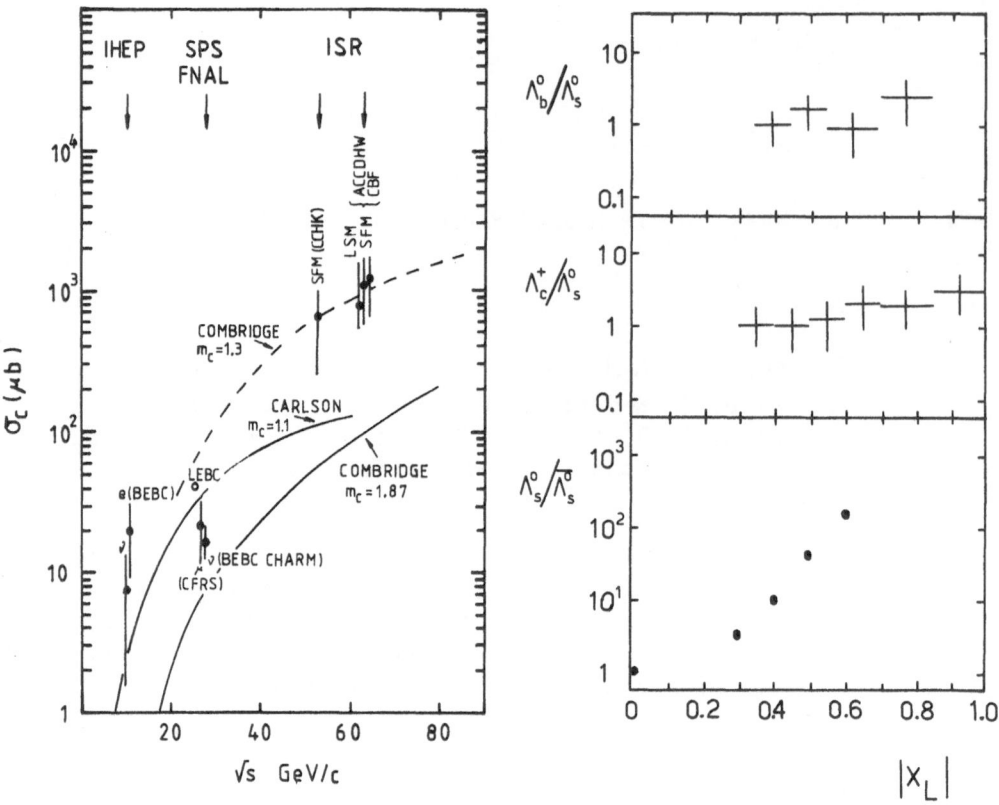

Fig. 1   Charm production cross-
         section as a function
         of √s

Fig. 2   Quantitative comparison
         between Λ_b^0, Λ_c^+, Λ^0, and
         $\overline{\Lambda}^0$ longitudinal-momentum
         distributions

iii) The "leading-particle effect" plays a very important role also
     in "heavy flavour" production[12,13], as arises from the Λ_b^0 and
     Λ_c^+ longitudinal momentum distributions that are quite flat as
     a function of $x_L = 2p_L/\sqrt{s}$. (Figure 2 shows the Λ_b^0, Λ_c^+ x_L-
     distribution compared with the Λ^0 "leading" and the $\overline{\Lambda}^0$ non-
     leading x_L distributions.)

2.   EXPERIMENTAL SET-UP AND HEAVY-FLAVOUR TRIGGER

     The semileptonic branching ratio of heavy-flavoured particles
is rather high (∿ 20%), and it is thought to be one of the main
sources of "prompt" leptons with high transverse momentum ($p_T >$
> 0.5 GeV/c). If this is the case, a "single electron trigger"
turns out to be a good signature of heavy-quark production. Even if
the direct identification of the parent particle is very difficult,
because of the presence of a neutrino in the final state, which
escapes detection, one can take advantage of the associated production of heavy flavours, and look, in the same event, for the other

heavy-flavoured partner, decaying, in most cases, hadronically. The search for this associated state can be performed via an invariant mass analysis among the particles in the final state. A wide acceptance detector is needed to apply this technique, and the Split Field Magnet (SFM) fulfils this requirement.

    The most important drawback of this method is represented by the combinatorial background in the invariant mass spectrum, together with the small branching ratios of the exclusive decay channels:

- The combinatorial background arises from the fact that all the reconstructed tracks have to be taken into account in the invariant mass spectrum, with all mass assignments, since one cannot pick up the secondary vertex of the decay, or do particle identification (as is the case for 90% of the solid angle, using the SFM).

- Neutral particles are not detected by the SFM, so the only decay channels we can look for are those for which all decay products are charged.

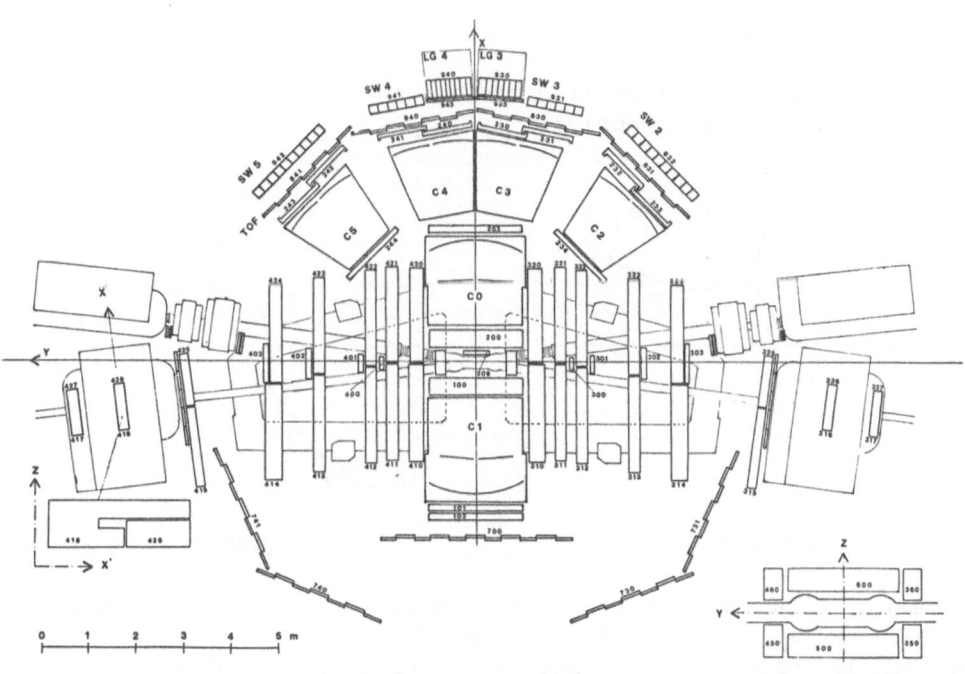

Fig. 3   Top view of the SFM detector showing the MWPCs, the time-of-flight (TOF) system, the electromagnetic shower detectors (SW, LG), the gas threshold Čerenkov counters (C), and the dE/dx chamber (209).

The main features of the experimental set-up, shown in Fig. 3, were:

i) A powerful electron detector in the 90° region[14], consisting of:
   a) electromagnetic shower detectors (EMSD)[15] and Cherenkov threshold counters for $e^{\pm}$ versus charged hadrons ($h^{\pm}$) discrimination, giving a $\leq$ 2% contamination of charged hadrons in the final electron sample;
   b) a little multiwire proportional chamber (209)[16], with analogue readout, placed very near to the intersection region, and used to reduce the background of e^+e^- pairs coming from neutral hadron decays; the contamination in the final sample of prompt electron candidates due to this source was $\sim$ 50%.

ii) A time-of-flight (TOF) system of counters[17], able to give $\pi/K/p$ separation up to $\sim$ 1.5 GeV, but covering only 10% of the solid angle.

iii) The SFM multiwire proportional chamber (MWPC) system[18] for track reconstruction and momentum determination over 90% of the solid angle ($\sim$ 75% if an accuracy of $\Delta p/p <$ 30% in momentum measurement is required).

3. CHARM PRODUCTION

The following diagrams illustrate the decay mechanism we have been looking for in our charm analysis

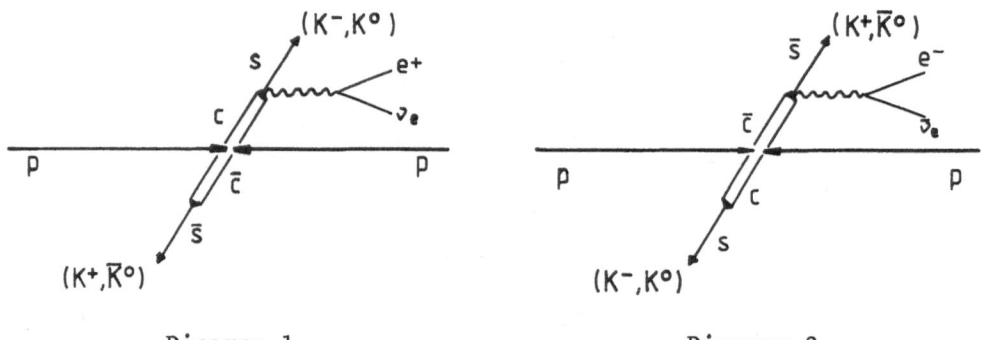

Diagram 1   Diagram 2

There is an unambiguous correspondence between the triggering lepton charge sign and the flavour-antiflavour determination; the c quark semileptonic decay produces an e^+ (diagram 1) and the associated hadronically decaying particle will be an anticharmed one; vice versa in the case of e^- triggering ($\bar{c}$ semileptonic decay), only charmed particles are expected to be found in the event. This correspondence is very important because it will be exactly reversed for beauty production. When changing the charge sign of the triggering lepton, going from the "right trigger" to the "wrong trigger" (in the sense of the previous remark) any observed signal for a given charm-decay channel should disappear. We have checked that this happens for all the channels we have been looking for[19]. We will now limit the discussion to those charm signals where statistical signifi-cance allows a more detailed study of their production distribution.

3.1 Λ_c^+ production

The Λ_c^+ production has been studied via the decay mode

$$\Lambda_c^+ \rightarrow pK^-\pi^+$$

in the reaction

$$pp \rightarrow e^- + \Lambda_c^+ + X .$$

The p, K^-, π^+ mass assignments were done on the basis of the following criteria:

  i) p: the positive fastest particle in the event with $|x_L| \geq 0.3$;
 ii) K^-: any negative track with $|y| \geq 1$ and not identified by the TOF system as a π^- or a $\bar{p}$;
iii) π^+: any positive track with $|y| \geq 1$ and not identified by the TOF system as a K^+ or p.

The condition (i) selects a proton sample with $\leq 20\%$ π^+ conta-mination[20]; the requirements of $|x_L| \geq 0.3$ for the proton and $|y| \geq 1$ for K^-, π^+, restrict the kinematical region available for Λ_c^+ detection to $|x_L(\Lambda_c^+)| \geq 0.30$, i.e. to the "forward" cone in phase space. All the p, K^-, and π^+ tracks were required to have a $\Delta p/p \leq \leq 30\%$ accuracy in momentum reconstruction and to fit the interaction vertex within ±5 cm. The presence of a "leading system", in the rapidity hemisphere opposite to the Λ_c^+ was also required $\left[|\Sigma_i \; x_{Li}| < \right.$ < 0.1 or $|\Sigma_i \; x_{Li}| > 0.5$, where i ranges over all the tracks such that $\left. (x_L)_i \cdot (x_L)_{\Lambda_c^+} < 0 \right]$.

Figure 4 shows the final $pK^-\pi^+$ invariant mass spectra for events associated with the e^- (Fig. 4a) and with the e^+ (Fig. 4b) triggers. A 4 standard deviation peak is present in the Λ_c^+ region associated with the e^- trigger. The width of this peak, compatible with the experimental resolution, and the absence of a similar peak in the sample of events triggered by an e^+, identifies it as a Λ_c^+ signal.

To get some more information about the Λ_c^+ production features, the following method has been used. In the $pK^-\pi^+$ invariant mass spectrum two mass regions have been defined:

a) an "IN" region, including the mass range where the Λ_c^+ signal is observed $\left[2.28 \leq m \; (pK^-\pi^+) < 2.38 \; \text{GeV}/c^2 \right]$;

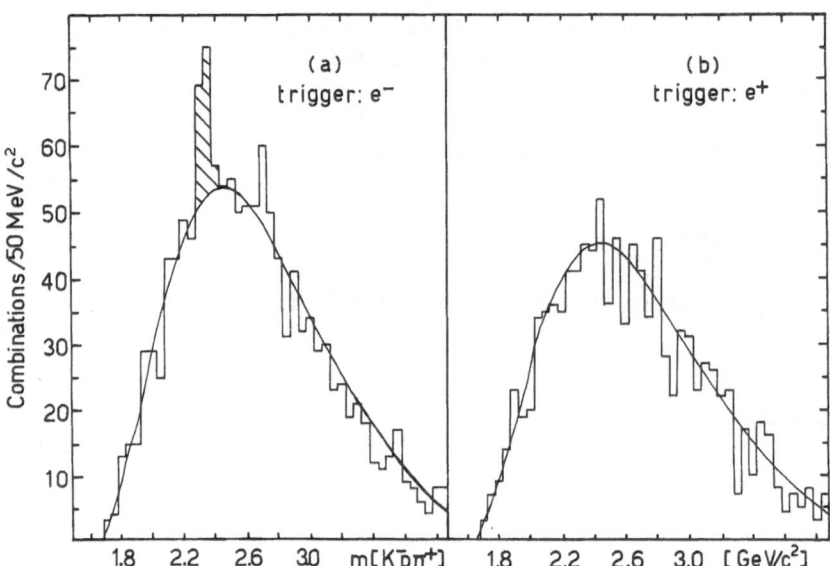

Fig. 4   Mass distribution of pK⁻π⁺ combinations, selected as des-
         cribed in the text, for events triggered (a) by an e⁻ and
         (b) by an e⁺. The solid-line fit represents the background
         shape as determined by the e⁺ triggered events.

b) an "OUT" region, including the mass range $[2.18 \leq m\ (pK^-\pi^+) <$
$< 2.28\ GeV/c^2]$ immediately below and the mass range $[2.38 \leq$
$\leq m\ (pK^-\pi^+) < 2.48\ GeV/c^2]$ immediately above the Λ_c^+ peak.

The longitudinal and transverse momentum spectra for the "IN"
$pK^-\pi^+$ combinations contain two components that cannot "*a priori*"
be separated; one is due to the Λ_c^+ events, and the other to the
background below the peak. The latter should have the same shape
as the corresponding distribution worked out for the OUT region,
so that the shape of the former can be obtained as the IN-OUT difference (with the right normalization).

Figure 5 shows the Λ_c^+ longitudinal momentum distribution ($x_L =$
$= 2p_L/\sqrt{s}$) obtained in this way. The best fit to the data

$$dN/dx_L \propto (1 - x_L)^\alpha$$

gives $\alpha = 0.40 \pm 0.25$. This quite flat x_L-distribution confirms that
for Λ_c^+ too, as for Λ^0 production, a strong leading effect character-
izes, in pp interactions, the baryon production, whereas the same effect is absent in antibaryon production[12]. We will have a further
proof of this fact from the analysis of the Λ_b^0 production.

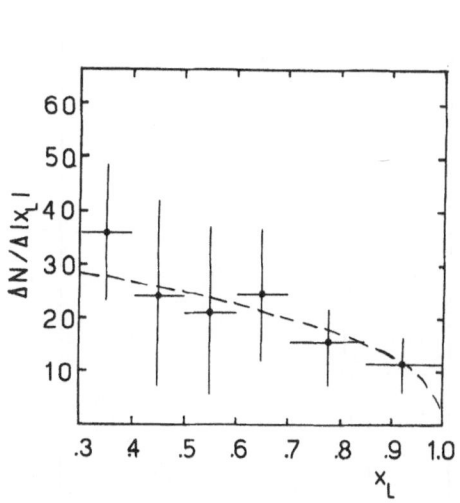

Fig. 5   Experimental x_L distri-
          bution of the Λ_c^+ events.
          The dashed line is the
          best fit with $dN/dx_L \propto$
          $\propto (1 - x_L)^\alpha$.

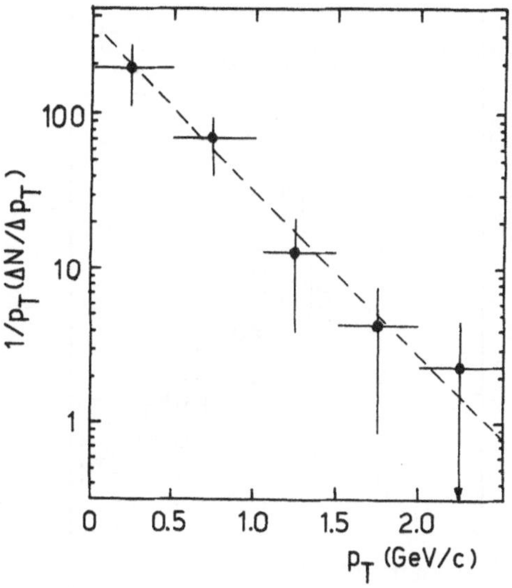

Fig. 6   Experimental p_T distri-
          bution of the Λ_c^+ events.
          The dashed line is the
          best fit with
          $(1/p_T)(dN/dp_T) \propto e^{-bp_T}$.

The transverse momentum distribution of the Λ_c^+ (Fig. 6) can be fitted with the function

$$\frac{1}{p_T}\,\frac{dN}{dp_T} \propto e^{-bp_T}$$

obtaining $b = 2.5 \pm 0.4$. The resulting average transverse momentum is

$$\left\langle p_T(\Lambda_c^+) \right\rangle = 0.8 \pm 0.12 \text{ GeV/c }.$$

3.2 D^+ production

D^+ production has been studied via the analysis of the reaction

$$pp \to e^- + D^+ + X$$

looking for the D^+ decay channel

$$D^+ \to K^- \pi^+ \pi^+$$

and identifying the final state as follows:

 i) K^- identified with 90% C.L. by the TOF counters;
ii) π^+ any positive track with $|x_L| < 0.3$ and not identified by the
    TOF counters as a K^+ or a p.

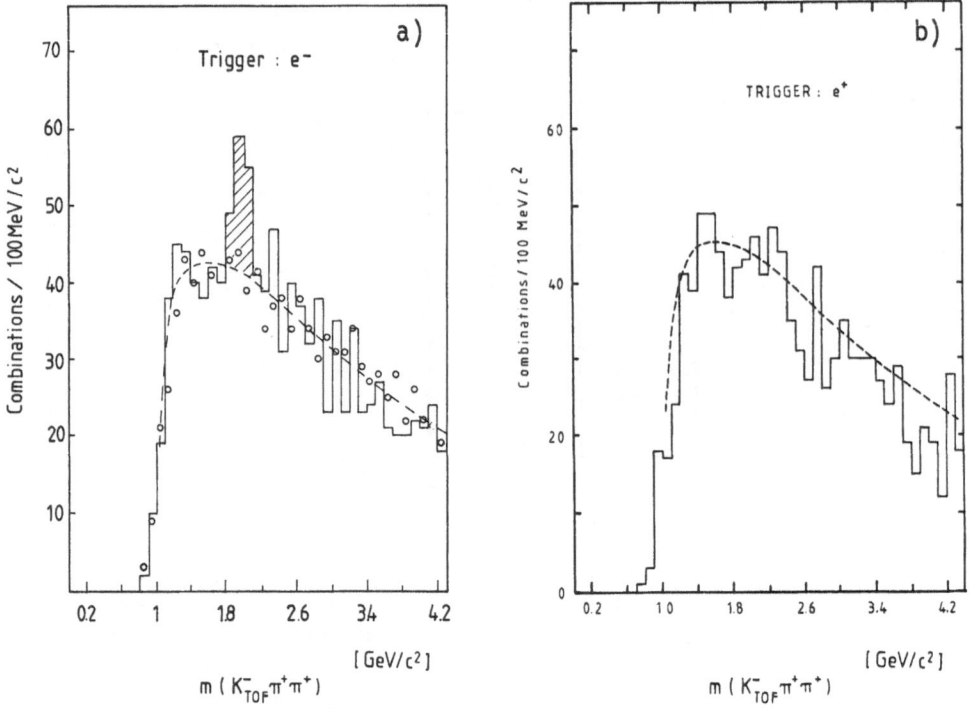

Fig. 7   Invariant mass spectrum of $K^-\pi^+\pi^+$ combinations triggered
          (a) by an e^- and (b) by an e^+, selected as specified in the
          text.   The circles and the dashed-line curve show the distri-
          bution and fit to the event-mixing background.

        A cut on the $K^-\pi^+\pi^+$ system transverse momentum at $p_T > 0.7$ GeV/c
has been applied in order to increase the signal/background ratio.
Figure 7a shows the corresponding $K^-\pi^+\pi^+$ mass spectrum from events
triggered by an e^-. An excess of 39 ± 11 events is present at the
D^+ mass, over the dashed curve representing the background. This
background shape is the best fit to the $K^-\pi^+\pi^+$ spectrum obtained
with the "event-mixing technique". In Fig. 7b this background is
compared with the $K^-\pi^+\pi^+$ spectrum obtained using e^+ as the triggering lepton. There is, as expected, no D^+ signal in this case, and
the two spectra fit well with each other.

        With the same "IN-OUT" technique as explained for the Λ_c^+ signal
the transverse and longitudinal momentum distributions of D^+ were
studied (Figs. 8, 9). For the transverse momentum distribution the
best fit to the data with the function:

$$\frac{1}{p_T}\frac{dN}{dp_T} \propto e^{-bp_T}$$

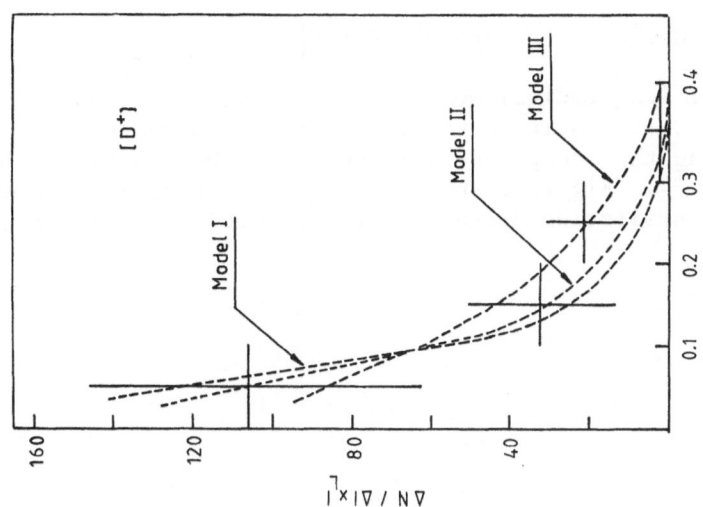

Fig. 9   Experimental x_L distribution of the D^+ events. The dashed lines are the Monte Carlo prediction on the basis of three possible models for D^+ production [Model I: $E(d\sigma/dx_L) \propto (1-x_L)^3$; Model II: $d\sigma/dy = const$; Model III: $d\sigma/dx_L = const.$].

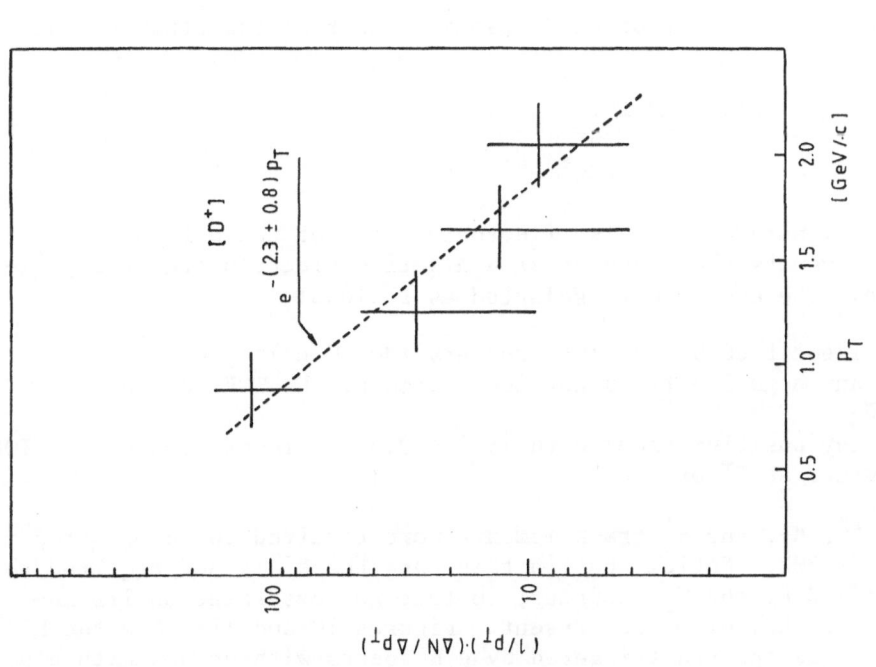

Fig. 8   Experimental p_T distribution of the D^+ events. The dashed line is the best fit.

gives a value (Fig. 8):

$$b = 2.3 \pm 0.8$$

in good agreement with the value obtained for Λ_c^+.

In so far as the longitudinal momentum distribution is concerned, it should be noticed that condition (i) limits the K^- momentum to $p \lesssim 1.5$ GeV, modifying the acceptance for the parent D^+, in particular depleting the high x_L's. Figure 9 shows the x_L distribution compared with the Monte Carlo prediction based on three possible distributions for D^+:

a) $E \dfrac{d\sigma}{dx_L} \propto (1 - x_L)^3$

b) $\dfrac{d\sigma}{dy} = \text{const.}$

c) $\dfrac{d\sigma}{dy} = \text{const.}$

The K^- momentum condition squeezes the three curves in such a way that no discrimination is possible among these production models.

3.3 D^0 production

Some more information on D-mesons comes from the study of the reaction:

$$pp \to (e^- + K^+) + D^0 + X$$
$$\,\rule{0pt}{1em}\hookrightarrow K^-\pi^+$$

where the signature of the semileptonic decay of the anticharmed particle is given by the presence of a negative electron and of a positive kaon. The events were selected as follows:

  i) K^+: identified by the TOF counters (90% C.L.);
 ii) K^-: any negative track not identified by the TOF counters as π^- or $\bar{p}$;
iii) π^+: any positive track with $|x_L| \leq 0.3$ not identified by the TOF counters as K^+ or p.

The K^+, K^-, and π^+ track momenta were required to be measured with $\Delta p/p \leq 30\%$. Notice that in this case the K^- is not required to be identified by the TOF counter, so that no constraint on its momentum or on that of D^0 is present. Figures 10 and 11a show the $K^-\pi^+$ invariant mass spectra triggered by e^-K^+ pairs without and with a p_T cut on the $K^-\pi^+$ mass ($p_T > 0.7$ GeV). The dashed line represents a

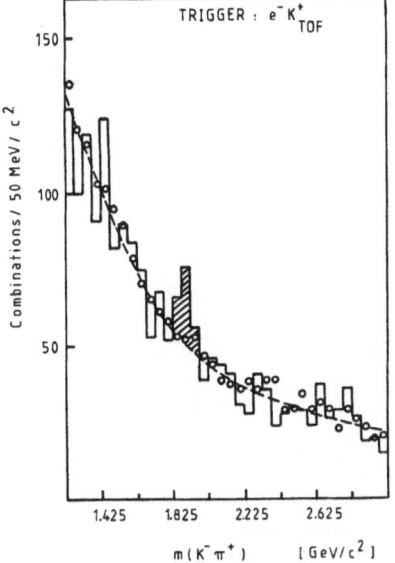

Fig. 10

Invariant mass spectrum of $K^-\pi^+$ combinations in events triggered by an e^-K^+ pair selected as specified in the text. The circles and the dashed-line curve show the distribution and fit to the event-mixing background.

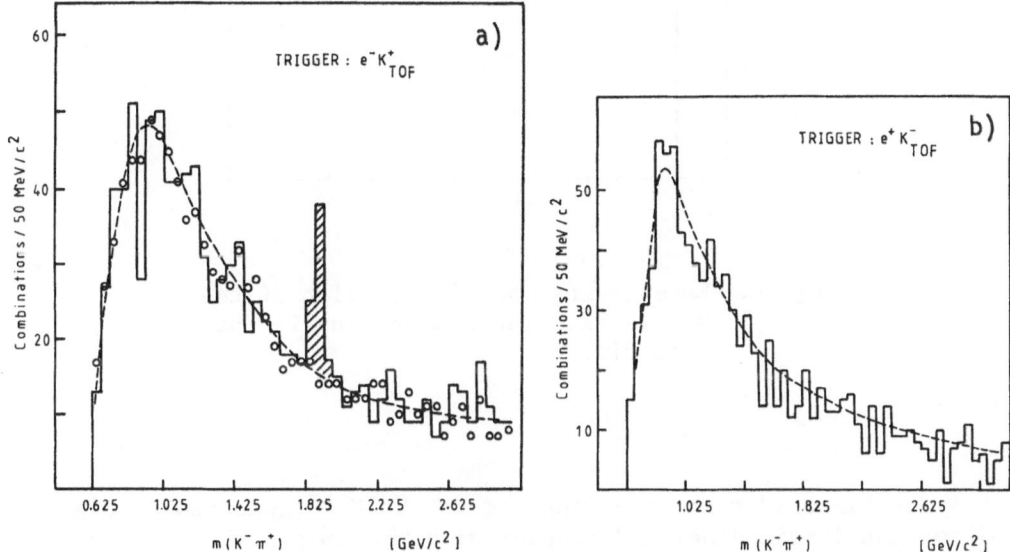

Fig. 11 a) Same as Fig. 10, when the requirement $p_T(K^-\pi^+) > 0.7$ GeV/c is also applied; b) As (a) but in events triggered by e^+K^-.

fit to the $K^-\pi^+$ spectrum obtained with the event-mixing technique (circles). The same background is compared with the $K^-\pi^+$ spectrum obtained with the "wrong trigger" e^+K^-, where no signal appears, as expected (Fig. 11b).

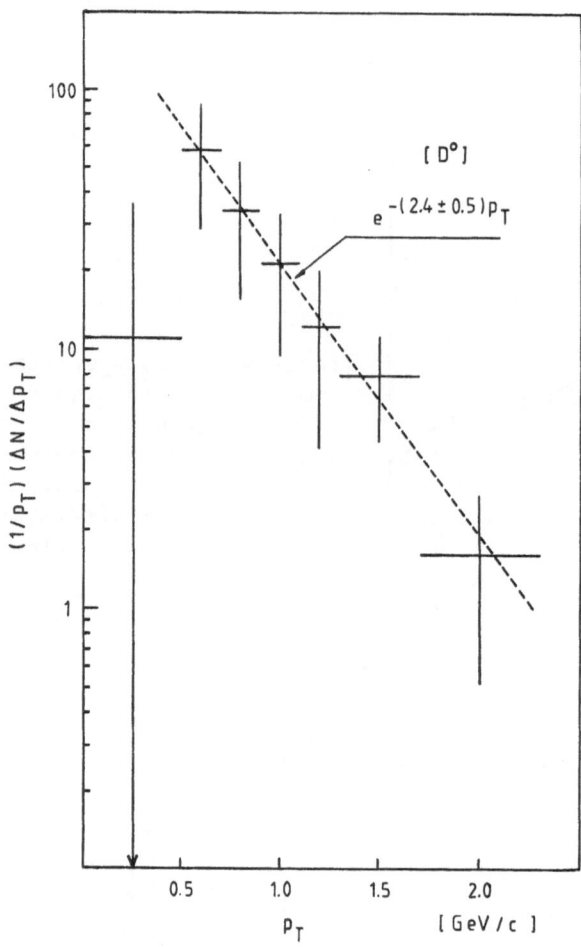

Fig. 12 Experimental p_T distribution of the
D^0 events. The dashed line is the
best fit.

The "IN-OUT" technique, applied to the $K^-\pi^+$ spectrum of Fig. 10,
allows the longitudinal and transverse momentum distributions for the
D^0 to be drawn (Figs. 12, 13). Here again the same slope, within
errors, found for D$^+$ and Λ_c^+ fits the D^0 data of Fig. 12. In Fig. 13
the longitudinal-momentum comparison with the Monte Carlo prediction
for three extreme possible x_L distributions shows that a "flat-x_L"
distribution is highly disfavoured, whereas nothing can be said about
the two other models (defined as in the D$^+$ case). The only conclusion which can be drawn is that Λ_c^+ is produced in the "forward" direction (dσ/dx_L $\simeq$ flat), whereas the trend for D production seems to
favour a rather "central" production.

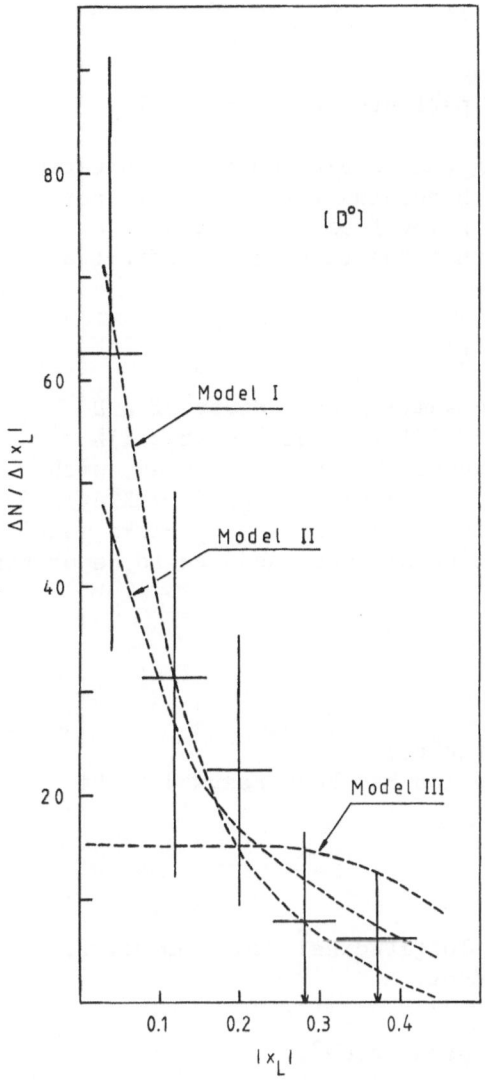

Fig. 13

Experimental x_L distribution of the D^0 events. The dashed lines represent the Monte Carlo prediction on the basis of three possible models for D^0 production [Model I: $E(d\sigma/dx_L) \propto (1 - x_L)^3$, Model II: $d\sigma/dy = $ const.; Model III: $d\sigma/dx_L = $ const.].

4. CHARM CROSS-SECTION ESTIMATES

Let me briefly comment on the inclusive charm cross-section estimates that these signals lead to. These estimates are strongly dependent on the assumptions one has to make to compute them. These assumptions are:

i) for the triggering particle decaying semileptonically:
   a) the mass,
   b) the production distribution,
   c) the decay modes,
   d) the decay dynamics;

  ii) for the observed charmed particle:
      a) the production distribution,
      b) the decay dynamics;
iii) any correlation between the two partners of the c,c̄ pair.

All these assumptions suggest keeping the results only as indicative; moreover, one has to fold in all the branching ratios needed (not always known with high accuracy) for going from the exclusive channel to the inclusive production, so that the results are affected by high errors, typically $\gtrsim$ 50%).

    We made the following assumptions:

i) The electron always comes from the semileptonic decay of a D meson, with a branching ratio $(D \to e^-X)/(D \to all) + (8 \pm 1)\%$ [21]. The decay proceeds via the two channels $D \to Ke\nu$, $D \to K^*e\nu$, with relative branching ratios $(D \to Ke\nu)/(D \to eX) = 0.6$, $(D \to K^*e\nu)/(D \to eX) = 0.4$. The decay dynamics is given by a K_{ℓ_3}-like matrix element. The production distribution has been assumed to be of the form:

$$E \frac{d^3\sigma}{dp^3} \propto f\left(\frac{y}{y \text{ max}}\right)(e^{-bp_T})$$

for both charmed and anticharmed particles;
  ii) the transverse momentum distribution has been assumed to be

$$\frac{1}{p_T} \frac{d\sigma}{dp_T} \propto e^{-2p_T} ,$$

iii) three extreme possibilities for longitudinal momentum distribution have been taken into account:

$$E \frac{d\sigma}{dx_L} \propto (1 - x_L)^3, \qquad \text{"central production"};$$
$$\frac{d\sigma}{dy} = \text{const.}, \qquad \text{"intermediate production"};$$
$$\frac{d\sigma}{dx_L} \simeq \text{const.}, \qquad \text{"forward production"}.$$

    For the hadronic decay of the observed particle the following assumptions have been made:

 i) decay distribution following Lorentz invariant (L.I.) phase space;
ii) branching ratios as from the literature[22]

$$(\Lambda_c^+ \to pK^-\pi^+)/(\Lambda_c^+ \to all) = (2.2 \pm 1.1)\%$$
$$(D^+ \to K^-\pi^+\pi^+)/(D^+ \to all) = (6.3 \pm 1.5)\%$$
$$(D^0 \to K^-\pi^+)/(D^0 \to all) = (3.0 \pm 0.6)\% .$$

Furthermore, no dynamic correlation has been assumed between the two charm quarks, which is an extreme assumption, but it is the simplest one, taking into account that no results are available on this point. The results can be summarized as in Tables 1, 2, and 3.

From D^0 longitudinal momentum investigation we have seen (Section 3.3) that a "forward" $D\bar{D}$ production mechanism is highly disfavoured, and that a "central" production is compatible with the data, with such a distribution we obtain the lowest cross-section estimate:

$$\sigma(D\bar{D}) \simeq 850 \ \mu b \ .$$

Table 1. $\Lambda_c^+\bar{D}$ cross-section estimates obtained
with different production models

Λ_c^+ model	$\bar{D}$ model	σ_{tot} (μb)
a) $E(d\sigma/dx_L) \propto (1 - x_L)^3$	$E(d\sigma/dx_L) \propto (1 - x_L)^3$	4200
b) $d\sigma/dy = const.$	$d\sigma/dy = const.$	750
c) $d\sigma/dx_L = const.$	$d\sigma/dx_L = const.$	1125
d) $d\sigma/dx_L = const.$	$E(d\sigma/dx_L) \propto (1 - x_L)^3$	184
e) $\begin{cases} \text{measured} \\ d\sigma/dx_L \text{ and} \\ d\sigma/dp_T \end{cases}$	$E(d\sigma/dx_L) \propto (1 - x_L)^3$	245

Table 2. $D^+\bar{D}$ cross-section estimates obtained
with different production models

D^+ model	$\bar{D}$ model	σ_{tot} (μb)
a) $E(d\sigma/dx_L) \propto (1 - x_L)^3$	$E(d\sigma/dx_L) \propto (1 - x_L)^3$	305
b) $d\sigma/dy = const.$	$d\sigma/dy = const.$	730
c) $d\sigma/dx_L = const.$	$d\sigma/dx_L = const.$	> 5000
d) $d\sigma/dx_L = const.$	$E(d\sigma/dx_L) \propto (1 - x_L)^3$	1080

Table 3. $D^0\bar{D}$ cross-section estimates obtained
with different production models

D^0 model	$\bar{D}$ model	σ_{tot} (μb)
a) $E(d\sigma/dx_L) \propto (1 - x_L)^3$	$E(d\sigma/dx_L) \propto (1 - x_L)^3$	575
b) $d\sigma/dy$ = const.	$d\sigma/dy$ = const.	1290
c) $d\sigma/dx_L$ = const.	$d\sigma/dx_L$ = const.	> 5000
d) $d\sigma/dx_L$ = const.	$E(d\sigma/dx_L) \propto (1 - x_L)^3$	1000

The Λ_c^+ production is dominated by the "leading" effect. This fact, together with the previous remark about D mesons, suggests that, in $\bar{D}\Lambda_c^+$ pair production, the $\bar{D}$ is produced in a central way, whereas the Λ_c^+ is produced according to a flat x model. The $\bar{D}\Lambda_c^+$ cross-section estimate, according to models d and e of Table I, gives

$$\sigma(\Lambda_c^+ \bar{D}) \simeq 200 \ \mu b \ .$$

These values are very high when compared with lower $\sqrt{s}$ measurements, meaning that the energy-rise of the charm cross-section is steeper than expected. What happens at $p\bar{p}$ collider energy? At $\sqrt{s}$ = 540 GeV the heavy-flavour production should be even more abundant, and, hopefully, we could have more detailed information about the production mechanism.

5. BEAUTY PRODUCTION

The logic used for charm analysis can easily be applied also to the search for new particles with open beauty. From recent results at Cornell[23] it has been found that beauty-flavoured particles also have a high semileptonic branching ratio ($\sim$ 20%). The b,$\bar{b}$ decay chain which we are looking for is described by diagrams 3 and 4:

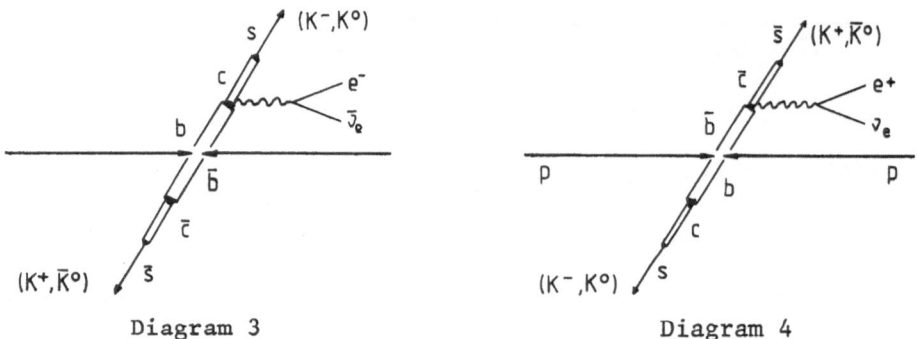

Diagram 3   Diagram 4

The semileptonic decay of an antibeauty-flavoured particle produces an e^+, the associated hadronic decaying state being a beauty-flavoured one. The situation is exactly reversed with respect to the direct charm production. Obviously a $\bar{b}$ hadronic decay should be looked for in association with a negative electron.

Since our previous studies on the production of the charmed baryon Λ_c^+ in pp collisions[3,9] had shown that this heavy baryon is produced according to a strong leading effect, i.e. with a flat distribution in $x_L = 2p_L/\sqrt{s}$, and that the identification of a proton as a positive particle with high x_L, is indeed a good way to trigger on the production of heavy-flavoured baryons, we have applied the same "leading" technique to the search for naked beauty states produced in pp collisions[6]. The lowest mass baryonic state with a beauty quark is the Λ_b^0 (udb). According to the Cabibbo-favoured mechanism, the beauty quark should decay into a charm quark, so that a charmed particle is expected to be present as a further signature, for example a D^0. Notice that this charmed particle is expected to be found in association with a positron, that is in a configuration "forbidden" by direct charm production (See Section 3).

To summarize, the reaction investigated was:

$$pp \rightarrow \Lambda_b^0 \qquad\qquad + (\bar{b}\ \text{state}) + \text{anything} \qquad\qquad (1)$$

$$\quad\ \ \llcorner\rightarrow pD^0\pi^- \qquad\qquad \llcorner\rightarrow e^+ + \text{anything}$$

$$\qquad\ \llcorner\rightarrow K^-\pi^+$$

in the kinematical region imposed by the "leading" conditions.

The following conditions have been applied to the sample of 3×10^4 positron events (corresponding to an integrated luminosity of $\mathcal{L} = 4.39 \times 10^{36}$ cm^{-2}) to study reaction (1)

  i) $p_T(e^+) \geq 0.8$ GeV/c;
 ii) the charged multiplicity of the "anything", n_{ch}, was required to be $n_{ch} \geq 4$, with at least one particle opposite in azimuth to the triggering positron;
iii) at least one positive particle with $x_L \geq 0.32$ and momentum accuracy $\Delta p/p \leq 0.30$ was required in the event, and this particle was assumed to be a proton;
 iv) the rapidity of the $(pK^-\pi^+\pi^-)$ system was required to be $|y| > 1.4$.

Condition (i) selects high transverse momentum positrons, as is expected for the lepton coming from the decay of a heavy b state. Condition (ii) with the same arguments selects highly inelastic events (a b,$\bar{b}$ pair creation requires an energy threshold of 10.5 GeV). Conditions (iii) and (iv) select events with a "leading baryon". We will call this set of conditions "set α".

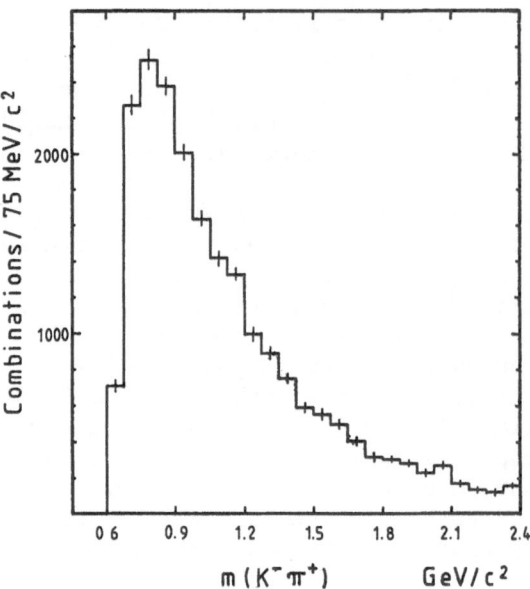

Fig. 14 The $K^-\pi^+$ invariant mass spectrum obtained from the events
         satisfying set-α conditions.

    Figure 14 shows the invariant mass spectrum for the system $K^-\pi^+$
obtained from the events satisfying the set α condition. Only particles with $\Delta p/p \leq 30\%$ and fitted to the interaction vertex were
used. To any negative track not identified as π^- or $\bar{p}$ was assigned
the K mass, any positive track not identified as K^+ or p was assigned
the π mass. As expected, no D^0 signal is visible in this spectrum.
(D^0 quark composition is $c\bar{d}$ and the positron itself gives the signature for a "charm" quark semileptonic decay so that D^0 production
is forbidden with an e^+ trigger in direct charm production.) Anyway
this spectrum should contain some D^0, with the $b\bar{b}$ cross-section rate,
if the reaction in diagram 3 takes place. To select events containing
a D^0, a cut has been applied to the spectrum of Fig. 14, requiring

    $1.7 \leq m(K^-\pi^+) \leq 2.0$ GeV/c .

    Figure 15 shows the invariant mass spectrum for the $(pK^-\pi^+\pi^-)$
system, where the set α condition and the D^0 mass cut has been applied. A clear enhancement (29.4 ± 7.4 combinations)[*] is observed
in the mass range $5.35 \leq m[p(K^-\pi^+)\pi^-] \leq 5.5$ GeV/c^2. On the contrary,
if the D^0 mass cut condition is released, the invariant mass spectrum

---

[*] Note that while 29.4 is the number of combinations above the back-
    ground, ±7.4 is the statistical fluctuation of the background
    level under the observed enhancement. The same applies when dis-
    cussing the other signals presented in this paper.

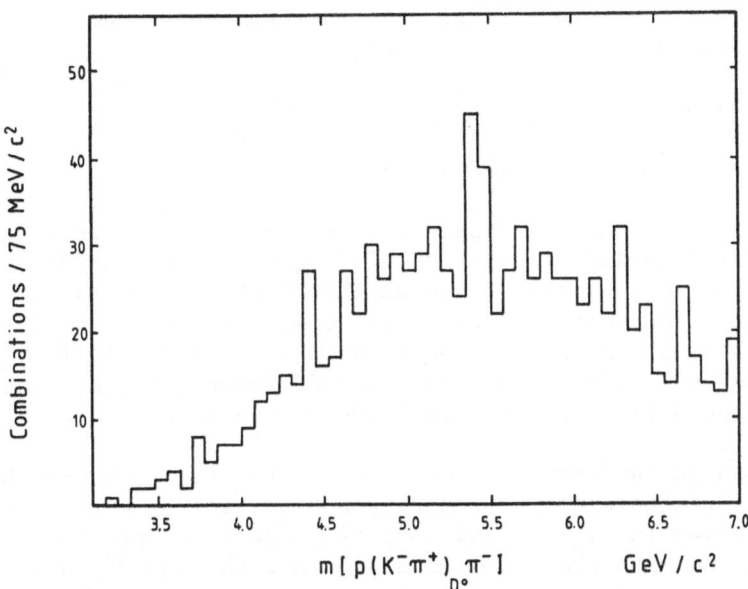

Fig. 15   The $p(K^-\pi^+)\pi^-$ invariant mass spectrum obtained from the
events satisfying set-α conditions and the wide D^0 cut.

of the system ($pK^-\pi^+\pi^-$) appears as shown in Fig. 16 and no enhancement is observed. The mass range of the enhancement agrees very well
with theoretical expectations[24,25]. A very important cross-check is
now the following: if the enhancement observed in Fig. 15 is really
due to the Λ_b^0 decay $\Lambda_b^0 \rightarrow pD^0\pi^-$, then, selecting the events on the
spectrum of Fig. 16 in the Λ_b^0 mass region $5.35 \leq m(pK^-\pi^+\pi^-) <$
< 5.5 GeV/c^2, and working out only for these events the $K^-\pi^+$ spectrum, the D^0 content should be enhanced and possibly a signal should
appear.

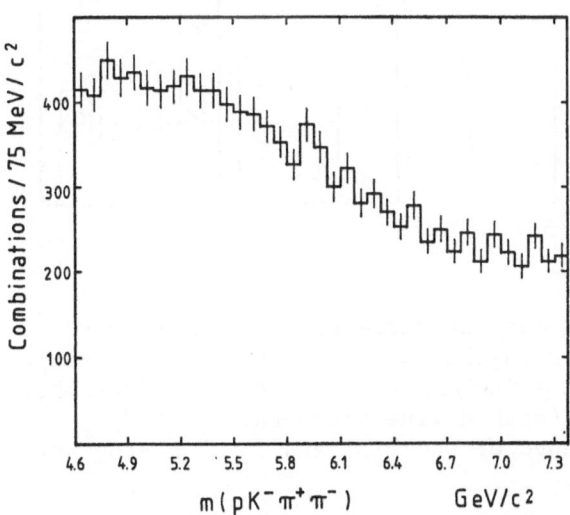

Fig. 16

The $pK^-\pi^+\pi^-$ invariant mass
spectrum obtained from the
events satisfying only set-α
conditions.

Let us define, in analogy with what has been done for Λ_c^+ two mass regions for the $pK^-\pi^+\pi^-$ system:

1) "IN" region $5.35 \leq m(pK^-\pi^+\pi^-) < 5.5$ GeV/c^2

2) "OUT" region $\begin{cases} 5.2 \;\; \leq m(pK^-\pi^+\pi^-) < 5.35 \text{ GeV/c}^2 \\ 5.5 \;\; \leq m(pK^-\pi^+\pi^-) < 5.65 \text{ GeV/c}^2 \;. \end{cases}$

Figure 17a shows the $K^-\pi^+$ spectrum for the events satisfying the set α conditions and with the $pK^-\pi^+\pi^-$ mass falling inside the "IN" mass region (solid line) or inside the "OUT" mass region (dashed line). An enhancement of (28 ± 5.5) combinations is indeed observed for the IN sample at a mass about the D^0 mass $1.725 \leq m(K^-\pi^+) <$ < 1.875 GeV/c^2. Figure 17b shows the difference between the solid and the dashed lines of Fig. 17a (IN-OUT spectrum).

We can go further, and check the correlation between this D^0 and the $pK^-\pi^+\pi^-$ enhancement by centring our initial D^0 cut on the mass value observed in Fig. 17 and computing again the $pK^-\pi^+\pi^-$ invariant mass spectrum with the set α condition and the new D^0 cut:

$$1.725 \leq m(K^-\pi^+) \leq 1.875 \text{ GeV/c}^2 \;.$$

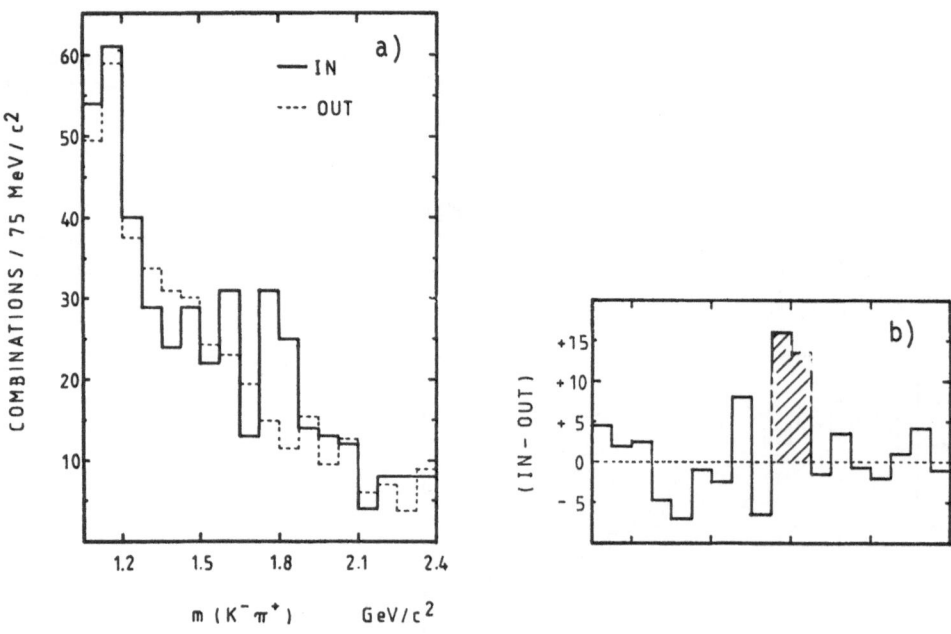

Fig. 17   a) The $K^-\pi^+$ invariant mass spectrum when the mass of the system $p(K^-\pi^+)\pi^-$ falls in the (5.35-5.5 GeV/c^2) mass range (solid-line histogram = "IN") or in two control regions above and below this range (dashed line histogram = "OUT").
b) Bin-to-bin difference of the "IN" and "OUT" histograms shown in (a).

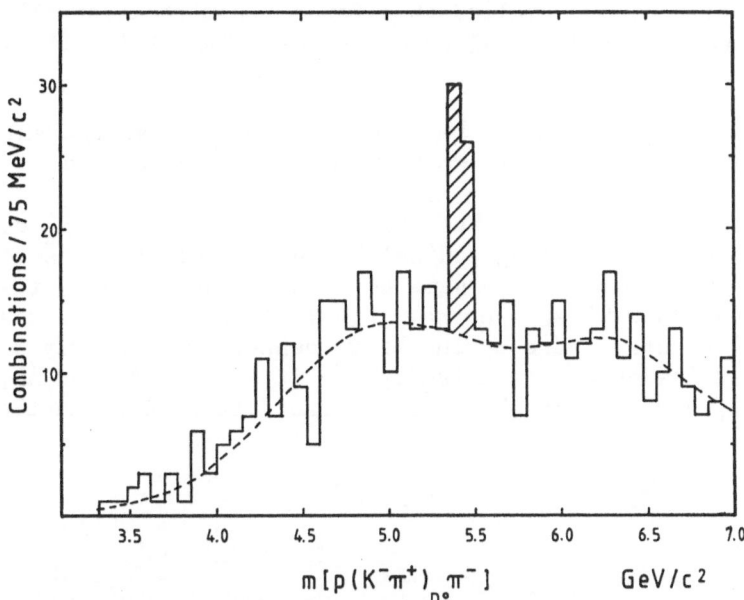

Fig. 18　The $p(K^-\pi^+)\pi^-$ invariant mass spectrum obtained from the
events selected by the set-α conditions when the mass $(K^-\pi^+)$
satisfies the narrow D^0 trigger.

Figure 18 shows the result obtained in this case. The signal is now
given by 30 ± 5 combinations with a 1 to 1 signal-to-background ratio,
with an improvement of a factor of 2 with respect to the spectrum of
Fig. 15.

　　The same analysis has been repeated for different samples of
events where we do not expect the Λ_b^0 signal, in order to be sure
that the effect was not due to instrumental uncertainties:

  i) mixing particles from different events;
 ii) using as the trigger particle a charged hadron instead of a
     positron;
iii) using as the trigger particle a negative electron.

Notice that in this last case we could expect a Λ_b^0 signal from the
reaction:

$$pp \to \Lambda_b^0 \qquad + (\bar{b}\ \text{state}) \qquad + \text{any} \qquad (2)$$

$$\begin{array}{l} \hookrightarrow pD^0\pi^- \qquad\qquad \hookrightarrow (\bar{c}\ \text{state}) \\ \qquad \hookrightarrow (K^-\pi^+) \qquad\qquad \hookrightarrow e^- + \text{any} \end{array}$$

However, Monte Carlo simulation shows that the transverse momentum
spectrum of the e^- in the final sample of reaction (2) is peaked to
much smaller p_T values, so that the acceptance for Λ_b^0 from reaction

(2) is a factor of $\sim$ 4 smaller than for Λ_b^0 from reaction (1), in such a way that we expect 8 ± 6 Λ_b^0 combinations, compatible with what we see.

The dashed line in Fig. 18 represents the background shape as it comes out from the analysis of these samples of events.

Even though the statistics are smaller than in the Λ_c^+ case we tried to look also for longitudinal momentum distributions for the Λ_b^0 [26]. The technique is the same as used in the Λ_c^+ case, i.e. to work out the "IN-OUT" x_L spectra. Figure 19 shows the x_L distribution for Λ_b^0 events. Here again the x_L behaviour is nearly flat for the heaviest baryon of the Λ family, as it was for Λ^0 and Λ_c^+, at least for $x_L \gtrsim 0.35$. The best fit to the data with a function

$$\frac{dN}{dx_L} \propto (1 - x_L)^\alpha$$

gives for α

$$\alpha = 0.87 \pm 1.26 \ .$$

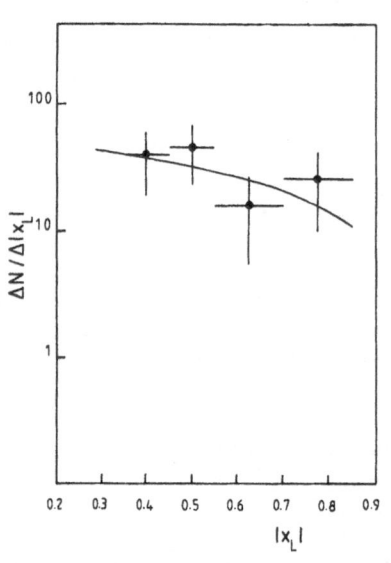

Fig. 19

The x_L distribution of the Λ_b^0

This can be understood in terms of the fact that the members of the Λ family carry two of the original quarks of the incident proton,

$$p \ \ = \left[(ud)u \right] \ ,$$
$$\Lambda^0 = \left[(ud)s \right] \ ,$$
$$\Lambda_c^+ = \left[(ud)c \right] \ ,$$
$$\Lambda_b^0 = \left[(ud)b \right] \ .$$

This again gives a serious hint that the same mechanism is at work irrespective of the flavour being produced.

An additional argument concerning the Λ_b^0 interpretation of the peak shown in Fig. 18 is given by comparing the positron transverse momentum distribution relative to the Λ_b^0 "IN" events, shown in Fig. 20a with the same spectrum for the "OUT" events (Fig. 20b). In both cases the dashed line shows the normalized inclusive distributions of the positrons.

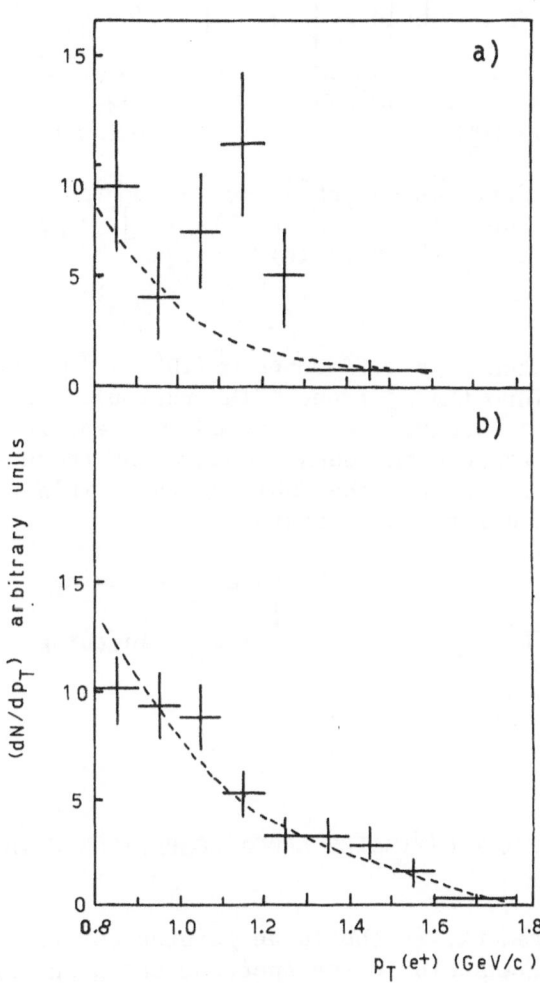

Fig. 20   a) The transverse momentum distribution of the positrons associated with the Λ_b^0 compared with the positron spectrum relative to all triggered events that are not Λ_b^0;   (dashed line). b) Same as (a) but for events with $m[p(K^-\pi^+)\pi^-]$ in the "OUT" mass region [dashed line as (a)].

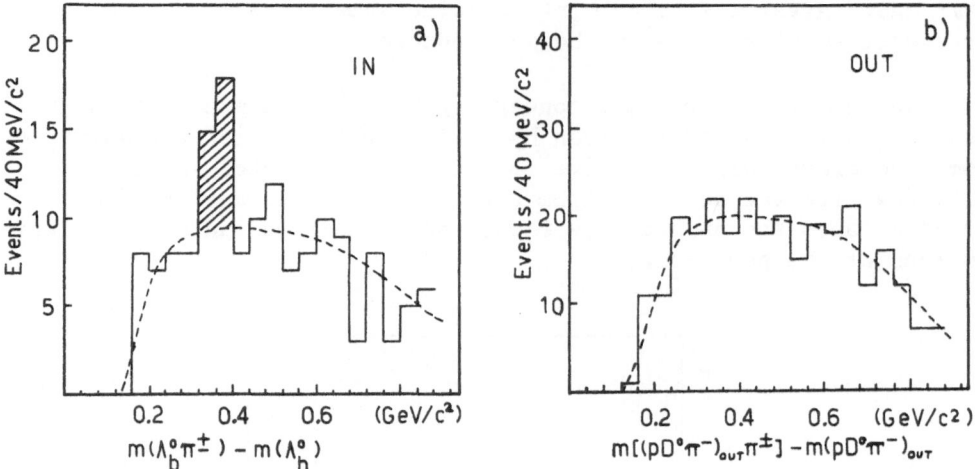

Fig. 21   a) Mass difference $m\left[\left(pD^0\pi^-\right)_{IN}\pi^\pm\right] - m\left[\left(pD^0\pi^-\right)_{IN}\right]$ compared
with the best fit to $m\left[\left(pK^-\pi^+\right)_{OUT}\pi^\pm\right] - m\left[p\left(K^-\pi^+\right)\pi^\pm\right]_{OUT}$
(dashed line) shown in (b).

Figure 21a shows the mass difference $m\left[\left(pD^0\pi^-\right)\pi^\pm\right] - m(pD^0\pi^-)$ when
the $pD^0\pi^-$ falls into the Λ_b^0 range ("IN" region). An enhancement of
14 ± 4.4 events is observed, which is not present in Fig. 21b where
the same mass difference has been computed for those events in which
the $(pD^0\pi^-)$ mass falls into the "OUT" region. This enhancement can
be interpreted as due to the reaction

$$pp \to \Sigma_b^\pm \qquad\qquad + (\bar{b}\ state) + anything \qquad\qquad (3)$$
$$\quad\ \ \downarrow\ \ \Lambda_b^0\ \pi^\pm \qquad\qquad \downarrow\ e^+ + anything$$
$$\qquad\quad \downarrow\ pD^0\pi^-$$
$$\qquad\qquad \downarrow\ K^-\pi^+$$

6. Λ_b^0 CROSS–SECTION ESTIMATES AND A COMPARISON WITH THE Λ_c^+ CROSS–SECTION

As a first remark, it should be pointed out that the number of
combinations per mass bin in the spectrum shown in Fig. 18 exceeds
the number of events by a small amount (25%). In order to work out
the cross-section for the observed signal[27], the number of combinations has, therefore, to be divided by 1.25.

As already shown in the Λ_c^+ case, the number of hypotheses we
have to assume in order to give a cross-section estimate is very
high, so let us proceed step by step in adding these assumptions.

The reaction under investigation is

$$pp \rightarrow \underbrace{\left[p(K^-\pi^+)\pi^- \right]}_{D^0} + e^+ + \text{anything} .$$

Let us calculate the partial cross-section $\Delta\sigma$ for this process, measured under the following conditions:

   i) $p(K^-\pi^+)\pi^-$ is a "leading system" in the sense specified in the text;
  ii) the positron has a transverse momentum cut at $p_T \geq$ ≥ 800 MeV/c;
 iii) the total multiplicity is limited by the condition on the "anything", $n_{ch} \geq 4$.

$\Delta\sigma$ can be interpreted as the sensitivity of our set-up, when illuminated by a final state such as in diagram (3) with conditions (i), (ii), and (iii). Then

$$\Delta\sigma = N(\Lambda_b^0)/\varepsilon \ell(pp) .$$

$N(\Lambda_b^0)$ is the number of Λ_b^0 events (25); ε is the total detection efficiency (11.9 ± 1.2)%; $\ell(pp)$ is the total integrated luminosity $(4.39 \times 10^{36} \text{ cm}^{-2})$. Therefore

$$\Delta\sigma = (3.8 \pm 1.2) \times 10^{-35} \text{ cm}^2 .$$

Let us extrapolate this result to the whole phase space in order to give an estimate for the total Λ_b^0 production cross-section. It is necessary to assume:

 i) *The nature of the $\bar{b}$ state produced in association with the Λ_b^0.*
    We know in fact that it is decaying semileptonically, but we do not know whether it is a meson or a baryon and its mass. Since an antibaryon needs three antiquarks to be created and since the pp experimental results favour baryon-antimeson production rather than baryon-antibaryon production, we will assume the e^+ to originate from the decay of an antibeauty-flavoured meson which we will call $M_{\bar{b}}$.
ii) *The longitudinal- and transverse-momentum production and decay distribution of the Λ_b^0 and of the $M_{\bar{b}}$.* For $M_{\bar{b}}$ we assumed, on the basis of charm studies[10,11] a central production mechanism such as

$$\left(E\frac{d^3\sigma}{dp^3} \right)_{M_{\bar{b}}} \propto (1 - x_L)^3 \, e^{-2.5 p_T} ,$$

with a $K_{\ell 3}$-like decay matrix for the channel $M_{\bar{b}} \rightarrow De^+ \nu_e$. For the Λ_b^0, following the Λ_c^+ results, we have chosen

$$\frac{1}{p_T} \frac{d\sigma}{dp_T} \propto e^{-2.5p_T}$$

$$\frac{d\sigma}{dx_L} = \text{const.},$$

with three possible decay mechanisms:

I) The leading baryon conditions

$$x_L(pD^0\pi^-) \geq 0.32 \ , \ |y(pD^0\pi^-)| \geq 1.4 \ .$$

Obviously the result found according to this model will not be the total cross-section. It will be the partial cross-section for the experimentally available phase-space region.

II) A minimum "leading baryon" condition $p_{proton} \geq p_{D^0}$, $p_{proton} > p_{\pi^-}$.

III) Pure L.I. phase space.

iii) *The correlation, if any, between b and $\bar{b}$.* We assumed, as in the charm case, no correlation.

Furthermore, the following branching ratios are needed:

$B_1(D^0 \rightarrow K^-\pi^+) = (D^0 \rightarrow K^-\pi^+)/(D^0 \rightarrow \text{all}) = (3.0 \pm 0.6)\%$
$B_2(\bar{b} \text{ state} \rightarrow e^+ + \text{any}) = (M_{\bar{b}} \rightarrow e^+ + \text{any})/(M_{\bar{b}} \rightarrow \text{any}) = (13 \pm 6)\%$ [23]
$B_3(\Lambda_b^0 \rightarrow pD^0\pi^-) = (\Lambda_b^0 \rightarrow pD^0\pi^-)/(\Lambda_b^0 \rightarrow \text{all}) = \text{unknown} \ .$

Our estimates will be given for the $\Lambda_b^0 M_{\bar{b}}$ production cross-section times the unknown branching ratio B_3

$$\Delta\sigma_b(\ \text{I}\ ) \times B_3 = \left(2.7 \ ^{+1.8}_{-1.1}\right) \times 10^{-30} \ \text{cm}^2$$

$$\sigma_b(\ \text{II}\ ) \times B_3 = \left(8.2 \ ^{+6.0}_{-2.8}\right) \times 10^{-30} \ \text{cm}^2$$

$$\sigma_b(\text{III}) \times B_3 = \left(27 \ ^{+17}_{-11}\right) \times 10^{-30} \ \text{cm}^2 \ .$$

The large uncertainties and the critical assumption on the production and decay model do not allow these to be taken as absolute estimates; anyway, a very interesting comparison can be made with the charm cross-section measured under the same hypothesis. In this way the ratio σ_b/σ_c of the beauty to the charm cross-section should be less dependent on any assumption. The fact that both Λ_c^+ and Λ_b^0 have been detected, in the same set-up, by the same experiment, with the same method for data reduction increases confidence in the comparison.

The Λ_c^+ cross-section estimates, worked out with the same production models I, II, and III, are

$$\Delta\sigma_c(\ \ I) = \left(77 \begin{array}{l} +57 \\ -28 \end{array}\right) \times 10^{30}\ cm^2$$

$$\sigma_c(\ II) = \left(100 \begin{array}{l} +68 \\ -36 \end{array}\right) \times 10^{30}\ cm^2$$

$$\sigma_c(III) = \left(190 \begin{array}{l} +140 \\ -68 \end{array}\right) \times 10^{30}\ cm^2\ .$$

Now $\sigma_b \times B_3/\sigma_c$ allows a direct investigation of the dependence on B_3 of the beauty to charm cross-section ratio.

Figure 22 shows now, for model I, where the assumptions are less critical (it refers to partial cross-section estimates in the kinematical region really investigated) the B_3 behaviour as a function of σ_b/σ_c. It can be pointed out that, within 1.5σ a realistic value of $B_3 \simeq 2\%$ is compatible with

$$\sigma_b/\sigma_c = \frac{m_c^2}{m_b^2} \simeq 1/8\ ,$$

which is in agreement with some theoretical estimates[28,29] of the mass coupling law for heavy-flavour production;

$$\sigma_f \propto \frac{1}{m_f^2}\ .$$

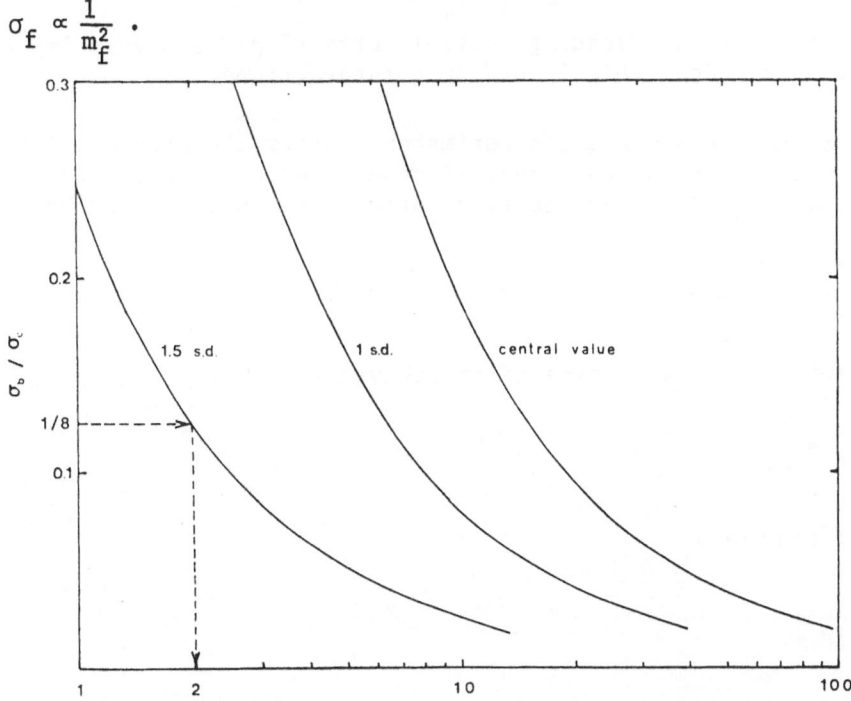

Fig. 22 σ_b/σ_c versus $B_3(\Lambda_b^0 \to pD^0\pi^-)$ relative to model I (as defined in the text).

CONCLUSIONS

The production of heavy flavours in hadronic collision has been investigated at $\sqrt{s}$ = 62 GeV.

For charmed particles the associated production of two mesons and one baryon has been observed.

In the same experiment, the associated production of a beauty baryonic state has been observed as well.

For charm meson states the results indicate a rather central production mechanism such as

$$E \frac{d\sigma}{dx_L} \propto (1 - x_L)^3 \text{ or } \frac{d\sigma}{dy} \text{ const.}$$

to be favoured.

The same production mechanism appears to be at work both for charm and beauty baryonic states, such as

$$\frac{d\sigma}{dx_L} = \text{const. for the baryon}$$

suggesting that the "leading particle effect" plays a very important, flavour-independent role in hadronic interactions.

The charm cross-section estimates confirm the previous ISR results, where very abundant heavy-flavour production was suggested, and the beauty-to-charm cross-section ratio is compatible with being

$$\sigma_b/\sigma_c = \frac{m_c^2}{m_b^2} \simeq 1/8 \; ,$$

in agreement with a cross-section behaviour of

$$\sigma_f = \frac{1}{m_f^2}$$

for any flavour f.

REFERENCES

1) F. Muller, Hadroproduction of charmed particles, Invited talk
   given at the 4th Warsaw Symposium on Elementary Particles
   Physics, Kazinsierz, Poland, May 1981, CERN/EP/0457R/RN/ed,
   June 1981, and references quoted therein.
2) D. Treille, Proc. 10th Int. Symposium on Leptons and Photon
   Interactions at High Energies, Bonn, 1981 (ed. W. Pfeil)
   (Universität Bonn, Bonn, 1981), p. 750.
3) M. Basile, G. Cara Romeo, L. Cifarelli, A. Contin, G. D'Ali,
   P. Di Cesare, B. Esposito, P. Giusti, T. Massam, R. Nania,
   F. Palmonari, G. Sartorelli, G. Valenti, and A. Zichichi,
   Nuovo Cimento 63A:230 (1981).
4) M. Basile, G. Cara Romeo, L. Cifarelli, A. Contin, G. D'Ali,
   P. Di Cesare, B. Esposito, P. Giusti, T. Massam, B. Nania,
   F. Palmonari, G. Sartorelli, G. Valenti, and A. Zichichi,
   Nuovo Cimento 65A:457, 1981.
5) M. Basile, G. Cara Romeo, L. Cifarelli, A. Contin, G. D'Ali,
   P. Di Cesare, B. Esposito, P. Giusti, T. Massam, R. Nania,
   F. Palmonari, G. Sartorelli, G. Valenti, and A. Zichichi,
   Nuovo Cimento 67A:40, 1982.
6) M. Basile, G. Bonvicini, G. Cara Romeo, L. Cifarelli, A. Contin,
   G. D'Ali, P. Di Cesare, B. Esposito, P. Giusti, T. Massam,
   R. Nania, F. Palmonari, G. Sartorelli, G. Valenti, and
   A. Zichichi, Nuovo Cimento Lett. 31:97 (1981).
7) M. Basile, G. Cara Romeo, L. Cifarelli, A. Contin, G. D'Ali,
   P. Di Cesare, B. Esposito, P. Giusti, T. Massam, F. Palmonari,
   G. Sartorelli, G. Valenti, and A. Zichichi, Nuovo Cimento
   62A:14 (1981).
8) M. Basile, G. Cara Romeo, L. Cifarelli, A. Contin, G. D'Ali,
   P. Di Cesare, B. Esposito, P. Giusti, T. Massam,
   F. Palmonari, G. Sartorelli, G. Valenti, and A. Zichichi,
   Nuovo Cimento Lett. 30:481 (1981).
9) M. Basile, G. Cara Romeo, L. Cifarelli, A. Contin, G. D'Ali,
   P. Di Cesare, B. Esposito, P. Giusti, T. Massam,
   F. Palmonari, G. Sartorelli, G. Valenti, and A. Zichichi,
   Nuovo Cimento Lett.30:487 (1981).
10) M. Basile, G. Cara Romeo, L. Cifarelli, A. Contin, G. D'Ali,
    P. Di Cesare, B. Esposito, P. Giusti, T. Massam, R. Nania,
    F. Palmonari, G. Sartorelli, G. Valenti, and A. Zichichi,
    Nuovo Cimento Lett. 33:17 (1982).
11) M. Basile, G. Cara Romeo, L. Cifarelli, A. Contin, G. D'Ali,
    P. Di Cesare, B. Esposito, P. Giusti, T. Massam, R. Nania,
    F. Palmonari, G. Sartorelli, G. Valenti, and A. Zichichi,
    Nuovo Cimento Lett. 33:33 (1982).
12) M. Basile, G. Cara Romeo, L. Cifarelli, A. Contin, G. D'Ali,
    P. Di Cesare, B. Esposito, P. Giusti, T. Massam, R. Nania,
    F. Palmonari, V. Rossi, G. Sartorelli, M. Spinetti,
    G. Susinno, G. Valenti, L. Votano, and A. Zichichi, Nuovo
    Cimento Lett. 32:321 (1981).

13) M. Basile, G. Cara Romeo, L. Cifarelli, A. Contin, G. D'Ali,
    P. Di Cesare, B. Esposito, P. Giusti, T. Massam, R. Nania,
    F. Palmonari, V. Rossi, G. Sartorelli, M. Spinetti,
    G. Susinno, G. Valenti, L. Votano, and A. Zichichi, Nuovo
    Cimento 66A:129 (1981).

14) M. Basile, G. Cara Romeo, L. Cifarelli, A. Contin, G. D'Ali,
    P. Di Cesare, B. Esposito, P. Giusti, T. Massam,
    F. Palmonari, G. Sartorelli, G. Valenti, and A. Zichichi,
    Nuovo Cimento 65A:421 (1981).

15) M. Basile, G. Cara Romeo, L. Cifarelli, A. Contin, G. D'Ali,
    P. Giusti, T. Massam, F. Palmonari, G. Sartorelli,
    G. Valenti, and A. Zichichi, Nucl. Instrum. Methods 163:93
    (1979).

16) H. Frehse, F. Lapique, M. Panter, and F. Piuz, Nucl. Instrum.
    Methods 156:87 (1978).
    H. Freshe, M. Heiden, M. Panter, and F. Piuz, Nucl. Instrum.
    Methods 156:97 (1978).

17) M. Basile, G. Cara Romeo, L. Cifarelli, A. Contin, G. D'Ali,
    P. Di Cesare, B. Esposito, L. Favale, P. Giusti, T. Massam,
    F. Palmonari, G. Sartorelli, G. Valenti, and A. Zichichi,
    Nucl. Instrum. Methods 179:477 (1981).

18) R. Bouclier, G. Charpak, E. Chesi, L. Dumps, H.G. Fischer,
    H.J. Hilke, P.G. Innocenti, G. Maurin, A. Minten,
    L. Naumann, F. Piuz, J.C. Santiard, and O. Ullaland, Nucl.
    Instrum. Methods 115:235 (1974); R. Bouclier, R.C.A. Brown,
    E. Chesi, L. Dumps, H.G. Fischer, P.G. Innocenti, G. Maurin,
    A. Minten, L. Naumann, F. Piuz, and O. Ullaland, Nucl.
    Instrum. Methods 125:19 (1975).

19) M. Basile, G. Cara Romeo, L. Cifarelli, A. Contin, G. D'Ali,
    P. Di Cesare, B. Esposito, P. Giusti, T. Massam, R. Nania,
    F. Palmonari, G. Sartorelli, G. Valenti, and A. Zichichi,
    EP Internal Report 81-02, 23 June 1981.

20) P. Capiluppi, G. Giacomelli, A.M. Rossi, G. Vannini, and
    A. Bussière, Nucl. Phys. B70:1 (1974).

21) W. Bacino, R. Burns, P. Condon, P. Cowell, A. Diamant-Berger,
    G. Donaldson, M. Duro, T. Ferguson, A. Hall, G. Irwin,
    J. Kirkby, J. Kirz, F. Merrit, L. Nodulman, W. Slater,
    H. Ticho, and S. Wojcicki, Phys. Rev. Lett. 43:1073 (1979).

22) G.A. Trilling, Phys. Rep. 75:57 (1981).

23) C. Bebek, J. Haggerty, J.M. Izen, W.A. Loomis, F.M. Pipkin,
    J. Rohlf, W. Tanenbaum, R. Wilson, A.J. Sadoff, D.L. Bridges,
    K. Chadwick, P. Gauci, H. Kagan, R. Kass, F. Lobkowicz,
    A. Melissinos, S.L. Olsen, P. Poling, C. Rosenfeld,
    C. Rucinski, E.H. Thorndike, G. Warren, D. Bechis,
    J.J. Mueller, D. Potter, F. Sannes, P. Skubic, R. Brody,
    A. Chen, M. Goldberg, N. Horwitz, J. Kandaswamy, H. Kooy,
    P. Lariccia, G.C. Moneti, M.S. Alam, S.E. Csorna,
    R.S. Panvini, J.S. Poucher, D. Andrews, K. Berkelmann,
    R. Cabenda, D.G. Cassel, J.W. De Wire, R. Ehrlich,
    F. Ferguson, T. Gentile, M.G.D. Gilchriese, B. Gittelman,

D.L. Hartill, D. Herrup, M. Hertzlinger, D.L. Kreinick,
N.B. Mistry, E. Nordberg, R. Perchonok, R. Plunkett,
K.A. Shinsky, R.H. Siemann, A. Silverman, P.C. Stein,
S. Stone, R. Talman, H.G. Thonemann, and D. Weber, preprint
CLNS-80/4/5, 1980.

24) D. Stanley and R. Robson, Phys. Rev. Lett. B45:235 (1980).

25) A. Martin, Phys. Lett. 103B:51 (1981).

26) M. Basile, G. Bonvicini, G. Cara Romeo, L. Cifarelli, A. Contin,
G. D'Ali, P.Di Cesare, B. Esposito, P. Giusti, T. Massam,
R. Nania, F. Palmonari, G. Sartorelli, G. Valenti, and
A. Zichichi, Nuovo Cimento 65A:408 (1981).

27) M. Basile, G. Bonvicini, G. Cara Romeo, L. Cifarelli, A. Contin,
G. D'Ali, P. Di Cesare, B. Esposito, P. Giusti, T. Massam,
R. Nania, F. Palmonari, G. Sartorelli, G. Valenti, and
A. Zichichi, Nuovo Cimento 65A:391 (1981).

28) G. Gustafson and C. Peterson, Phys. Rev. Lett. B67:81 (1977).

29) A. Martin, Phys. Lett. 100B:511 (1981).

DISCUSSION

B. Reay: You see for your Λ_b^0 signal a FWHM of 150 MeV/c². Do you put the vertex into your mass fit, and is the width of the Λ_b^0 compatible with your mass resolution?

Answer : Our vertex resolution is such that we are quite unable to distinguish between primary interaction vertex and b-decay vertex, (to such an extent that all the tracks are required to fit the interaction vertex ±5 cm in order to be taken into account for Λ_b^0 invariant mass computing). In other words, our Λ_b^0 are coming from the primary interaction vertex, as we can define it. (The mean value of the distance of our reconstructed tracks from the fitted vertex is d ≃ 1.5 cm.) In so far as concerns the Λ_b^0 mass resolutio we have to stress the fact that, owing to the inhomogeneous magneti field of the SFM, we cannot quote a value for the mass resolution valid for the whole detector. 150 MeV/c² turns out to be compatibl with the resolution of a mass of ∿ 5.5 GeV/c² decaying into four bodies in the kinematical region where we see the Λ_b^0 products. Thi result has been obtained via a Monte Carlo simulation of the Λ_b^0 pro duction.

R. Baltay: What is the statistical significance of the Λ_b effect without the "wrong mass" D^0 cut, i.e. from the $pD^0\pi^-$ mass plot with the original 1.7 to 2.0 GeV/c² cut on the D^0?

Answer . Approximately 4.0 standard deviations; we see 29.4 ± 7.4 Λ_b^0 combinations over the computed background with the non-optimized D^0 cut.

R. Bizzarri: The value you quote for the branching ratio B_3 = = $(\Lambda_b^0 \to pD^0\pi)/(\Lambda_b^0 \to all) \simeq 2\%$ does not seem to me to fit with the table you have shown. Could you comment please?

Answer : The meaning of the value we quote for B_3 is the following, as shown by Fig. 22. The Λ_b^0 cross-section estimate, in terms of the unknown branching ratio of $\Lambda_b^0 \to pD^0\pi^-$, turns out to be compatible, within 1.5σ, with a value B_3 = 2%, if the ratio of beauty to charm cross-section is σ_b/σ_c = 1/8 as suggested by the ratio $m_c^2/m_b^2 \simeq 1/8$.

V. Korbel: There is a basic difference between the observed longitudinal momentum spectra of D mesons and of Λ_c^+. In the case of the D you need to identify positively the kaon. You can do this only in the low-momentum range with your TOF system. This leads to the observation of low-momentum D mesons, whereas you observe Λ_c^+ in the high-momentum region. Could this affect your conclusions on the D, Λ_c^+ longitudinal momentum distribution?

Answer : What you say is correct for the $D^+ \to K^-\pi^+\pi^+$ signal, where we can say nothing about the production distribution (the Monte Carlo predictions for different longitudinal momenta are nearly degenerate in Fig. 9). The reason for this is exactly the distortion of the acceptance introduced by the TOF momentum limitation on kaons. For the $D^0 \to K^-\pi^+$ signal, the K^- is not positively identified by TOF (we require a K^+ coming from the semileptonic decay of the triggering particle $\bar{D} \to e^-K^+X$ to be positively identified) so that no limitation on the $K^-\pi^+$ momentum is imposed. As can be seen from Fig. 13, in this case we can distinguish to a certain extent between the different types of production distributions, and a "forward" D^0 production turns out to be highly disfavoured. For Λ_c^+ our acceptance is only limited by $x_L > 0.3$ and we base our conclusions on the dN/dx_L distribution in the region $x_L(\Lambda_c^+) > 0.3$.

J. Appel: What is the effect of forcing the $K^-\pi^+$ mass to be the "real" D^0 mass in the Λ_b^0 events? Is the peak smeared out at all?

Answer : We did in fact try to force our D^0 mass at the value $m(D^0) = 1.865$ GeV/c^2. We can do this in two ways:

i) The error in the D^0 mass is due to an error in the momentum measurements. In this case we can correct the momenta of K^-, π^+ in order to get their invariant mass shifted upwards. The effect on the $p(K^-\pi^+)\pi^-$ spectrum is a shift of the Λ_b^0 peak by ~ 60 MeV/c^2.

ii) The error in the D^0 mass is due to the $K^-\pi^+$ relative angle. We can correct the director cosines and again the result is a shift of the Λ_b^0 mass by ~ 100 MeV/c^2, but the enhancement is still present, so that we can quote a further systematic uncertainty on the Λ_b^0 mass estimate:

$$0.535 < m(\Lambda_b^0) \lesssim 5.6 \text{ GeV/c}^2 .$$

R. Bizzarri: You seem to have some problems with the mass scale for the D^0; could you please comment on this point?

Answer : Our experiment is not an emulsion experiment, so that we have rather big mass uncertainties compared with their measurements. Rather than on a precise mass measurement, our identification of the D^0 is based (i) on the signature given by the e^+ (which excludes a direct charm production), and (ii) on the fact that we expect a D^0 signal to appear if we are looking at a Λ_b^0 (and it appears). Moreover, if the enhancement we see is a D^0 and if we centre our trigger on the $K^-\pi^+$ excess we expect the Λ_b^0 to be enhanced (and this happens).

In so far as the mass scale is concerned, there are at SFM, very local systematic effects due to the highly inhomogeneous magnetic field (known only from interpolation between a given number of discrete measured points); a mass shift can also appear because of some slight misalignment of the chambers; so we do not stress any conclusions about the mass values, as it was clear from the answer to the last question.

THE MASSES OF HEAVY FLAVOURED HADRONS

André Martin

Theoretical Physics Division
CERN
1211 Geneva 23

You have heard, in other talks, discussions on the production mechanisms of hadrons with visible heavy flavour. Here, I just want to present a method to estimate what will be the masses of these objects, especially the beautiful mesons and baryons. However, naturally, once the $t\bar{t}$ system is discovered the method will also apply to calculate masses of objects with visible top.

For us a baryon with heavy flavour Q will be (qqQ), and a meson will be $(\bar{q}Q)$. q will be an ordinary quark u or d. Q will be t, b, c and even s (which we dare to treat as heavy because of the success of the description of the $s\bar{s}$ and $c\bar{s}$ states in this way). One can think of other systems like (qQQ) (the Ξ is of this type !) or (QQQ) like the Ω^-, but the production of these things, when Q = c or b seems very unlikely and will not be observed before a very long time presumably.

The most crude prediction of masses one can make is based on the crude additive quark model. For instance, if we take $m_s = \frac{1}{2}M\phi = 0.51$ GeV we predict $m_{\Omega^-} = 1.53$ instead of the experimental number 1.672. So if you do not want an accuracy better than about 150 MeV, this is good enough (the spin effects, however, cannot be disregarded ! The mass differences $\rho-\pi$, $\Delta-N$ show this very clearly).

The next, much more sophisticated thing you can do is to manufacture a potential between quarks or between quark and antiquark, which fits existing hadrons and try to use this potential to predict the masses of beautiful or truthful hadrons. One can constraint the fit further by demanding that in the three-quark system only two-body forces are important and that $V_{qq'} = \frac{1}{2}V_{q\bar{q}'}$, a property which is

strictly true for the one-gluon exchange part of the force and which
has the merit of giving results insensitive to the particular choice
of quark masses made. Specifically, if we neglect the change in the
kinetic energy we can make the replacements

$$m'_{q_i} = m_{q_i} + \Delta$$

$$V'_{q_i \bar{q}_j} = V_{q_i \bar{q}_j} - 2\Delta$$

$$V'_{q_i q_j} = V_{q_i q_j} - \Delta \, ,$$

and these changes preserve the above-mentioned relation between qq
and q$\bar{q}$ forces. As an example, Richard [1] using a potential which
fits the b$\bar{b}$, c$\bar{c}$, c$\bar{s}$ and s$\bar{s}$ systems [2] has obtained M_{Ω^-} = 1665
(experiment: 1672). Bhadury et al. [3] and Stanley and Robson [4]
have made over-all fits of all mesons and baryons with a universal
potential and make predictions like, for instance, m_{B^*} = 5.344,
m_B = 5.308, m_{Λ_b} = 5.574, m_{Σ_b} = 5.846, $m_{\Sigma_b^*}$ = 5.862 MeV. Their approach
is sometimes criticized because of the large number of parameters in
their potentials. This seems unavoidable if one wants to treat spin
effects in a non-perturbative way and take into account annihilation
diagrams to explain the impurity of the quark content of mesons
[such problems do not occur if one restricts oneself to heavy quarks
as was done in Ref. 1)].

Anyway there is a need for a less model-dependent method of
prediction of the masses of beautiful and truthful hadrons. What we
want to point out is that if one believes in flavour-independence of
the forces and if one believes that there is some effective, but
unknown, Hamiltonian for the quark-antiquark, and three quark systems
containing one heavy quark, one can make some general predictions.

Because of flavour-independence the heavy quark mass is the only
characteristic of the heavy quark. It will appear in the kinetic
energy as $p^2/2M_Q$, in the spin-spin forces as $(1/m_Q)f(r_i,\vec{\sigma}_i)$. [For
instance, in the case of a contact term we have just

$$c \quad \frac{1}{m_Q m_{q_i}} (\vec{\sigma}_Q \cdot \vec{\sigma}_{q_i}) \; \delta(\vec{r}_Q - \vec{r}_{q_i})$$

neglecting terms of the order of $(1/m_Q)^2$ which are under control.]
The same is true for spin-orbit terms, which anyway are unimportant
in the ground states of hadrons.

In short, the Hamiltonian of the $Q\bar{q}$ or Qqq system is essentially of the form

$$H = \frac{1}{m_Q} H_1 + H_2$$

Notice that this does not say that light quark motion is not relativistic ! Anyway when we see this form we can use an elementary theorem which is

If $H = A + \lambda B$, the <u>ground state</u> energy of the system $E(\lambda)$ is a concave function of λ, i.e.,

$$\frac{d^2 E}{d\lambda^2} \leq 0.$$

So the binding energy of the Qqq system (with given external quantum numbers like spin, isospin, etc.) will be a concave function of $1/m_Q$. In other words, if

$$M(Q\bar{q}) \quad \text{or} \quad M(Qqq) = A + Bm_Q + \frac{C}{m_Q} + \frac{D}{m_Q^2}$$

the theorem says that D is negative.

It is perhaps slightly daring to treat strange quarks as heavy quarks but the success in the treatment of the $s\bar{s}$ and $c\bar{s}$ systems seems to justify it [2]. Then the concavity property allows us to get, for instance, an upper bound on $M(B^*)$ from $M(D^*)$ and $M(K^*)$. The only extra ingredient needed is a set of quark masses. Though quark mass differences change very little from one good fit to another good fit, there is some freedom in the choice of the absolute value of the masses. We have found, however, that taking two sets of masses gives very little change in the predictions. If we take the set we proposed [2] to fit simultaneously $b\bar{b}$, $c\bar{c}$, $c\bar{s}$, $s\bar{s}$:

$$m_b = 5.174, \quad m_c = 1.8, \quad m_s = 0.518 \text{ GeV},$$

we obtain from the known masses of K, K^*, D, D^*, Λ, Σ, Λ_c, Σ_c :

$$m_B \leq 5.263 \text{ GeV}, \quad m_{B^*} \leq 5.340 \text{ GeV},$$

$$m_{\Lambda_b} \leq 5.629 \text{ GeV}, \quad m_{\Sigma_b} \leq 5.826 \text{ GeV}.$$

These inequalities are interesting because they are seen to be very constraining in the meson case. Indeed, we get

$$m_B + m_{B*} < 10.603 \text{ GeV}.$$

According to CUSB [5] the Υ''', with a mass of 10.580 GeV, is lighter than $m_B + m_{B*}$ because they see no signal from the decay

$$m_{B*} \to m_B + \gamma,$$

at the Υ''' energy.

If this is true this means that the upper bound we get is less than 25 MeV away from the experimental value.

We are thus led to ask ourselves the question whether formulae of the type

$$M = A + m_Q + B/m_Q$$

do not represent reasonable interpolation formulae.

This can be checked on potential models both for the two-body system and for the three-body system. Our limited "numerical experiments" indicate that by doing this one never makes a mistake of more than 50 MeV.

In other works we expect

$$5.213 \leq m_B \leq 5.263 \text{ GeV}, \quad 5.290 < m_{B*} < 5.340 \text{ GeV},$$

$$5.581 \leq m_{\Lambda_b} \leq 5.629 \text{ GeV}, \quad 5.776 \leq m_{\Sigma_b} \leq 5.286 \text{ GeV}.$$

To end up with still less rigorous considerations, let us suggest to predict vector-pseudoscalar mass differences by formulae of the type

$$M(Q\bar{q})_{1^-} - M(Q\bar{q})_{0^-} = A/mQ + B/m_Q^2.$$

Taking into account deviations from this formula, we predict

$$50 < M_{B*} - M_B < 57 \text{ MeV}.$$

Similarly, numerical experiments indicate that the $\Sigma_Q - \Lambda_Q$ mass difference, where

$$\Sigma_Q = \left[(qq)_{J=1} \ Q \right]_{J=\frac{1}{2}}$$

and

$$\Lambda_Q = \left[(qq)_{J=0} \ Q \right]_{J=\frac{1}{2}}$$

is well represented by

$$\Sigma_Q - \Lambda_Q = A + B/m_Q.$$

In this way, from $\Sigma - \Lambda = 77$ MeV and $\Sigma_c - \Lambda_c = 166$ MeV, one predicts

$$\Sigma_b - \Lambda_b = 197 \pm 10 \text{ MeV}.$$

On the other hand, we expect that the mass difference $\Sigma_Q^* - \Sigma_Q$, where

$$\Sigma_Q^* = \left[(qq)_{J=1} \ Q \right]_{J=\frac{3}{2}} \quad ,$$

vanishes for $m_Q \to \infty$ because the hyperfine interaction of the heavy quark with the light quarks vanishes, so that $A = 0$.

From $\Sigma^* - \Sigma = 190$ MeV, we expect

$$\Sigma_c^* - \Sigma_c = 55 \text{ MeV} \quad \text{and} \quad \Sigma_b^* - \Sigma_b = 19 \text{ MeV}.$$

Notice that since Σ_b is allowed to decay strongly via $\Sigma_{c,b} \rightarrow \Lambda_b + \pi$, with an appreciable phase space available, the widths of Σ_b and Σ_b^* exceed their splitting, and we expect, as pointed out by Lipkin [6], interesting interference phenomena in the angular distribution of the decay products as a function of energy.

REFERENCES

1) J.M. Richard - Phys.Letters 100B (1981) 515.
2) A. Martin - Phys.Letters 100B (1981) 511.
3) A. Bhadury et al. - Phys.Rev.Letters 44 (1980) 3180.
4) D.P. Stanley and D. Robson - Phys.Rev. D21 (1980) 3180; Phys. Rev.Letters 45 (1980) 235.
5) J. Lee-Franzini and P. Franzini - Physics Reports 81 (1982) 240.
6) H.J. Lipkin - Private communication.

TESTS OF THE t QUARK MASS FROM B MESON DECAYS

Bruce A. Campbell and Patrick J. O'Donnell

Dept of Physics and Scarborough College
University of Toronto
West Hill, Ontario, Canada   M1C 1A4

ABSTRACT

For a top quark of large mass certain processes may no longer be suppressed sufficiently by the G.I.M. mechanism. We examine the induced decays $b \to s + \gamma$, $B_s \to \tau^+ \tau^-$ and $B_s \to \gamma\gamma$ and the mixings of $B_d(\overline{B}_d)$ and $B_s(\overline{B}_s)$ mesons; the $b \to s + \gamma$ decay and the neutral meson mixings give sensitive tests of the top quark mass.

In the Kobayashi-Maskawa[1] (KM) six quark (and lepton) model the tree level couplings of the neutral gauge bosons (γ and Z) are naturally flavour diagonal in the basis of quark mass eigenstates. Furthermore this extended Glashow-Iliopoulos-Maiani[2] (GIM) cancellation implies that flavour changing neutral current couplings induced at the one loop level will vanish if the virtual quarks appearing in the loop are degenerate in mass (and similarly for induced mixings of neutral mesons such as $K^\circ \leftrightarrow \overline{K}^\circ$[3]). As such, flavour changing neutral currents and induced neutral meson mixings provide a measure of the mass splitting between the virtual quarks involved.

In this talk we will examine the constraints one may put on the mass of the as yet unobserved top(t) quark from limits on the mixing or neutral current decay of B mesons. This follows the approach[3] used to infer limits on the charmed quark mass from the decays and mixing of neutral kaons. Examination of these kaon processes in the six quark KM model has recently been undertaken[4,5]; these authors have extended previous treatments by calculating the induced vertices for arbitrary intermediate quark mass, whereas previous work[6] considered the limit $M_Q/M_W \ll 1$.

This last point is important in view of the fact that theoretical constraints on the mass of the top quark are so weak. There is a restriction[7] on the mass splitting of weak SU(2) fermion doublets from the radiative corrections to the ratio of vector boson masses (with isodoublet Higgs only) which give:

$$\frac{M^2_{W^\pm}}{M^2_{Z^o}} \simeq \cos^2\theta_W \left[1+c\frac{G_F}{8\sqrt{2}\pi^2}\left(\frac{2M_1^2M_2^2}{(M_1^2-M_2^2)}\ln\left(\frac{M_1^2}{M_2^2}\right)+M_1^2+M_2^2\right)\right] \qquad (1)$$

where the colour factor c is 1(3) for leptons (quarks). Present data[8] then implies that $M_t < 0(400\text{GeV})$. A large t quark mass$\geq 0(108)$ GeV also tends[9] to destablize the vacuum of the Weinberg-Salam model; the addition of further heavy Higgs (or vector) bosons tends to stabilize the desired minimum of the effective potential - hence this type of argument may be vitiated by proliferation in the Higgs sector. In sum, there is no good a priori theoretical reason to rule out a top quark mass up to 0(400) GeV; so it is important to retain the full mass dependence in our study.

Constraints on the top mass in rare kaon processes are intricately related to determination of the KM parameters which are also inferred from these same processes. This, combined with the uncertainty in the hadronic matrix elements involved makes it difficult to isolate the value of the t mass in these processes. We wish to extend this analysis to the decays and mixings of b quarks; when combined with the constraints from K decays, these processes present direct and experimentally feasible tests of the t quark mass. There have been analyses of the mixing of neutral B mesons previously; however these utilized expressions for the mixing matrix element only valid for $M_t \ll M_W$[10,12].

Consider first those decays of the b quark which involve the emission of a neutral gauge boson: viz., $b \to s + \gamma$; $b\bar{s} \to \tau^+\tau^-$; $b\bar{s} \to \gamma\gamma$. We estimate the branching ratios by dividing the computed width for the decay in question by the total B decay width which we take[11] to be

$$\Gamma(B) \simeq \Gamma(b \to c + W^- \to c + X)$$

$$\simeq \frac{G_F^2 M_B^5}{192\pi^3}2|V_{bc}|^2 \simeq \frac{G_F^2 M_B^5}{192\pi^3}2(s_2^2+s_3^2+2s_2s_3\cos\delta) \qquad (2)$$

where V_{bc} is the (b,c) entry in the KM[1] mixing matrix,

$$V = \begin{array}{c} u \\ c \\ t \end{array} \begin{bmatrix} \overset{d}{c_1} & \overset{s}{s_1 c_3} & \overset{b}{s_1 s_3} \\ -s_1 c_2 & c_1 c_2 c_3 - s_2 s_3 e^{i\delta} & c_1 c_2 s_3 + s_2 c_3 e^{i\delta} \\ -s_1 s_2 & c_1 s_2 c_3 + c_2 s_3 e^{i\delta} & c_1 s_2 s_3 - c_2 c_3 e^{i\delta} \end{bmatrix} \qquad (3)$$

with $s_i = \sin\theta_i$; $c_i = \cos\theta_i$; $0 \le \theta_i \le \pi/2$ $i=1,3$. The constraints on the mixing angles have been considered in a number of recent papers. From β- and μ-decay the limit is $|V_{ud}| = \cos\theta_1 = 0.973 \pm 0.0025$. The semi-leptonic decays of Σ, Λ, Ξ yield $|V_{us}| = \sin\theta_1 \cos\theta_3 = 0.219 \pm 0.002$ which implies $\sin\theta_3 < 0.5$ and $|V_{ub}| = \sin\theta_1 \sin\theta_3 = 0.06 \pm 0.06$. CP violation in the $K^\circ - K^\circ$ system has been used to further constrain the mixing angles but unfortunately the constraints are not too certain. If s_3 is small then it appears that $s_2 s_3 s_\delta \sim 10^{-3}$ which can be satisfied with $\delta \sim 0$ (quadrant I solution) or $\delta \sim \pi$ (quadrant II solution). For very small values of s_3 it is possible to have δ taking on a value anywhere in the range from zero to π. The quantity $|V_{cs}|^2 = |c_1 c_2 c_3 - s_2 s_3 e^{i\delta}|^2$ appears to have a lower bound of 0.6. The quantity $|V_{bu}|^2 / |V_{bc}|^2 = |s_1 s_3|^2 / |c_1 c_2 s_3 - s_2 c_3 e^{i\delta}|^2$ has an upper limit of 0.4 from the CLEO results. In Eq. (2) we use $\theta_i \ll 1$ to retain only leading terms in $\sin\theta_i$, and use the explicit solutions of ref. 13 in the numerical evaluation.

Consider first the decay $b(p_1) \to s(p_2) + \gamma(q,\epsilon)$. This will proceed via an induced dipole transition matrix element.

$$M = \frac{G_F}{2\sqrt{2}} \left(\frac{e}{2\pi^2} \right) \sum_i V_{ib} V_{is}^* F_2^i q^\mu \epsilon^\nu \bar{s}(p_2) \sigma_{\mu\nu} (M_b R + M_s L) b(p_1) \qquad (4)$$

where $L(R)$ are the left (right) projection operators, the sum is over intermediate quark states, we have dropped terms suppressed by powers of the initial or final quark masses (assumed small) and[4,5]

$$F^i(x) = Q_i \left\{ \left[-\tfrac{1}{4} \frac{1}{(x-1)} + \frac{3}{4} \frac{1}{(x-1)^2} + \frac{3}{2} \frac{1}{(x-1)^3} \right] x - \frac{3}{2} \frac{x^2}{(x-1)^4} \ln x \right\}$$
$$- \left[\tfrac{1}{2} \frac{1}{(x-1)} + \frac{9}{4} \frac{1}{(x-1)^2} + \frac{3}{2} \frac{1}{(x-1)^3} \right] + \frac{3}{2} \frac{x^3}{(x-1)^4} \ln x \qquad (5)$$

where $Q_i = 2/3$ is the charge of the intermediate quark, and $x = M_i^2 / M_W^2$. The resulting decay width is then

$$\Gamma = \frac{\alpha G_F^2}{32\pi^4} M_b^5 \left(1 - \frac{M_s}{M_b} \right)^3 \left(1 + \frac{M_s}{M_b} \right) \left| \sum_i V_{is} V_{ib}^* F_2^i \right|^2 . \qquad (6)$$

An inspection of the KM mixing matrix shows that the $b \to s + \gamma$ decay is favoured over $b \to d + \gamma$. The branching ratio for this process is shown in Figure 1.

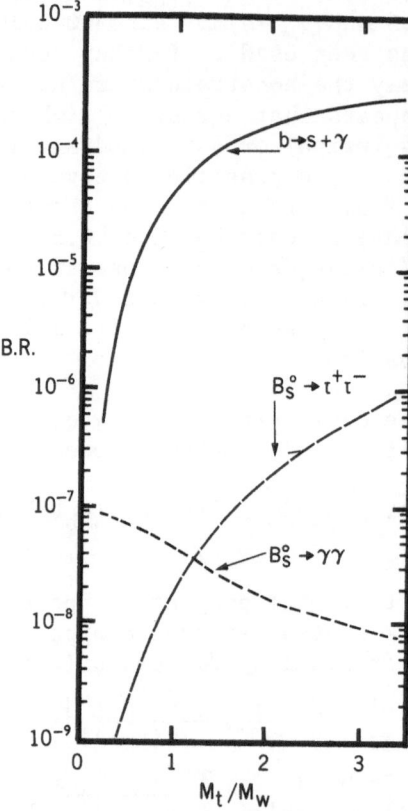

Figure 1.

Two other decays,[14] $B_s^o \to \tau^+ \tau^-$, and $B_s^o \to \gamma\gamma$ are shown also in the figure. The branching ratios appear to be too small for observation at present. If the t quark is massive the decay $b \to s + \gamma$ should give a measurable rate. By putting upper bounds on the branching ratio for this decay mode one will be able to bound the mass of the t quark. Furthermore in the parton view a decaying B meson at rest should give a photon recoiling against a strange quark "jet".

It is this result which gives a distinctive experimental signature. Although the multiplicity in B decays involving strange particles may be relatively large such events will be accompanied by photons which are non monochromatic and of low energy. These processes will have branching ratios much <u>smaller</u> than those presented in the figure since they will involve higher order processes involving gluons. By contrast the higher energy photons, whose energy is about $M_B/2$, will appear back to back with a strange quark "jet" which will have dominant <u>few</u> body modes such as K*(890), Q(1400), K*(1400) or K*(1890). In this energy regime there will be a monochromatic photon and a $K\pi$ or $K\pi\pi$ state reconstructing to form the resonances. The width of the resonances will give the resolution of the photon energy. Specifically the energy (ω) and resolution of the outgoing photon should be as follows: for the K*(890) final state, $\omega = 2.42 \pm .05$ GeV; for the K*(1430) final state, $\omega = 2.3 \pm .1$ GeV (with approximately the same energy and a slightly higher spread in the $K\pi\pi$ modes corresponding to Q production); for the K*(1800) final state $\omega = 2.18 \pm .13$ GeV. Wave function overlap effects may reduce the rate appreciably from the parton model estimates, however.

Although we have considered the standard six quark model we should note that five quark models which have no t quark have recently been examined.[15,16] If all the right handed quarks are in singlets there is a lower bound[15] for $\Gamma(B \to Xe^+ e^-)/\Gamma(B \to Xe^+ \nu) > 0.12$. This quantity has now been measured[17] and is less than 0.08 thus ruling out this model. If the five quarks have a right handed doublet (c,b) then a branching ratio comparable to that shown in fig. 1 may arise.

The analysis of mixings in the neutral B_q^o meson system proceeds[10-12] by direct analogy to that of the K meson system. However the phenomenology of mass mixing and CP violation need not have the same small ε (CP) violation limit.[18] Nevertheless the overlap between CP violation and mass mixing seems to be small and it has been argued that it is unlikely that CP violation in B-$\bar{B}$ mixing will be observed. The CP conserving piece of the mixing matrix element for the B_q meson is given by

$$\Delta M = 2\mathrm{Re}\langle \bar{B}_q^o | -L_{eff} | B_q^o \rangle$$
$$= \frac{2}{3} \frac{G_F}{\sqrt{2}} \frac{\alpha}{4\pi \sin^2\theta_W} f_{B_q^o}^2 M_{B_q^o} \mathrm{Re}(E_{B_q^o}) \qquad (7)$$

where $E_{B_q^o} = \sum_{j,k} V_{jq}^* V_{jb} V_{kq}^* V_{kb} \overline{E}(x_j, x_k)$, with

$$
\begin{aligned}
\overline{E}(x_j, x_k) = -x_j x_k \Bigg\{ & \frac{1}{(x_j - x_k)} \left[\frac{1}{4} - \frac{3}{2} \frac{1}{(x_j - 1)} - \frac{3}{4} \frac{1}{(x_j - 1)^2} \right] \ln x_j \\
& + \frac{1}{(x_k - x_j)} \left[\frac{1}{4} - \frac{3}{2} \frac{1}{(x_k - 1)} - \frac{3}{4} \frac{1}{(x_k - 1)^2} \right] \ln x_k \\
& - \frac{3}{4} \frac{1}{(x_j - 1)(x_k - 1)} \Bigg\}
\end{aligned}
\tag{8}
$$

where $x_{j,k} = M_{j,k}^2 / M_W^2$ with j,k the virtual quarks. In these expressions we have retained the full dependence[4-6,10] on the intermediate quark mass including both gauge and Higgs exchange contributions to the box diagrams. Renormalization effects from the strong chromodynamic interaction are expected[11,18] to be small and are ignored. To leading order in s_1 the mixing functions are

$$
\begin{aligned}
\mathrm{Re}(E_{B_d^o}) = & \; s_1^2 (s_3^2 + 2 s_2 s_3 \cos\delta + s_2^2 \cos 2\delta) \overline{E}(x_c, x_c) \\
& - 2 s_1^2 (s_2 s_3 \cos\delta + s_2^2 \cos 2\delta) \overline{E}(x_c, x_t) \\
& + s_1^2 s_2^2 \cos 2\delta \overline{E}(x_t, x_t) \; .
\end{aligned}
\tag{9}
$$

$$
\begin{aligned}
\mathrm{Re}(E_{B_s^o}) = & \; (s_3^2 + 2 s_2 s_3 \cos\delta + s_2^2 \cos 2\delta) \{ \overline{E}(x_c, x_c) \\
& - 2 \overline{E}(x_c, x_t) + \overline{E}(x_t, x_t) \} \; .
\end{aligned}
\tag{10}
$$

This mixing will have an effect on the decays of B_q^o meson pairs produced in $e^+ e^-$ annihilation[19,20] with the most experimentally distinctive characteristic of mixing occurring when both members of the $B_q^o(\overline{B}_q^o)$ pair undergo primary decay semileptonically. The fraction of same sign dileptons produced is[19,20]

$$
\begin{aligned}
R &\equiv \frac{N^{++} + N^{--}}{N^{++} + N^{--} + N^{+-} + N^{-+}} \\
&= \frac{[4(\Delta M)^2 + (\Delta\Gamma)^2][8\Gamma^2 + 4(\Delta M)^2 - (\Delta\Gamma)^2]}{32[\Gamma^2 + (\Delta M)^2]^2}
\end{aligned}
\tag{11}
$$

where Γ is the width of the B_q^o meson and $\Delta\Gamma$, which is the difference in width of the CP even (odd) eigenstates of the B_q^o system, is small compared to Γ.

There are two physically distinct sets of KM parameters[18,21,22] corresponding to $\delta \sim 0$ (quadrant I) or $\delta \sim \pi$ (quadrant II). This gives

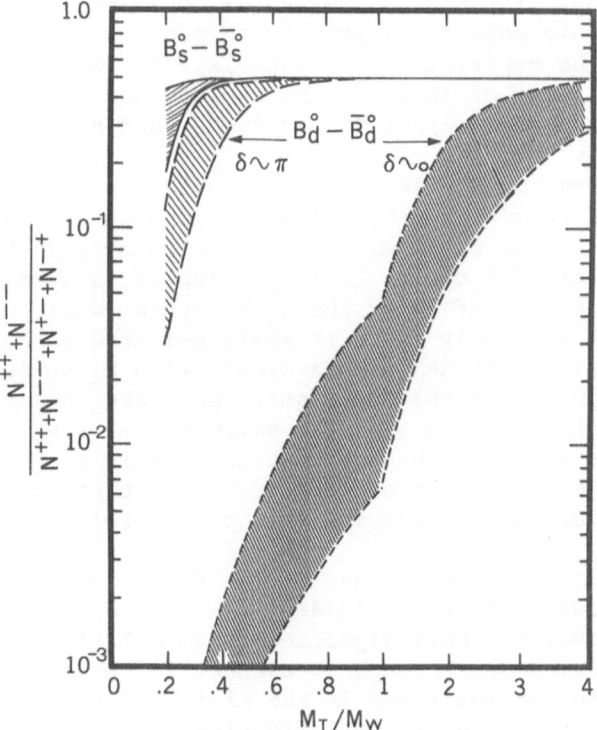

Figure 2.

a severe ambiguity for the ratio R in the case of B_d^o (B_d) mixing but to leading order in s_i, does not affect the ratio R for B_s^o (B_s^o) In Fig. 2 we show the ratio R computed using the vacuum insertion approximation for ΔM and taking a range of values for the pseudoscalar decay constants f covering the estimates of the papers in ref. 23. From Fig. 2 we see that R can be sensitive test for a t quark if it has mass in the lower end of the possible mass spectrum in the case of B_s^o (B_s^o) mixing. (To leading order in s_i the quadrant ambiguity does not affect the branching ratio calculation of b→s+γ).

In summary, prospects for determining the mass of the t quark from rare decays of B mesons are as follows. We have found that b→s+γ is the only rare decay process to proceed at a sufficiently large rate to be experimentally feasible. It should be useful for exploring the range of larger quark masses, and should have distinctive experimental signatures B→K*γ→Kπγ and B→Qγ→Kππγ.

For the same sign dilepton fraction R from neutral B_q^o meson pairs produced in e^+e^- annihilation the results depend on the companion quark q. For $B_s^o \bar{B}_s^o$ mixing R rises quickly with t quark mass getting to within 10% of its maximum value of .5 by a t-quark mass of 0.4 M_W; so values of R for the $B_s^o \bar{B}_s^o$ system will be useful tests of the t quark mass only if it is small compared to M_W. For the $B_d^o \bar{B}_d^o$ system mixing turns on more slowly with M_t so in principle R for this system will probe the t quark mass over a larger range; however predictions for R in this system suffer a severe quadrant ambiguity, and will only be useful once that is resolved (conversely, if one knew the t quark mass from direct measurement this would be a convincing method of resolving this ambiguity).

For all of these processes, if the KM parameters do have δ in the second quadrant then the leading mixing and decay terms $s_2^2+s_3^2+2s_2s_3 \cos\delta$) can have significant cancellations. One must know the KM angles well enough to account for this (or possibly include all the smaller terms in the mixing factors). Indeed the approach advocated here is most effective when combined with studies (for arbitrary t mass) of (rare) K and D decays to pin down the KM parameters as well as the top quark mass.[24]

REFERENCES

1. M. Kobayashi and T. Maskawa, Prog. Theor. Phys. 49, 652 (1973).
2. S.L. Glashow, J. Iliopoulos, and L. Maiani, Phys. Rev. D2, 1285 (1970); see also S.L. Glashow and S. Weinberg, Phys. Rev. D15, 1958 (1977); E.A. Paschos, Phys. Rev. D15, 1966 (1977).
3. A.I. Vainstein and I.B. Kriplovich, J.E.T.P. Lett. 18, 83 (1973); E. Ma, Phys. Rev. D9, 3103 (1974); M.K. Gaillard and B.W. Lee, Phys. Rev. D10, 897 (1974); M.K. Gaillard, B.W. Lee, and R.E. Schrock, Phys. Rev. D13, 1674 (1976).

4. T. Inami and C.S. Lim, Prog. Theor. Phys. 65, 297 (1981).

5. E. Ma and A. Pramudita, Phys. Rev. D22, 214 (1980), Phys. Rev. D24, 1410 (1981), also preprint UH-511-428-80 (1980).

6. For references see Inami and Lim, reference (4).

7. M.J.G. Veltman, Nucl. Phys. B123 (1977).

8. For a review see J.E. Kim et al. Pennsylvania preprint UPR-158T

9. N.V. Kraznikov, Sov. J. Nucl. Phys. 28, 549 (1978); P.Q. Hung, Phys. Rev. Lett. 42, 783 (1979); H.D. Politzer and S. Wolfram, Phys. Lett. B82, 242 (1979), Erratum B83, 421 (1979); P.J. O'Donnell, Z. Physik C5, 43 (1980).

10. J.S. Hagelin, Phys. Rev. D23, 119 (1981).

11. J. Ellis, M.K. Gaillard, D.V. Nanopoulos and S. Rudaz, Nucl. Phys. B131, 285 (1977).

12. A. Ali, and Z.Z. Aydin, Nucl. Phys. B148, 165 (1979); E. Ma, W.A. Simmons, and S.F. Tuan, Phys. Rev. D20, 2888 (1979); J.S. Hagelin, Phys. Rev. D20, 2893 (1979).

13. R.E. Shrock, S.B. Treiman, and L.L. Wang, Phys. Rev. Lett. 42, 1589 (1979); L.L. Wang, Brookhaven preprint (1980).

14. See B.A. Campbell and P.J. O'Donnell, Phys. Rev. D25, 1989 (1982) for further details.

15. G. Kane and M. Peskin, preprint UMHE81-51.

16. R. Decker and E.A. Paschos, preprint DOTH 81/09.

17. G. Monetti, these proceedings.

18. For summary and references see J.S. Hagelin, Proc. Cornell B-Decay Workshop, 1981 (to be published).

19. A. Pais and S.B. Treiman, Phys. Rev. D12, 2744 (1975).

20. L.B. Okun, V.I. Zakharov, and B.M. Pontecorvo, Lett. Nuovo Cimento 13, 218 (1975).

21. V. Barger, W.F. Long, and S. Pakvasa, Phys. Rev. Lett. 42, 1585 (1979); S. Pakvasa, S.F. Tuan, and J.J. Sakurai, Hawaii preprint UH-511-427-80 (1980).

22. B. Gaisser, T. Tsao, and M. Wise, Ann. Phys. (N.Y.) 132, 66 (1980).

23. M. Claudson, Harvard preprint; V.S. Mathur and T. Yamawaki, Rochester preprint. H. Krasemann, Phys. Lett. 96B, 397 (1980).

24. A.J. Buras, Phys. Rev. Lett. 46, 1354 (1981).

BEAUTY EXPERIMENTS AT THE SPS

Paul Musset

EP Division

CERN, 1211 Geneve 23, Switzerland

INTRODUCTION

Several experiments, one already in the course of analysis and others to be done in the near future, have been designed to search for beauty at the SPS. They are of a hybrid character, combining the capabilities of emulsion with a selective signature of the beauty production in emulsion. The selection is done by an on-line trigger and by the off-line analysis.

The first experiment in this line of research, NA19[1] was run for data taking and emulsion exposure in 1980, and the analysis is presently under completion. Three other experiments have been proposed P159[2], P166[3], and P167[4], which have various features, and are planned to run in the years 1983 - 1984.

MOTIVATION

The main aims of these experiments are the direct observation of B-particles and of their decays, the measurement of their life-times - possibly different for $B^{\pm}$ and for B^0, and the measurement of the branching ratio $b \rightarrow c/b \rightarrow u$.

In comparison with charm physics, the study of the B-particles develops slowly. One year after the theoretical prediction of charm , the first short-lived particles were discovered in cosmic rays . Subsequently, the J/ψ[7] the dilepton[8] and the $\mu e V^{\circ}$[9] events in neutrino interactions were observed, before charmed particles were aboundantly produced and studied at SLAC[10]. On the contrary, the theoretical prediction of beauty in 1973[11] waited the year 1977[12] for the observation of the T at FNAL, and only recently[13] particles with open beauty were indirectly observed at CESR[14]. The hadronic production of beauty at the ISR is still the subject of contradictory results[14,15] and we are waiting for a second generation experiment.

The difficulties in the study of beauty are the presumably short lifetime and very small production cross section.

To produce B-particles, photons are not quite energetic (or intense enough) in present beams. The $\bar{\nu}$ production of beauty is still too small at the present accelerator energies to provide the basis of a realistic experimental programme. Reactions induced by hadrons, mainly pions, will be considered as the possible source of B-particles in the experiments to be described here.

The physical parameters which can be reached in the study of beauty are expressed, in the framework of the standard six quark--model, in terms of mixing angles of the Kobayashi-Maskawa matrix. In this model, the particles inside the left-handed doublet of $SU(2) \otimes U(1)$ can be expressed as combinations of the d, s, and b quarks.

d'		c_1	$s_1 c_3$	$s_1 s_3$	d
s'		$-s_1 c_2$	$c_1 c_2 c_3 + s_2 s_3 e^{i\delta}$	$c_1 c_2 s_3 - s_2 s_3 e^{i\delta}$	s
b'	=	$-s_1 s_2$	$c_1 s_2 c_3 - c_2 s_3 e^{i\delta}$	$c_1 s_2 s_3 + c_2 c_3 e^{i\delta}$	b

In this expression, s_1 is the sinus of the Cabibbo angle $s_1 = 0,23$; s_3 is limited by the rate of the strange particle decays $s_3 = 0.28^{+0.28}_{-0.21}$ and s_2 is limited by the $K^0_L - K^0_S$ mass difference. $S_2 < 0.36$, δ is the CP violating phase $\delta > 10^{-3}$.

In the approximation of all mixing angles being small ($c_1 \simeq c_2 \simeq c_3 \simeq 1$), the relevant terms are the two upper terms in the last column, which become $s_1 s_3$ and $s_3 - s_2 s_3 e^{i\delta}$ respectively. Then the measurable quantities τ_B and $b \to c/b \to u$ are expressed[16] as:

$$\tau_B \simeq \frac{4 \cdot 10^{-5}}{s_2^2 + s_3^2 + 2 s_2 s_3 \cos\delta}$$

$$\frac{b \to c}{c \to \mu} \simeq \left(\frac{s_3 - s_2 e^{i\delta}}{s_1 s_3}\right) \quad F \quad \left(\frac{m_c^2}{m_b^2}\right)$$

Note also the prediction that the ratio

$$\frac{\tau_{B^\pm}}{\tau_{B^0}} > 1 \quad \text{(see for ex. Kobayashi and Yamazaki[17]).}$$

With our present knowledge on the limits of the mixing angle, τ_B is currently estimated to be within the range 10^{-14}s - 10^{-13}s, whereas the range 10^{-15}s - 10^{-12}s is not completely excluded. The decay b → c is believed to be dominating over the decay b → u.

PRODUCTION CROSS SECTION

In order to design a beauty experiment, one has to assume values for the production cross section, for its energy dependence and for its dependence on the nature of the primary, π or p. Theoretical estimates of the cross sections differ widely according to models[18]. Simply speaking, the first QCD estimates lead to very small values, typically of the order of the nb, but presently more complete calculations have the tendency to provide higher values. Starting from a particular hypothesis, the gluodominance mechanism calculations, invoked to explain a possibly large (of the order of the μb), cross section[19], now seems to be ruled out by experimental limits[20]. There exist typically two types of diagrams in QCD for the b-production (Fig.1). The first one is called flavour creation: $q\bar{q} → b\bar{b}$, and leads to a central production, i.e. with small x-values. It operates via quark fusion or via gluon fusion. The second one is called flavour excitation qb → qb, and it leads to a diffractive production, i.e. it occurs at all x-values, and it has a characteristic threshold behaviour at low energy.

These mechanisms are likely to be very similar to the case of charm, but both the theoretical and the experimental situations for charm production are unclear and it is not possible to found speculations about beauty production on the case of charm. In particular, in charm experiments, only a partial covering of the allowed kinematical region is realized, and one is not able to extrapolate the results from one experiment to the other. There are still order-of-magnitude uncertainties in the knowledge of the charm production cross-section by hadrons.

Table I gives some examples of theoretical predictions of the production cross-sections for π at 360 GeV, together with the experimental limits for protons and pions.

Let us first remark that to extrapolate from existing data around 200 GeV π to the presently maximum available energy 360 GeV π, the various theories predict an increase in the range 3-6. Note also that in general, $\sigma_\pi > \sigma_p$ in these theories, and this is further supported by the experimental result[21] that $\pi^+ → T/p → T = 30$ at 380 GeV.

Our conclusion will be that σ is certainly small, but still uncertain by more than one order of magnitude from a few nb to 50 nb.

Beauty Production Mechanisms

Flavor Creation

$$q\,\bar{q} \longrightarrow b\,\bar{b}$$

Central ,   x small

Quark fusion   Gluon fusion

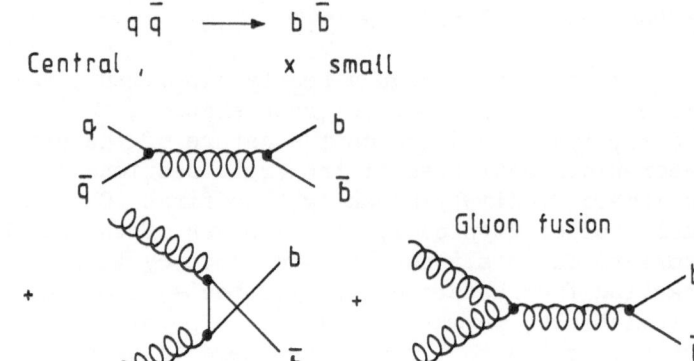

.Flavor Exitation

$$q\,b \longrightarrow q\,b$$

Diffractive,   all x (threshold.)

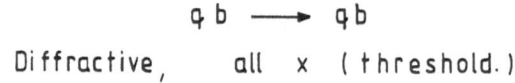

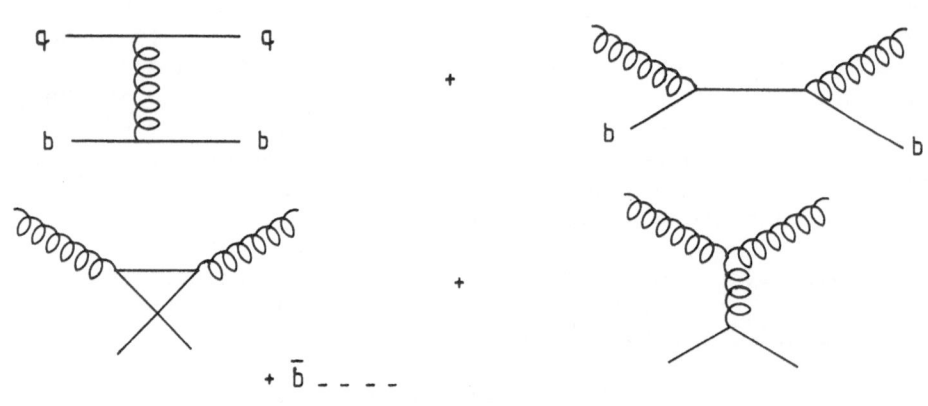

+ $\bar{b}$ _ _ _ _

Gluo Dominance

$\cong$ VDM for gluon

Figure 1

TABLE I: Theoretical Predictions

Author	σ	Remark
B.L. Combridge	≤1 nb	QCD
R. Winder-C. Michael	>1 nb	"
T.S. Tu	∿6 nb	"
Y. Afek, C. Leroy, B. Margolis	∿10 nb	"normalized" on production
Fritzsch	∿0.6 μb	gluo-dominance model
G. Guftafson, C. Peterson	∿1% of C	diffractive
V. Barger, F. Halzen, B. Margolis	∿0.16 μb	mainly diffractive based on Λ_c at ISR.

Experimental Limits

Primary	Experiment	Energy	σ limit (90% CL)
p	FNAL	400 GeV	50 nb
π	NA3	280 GeV	25 nb
π	CIP	225 GeV	10 nb

GENERAL FEATURES OF THE HYBRID BEAUTY EXPERIMENTS

In general, the experiments under discussion make use of a beam with the highest possible energy E, because of the rise of σ with E.

The B-events produced in the reaction $\pi + N \rightarrow B + \bar{B} + X$ are characterized by the observation of the cascade chain $B \rightarrow C \rightarrow X$; a multiple decay vertex-up to 4-events can be observed. The B-decays are always observed in emulsion, whereas the subsequent C-decays may be observed in emulsion or in an ad-hoc telescope placed behind the emulsion.

Since the scanning and the analysis of the events in emulsion is a time-consuming procedure and due to the small cross section, the trigger and the off-line selection of events has to be very selective, of the order of 10^4-10^5. Nevertheless, the acceptance has to be kept as large as possible, in order to minimize the amount of emulsion used in the experiment.

In the P159 and P167 experiments, it is intended to use the

Ω' spectrometer to identify and measure secondaries. In the other
experiments, this is precluded by the presence of a muon filter.
Triggers are based on muonic decays in NA19 and P166, on charm
decays in P159 and on the yield of kaons in P167.

All the experiments described here use emulsion to detect the
primary and the decay vertices. This choice is dictated by the high
resolving power of emulsion, i.e. 0.5μ and the high grain density,
i.e. 2000 grains/cm along the track. This is to be compared to
corresponding values for holographic bubble chambers, i.e. 10 μ for
the bubble size and 200 bubbles/cm. Remember that with a typical
beauty lifetime of $5 \ 10^{-14}$s, the corresponding impact parameter is
15 μ. Since at least 3 decay vertices are used to sign a B-event,
the efficiency has to be high for each decay, and this explains
the choice for emulsion.

The Na19, P166 and P167 experiments use thick emulsion, in
order to directly observe C-decays as well as B decays. On the
contrary the P159 experiment uses a thin emulsion, the charm decay
telescope being placed immediately after the emulsion. Thick emulsion has the advantage of providing a larger range for the explora-
tion of lifetime and the direct observation of the C-decays. Thin
emulsion produces less multiple scattering and permits the C-decay
trigger.

Although the selection of events is very restrictive, the
experiments collect a large number of candidates from background,
typically a few tens of thousands and the corresponding scanning
and measuring effort has to be reduced by automation. On-line com-
puters, TV screens, displays and motorized displacements are
installed in Nagoya, Bari, Brussels, CERN, Genova, Milano, Roma etc.
Two types of exposure are used: the "vertical" one, that is to say
perpendicularly to the incident beam and the "horizontal" one,
parallel to the beam. The first one permits a higher track density,
typically $\sim$3000 tr/mm^2, compared to 1500 tr/mm^2 for horizontal
emulsion, but the second one is certainly more efficient in the
search for very short lifetimes, because of the obscuration of the
vertex region by the secondary tracks, when seen from the direction
of the beam in the case of perpendicular emulsion.

In addition to the emulsion, the experiments have a beam
hodoscope, an emulsion stage, a vertex detector, and a spectrometer
or a muon analyzer.

BEAM HODOSCOPES

In the NA19 experiment (Fig. 2), the beam particle was measured
by a set of 6 MWPC chambers[22] placed near the emulsion plus 3 MWPC
chambers far upstream. The chambers had cathode read-outs and the
track position was measured by the centroid method[23]. A calibration

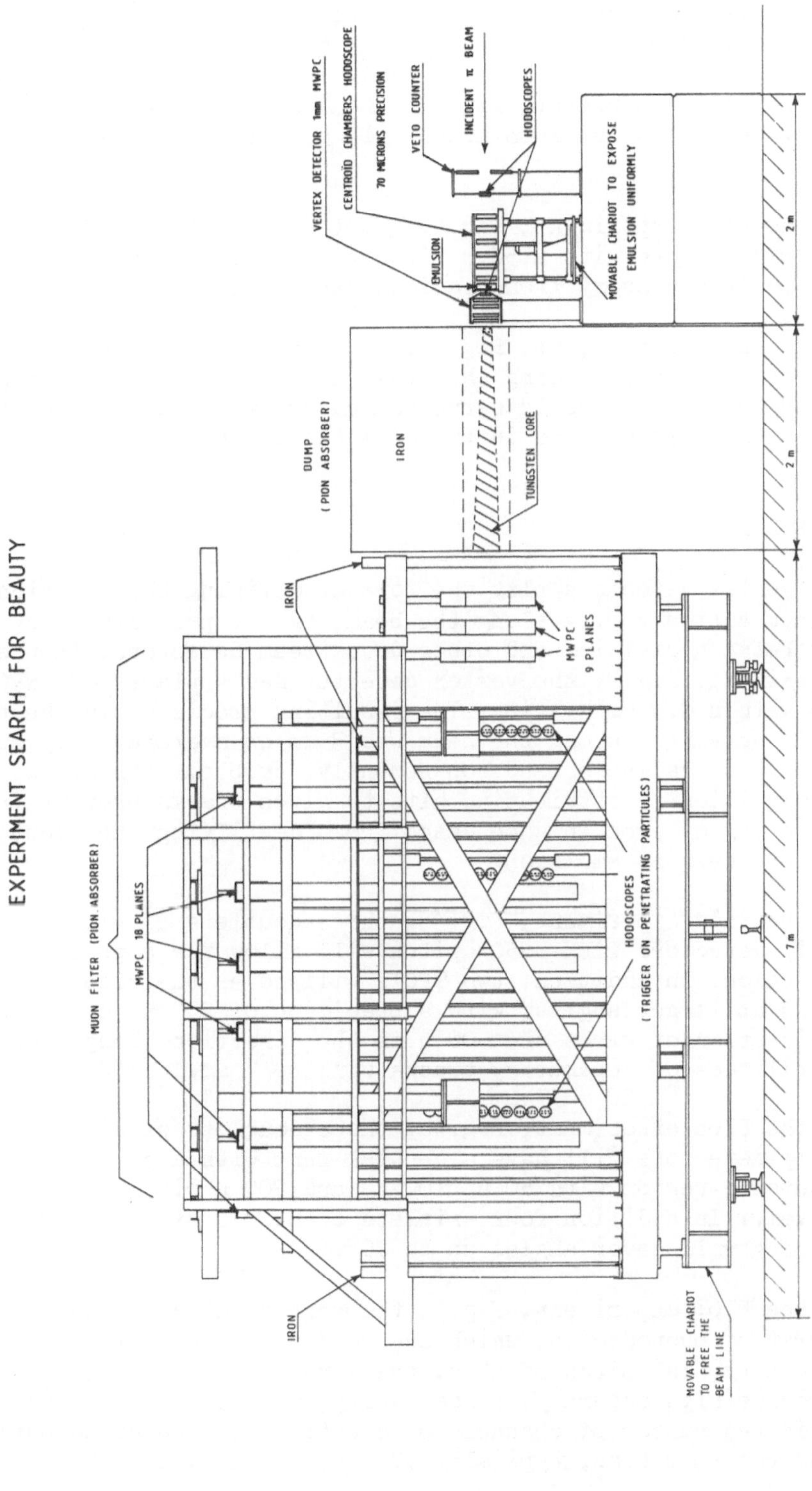

EXPERIMENT SEARCH FOR BEAUTY

VERTEX DETECTOR 1mm MWPC

CENTROID CHAMBERS HODOSCOPE

70 MICRONS PRECISION

VETO COUNTER

INCIDENT π BEAM

HODOSCOPES

EMULSION

MOVABLE CHARIOT TO EXPOSE
EMULSION UNIFORMLY

DUMP
(PION ABSORBER)

IRON

TUNGSTEN CORE

MUON FILTER (PION ABSORBER)

IRON

MWPC 18 PLANES

IRON

MWPC
9 PLANES

HODOSCOPES
(TRIGGER ON PENETRATING PARTICULES)

MOVABLE CHARIOT
TO FREE THE
BEAM LINE

2 m

2 m

7 m

Figure 2: General Layout NA 19

was made by comparing beam tracks observed in the chambers with
tracks observed in a vertical emulsion. The measured dispersion is
~70μ in the two directions perpendicular to the beam.

In the P159 experiment, Fig. 3, the beam hodoscope will be
made of two planes of silicon microstrip detectors with a 100 μ
pitch.

In the P166 experiment, Fig.4, a set of six planes of silicon
microstrip detectors with a 50 μ pitch will be used. This redundancy
was felt to be necessary from previous experience with NA19.

In the P167 experiment, Fig. 5, no beam hodoscopes are
planned for, the reason being that the vertex detector has a high
resolution, so that it would serve to measure the coordinates of
the interaction vertex also in the directions perpendicular to
the beam.

VERTEX DETECTORS

The vertex detectors play the role of defining the position
of the interaction vertex along the beam, and of providing the
matching between emulsion and other downstream detectors. In the
NA19 experiment, Fig.2, the vertex detector had 4 planes of MWPC
chambers with a 0.5 mm spacing. The resulting precision in the position of the vertex along the beam was 3 mm on average. Only one
projection was available, and consequently, problems associated
with ambiguities in attributing hits to given tracks prevent from
reconstructing or even in some cases from identifying the event
with the one seen in emulsion.

In the P159 experiment, Fig. 3, three doublets of silicon
microstrip detectors with a 50 μ pitch will allow for a precise
reconstruction. In addition, two TPC's will cover the acceptance
for tracks at larger angles, with a precision of 200 μ, or better,
in the direction of wires and 500 μ in the direction along pads,
the capabilities of separating tracks will be 7 mm.

In the P166 experiment, Fig.4a, three doublets of silicon
microstrip detectors will have a central part with a 50 μ pitch
and an outside region with 50 μ, 100 μ, and 200 μ pitches
respectively. In addition four triplets of MWPC's with a 1 mm
wire spacing will cover angles up to 20°.

In the P167 experiment, Fig.5, the vertex detector consists
of an array of photodiodes, which are read out sequentially by
CCD, with a typical pitch of 22 μ, the advantage of this technique
is to provide high accuracy, no ambiguity, and to operate with
only a limited number of channels of electronics. The disadvantage
is a long read-out time, typically 10^{-2} s, during which the

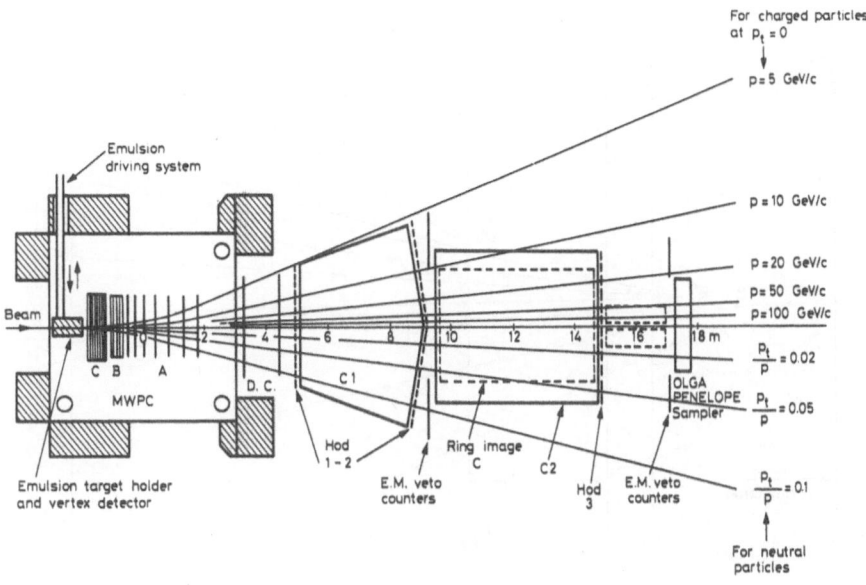

P159 Experiment - General Layout

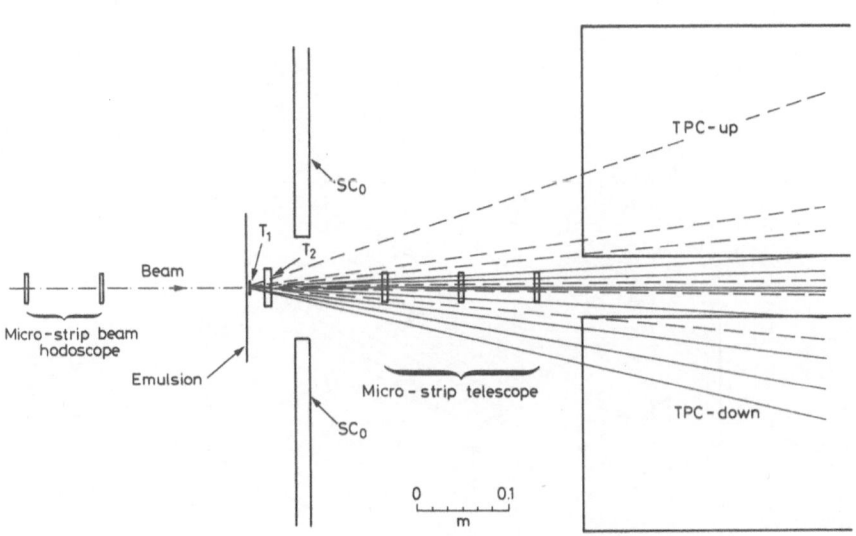

P159 Experiment - Vertex Detector

Figure 3a

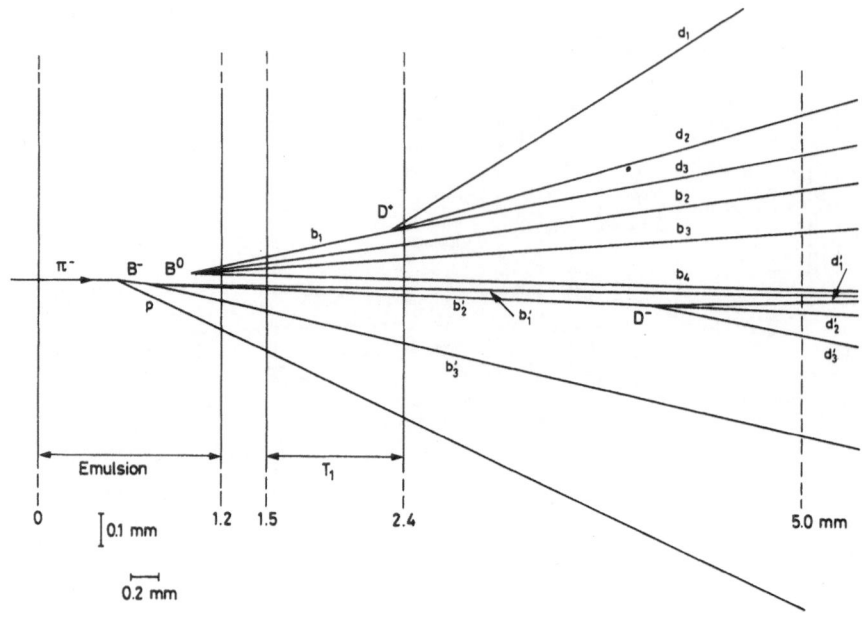

P159 Experiment — Simulated Beauty Event

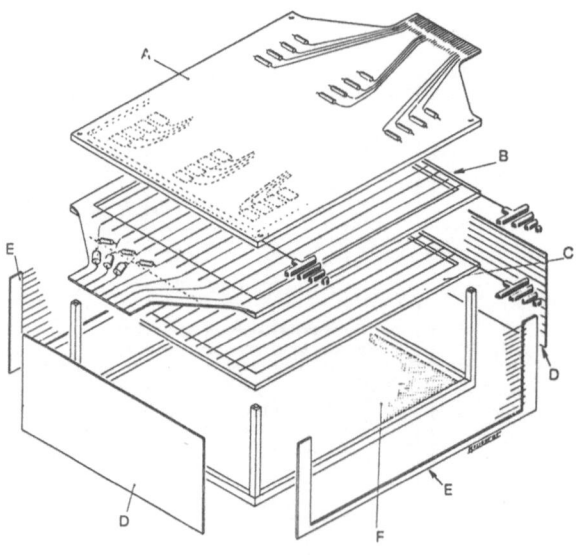

P159 Experiment — Time Projection Chamber

Figure 3b

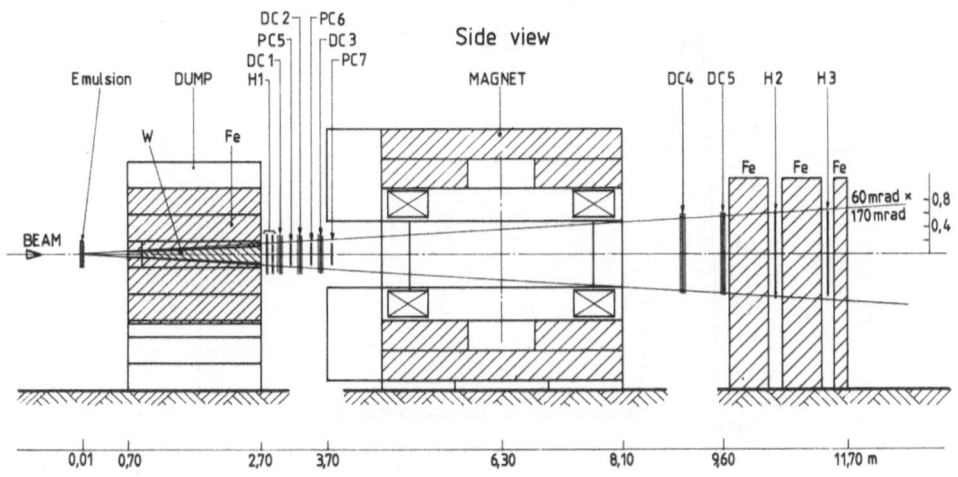

Experiment P166 — General Layout

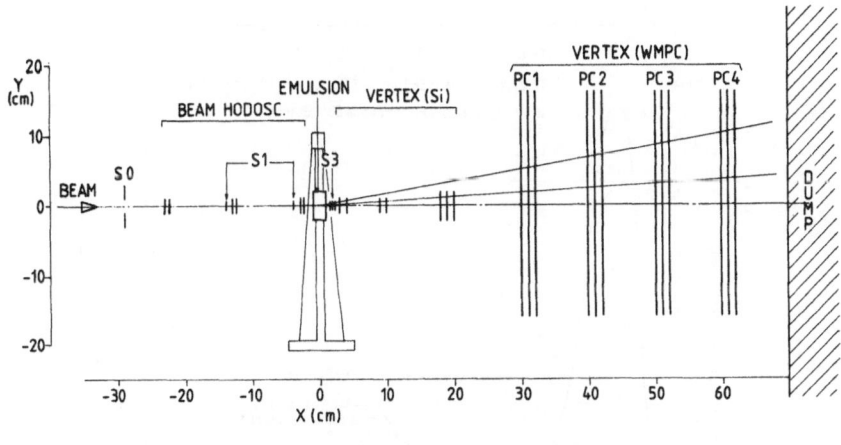

Side view

Experiment P166 — Front End

Figure 4a

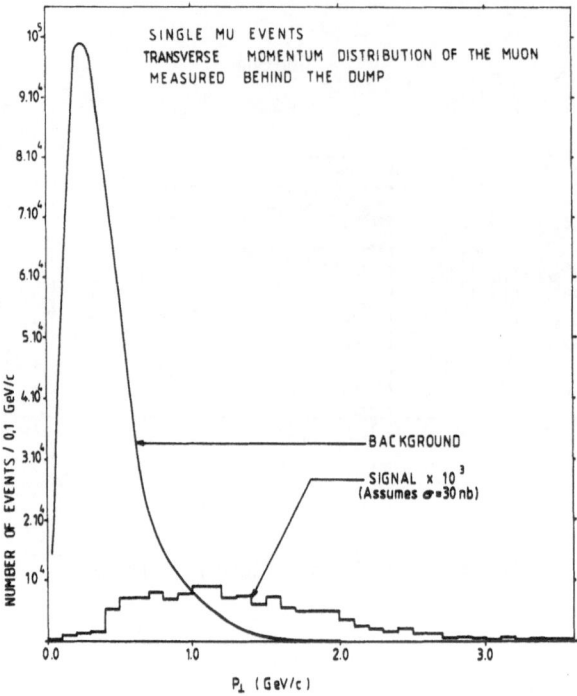

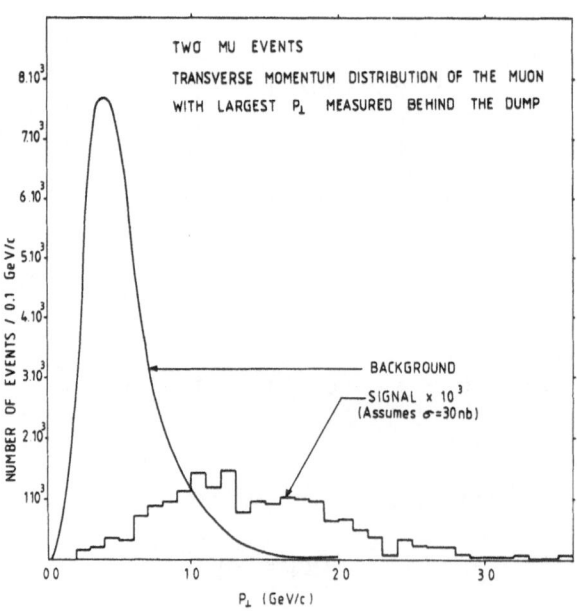

Experiment P166

Figure 4b

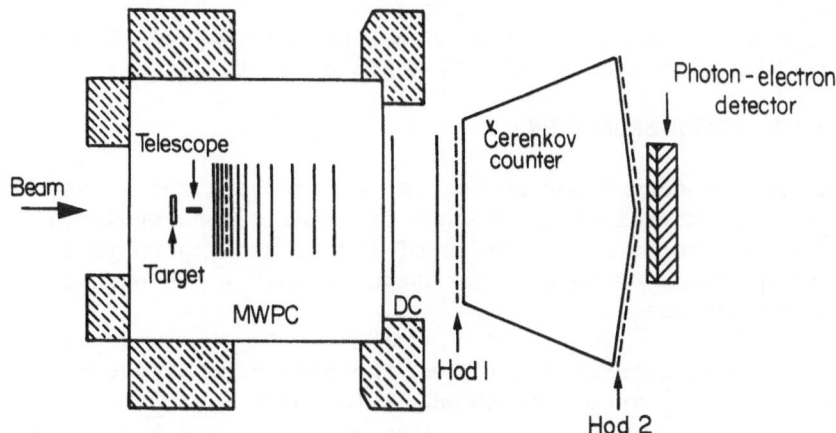

Experiment P167 — Genral Layout

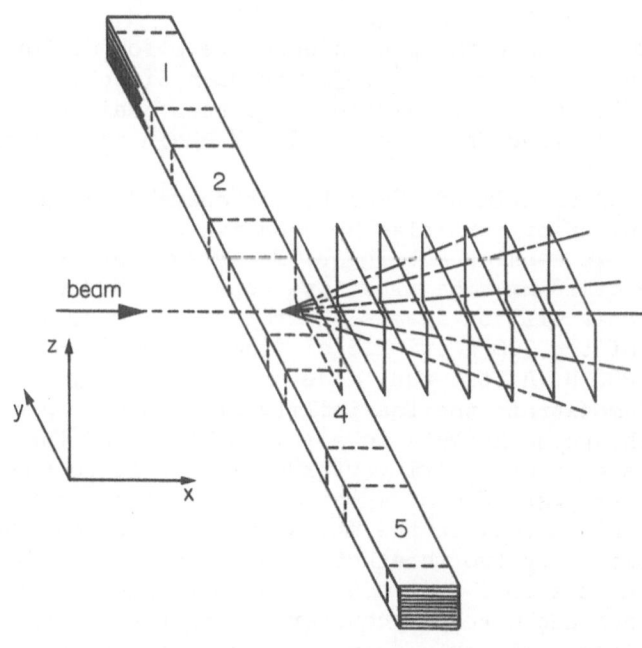

Experiment P167 — Emulsion and Vertex Detector

Figure 5

detector remains sensitive, so that tracks not simultaneous with
the trigger event have to be subtracted. Practically, the prototype
array is made of seven planes with a 1 cm spacing and it covers
8.5 x 12.5 mm with a total of 385 x 576 elements.

TRIGGER AND EVENT SELECTION

In these beauty experiments, the trigger and the event
selection have to reduce the number of interactions produced in
emulsions, typically of the order of 10^8 or more, down to a level
of a few thousands or tens of thousands events to be actually
scanned and analyzed.

In the NA19 experiment, Fig.2, the selection of events was
very restrictive, since only events with three muon candidates
possibly coming from the decay chains B → C → X and $\overline{B}$ → $\overline{C}$ → X were
looked for. The identification of muons was made after a dump of
2 m iron equipped with a tungsten core. The penetrating tracks
were reconstructed after the dump through a set of MWPC chambers:
3 x 3 planes with a 2 mm wire spacing and 6 x 2 planes with 4 mm
wire spacing and anode read-out. These chambers were interspersed
with iron absorbers. The scintillation hodoscopes placed in the
absorber provide the trigger for penetrating particles.

About 300 events with 3μ candidates were found. For every
candidate, a volume of 2.5 mm^3 (∿3σ in each direction) was scanned.
Today about 1/3rd of the candidates have been analyzed. No events
with possible vertices from the B → C → X chain were found.

In the P159 experiment, Fig.3b, the charm decay producing a
step in the number of particles between two sets of silicon
counters, will be used as a trigger. These two sets are separated
by a distance of 10 mm. The first one consists of five 100 μ thick
and 1 cm diameter silicon sheets separated by 100 μ, and the second
one consists of ten 300 μ and 2.5 cm diameter silicon sheets
separated by 200 μ. A test run, made at lower energy, was performed
in order to demonstrate the feasibility of such a trigger. The
analysis of the pulse heights in every counter, and the reconstruction of the second charm vertex will be used to enhance the
signal in the trigger. The Ω' spectrometer will be used in the
selection of candidates. In the spectrometer, the secondary tracks
will be identified by two threshold gas Cerenkov counters and by
the transition detector, from 4.5 to 45 GeV. This can be supplemented by a ring gas Cerenkov counter in the range from 45 to 80
GeV. A lead glass wall (or a scintillator sandwich wall) will
measure π^0's.

The volume to be scanned in emulsion for every candidate will
be 0.6 mm^3 and the number of candidates ≃ 30.000.

In the P166 experiment, Fig. 4b, the events will be selected by the presence of a muon with a p_T > 1.4 GeV or, in the 2 muons channel, by the presence of at least one muon with a p_T > 1.0 GeV. This was proven to be characteristic of beauty decays, as opposed to ordinary background. The transverse momentum p_T is measured behind a 2 m iron dump equiped with a tungsten core. Behind the dump, the penetrating tracks are measured in position and direction by a set of 10 drift chamber planes. The tracks are deflected by a 3 Tesla meter supraconductive magnet and measured behind the magnet by a set of 6 drift chamber planes. In addition, in order to eliminate ambiguities in the high-density region nine planes of MWPC's with a 1 mm wire spacing cover the forward cone immediately behind the dump. The events selected in that way are expected to be ≃ 9000 and of every event, a volume of 0.5 mm^3 has to be scanned and analyzed in the emulsion. A volume scan will be made, mainly for charged decays, and a scan "back" starting from the vertex detector information will be also useful for neutral decays.

In the P167 experiment, Fig. 5, the trigger is based on the result that the average number of charged K's, for $B\bar{B}$ events is 2.2. The trigger will require 3 K's in the Cerenkov of the Ω' spectrometer. The efficiency of such a trigger has been evaluated from the data to be ∿ 10%. The number of background interactions which produce this multikaonic trigger is difficult to evaluate. The hypothesis that the fraction of events with ≥ 3 K's is 1% leads to a background of a few thousands events. It is intended to study in the test run the background event rates with different kinds of multikaonic triggers.

CONCLUSION

The main characteristics of the experiments discussed in this paper have been presented and can be compared in Table II.

It is clear that, despite the presumably low cross section production and short lifetime, the new generation of experiments is designed to reach the sensitivity of the order of one well identified event per nanobarn. The nuclear emulsion well adapted to the study of particles with lifetimes in the range 10^{-14} to 10^{-12}s has been adopted by the four experiments. On the contrary, a variety of approaches characterize the triggers and the selection criterias.

There is every reason to believe that beauty will be directly observed in the next two years or so at the SPS. Assuming reasonable values for the parameters governing its production and decay, one should also be able to measure its main properties for the first time.

I would like to acknowledge very useful discussions with colleagues from the NA19, P159, P166 and P167 experiments.

TABLE II: Main characteristics of the beauty experiments at SPS.

Experiment	NA19	P159	P166	P167
Selection and analysis	3μ	Charm decay, Ω'	$1\mu\ p_T > 1.4$ / $2\mu\ p_T > 1.0$	3K in C, Ω'
Emulsion type	Horiz.	Vertic.	Horiz./Vertic.	Horiz.
Emulsion volume	50 l.	40 l.	20 l/40 l.	28 l.
Thickness	50 mm	1.2 mm	40 mm/20 mm	15' mm
Track density	1000+1000	2000	(750+750)/3.000	2000
Total no. of interactions	$2.5\ 10^8$	$2\ 10^8$	$4\ 10^8$	10^8
Candidates for scanning	300 x 30	30 000	8000	20 000
Scanning volume per event	2.5 mm^3	0.6 mm^3	0.5 mm^3	0.002 mm^3
Sensitivity $\overline{BBC}$ event/nb	0.03	0.2	1.5	0.75

REFERENCES

1. J.P. Albanese et al., An experiemnt to observe directly beauty
   particles selected by muonic decay in emulsion; CERN/SPSC 79-102,
   SPSC/P133, (1979).
2. M. Adamovich et al., An experiment for studying beauty production
   and lifetime in the upgraded Ω's spectrometer. CERN/SPSC/81-18,
   SPSC/P159, (1981).
3. J.P. Albanese et al., An experiment to observe directly beauty
   particles selected by muonic decay in emulsion and to estimate
   their lifetimes; CERN/SPSC/81-69, SPSC/P166 (1981).
4. M. Bocciolini et al., CERN/SPSC/81-72, SPSC/P167, (1981). Measurement of the lifetime of the beauty in the ' spectrometer by
   a high precision vertex detector and emulsion target.
5. S.L. Glashow, I. Illiopoulos, L. Maiani, Phys. Rev. D2 (1970)
   1285.
6. K. Niu et al., Progr. Theor. Phys. 46 (1971) 1644.
7. J.J. Aubert et al., Phys. Rev. Lett. 33 (1974) 1404;
   J.E. Augustin et al., Phys. Rev. Lett. 33 (1974) 1406.
8. A. Benvenuti et al., Phys. Rev. Lett. 35 (1975) 1199.
9. H. Deden et al., Phys. Lett. 58B (1975) 361.
10. See for ex. G.H. Trilling, LBL-12283, Feb.1981.
11. M. Kobayashi and T. Maskawa, Progr. Theor. Phys. 49 (1973) 652.
12. S.W. Herb et al., Phys. Rev. Lett. 39 (1977) 252.
13. D. Andrews et al., Phys. Rev. Lett. 45 (1980) 219;
    G. Cinocchiaro et al., Phys. Rev. Lett. 45 (1980) 222.
14. M. Basile et al., CERN/EP/81-38, Submitted to Nuovo Cimento
    Letters.
15. D. Drijard et al., CERN/EP/81-96, Submitted to Physics Letters B.
16. J. Ellis, M.K. Gaillard, D.V. Nanopoulos, R. Rudaz, Nucl. Phys.
    B131 (1977 285, see also: Fermilab Conf. 78/64 THY;
    M.K. Gaillard, SLAC Summer Institute (1978);
    R.E. Shrock and L.L. Wang, BNL-24930/PU-3043 (Aug. 1978);
    T.G. Rizzo, Phys. Rev. D20 (1979) 706;
    T. Kobayashi, Tsukuba preprints UTHEP-65, UTHEP-78
    V. Barger, W.Y. Keung and R.J.N. Phillips, DOE-ER/00881-190
    (March 81).
    V. Barger, W.F. Long and S. Paksava, UH-511.335-7a/UCN 600-881-90
    (March 81);
    B.D. Gaiser, T. Tsao and M.B. Wise, to appear in Annuals of
    Physics (1981);
    T. Kizukuri, Waseda preprint WU-HEP/81-4;
    M.K. Gaillard, B.W. Lee and J.L. Rosner, Rev. Mod. Phys. 47
    (1975) 277;
    M.K. Gaillard and L. Maiani, Cargese Inst. (1979) LAPP TH-09
    (1979).
17. T. Kobayashi and N. Yamazaki, UTHEP-65 rev.

18. B.L. Combridge, Physica Scripta vol. 20 (1979);
    J. Babcolk, D. Sivers and S. Wolfram, Phys. Rev. D18 (1978) 1902;
    T.S. Tu, TH 2653, CERN-TH (Oct. 1979);
    G.A. Ringland, workshop on the production of new particles
    Wisconsin Madison (1979);
    R. Winder and C. Michael, Liverpool LTH 5780, PO 850 (Jan.1980);
    V. Barger, F. Halzen and W.Y. Keung DOE-ER/DO 881.211 (May 1981),
    and private communications;
    Y. Afek, C. Leroy and B. Margolis, Phys. Rev. D22, 86 (1980);
    C. Peterson, G. Gustafson , private communication.
19. H. Fritzsch and K.H. Streng, Phys. Lett. 78B (1978) 447;
    H. Fritzsch, Phys. Lett. 86B (1979) 164.
20. T. Badier et al., CERN/EP/80-89, XV Rencontre de Moriond,
    Les Arcs (1980);
    A. Diamant Berger et al., Phys. Rev. Lett. 44 (1980) 507;
    F.N. Coleman et al., COO-3072-105.
21. G. Matthiae, Rivista del Nuovo Cimento, Vol. 4 n.3 (1981).
22. G. Charpak et al., Nuclear Instr. and Methods 148 (1978) 471.
23. F. Piuz et al., CERN/EF 80-11 Submitted to Nucl. Instr. and
    Methods.
24. E. Heijne et al., Nucl. Instr. and Methods 178 (80) 331.

HADRONIC PRODUCTION OF PROMPT MUONS*

Presented by R. N. Coleman

A. Bodek,[a] R. Breedon,[b] R. N. Coleman,[c] W. Marsh,[c]
S. Olsen, and J. L. Ritchie
University of Rochester, Rochester, New York

B. C. Barish, R. L. Messner,[d] M. H. Shaevitz,[e] and
E. J. Siskind[f]
California Institute of Technology, Pasadena, California

F. S. Merritt
University of Chicago, Chicago, Illinois

H. E. Fisk, Y. Fukushima,[g] and P. A. Rapidis
Fermilab, Batavia, Illinois

G. Donaldson and S. G. Wojcicki
Stanford University, Stanford, California

---

*Supported by the U.S. Department of Energy and the National Science
Foundation.

(a) Alfred P. Sloan Foundation Fellow.
(b) Present Address: CERN, Geneva, Switzerland
(c) Present Address: Fermilab, Batavia, Illinois
(d) Present Address: SLAC, Stanford, California
(e) Present Address: Columbia University, New York, New York
(f) Present Address: Brookhaven National Laboratory, Upton, New York
(g) Present Address: Laboratory for High Energy Physics, Tsukuba-gun,
                      Ibaraki-Ken 305, Japan

In an experiment performed at Fermilab (E-595) we have measured
the production of prompt single muons and dimuons in hadronic interactions. Data were taken with both 350 GeV protons and 280 GeV π^-'s
incident on an iron "beam dump" instrumented with scintillation counters. Density extrapolations were used to separate prompt muons from
muons originating from the decays of long lived particles such as π's,
K's and hyperons. Prompt dimuons were identified in a very large
acceptance muon identifier (see Figure 1). The target calorimeter
identified interactions and measured the total electromagnetic and
hadronic energy produced in the event ($\Delta E/E \sim 4\%$). The muon identifier
identified all muons produced with a laboratory momentum greater than
5 GeV/c. Muon energies below 15 GeV were determined from their range
in the muon identifier. Muon energies above 15 GeV were determined
by use of toroidal magnets placed at the downstream end of the detector. Data were taken with two different triggers. One required a
muon with $p_\mu > 8$ GeV/c (as determined from range in the apparatus)
while the other required a muon which traversed the entire muon identifier ($p_\mu > 20$ GeV/c). The $p_\mu > 8$ GeV/c trigger was used with beam
intensities which were typically 10^4/sec while the $p_\mu > 20$ GeV/c
trigger was run at intensities of $\sim 10^5$/sec. Note that only those
muons with $p_\mu > 15$ GeV/c penetrated enough of the toroids to enable a
sign determination.

Here, we report results from the $\underline{p_\mu > 20}$ GeV/c trigger only:

(1) Prompt single μ (p-Fe, ~1/2 data sample)
(2) Prompt 2μ + missing energy (p-Fe and π-Fe)

The prompt single muon results were reported earlier;[1] in this report
we emphasize the recent 2μ + missing energy results.[2,3] An analysis
of the complete data sample, $p_\mu > 8$ GeV/c and $p_\mu > 20$ GeV/c triggers
including both proton and pion induced data is in progress. In an
early test run of this experiment we found $\sigma(c\bar{c}) = 22 \pm 9$ µb/nucleon

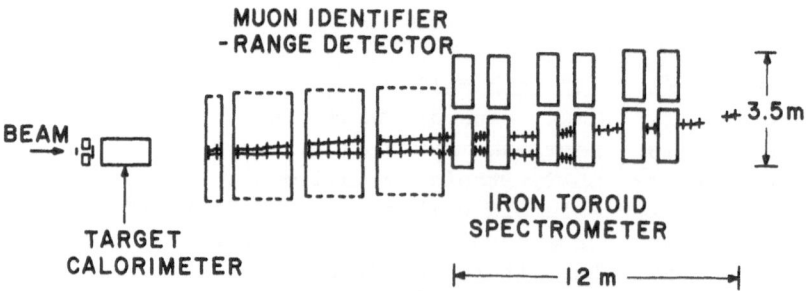

Fig. 1. Plan View of the Apparatus.

for 350 GeV p-Fe collisions using the $p_\mu > 8$ GeV/c trigger.[4] These
results are insensitive to the details of the production mechanism
(charm acceptance ≈40%). Our new data will provide a substantial
improvement in the low momentum region and also allow a comparison of
proton and pion induced prompt muon production.

Prompt Single Muon Results (350 GeV p-Fe)

In the $p_\mu > 20$ GeV/c data the μ's have traversed the toroid system and the sign of their charge is measured. In diffractive production from incident protons we expect μ^+'s from Λ_c^+ while μ^-'s
come from $\bar{D}$'s. Since in this picture we expect forward $\bar{D}$'s but not
forward D's, differences between μ^+ production and μ^- production might
be expected. Figure 2 shows the density extrapolation for the μ^+ and
μ^- sample. The steeper slope for the μ^+'s reflects the excess of π^+'s
over π^-'s in the forward direction. Note that both curves give the
same intercept at $1/\rho = 0$, corresponding to equal amounts of prompt
μ^+ and μ^- production.

There was a contamination in the single muon sample from asymmetric dimuon events because muons of momentum less than 5 GeV/c were not
identified. This background was subtracted with the aid of a Monte
Carlo calculation which was normalized to the observed number of identified dimuon events. The resulting prompt single muon distributions
versus momentum are shown in Figure 3. The curves in each figure
correspond to fits to the data using a central production model and a
diffractive model using the x distributions suggested in Ref. 5. The

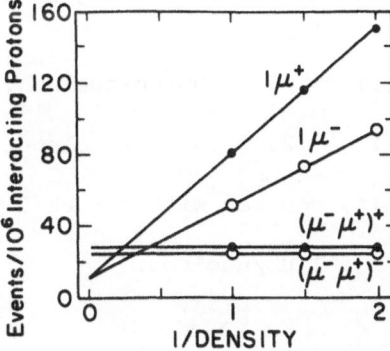

Fig. 2. Event rates vs. density for the μ^+ and μ^- data with $p_\mu >$
         20 GeV/c. The symbol $(\mu^-\mu^+)$ refers to dimuon events with
         a triggering μ^+.

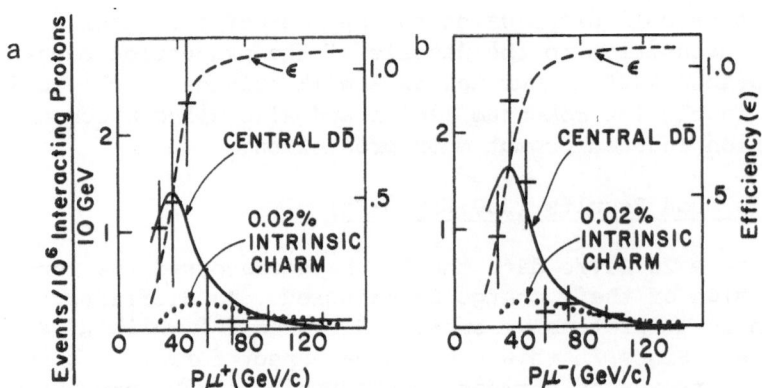

Fig. 3. Prompt single muon rates vs. momentum for μ^+ (a) and μ^- (b).
The rates are not corrected for trigger efficiency. The
dashed line is the efficiency. It can be greater than 1.0
because it includes resolution smearing effects. The solid
line is from the central $D\bar{D}$ production model.

diffractive curves are fit only to those points with $p_\mu > 50$ GeV/c.
From the central model, using

$$E \frac{d^3\sigma}{dp^3} \propto (1-x_D)^\alpha \, e^{-\beta P_T}$$

with $\alpha = 4.7 \pm 1.0$, $\beta = 2.5$(fixed), $B_\mu = .08$ and $A^{1.0}$ behavior we get

$$\sigma_{D\bar{D}} = 16 \pm 4 \ \mu b/nucleon.$$

For diffractive production we use the intrinsic charm model of
Brodsky, Hoyer, Peterson and Sakai.[5] Using this model with an $A^{2/3}$
dependence we get

$$\sigma \cdot BR(\Lambda_c \rightarrow \mu^+ + \ldots) = .27 \pm .08 \ \mu b/nucleon,$$

$$\sigma \cdot BR(\bar{D} \rightarrow \mu^- + \ldots) = .23 \pm .08 \ \mu b/nucleon.$$

Using only the μ^- result, coupled with the BR$(D \rightarrow X\mu\nu) = 8\%$ gives

$$\sigma_{\Lambda\bar{D}\,diffractive} = 3 \pm 1 \ \mu b/nucleon.$$

Comments:

i) Using an $A^{1.0}$ dependence here would <u>reduce</u> the quoted cross sec-
    tion by a factor of 3.8.

ii) This rate for diffractive production translates into a probability
    for intrinsic charm in the proton of about 0.02%. This should be
    contrasted to the 1% predicted by the proponents of this model,
    and suggested by recent ISR results.[6]

Prompt 2μ + Missing Energy (p-Fe and π-Fe)

In the p_μ > 20 GeV/c sample we find a number of dimuon events.
Of these some have a substantial amount of missing energy characteristic of parallel semileptonic charm decays with neutrinos carrying
away a significant amount of energy. These data are used both to obtain limits on diffractive charm[2] and bottom production as well as for
D^o-$\bar{D}^o$ mixing.[3]

In Figure 4, we show the total measured energy for those events
in the p_μ > 20 GeV/c trigger which contained a second muon with p_μ >
15 GeV/c. These results are shown both for 350 GeV protons and
278 GeV pions. The total measured energy corresponds to the energies
of the two muons $E_{\mu+}$ and $E_{\mu-}$, inferred from the magnetic measurements,
plus the total hadronic energy E_{had}, as determined by the target calorimeter. The solid points correspond to E_{TOT} measurements made using
randomly triggered interactions. Note an excess of events in each
sample for dimuon events on the low side of the curve. These events
are interpreted as coming from charm-anticharm production where both
the charm and anticharmed particle subsequently undergo semileptonic
decays. The energy deficiency corresponds to energy of the neutrinos.
The gaussian shape (no significant non-gaussian tails) of the curve
corresponding to unbiased triggers and the lack of an excess of events
with E_{TOT} > E_{beam} provides confidence to this interpretation.

Dimuon events with $E_{missing}$ > 45 GeV for the proton sample and
>40 GeV for the π^- sample were selected as candidates for double charm
decay. This corresponded to 59 $\mu^+\mu^-$ events for the proton sample and
154 events for the π^- sample. These numbers were corrected for possible backgrounds from π and K decay, Drell-Yan production of $\tau^+\tau^-$
pairs followed by two leptonic decays and leakage from ordinary dimuon
events. These corrections corresponded to a total of about 10% in
each data sample. (The number of same sign dimuons satisfying the
same cuts were 0 ± 1 for the protons sample and 3 ± 1.7 for the π^-
sample.

The acceptance for central $D\bar{D}$ production is small (~1%). On the
other hand, for charm production via diffractive processes a la the
intrinsic charm model of Brodsky et al[5] it is quite good. In this
model the x distributions of the Λ_c and $\bar{D}$ (with incident protons) and

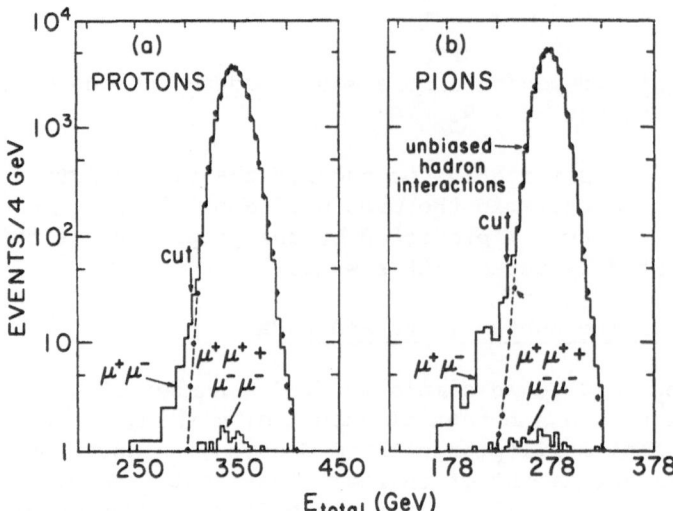

Fig. 4. Total energy distributions for accepted opposite sign and
         same sign dimuon events. The solid circles correspond to
         the observed energy distributions for unbiased hadron inter-
         actions taken simultaneously with the dimuon data; a) proton
         data, b) pion data.

D and $\bar{D}$ (with incident π^-'s) peak near x = 1/2. Here we find the
acceptance of our cuts to be about 16%. Thus those data are quite
sensitive to diffractive like production but are poorer than the
single muon data for determining $\sigma(c\bar{c})$ in the case of central production.

     Figure 5a (b) shows the distribution of events for the proton
(π^-) sample vs E_{lep}(= $E_{\mu+}$ + $E_{\mu-}$ + E_{miss} = E_{beam} - E_{had}). The dashed
curve shows the results of normalizing the Brodsky model to all of the
data and the solid curve shows the results of normalizing only to 3
points with E_{lep} > 160 GeV. Normalizing to the data with E_{lep} > 160
and assuming no central production contribution in this region, we
can estimate diffraction cross sections to be 2.5 μb for pN and 1.9 μb
for πN using an $A^{2/3}$ dependence. This corresponds to an intrinsic
charm content of the proton or pion of ~.02%. Note that an $A^{1.0}$

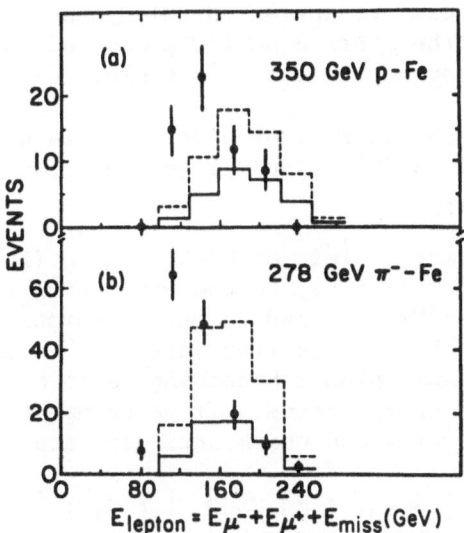

Fig. 5. The $E_{lep} = E_{\mu^+} + E_{\mu^-} + E_{miss} = E_{beam} - E_{had}$ distribution for
         a) proton, b) pion events with missing energy. The lines are
         predictions of the intrinsic charm model. The dashed line is
         normalized to the data, the solid curve is normalized to the
         data with $E_{lep} > 160$ GeV.

behavior would result in diffractive cross sections which are a factor of 3.8 smaller.

In the dimuon + missing energy sample we see 59 $\mu^+\mu^-$ and no $\mu^+\mu^+$
or $\mu^-\mu^-$ events for 350 GeV/c incident protons, and 154 $\mu^+\mu^-$ with one
$\mu^+\mu^+$ and two $\mu^-\mu^-$ events for 278 GeV/c incident pions.

Same sign dimuon events with missing energy can come from three
sources:

1) An additional π or K decay in association with a single muon
event (the single muon event can be from charm or pion decay).

2) $D^0-\bar{D}^0$ mixing, i.e., charm production followed by two semileptonic decays where in one decay a D^0 decayed to the wrong sign muon.

3) Bottom particle production in which one B particle decayed semileptonically and the other B particle decayed to charm followed by a semileptonic decay of the charm particle.

Backgrounds from π and K decay were determined experimentally to be $1.1 \pm .1$ and $1.9 \pm .9$ ($\mu^+\mu^+ + \mu^-\mu^-$) for the proton and pion data samples, respectively.

In order to extract a limit on D^o-$\bar{D}^o$ mixing (i.e., the relative probability that a D^o will decay to the wrong sign rather than the right sign muon, $P(D^o \to \bar{D}^o)$) we need to make assumptions about (a) the probabilities that a charm quark fragments to D^o's and other charm states, and (b) the semileptonic branching ratio of the D^o relative to those of the other charm states. If we assume that the fragmentations of the charm quark and charm antiquark are independent then

$$P(D^o \to \bar{D}^o) = \frac{1}{2} \left[\frac{N_{\mu^+\mu^+} + N_{\mu^-\mu^-}}{N_{\mu^+\mu^-}} \right] \left[\frac{\sum_0^3 f_i B_i}{f_0 B_0} \right]$$

where f_0, f_1, f_2, f_3 and B_0, B_1, B_2, B_3 are the fractions of D^o, D^+, F^+, Λ_c and their semileptonic branching ratios respectively. The available data on charm particle semileptonic decays are $B_0 < 4\%$ (90% CL) and $B_1 = 22.0^{+4.4}_{-2.2}\%$ from the DELCO group;[7] and $B_0 = 5.5 \pm 3.7\%$ and $B_1 = 16.8 \pm 6.5\%$ from the Mark II group.[8] Other information on semileptonic decays is available if we assume that the ratios of the semileptonic branching ratios are proportional to the lifetimes. An average of the available data on charm particle lifetimes[9] (in units of 10^{-13} sec) is

$$\tau(D^o) = 2.8 \pm 0.6, \quad \tau(D^+) = 8.1 \pm 1.4, \quad \tau(F^+) = 2.0^{+1.8}_{-0.8},$$

$$\tau(\Lambda_c) = 1.4^{+0.8}_{-0.4} \quad .$$

The ratios of the lifetimes are consistent with the data on semileptonic decays.

For the fractions f_i, $f_0(D^o) = 0.35$, $f_1(D^+) = 0.35$, $f_2(F^+) = 0.15$ and $f_3(\Lambda_c) = 0.15$ are assumed. Such production probably underestimates $f_0(D^o)$ since substantial D* production would favor D^o's over $D^\pm$'s. Assuming the above lifetimes are proportional to the semileptonic branching ratios we find an upper limit (90% CL) of $P(D^o \to \bar{D}^o)$ < 0.044.

In general, the fraction of wrong sign muons is related[10] to the difference in mass δm and the difference in inverse lifetimes $\delta\lambda$ for the CP eigenstates of the D^o

$$P(D^o\text{-}\bar{D}^o) = \frac{\delta m^2 + \frac{1}{4}\delta\lambda^2}{2\lambda^2 + \delta m^2 - \frac{1}{4}\delta\lambda^2} \leq 0.044$$

where λ is the average inverse lifetime of both D^O states. Using $\lambda = 0.36 \times 10^{13}$ sec we obtain the limit

$$\delta m \leq 1 \times 10^{12} \, \hbar \, sec^{-1} \, (= 6.5 \times 10^{-4} \, ev)$$

This can be compared with the measured K_L^O-K_S^O mass difference of $0.5 \times 10^{10} \, \hbar \, sec^{-1}$. Our data also imposes the limit $\delta\lambda/\lambda < 0.55$ on the difference in the inverse lifetimes of the CP eigenstates of the D^O.

The same sign dimuons can also be used to investigate diffractive bottom particle production in hadronic collisions. Using muon and neutrino distributions from CESR data,[11] semileptonic branching ratios for $B \rightarrow \mu X$ and $\bar{B} \rightarrow (D \rightarrow \mu X)X$ of 10% and 8%, respectively, and production distributions analagous to those for intrinsic charm,[5] we find upper limits (90% CL) for diffractive bottom production of 40 nb/nucleon for 350 GeV protons and 140 nb/nucleon for 278 GeV π^-'s. These upper limits assume an $A^{2/3}$ behavior. If $A^{1.0}$ is assumed, these limits should be reduced by a factor of 3.8.

We would like to thank Fermilab for its support. This work was supported by the U.S. Department of Energy and the National Science Foundation.

REFERENCES

1. J. L. Ritchie et al., Proceedings of the XVIth Rencontre de Moriond, Vol. II, 121 (1981), J. Than Tran Van, Editor.
2. A. Bodek et al., "A Study of Forward Production of Charm Particle Pairs in p-Fe and π^--Fe Interactions," University of Rochester preprint #UR-804, submitted to Physics Letters (1982).
3. A. Bodek et al., "Limits on D^O-$\bar{D}^O$ Mixing and Bottom Particle Production Cross Sections from Hadronically Produced Same Sign Dimuon Events," University of Rochester preprint #UR-802, submitted to Physics Letters (1982).
4. J. L. Ritchie et al., Phys. Rev. Lett. <u>44</u>, 230 (1980).
5. S. J. Brodsky et al., Phys. Lett. <u>93B</u>, 451 (1980); Phys. Rev. D23, 2745 (1981).
6. See, for example, D. Treille, these proceedings.
7. W. Bacino et al., Phys. Rev. Lett. <u>40</u>, 671 (1978); ibid <u>45</u>, 329 (1980).
8. R. H. Schinler et al., Phys. Rev. <u>D24</u>, 78 (1981).
9. For a review of charm particle lifetime see L. Foà, Lepton-Photon Conference, Bonn, 1981; K. Ushida et al., Phys. Rev. Lett. <u>45</u>, 1049, 1053 (1980); and N. R. Stanton, Hawaii Conference (1981).
10. R. L. Kingsley, Phys. Lett. <u>63B</u>, 329 (1976); R. L. Kingsley, F. Wilczek and A. Zee, Phys. Lett. <u>61B</u>, 259 (1976); L. B. Okun, V. I. Zakharov and B. M. Pontecorvo, Lett. Nuovo Cimento <u>13</u>,

218 (1975); E. Pachos, Phys. Rev. D15, 1966 (1977); F. Buccella and L. Oliver, Nuc. Phys. B162, 237 (1980).

11. C. Bebek et al., Phys. Rev. Lett. 46, 84 (1981); K. Chadwick et al., Phys. Rev. Lett. 46, 88 (1981); L. J. Spencer et al., Phys. Rev. Lett. 47, 771 (1981).

A STUDY OF CHARM PRODUCTION IN THE STRONG INTERACTIONS USING A PROMPT MUON TRIGGER

Presented by R. Ruchti

J. Bishop[3], N. Biswas[3], D. Buchholz[1], N. Cason[3], L. Cremaldi[1], L. Dauwe[3], S. Delchamps[1], R. Edelstein[2], R. Erichsen[3], C. Forsyth[2], K. Gamarnik[2], G. Ginther[2], J. Godfrey[3], D. Johnson[4], V. Kenney[3], A. Kreymer[2], R. Lipton[2], M. Huishum[1], J. McQuade[2], P. Mooney[3], R. Pemper[3], D. Potter[5], E. Rojek[3], J. Rosen[1], R. Ruchti[3], J. Russ[2], W. Sakumoto[1], M. Sarmiento[3], R. Schluter[1], W. Shephard[3], S. Sontz[1], L. Spiegel[2] and C. Winter[1]

[1] Physics Department, Northwestern University
    Evanston IL 60201, U.S.A.
[2] Physics Department, Carnegie Mellon University
    Pittsburgh PA 15213, U.S.A.
[3] Physics Department, Notre Dame University
    Notre Dame IN 46556, U.S.A.
[4] Fermi National Accelerator Laboratory
    Batavia IL 61510, U.S.A.
[5] Physics Department, Rutgers University
    New Brunswick NJ 08903, U.S.A.

Our group is studying charm production in 200 GeV/c π^- nucleon interactions via the process

$$\pi^- + Be \rightarrow C_1 + \bar{C}_2 + x$$
$$200\ GeV/c)$$

$$\downarrow \quad \downarrow \rightarrow e^-, \mu^-, K^+\pi^-, \ldots$$
$$\downarrow \rightarrow \mu^+$$

The experimental strategy is to trigger on a prompt muon from the semileptonic decay of one of the charm particles (C or $\bar{C}$), and to examine the decay of the associated state with open geometry and minimum bias[1].

The detector (Fig. 1) is a two-arm spectrometer, situated in

the M1-West beam line at Fermilab. Trigger muons are selected and
identified in an upward arm ($\theta_{vert} \gtrsim 40$ mr), consisting of a
tungsten-and-steel absorber and polarized iron (the return yoke of
the 40D48 spectrometer magnet). This arm is instrumented with
scintillation counter hodoscopes (M1, M2, M3, M4) and proportional
chambers to track the muon and measure the momentum. Particles are
detected in this arm if they have $P_T \gtrsim 0.4$ GeV/c and $p \gtrsim 5$ GeV/c
(Fig.2). By measuring the deflection of the muon in its traversal
through the polarized iron, the muon momentum is determined to
$\delta p/p \sim 20\%$. Special attention has been paid to minimize the decay
length for hadrons (π, K) in this arm by careful beam design and
absorber placement. The 200 GeV/c negative beam of $\leq 10^7$ particles
per pulse is strongly focused in the vertical plane at the target
(see Fig.3). The dimensions of the beam profile are: ±1 mm vertical
full width), ±1.5 cm (horizontal width). The upper edge of the beam
is kept 1 mm below the lip of the muon arm absorber, the first layer
of which is two absorption lengths of tungsten. In this geometry
the mean decay length is ~ 20 cm.

The forward arm of the detector, which subtends (±200 mr horizontally and -80 mr $<\theta_y< +40$ mr vertically) is a conventional magnetic
spectrometer consisting of proportional and drift chambers. Particle
momenta in this arm are determined to $\delta p/p \simeq 1\%$. Ancillary detectors
for particle identification include a 46 cell, N_2 gas Cerenkov
counter for π, K separation in the range $6 \lesssim p \lesssim 22$ GeV/c and a
liquid argon shower detector comprised of 24 radiation lengths of
lead and 800 separate readout channels for electron and photon
detection. The resolution of this latter device which contains an
0.5% admixture of methane in the argon is $\sigma(E)/E \simeq 20\%/\sqrt{E}$. Muon
identification is also provided in the forward arm by two hodoscopes
placed behind a downstream hadron dump.

Typical acceptance for the apparatus are (assuming central
production): in the upward arm, $\sim 5\%$ per muon from charm decay; in
the forward arm, $\sim 10\%$ for $K^-\pi^+$ decay of a D^o meson.

The trigger for the experiment consists of an incident beam
particle (unaccompanied by halo muons) in coincidence with at least
one detected particle which penetrates through the length of the
muon arm (BEAM · $\overline{HALO}$ · M1 · M2 · M3 · M4). The observed trigger
rate is $\sim 4 \times 10^{-5}$ triggers per interacting beam pion.

Using a 7% Be target and a typical beam flux of 8×10^6
π^-/spill yields 20 triggers per spill in our configuration. Of
these raw triggers, $\sim 32\%$ yield tracked particles in the muon arm
in the off-line analysis. (The remainder consist of upstream
showers, magnet pole-tip scatters, accidentals, etc.).

The fraction of the triggers which are from prompt muons was
determined by removing various amounts of absorber from the muon

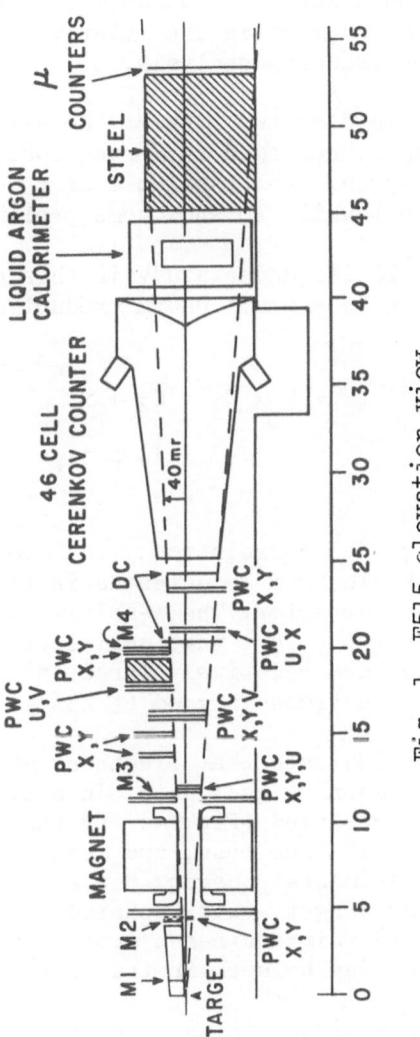

Fig. 1. E515 elevation view.

arm to increase the decay component. Using these measured rates
to extrapolate to zero decay length, it was observed that ∿20% of
the triggers are from prompts. Since the electromagnetic cross
section ($\sigma_{em} \simeq 10$ µb) and charm cross section ($\sigma \simeq 15$ µb) are
comparable at our energy ($\sqrt{s} \simeq 20$), we expect that roughly half of
the prompt muon yield is due to charm, or ∿10% of the raw triggers.
Since charm decays should lead to tracked muons in the trigger arm,
we expect a further improvement in the relative percentage of charm
after the offline muon tracking analysis.

Thus far there have been two runs of the experiment with a
total of ∿450 K triggers recorded in Spring 1980, and 2×10^6
triggers recorded in Spring 1981. Analysis of these data is in
progress, but it is worthwhile to show some preliminary results.

Of the event topologies under study in this experiment, the
reaction most sensitive to a small charm production cross section
is:

$$\pi^- + N \rightarrow C_1 + \overline{C}_2 + x$$
$$\left| \qquad \right|_{\rightarrow e^-, \ \mu^-}$$
$$\left|_{\rightarrow \mu^+, \ e^+}\right.$$

in which one looks for $\mu^{\pm}e^{\mp}$ correlations. The muon is detected in
the upward arm and the electron is detected in the forward-arm
spectrometer and shower detector. The sensitivity to small cross
sections is derived from the fact that semi-leptonic branching
ratios for charm states are relatively large and that the experimental acceptance for electrons is good ($\geq 25\%$).

Figures 4a and 4b indicate that electrons and π^0 mesons are
observed in the prompt muon trigger data. In addition, the decay
mode $K_S^0 \rightarrow \pi^0\pi^0$ is also observed (Fig. 4c) by taking pairs of identified π^0, assigning their true mass, and demanding that the K_S^0
decay occur at the experimental target. Since the decays actually
occur downstream of the target, this technique underestimates the
opening angle of the π^0 pair and hence artificially lowers the
K_S^0 invariant mass - as can be seen in the $\pi^0\pi^0$ mass spectrum.

To look for a charm signal in the µe event sample, one
compares the yield of observed $\mu^{\pm}e^{\mp}$ pairs to the yield of $\mu^{\pm}e^{\pm}$
pairs. (Charm contributes only to the $\mu^{\pm}e^{\mp}$ yield in the absence
of $D^0\overline{D}^0$ mising[2], whereas conventional backgrounds contribute to
both topologies). An excess of opposite-sign pairs over like-sign
pairs is them indicative of charm production.

The yield of µe pairs (based on the raw data sample of ∿450 K
triggers taken during Spring 1980) is presented in Table I. Also
shown are the µe yields corrected for charge asymmetry observed

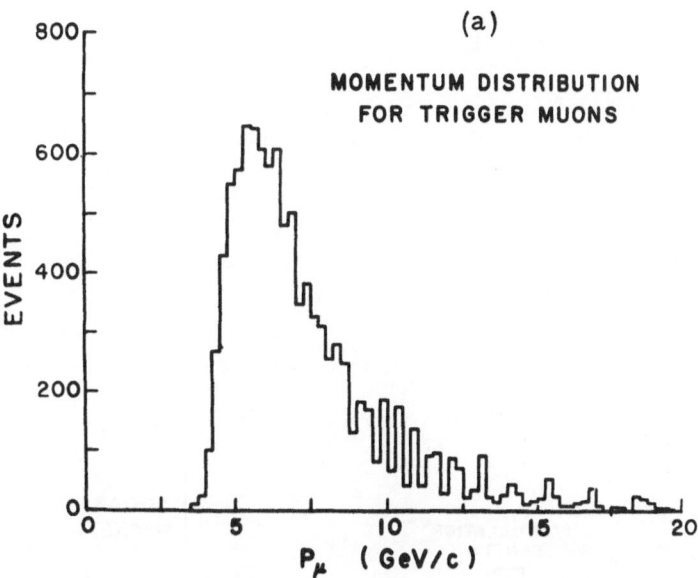

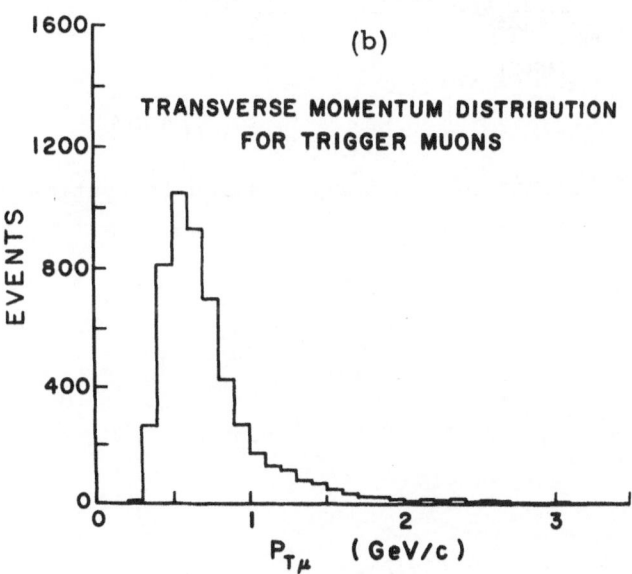

Fig. 2. Momentum and transverse momentum distri-
         butions for tracked muons in the upward
         (trigger) arm of the spectrometer.

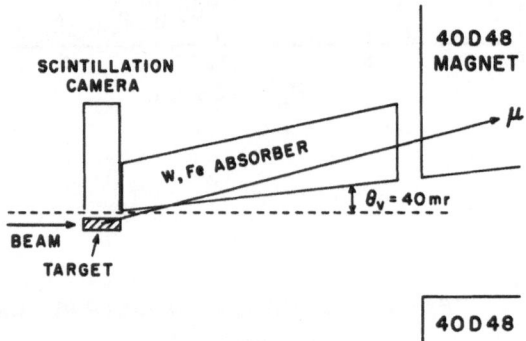

Fig. 3. Schematic of the target region (elevation view).

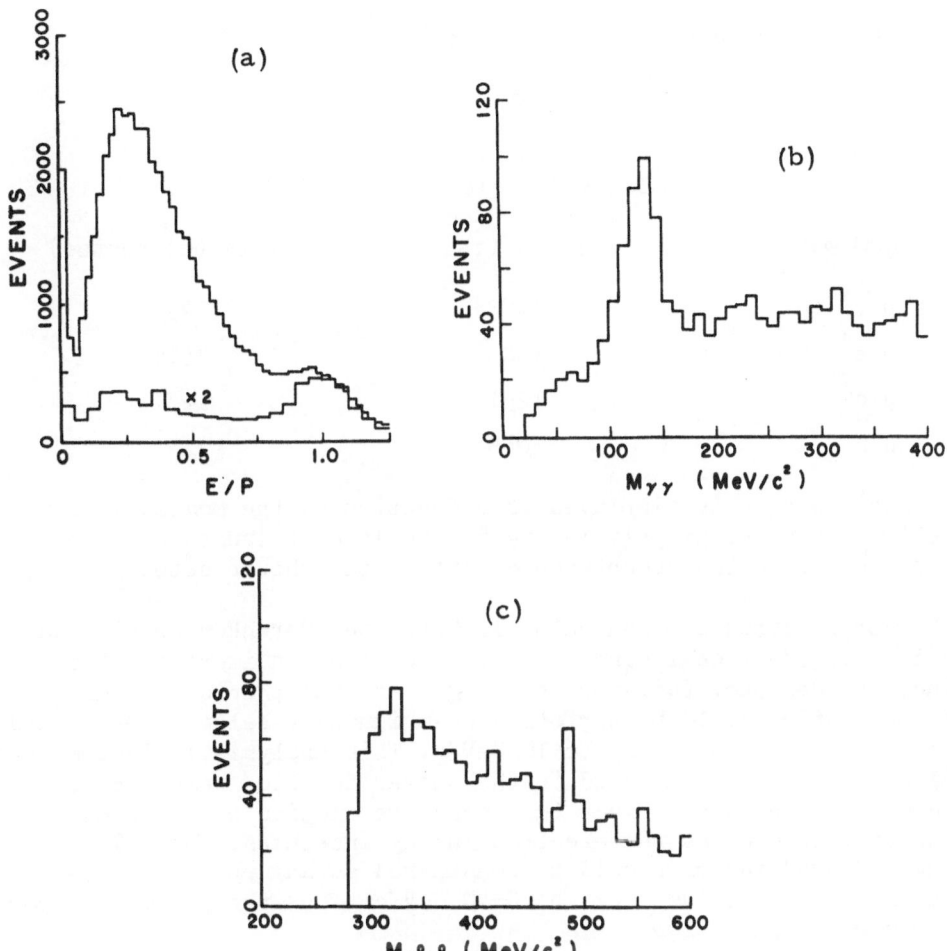

Fig. 4.   (a) The plot of E/P for particles in the forward
          arm, where E is measured in the liquid Argon
          detector, P in the magnetic spectrometer. The
          upper distribution is all charged tracks, the
          lower with cuts favorable to electromagnetic
          showers.
          (b) The γγ invariant mass spectrum.
          (c) the π⁰π⁰ invariant mass spectrum (see text).

in the events due in part to beam/spectrometer misalignment and detection inefficiencies. The excess yield of corrected, opposite--sign pairs is obtained as follows:

$$\text{Excess Yield} = N(\mu^+ e^-) + N(\mu^- e^+) - N(\mu^+ e^+) - N(\mu^- e^-)$$

$$= 534 \pm 74 \text{ events.}$$

Table I: μe Yields

Topology	Yield (raw)	Yield (corrected)
$\mu^+ e^-$	1483	1498
$\mu^- e^+$	1303	1511
$\mu^+ e^+$	1251	1251
$\mu^- e^-$	1045	1224

This excess is displayed as a function of the momentum of the electron from the μe pair in Fig.5. The lack of events below 5 GeV/c is due to the acceptance cutoff of the shower detector.

For electron momenta below 10 GeV/c the Cerenkov counter was used as an additional constraint in the electron identification. Since the Cerenkov radiator was N_2 gas at atmospheric pressure, π/e separation could be performed unambiguously below ~ 6 GeV/c and on a statistical basis up to 10 GeV/c. This analysis indicated that 50% of electrons identified in the shower detector were due to hadron "feedthrough". Since the energy resolution of the shower detector improves as the electron energy increases, the 50% "feedthrough" factor should be considered as an upperbound over the extended momentum range of 0-40 GeV/c. Thus the yield of opposite sign μe pairs is:

$$267 \pm 52 \text{ events} \leqslant \text{excess yield} < 534 \pm 74 \text{ events}$$

which translates into a production cross section of

$$16 \text{ } \mu b \leqslant \sigma_{c\bar{c}} < 32 \text{ } \mu b/\text{nucleon.}$$

Although a detailed model calculation has not yet been performed, the μe data are strongly suggestive of central production for charm.

At present, an intensive analysis of the Spring 1981 data is underway (a factor of 5 increase in statistics), to further refine the μe analysis and to search for hadronic weak decay modes of charm states. Pivotal in this analysis are good mass resolution and

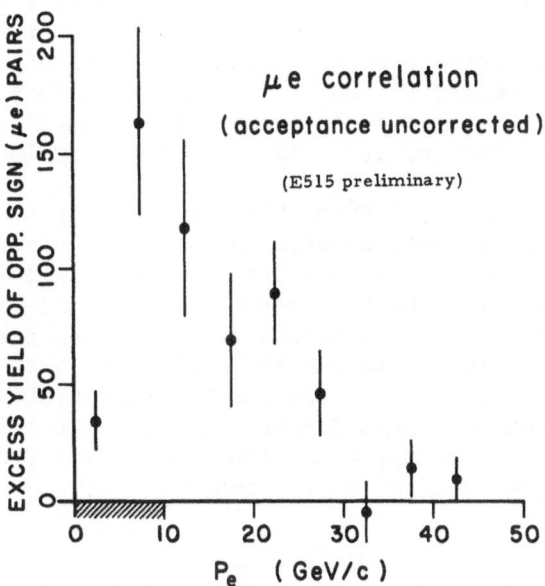

Fig. 5. Excess yield of $\mu^{\pm}e^{\mp}$ pairs as a function of the momentum of the electron of the μe pair. Acceptance for electrons deteriorates below 5 GeV/c. Cross hatched region indicates momentum range where Čerenkov constraint applies (see text).

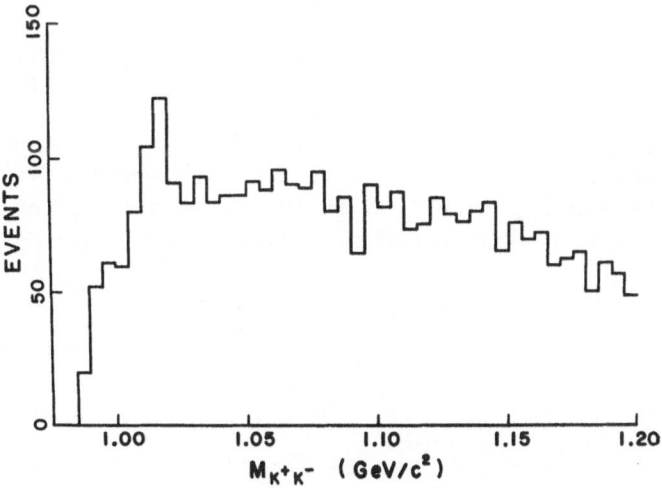

Fig. 6. The K⁺K⁻ invariant mass spectrum with both kaons identified in the Čerenkov counter.

kaon identification. An early observation in this study has been the decay $\phi \to K^+K^-$, with both kaons identified in the Cerenkov counter (Fig. 6). These ϕ mesons are observed to accompany trigger muons of either charge, μ^+ and μ^-. If we assume that these ϕ mesons are associated with the π-decay (non-prompt) component of the muon trigger, the ϕ production level is $\sigma_\phi \simeq 300$ μb/nucleon.

Finally, during the Spring 1981 run, we tested an active target with high interaction rate capability, in conjunction with the spectrometer. This device is a triggerable scintillation camera system which will allow us to observe particle lifetimes $\tau \gtrsim 10^{-13}$ sec. (For details, see D. Potters's lecture elsewhere in these proceedings)[3]. We plan to take $\gtrsim 100$ K pictures with this target using the prompt muon trigger and spectrometer system beginning in January 1982. From this data sample, we expect to obtain $\lesssim 10$ K events associable to prompt muons from charm, and from these to reconstruct several hundred charm decays.

REFERENCES

1. For further discussion of this and other trigger schemes see
   R. Ruchti, "Hadronic Charm Production",
   XII International Symposium on Multiparticle Dynamics,
   Notre Dame, USA, (June 1981), W. Shephard and V. Kenney, eds.,
   and R. Lipton, "Lepton Triggers for Heavy Quarks",
   Physics Opportunities at the Fixed Target Tevatron, (July 1980),
   G.L. Kane and N.M. Gelfand, eds.
2. $D^0\overline{D}^0$ mixing is assumed to be small (see J. Appel in these pro-
   ceedings). The present best limt is < 11% (90% CL) from E400 at
   Fermilab. (J. Butler private communication).
3. Additional discussion of the Scintillation Camera may be found
   under D. Potter, "The Scintillation Camera and the Microchannel
   Plate", Physics Opportunities at the Fixed Target Tevatron,
   (July 1980), G.L. Kane and N.M. Gelfand, eds.

TRIGGERING FOR CHARM, BEAUTY AND TRUTH

Jeffrey A. Appel

Fermi National Accelerator Laboratory
Batavia, Illinois 60510

INTRODUCTION

As the search for more and more rare processes accelerates, the need for more and more effective event triggers also accelerates. In the earliest experiments, a simple coincidence often sufficed not only as the event trigger, but as the complete record of an event of interest. In today's experiments, not only has the fast trigger become more sophisticated, but one or more additional level of trigger processing precedes writing event data to magnetic tape for later analysis. Future search experiments will certainly require further expansion in the number of trigger levels required to filter those rare events of particular interest.

VARIETY OF TRIGGER LEVELS

With the large variety of triggered devices appearing in experiments today, it is useful to class them according to the time required for a decision at each level. The time associated with each level limits the kind of hardware available to make that decision. The hardware in turn limits the kind of calculation or logic which can be economically implemented at that level. Table I shows the correlation between decision times and hardware characteristics.

In the current climate, experiments tend to emphasize one or another of the trigger levels in combination with the first, fast

Table I

TRIGGER LEVELS

DELAY OR DECISION TIME	N PER % DEAD TIME	CHARACTERISTICS
100ns = 10^{-7}s	10^5	FAST PRETRIGGER-PIPELINED DC LOGIC
10μs = 10^{-5}s	10^3	ECL HARDWIRED LOGIC PROGRAMMABLE HARDWIRED COMPUTERS
1ms = 10^{-3}s	10	BIT-SLICED MICROPROCESSOR MICROCOMPUTER
100ms = 10^{-1}s	1/10	MINICOMPUTER FILTER

pre-trigger. At Fermilab, this emphasis has been centered primarily on hardwired devices. At the CERN SPS, there has been a greater emphasis on microprocessor devices. Finally, at the various colliding beam facilities, the microcomputer system has found most favor. An excellent summary of the many devices in use today is provided in the proceedings[1] of the Microprocessor Conference at CERN. The existing trigger processors have been selected to reduce the rate of triggers such that the dead time does not seriously limit the physics of the experiment. As the search for Beauty and Truth continues, the ability of any one or two levels of trigger to achieve this feat will be reduced, and we will see the multiplicity of trigger levels growing in experiments. At the same time, the size of collaborations and the subdivision of labor on experiments is increasing. This will place a premium on the ability to maintain a coherent trigger program throughout the various levels. The program will include integration of the detector design with the trigger philosophy. This integration has been almost second nature in the case of the fast pretrigger. However, there has been an increase in the generality and related cost of higher level devices as the logical distance from the detector has grown.

In addition to the increase in trigger level multiplicity, it is possible to see now the coming increase in the local storage of data in front-end processors, a system of distributed intelligence to work on the locally stored data, and an increase in the parallelism of such processes. Each of these three developments is being discussed in detail in current designs and is basic to such explicit next generation hardware as the FASTBUS system[2].

TPS TRIGGER PROCESSOR

The Tagged Photon Spectrometer trigger processor provides an example of what can be achieved in a modular hardwired trigger system. Since it is less familiar to a European audience, some of its features and the lessons it provides will be described. As used in the Tagged Photon Spectrometer, the processor reduces the fast pre-trigger rate of 2000/second to less than 100/second. An average event requires ten microseconds for complete processor analysis and results in a 2% deadtime. Input to the processor includes three ten-bit addresses, one from each of three evenly spaced cylindrical PWC's which surround a liquid hydrogen target.

In addition, there are six channels of 8-bit timing and pulse height information from each of 15 sectors surrounding the target. Each sector has four layers of scintillator. Combining these data with similar information from the photon tagging system allows identification of recoiling protons from events with forward effective mass between 2 and 11 GeV/c^2.

The processor (with the logic indicated in Table II) begins with a hardwired "do loop" searching for "tracks" where the average of the inner and outer hit PWC wire addresses matches a hit from the middle PWC. Given such a track, a memory look up unit is used to project the track information into the first layer of scintillator. An additional "do loop" finds the sector (one of 15) in which the end-to-end signal timing corresponds to the track location. Given the appropriate sector, the two DE/DX pulse heights (7 bits each) are combined with position information (6 bits) to "calculate" an attenuation and photomultiplier saturation corrected DE/DX (5 bit output). The "calculation" consists of using the input bit pattern to provide the address in a memory whose stored data is the desired output. The information in the memory lookup modules has been precalculated and preloaded so that the results on-line are available in less than 50 nanoseconds. The DE/DX information from two layers is combined with 3 bits of track angle information to "calculate" a predicted kinetic energy (5 bits) and two bits of information, one identifying consistency with a proton input pattern, the other for a pion pattern.

Table II

TPS TRIGGER PROCESSOR LOGIC
(SIMPLIFIED)

FIND TRACK IN PWC's USING 10 BIT ADDRESSES

Do Loop: $(\text{INNER} + \text{OUTER})/2 \stackrel{?}{=} \text{MIDDLE}$

FIND SECTOR ASSOCIATED WITH TRACK

Do Loop: $Z_{\text{PROJECTION}} \stackrel{?}{=} \text{EET}_{\text{1ST LAYER}}$

→1 OF 15 φ SECTORS

CORRECT SCINTILLATION PULSE HEIGHT

LOOK UP: DE/DX ← PM Saturation Attenuation

(7 BITS) P.H. →
(6 BITS) Z → [ ] → DE/DX (5 BITS)
50NS

ESTIMATE KE AND PARTICLE ID

LOOK UP:

(5 BITS) DE/DX$_1$ → [ ] → KE (5 BITS)
(5 BITS) DE/DX$_2$ → → π (1 BIT)
(3 BITS) CSC θ → → P (1 BIT)
50NS

CALCULATE

LOOK UP:

(5 BITS) KE →
(5 BITS) TAN θ → [ ] → M$_X$ RANGE (4 BITS)
(4 BITS) E$_γ$ →
50NS

The modularity of the "do loop" controller and memory lookup modules, among others, have led to their use in an entirely different application finding tracks in a forward spectrometer in another experiment at Fermilab.

In addition to the useful modularity of the TPS system, the division of effort between arithmetic units and memory lookup devices is worth noting. Even with the low prices of highly integrated memory chips, the calculations involving numbers requiring 10 bits are more appropriate for arithmetic units (e.g., the track calculation). On the other hand, when a smaller number of bits is sufficient for a number of parameters, even complicated transcendental functions can be precalculated for use in memory lookup devices (e.g., the identification of DE/DX patterns consistent with a recoiling proton). The ability to combine both types of techniques in one system is one of the strengths of the TPS trigger processor.

A final lesson of the TPS experience is that the integration of the apparatus design with the details of the trigger processor capability save effort and cost. The equal spacing of PWC layers in the tracking apparatus is one example of this. However, one less obvious example which is even more important is the uniformity of similar information channels. For example, the attenuation and gain of the 15 scintillator—photomultiplier—ADC combinations in each layer are expected to be the same. The equality of the ADC gain times photomultiplier times light output for each channel must be established early and maintained throughout the course of the experiment. This places a premium on the understanding and the stability of the detector from the earliest possible time.

TRIGGER POSSIBILITIES FOR C, B, AND T

Quite a number of triggering requirements have been proposed in the search for Charm, Beauty and Truth. A nearly complete list of such trigger requirements appears in Table III. Those which have already been used in experiments are indicated with an asterisk in the Table. A number of the trigger requirements which have yet to be used deserve consideration for the future. Many of them become even more effective as the mass of the heavy quark increases. As indicators of Beauty and Truth, they may become more attractive than they have been heretofore.

An example of the mass dependence of the trigger requirement is the minimum momentum transfer, t_{min}, needed for diffractive produc- tion of a heavy forward

Table III

TRIGGER POSSIBILITIES

A. TARGET RECOIL
  1. HIGH FORWARD MASS
    * A. MISSING MASS À LA TPS
    B. $T > T_{MIN}$ OR RECOIL KE $> KE_{MIN}$
  2. COHERENCE OF SCATTERING FROM NUCLEUS
    A. FOR LATER RECONSTRUCTION
    B. PRIMAKOFF TRIGGER FOR $\eta_{C,B,...}$

B. DECAY PRODUCT
  1. HIGH P_T
    * A. LEPTONS
    B. CHARGED PARTICLES
  2. DECAY CHAIN
    A. $K^{\pm}$ IN CERENKOV/MOMENTUM CORRELATION
    B. K^0_S, Λ^0 - DOWNSTREAM V OR ΔQ
    * C. $\pi^{\pm}$ FROM $D^* \rightarrow D$
            $B^* \rightarrow B$
    D. ΔQ'S NEAR TARGET (E.G., CHARM TRIGGER FOR BEAUTY)
    * F. ν - MISSING ENERGY

C. EVENT TOPOLOGY
  *1. MULTIPLICITY
  2. $\Sigma|P_T|$, ΣP_T^2, ΣW^2 FROM CALORIMETERS

mass, M.   The minimum momentum transfer is directly related to a minimum kinetic energy for the recoiling target particle of mass, m_r. For example, a minimum kinetic energy of 30 MeV is required for the pair photoproduction of naked Beauty with a hydrogen target.   The minimum energy requirement is inversely proportional to the atomic weight of $(m_r = A)$ a nuclear target. Such minimum kinetic energies are easy to detect in real time, either by placing DE/DX counters around the target or by use of active target elements.

The opposite requirement of no measureable recoil can be used when the Primakoff effect is the sought after production mechanism. This is a potentially valuable technique for observing the photoproduction of η_b, assuming the two photon decay is large enough.

High transverse momentum has been a popular trigger for Charm, especially leptons with high p_T. However, this type of trigger may become less useful as one goes to higher mass and a typically larger number of decay products, n_d. The typical transverse momentum for a particle of mass M is M/n_d. On the other hand, if one can add the transverse momentum of all of the particles from a given parent or even all the particles produced in the interaction, there will still be a dependence on the effective mass of such a forward going state. Calorimeters can be used for such purposes with relative ease.   A signal approximately proportional to transverse momentum can be obtained by weighting individual calorimeter elements in a sum by their distance from the beam line.   It is also possible to consider non-linear expressions of this type.   A signal proportional to p_T^2 is of interest here, for example. It is trivial to have the weighting proportional to the square of the distance from the beam line. The requirement of a signal proportional to the square of the energy deposit is less trivial.   However, an amplifier[3] whose output is proportional to the square of the input voltage placed in the summing circuit of a trigger system can easily produce signals proportional to p_T^2. One more elaborate quantity might be the square of the effective width of showers produced from a single parent. Such a term would be proportional to the mass squared of the parent particle.   Although such techniques may become popular with physicists studying jets, their applicability in the search for Charm, Beauty and Truth should not be overlooked.

Finally, we seem to be at the threshold of the use of active targets for observing changes in ionization corresponding to the charged decay of heavy particles. Such techniques are difficult to implement for particles with very short lifetimes because of the combined effects of Landau fluctuations of ionization and nuclear breakup products masking the step in ionization near the interaction point.   On the other hand, the existence of a coherently recoiling nucleus may be useful in identifying events for which it will be possible to reconstruct tracks near to the interaction point, either in a next level of trigger processing or off-line.

CONCLUSION

In the past, thinking has tended to concentrate on a single triggering concept. Any number of the trigger possibilities listed in Table III could be rejected on such a basis as being insufficiently sensitive to the decay of interest. On the other hand, as multiple levels of triggering become more prevalent, they may be very useful in reducing the throughput to a higher level of trigger processing. Thus, the deadtime for the experiment may be kept at an acceptable level. Without enhancing the signal of interest, one trigger stage may be very useful in suppressing background. The design decision about trigger processing depends on the signal over background improvement at each stage of the process, consistent with deadtime and total number of events of interest. Thus, the future is likely to bring a greater variety of trigger requirements as well as trigger levels in each experiment.

REFERENCES

1. Proceedings, Topical Conference on the Application of Microprocessors to High-Energy Physics Experiments, CERN, May 4-6, 1981; CERN 81-07.

2. FASTBUS Modular High Speed Data Acquisition System for High Energy Physics and Other Applications, Tentative Specification, U.S. NIM Committee, 1980. A final specification is expected in the summer of 1982.

3. Cordon Kerns, Fermilab Physics Section, has prototyped such a device.

MUON TRIGGERS IN SEARCH FOR CHARM AND BEAUTY IN HYBRID EMULSION EXPERIMENTS

Giorgio Romano

Istituto di Fisica "G.Marconi" - Università di Roma
Istituto Nazionale di Fisica Nucleare - Sezione di Roma
Piazzale A.Moro, 2   00185 Roma (Italy)

INTRODUCTION

The estimated production rates of new flavours in the interaction of high energy (350-400 GeV) hadrons in emulsion are

$C\bar{C}$ 1/200   interactions

$B\bar{B}$ 1/200000 interactions

whilst, in order to maintain the scanning effort within a reasonable level, but also to have good and controlled efficiency and to avoid background problems, it is desirable to have finding rates not exceeding ~1/100 in the case of pioneering, low statistics experiments, and ~1/10 in refined, high statistics experiments. Hence, emulsion experiments in this field must be hybrid and a selective trigger has to be applied.

Among the possible on-line and/or off-line triggers, those based on the presence of one or more muons seem particularly promising due to the sizeable branching ratio of the new flavours into leptons and to the good selection power against background. An obvious drawback is the presence of an undetected neutrino in the final state, and this prevents any univoque mass assignement; however, a topological evidence could be sufficient if the trigger is selective enough, and, in hadroproduction, one particle of the pair is unbiased in the case of a single muon trigger.

MUON TRIGGERS

The easiest way to trigger on muons (even on-line) is to place a dump behind the target: as close as possible in order to avoid π's and K's decays. The dump introduces a momentum cut-off that could be as low as few GeV/c; a magnetic analysis of the particles surviving behind the dump allows a momentum measurement, and the transverse momentum with respect to the beam can be also determined (typical error down to $\Delta p_\perp \sim 0.1$ GeV/c, mainly due to multiple scattering).

In the search for charmed and beauty particles, different kinds of triggers can be reasonably envisaged: the easiest require just the presence of a certain number of muons (up to two for charm, up to four for beauty); the addition of a transverse momentum cut-off further enhances the signal with respect to background.

The following calculations are mainly based on the results obtained with the dump used in the experiment NA19 at CERN and scheduled for P166. It consists in 2 m of iron with a tungsten core, followed by several iron modules interlayed with suitable detectors (scintillator hodoscopes, wire chambers, etc.); momenta will be determined with a superconducting air gap magnet of 3 T·m bending power placed between two sets of drift chambers.

SEARCH FOR CHARMED AND BEAUTY PAIRS

Background triggers are due to muons produced in the primary interaction (mainly Drell-Yan) or resulting from short lived particles (charm decays are a source of background in a search for beauty) or from long lived particle decays (mainly π's and K's). The last source is minimized by approaching the dump as close as possible to the target, but the need of some kind of analysis makes it difficult to leave less than about 0.5 m free space between target and dump.

Table 1. Trigger selection, loss of signal and background as a function of the number of muons.

n° of μ's	T	F (charm)	R	F (beauty)	R
0	1	1	200	1	200000
1	10^{-2}	7×10^{-2}	30	10^{-1}	20000
2	2×10^{-4}	2×10^{-3}	20	1.5×10^{-2}	2000
3	2×10^{-6}	–	–	1.2×10^{-3}	300
4	3×10^{-8}	–	–	3.0×10^{-5}	200

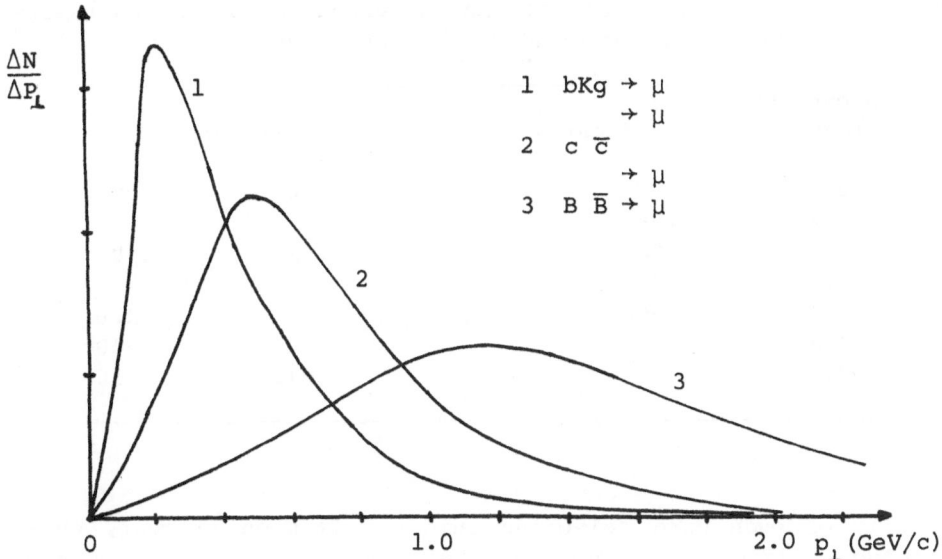

Fig.1.   Sketch of the transverse momentum distribution of muons
          originating from background, charmed and beauty particles.

     Table 1 shows the order of magnitude of the trigger selec-
tion T, of the accepted fraction of production cross section F
and of the background to signal ratio R as a function of the
number of detected muons. The acceptance has been included,
whereas in the beauty signal the further requirement of at least
one chain B → C was added in order to minimize topological background. A trigger is defined as (beam) + (interaction in target)
+ (required number of muons). The following branching ratios
were used: charm →μ (average): 10%; beauty →μ : 15%; beauty →
charm: 65%. It is seen that for charmed particles the value of R
is acceptable for μ and there is no real gain when going to
2 μ's because then F decreases very much. As background in these
conditions is not expected be a problem, even on a topological
base, it could be possible to perform a systematic study of
charmed particles, in particular to derive their semileptonic
branching ratios.

     For a beauty search R is suitable only in the 3 or 4 μ's
channel, but the present estimate of the production cross
section would make the signal marginal (if any) even in an
experiment aiming to collect (10^8 to 10^9) interactions.

     As stated above and sketched in Fig. 1, charmed and beauty
particles produce, on average, muons with much higher transverse
momenta than background, and thus a trigger requiring a low
number of muons (1 or 2) with a suitable $p_\perp$ cut-off could be

Table 2.   Fraction of signal and background on charm and beauty
            as a function of $p_\perp$ in the $1\,\mu$ channel.

$p_\perp$ cut-off (GeV/c)	F (charm)	R	F (beauty)	R
0.	.07	30	.10	20000
0.3	.06	12		
0.5	.04	6	.09	2000
0.7	.023	5.5		
1.0	.006	5.2	.07	220
1.5			.04	40
2.0			.012	30

equally or even more selective than a multimuon trigger, while
keeping a larger fraction of the signal. Of course, a more
sophisticated detection and measurement apparatus is needed in
this case.

Table 2 shows F and R as a function of $p_\perp$ in the $1\,\mu$ channel
for both charmed and beauty particles. It is seen, for instance,
that in a charmed particle search a cut-off at $p_\perp > 0.5$ GeV/c
would give an absolute rate comparable with that obtained with
high energy photons and a finding rate better than that observed
with neutrinos. A suitable cut-off for B particles is in the
range of $p_\perp$ between 1 and 1.5 GeV/c.

Further improvements, like those proposed in P166, are
still possible, namely

a)   an angular cut-off against small emission angles, which can
     be applied on-line (with $\alpha > 30$ mr the trigger rate is
     expected to decrease by a factor ~ 3, while the beauty signal
     with $p_\perp > 1$ GeV/c would remain almost unchanged);

b)   a $p_\perp$ cut-off on multimuon events (2 μ's with largest $p_\perp > 1$
     GeV/c would increase the number of beauty events by $\sim 20\%$).

POSSIBILITY OF TRIGGER FOR SHORT LIVED STATES

M. di Corato

Istituto di Fisica dell'Università and Sezione INFN

Milano

The subject of this talk is the possibility of using for the trigger the change in secondary track multiplicity which is due to a multi-charged decay of one of the particles produced in the interaction. The process under study is the associated production of two charmed states one of which has a decay length of the order of 1 cm, the other decay length being shorter. The target used is a silicon detector telescope, made of 10 layers 300 μ thick put at a distance of 100 μ.

We accept all interactions produced in the first 7 sheets. We measure the multiplicity through the ionization produced in the last 3 sheets (with the rejection of the highest one in order to reduce the fluctuations). Then we measure the ionization at a distance of two "long" decay lengths (2 cm) with a second telescope of three silicon sheets. We select the events where the second multiplicity is at least 1.5 units higher than the first.

It has been shown with a Montecarlo simulation that, for a detector noise of 20 keV (standard deviation) and interactions with charged multiplicity 2 or 4, the acceptance is 90% for events where at least one of the two charmed states decays between the two telescopes with a change in charged multiplicity of 2 or more, The problem is how big is the background.

There are two main background sources:
1) secondary interactions
2) γ conversion.

1) The probability of interaction of one particle in 300 μ Silicon is 300 μ/30 cm = 10^{-3}. If we take 3 particle interactions and 3 silicon sheets, this probability becomes 1%. If we accept only coherent interactions (with a veto counter around the target) it becomes ∿ .5%.

So for coherent interaction the background from interaction

is about

$$\sigma_{\text{inter BKGR}} \sim 10 \text{ mb} \times .5 \times 10^{-2} = 50 \text{ } \mu b$$

2)   For each π^0 produced in the interaction the probability of γ conversion in 3 silicon sheets is $2 \times 3 \times 300 \text{ } \mu/10 \text{ cm} = 2\%$.
     So for each π^0 the background is

$$\sigma_{\pi^0 \text{ BKGR}} = 10 \text{ mb} \times 2 \times 10^{-2} = 200 \text{ } \mu.$$

The conclusion is that the trigger through the change in multiplicity cannot be the only trigger. However added to other requirements (K and/or μ's) it can give a substantial contribution to the selection of the events.

TRIGGER ON CHARM AND BEAUTY FOR A HYBRID EXPERIMENT IN THE Ω' SPECTROMETER

L. Rossi

Istituto Nazionale di Fisica Nucleare and Istituto di
Fisica dell'Università, Genoa, Italy

INTRODUCTION

This paper reports on the trigger strategy that will be applied
in the WA71 experiment at CERN. Calculations on the expected signals and backgrounds are presented together with experimental results on the selection criterion.

SUMMARY OF THE EXPERIMENTAL METHOD

Our idea[1] is to use the knowledge of the charmed particles
lifetimes in order to select them.

A specially designed silicon counter detector[2] allows a selective trigger on charm decays. Using this detector (see Fig. 1), the
multiplicity of an event produced in a high-energy interaction is
measured at two distances, say $X_1 = 1$ mm, $X_2 = 10$ mm, from a thin
sensitive target (i.e. nuclear emulsion perpendicular to the incident beam). If the second multiplicity value is increased by 2,4
or more with respect to the first one, then a decay may have occurred
inside the X_2-X_1 region, the event is triggered and will be searched
later in the emulsion target.

The information on the charged multiplicity is obtained as follows: the total amount of energy released by the secondaries of a
high-energy interaction in two ion-implanted silicon telescopes
[T1: 5 elements each one 100 μm thick, 100 μm spaced, $\emptyset$ 10 mm and
T2: 10 elements each one 200 μm thick, 200 μm spaced, $\emptyset$ 23 mm] is
measured via a set of low noise fast amplifiers[3]. The comparison

of the energy released in T2 and in T1 (available within 200 nsec
from the occurrence of the event) is the main ingredient of the trigger; the information of the energy released in each silicon sheet
is stored and can be used in the analysis stage to improve the selection criteria.

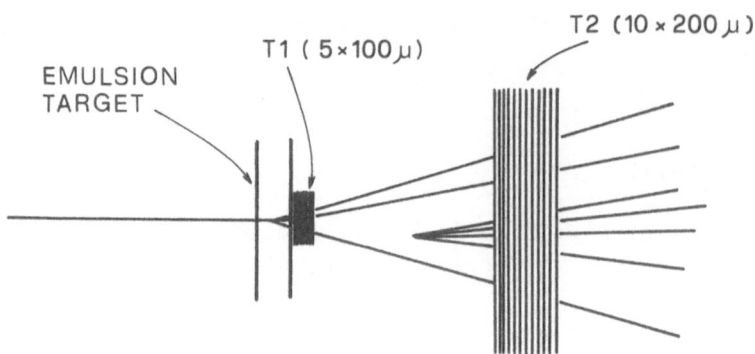

Fig. 1: Sketch of the silicon telescope system described in the
        text

The telescopes system we designed is then able to trigger on short-
lived (mainly charmed) particles decaying in the empty space between
T2 and T1. This opens up the possibility of extracting the Beauty
particles yield from the hadronic background looking at the dominant
(B_r > 50%) decay mode

$$B \rightarrow C + X$$

with a consequent reduction of the signal over noise ratio from
$\sim 10^{-5}$ to $\sim 10^{-3}$.

EXPERIMENTAL RESULTS ON THE METHOD

To check this idea we performed a test experiment[4] in Ω', with
a 35 GeV/c unseparated hadron beam hitting a 1 mm thick Cu target
in front of the telescopes. The accuracy in multiplicity determination of the silicon telescopes has been measured[5] comparing the
pulse height distribution of the telescopes for various multiplicities as counted in the Ω' spectrometer after the track reconstruction procedure (see Fig. 2).

These data fit the predictions of the Monte Carlo calculations
and let us be confident in our capability to extrapolate from
35 GeV/c to 350 GeV/c incident momentum.

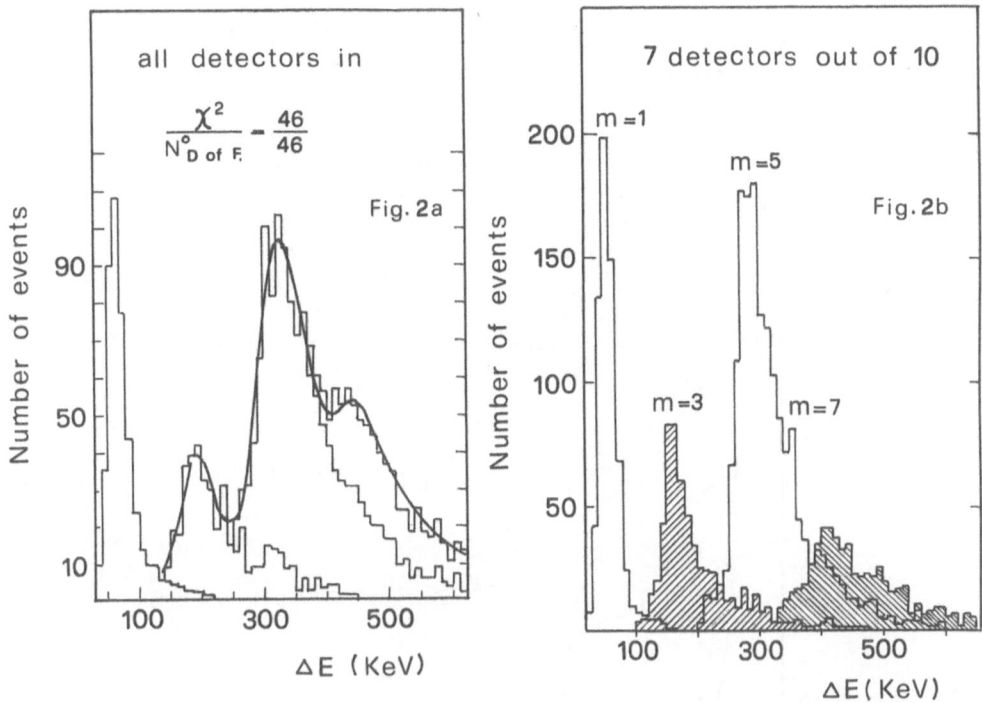

Fig. 2: a) T2 pulse height mean value distribution for multiplicities 1, 3, 5, 7 as measured in Ω'. b) The same distribution after the extraction of the three highest pulses of each event (corrected mean).

BEAUTY AND CHARM EFFICIENCY EVALUATION AND NOISE SOURCE ANALYSIS

We made a detailed Monte Carlo study of the telescope behaviour when a high momentum (π^-: 350 GeV/c) particle interacts in the emulsion target in front of the telescopes (see Fig. 1). We started from the experimental Landau distribution of the energy released by a minimum ionizing particle (m.i.p.) in each silicon sheet[5], we considered the multiplicity and angular distribution (as measured in similar conditions in an emulsion experiment) for minimum ionizing, gray and black tracks and we, finally, took into account the various noise sources (electronic noise of the amplifiers, K_S^0 and Λ^0 decays, γ conversions, etc.).

We optimized the telescope geometry for charm detection and we studied the signal and noises versus the difference in multiplicity as measured in T2 and T1. The results are shown in Fig. 3.

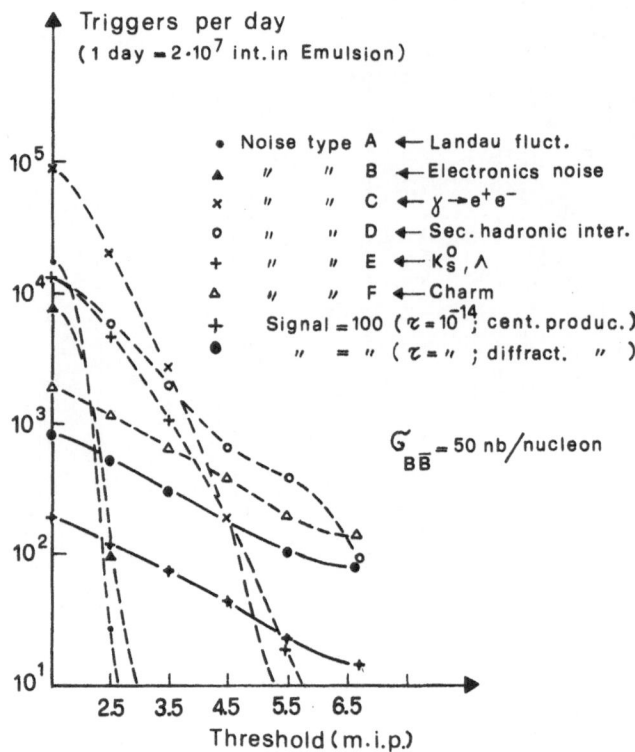

Fig. 3:   Beauty signal and noises dependence on threshold between
          the energy released in 7 out of 10 elements of T2 and in
          5 (out of 5) of T1 telescope. The threshold is expressed
          in minimum ionizing particles mean energy release.

The B curve takes already into account all the various reduction
factors (spectrometer efficiency, matching efficiency, etc.).

    Some of the noises contributions can be identified and strongly
depressed in the analysis stage (typically the secondary hadronic
interactions and the K_S^0, Λ^0 decays), some are less sensitive even
to a more refined analysis (typically the γ conversions). Finally:
$B\bar{B}$ efficiency $\sim 1\% \rightarrow \sim 15$ $B\bar{B}$ events unambiguously identified for
the full data taking period (2.0×10^8 interactions in emulsion);
$D\bar{D}$ efficiency $\sim 1\% \rightarrow \sim 10^4$ $D\bar{D}$ events $\rightarrow 1.5$-2.0×10^3 charm particles
decaying in emulsion $\rightarrow \sim 500$ charm particles fully reconstructed.

    This sample should be searched for amongst 3-5 $\times 10^4$ stars in
emulsion.

CONCLUSION

An experiment[6] will be run at CERN in 1983 to look for beauty particles hadroproduced in emulsion. The trigger selection will be based on the B → C chain and on the subsequent C decay. The signature for a C decay will be a "multiplicity jump", that should be detected by two silicon telescopes appropriately designed.

REFERENCES

1. G. Diambrini-Palazzi and L. Rossi, CERN/SPSC/79-92, SPSC/I 123, Part I, September 1979.
2. S. Benso et al., "Silicon telescopes as charm decay detectors", submitted to Nuclear Instruments and Methods.
3. M. Artuso et al., "High rate pulse processing systems for silicon active targets", 10th Int. Symp. on Nucl. Electr., 10-16 April 1980, Dresden, DDR.
    P.F. Manfredi, these proceedings.
4. CERN-Genoa-Milan-Moscow Collaboration, CERN/SPSC/80-02, SPSC/P 137, January 1980.
5. CERN-Genoa-Glasgow-Lancaster-Manchester-Milan-Moscow-Rutherford-Sheffield Collaboration, CERN/SPSC/81-17, SPSC/M 281, February 1981.
6. CERN-Genoa-Milan-Moscow-Rome-Paris Collaboration, CERN/SPSC/81-18, SPSC/P 159, February 1981.
    See also, P. Musset, these proceedings.

CONCLUSIONS

L. Montanet

CERN
1211 Genève 23
Switzerland

The aim of these conclusions is not to attempt a complete review
of all the talks given at this very active, very informative, very
fruitful workshop, but to present, in a somewhat subjective way, a
viewpoint on the highlights. The topics I have selected in these
conclusions are those which have generated the most interesting
discussions during this workshop and can be related in an obvious
way to the contributions reproduced in these proceedings. For simplicity, I shall not give explicit references and shall not reproduce
the diagrams, figures and histograms which can be found in this book.

Needless to say, the study of the properties of heavy flavours
requires new and sophisticated techniques. We are far from the "two-
body problem", which is the only problem we know how to solve well,
both theoretically and experimentally. Except at the e^+e^- collider
machines, these heavy flavour particles are produced in complex
interactions, and with small cross-sections (typically 10%, 1% and
0.1% of the total cross-sections, for lepto, photo and hadro-
production, respectively). These heavy flavour particles decay into
several channels, with no dominant, typical one, and therefore do
not offer an easy way for their selection. Typical life-times for
the ground states are of the order of 10^{-13}s, leading to mean
decay lengths of the order of a few hundred microns. New techniques
have been developed: high-resolution streamer and bubble chambers,
live-targets making use of semiconductor detectors, as developed,
in particular, by the Milano school. One problem is to develop
these new high-resolution detectors to "see" the decay vertex of the
charm or beauty particle and get a clean sample of events. Another
problem is to use these high-resolution techniques to trigger on

573

the decay by detecting a decay product which misses the production
vertex by a small quantity (the impact parameter). It may well be
that two techniques will have to be combined: an emulsion or a
holographic bubble chamber, to see the decay vertex, the pictures
being selected by a set of semiconductor detectors forming a high-
resolution trigger.

The most elusive heavy flavour remains the t quark.

The results, at Petra (for M_t < 17 GeV) are very eloquent:
the ratio $R = \sigma_{tot} / \sigma_{\mu\mu}$ gives negative results. No narrow structures
are observed (with $\Gamma_{ee}^{\mu\mu}$ > 0.61 KeV) the event shape is consistent
with no t (< T > = 1, instead of 0.65 if t were present), no direct
leptons indicative of the t are observed. However, results from
CESR put the topless models in trouble: no excess of baryon or
dilepton is observed in the b-decay, as some topless models would
require. It seems clear that we have to wait, here, for higher energy
e^+e^- colliders. Meanwhile, CESR is operating as a beauty factory,
yielding 250 B a day! The trick is to set the beam energy in order
to produce the $\gamma(4s)$ which corresponds to the $B\bar{B}$ threshold, as its
width indicates (Γ_{obs} = 19.1 $\pm$ 1.8 MeV, $\Gamma_{res} \sim$ 10 MeV).

Then one observes an abnormal multiplicity

$$n_{ch} = 11.6 \pm 0.4 \ (8.2 \pm 0.4 \text{ for the continuum}),$$

an event shape which is different of the shape of the continuum.
It tells us that the mass of the B-meson is

$$M(B) = (5257 \pm 8) \text{ MeV}.$$

With this B factory, CESR has already produced an impressive
series of results on the properties of the B:

$$B \rightarrow \mu\nu X \quad : \quad 0.100 \pm 0.013 \pm 0.021$$

$$B \rightarrow e\nu X \quad : \quad 0.136 \pm 0.021 \pm 0.017$$

$$b \rightarrow s / b \rightarrow u \text{ large, etc.}$$

It seems very hard to complete with this factory.

However, the e^+e^- machines cannot produce easily heavy flavour
baryons, and to study the Λ_c and the Λ_b, it may be wise to fall
back on the hadron interactions. Indeed, observations made at the
ISR at CERN have lead to the conclusion that the charm and beauty
production may be large at high energies ($\sqrt{s}$ = 60 GeV). Without going
into a detailed discussion, let me underline the large difficulties

one encounters to extract from complex interactions a signal which
may represent only a small fraction, with an essentially unknown
branching ratio, of the charm or beauty particle. By necessity, one
is forced to look through a narrow phase space window. For instance,
to look for a Λ_b signal, one of the ISR collaborations has shown that
one must perform the following cuts:

$$p_t \ (e^+) \ > 0.8 \ \text{GeV (to enrich the signal in charm and beauty)}$$

$$|x|_{proton} > 0.32 \qquad \text{(to privilège the baryon decays)}$$

$$|y|_{pK\pi\pi} \ > 0.4$$

$$n_{ch} \geqslant 4$$

$$K^-\pi^+ \ \text{in "D"}$$

With all these cuts, and knowing so little on the branching
ratios and the production mechanisms, it is hard to evaluate the
inclusive charm or beauty cross-sections.

No doubt that charm is observed at the ISR (D^o, D^+, Λ_c) and
that the observations form a relatively coherent picture, with a
remarkable flat x distribution for the Λ_c and a more central
production for the D's. It may even be the best machine to study
the Λ_b if the cross-sections are too small at $\sqrt{s}$ = 20 GeV and the
event complexity too large at $\sqrt{s}$ = 540 GeV, but to estimate the
cross-sections, it seems more reliable, for the time being, to refer
to the measurements made on the e/π ratio. These measurements, made
by several groups, indicate that the ratio e/π is an order of magnitude larger at $\sqrt{s}$ = 60 GeV than at $\sqrt{s}$ = 20 GeV. Interpreted in
terms of charm and beauty cross-sections, they yield

$$\sigma_c \sim 300 \ \mu\text{b and}$$

$$\sigma_b \sim \ \ 4 \ \mu\text{b.}$$

Even if these evaluations are smaller than some others, made
from the observations of peaks in mass spectra, they are still
considerably larger than those made at SPS-Fermilab energies, which
cluster around 30 - 50 μb. None of the current models for hadro-
production gives a good interpretation of these observations, and
it will be probably necessary to introduce several mechanisms to
give simultaneously a good picture of the cross-sections and
longitudinal momentum distributions observed from SPS to ISR
energies. In this respect, interesting results have already been
obtained on the correlations between the charm particle produced

and between the charm particle and incident particle quark content
(leading particle effects).

Another field of physics discussed at the workshop was the decay
of the charm particles. It was first believed that all the charm
particles have the same life time, with the dominance of the quark
spectator diagram. The first experimental results did not confirm this
prediction, and gave indications in which the life-time ratio between
$D^{\pm}$ and D^{o} could be as large as 10/1. New results, although not all in
perfect agreement, indicate that the truth may be somewhere in between
with a ratio $\tau(D^{\pm})/ \tau(D^{o})$ close to 3 (and still with large errors,
$\pm$ 1.5). The "world" average life-times, after this workshop, are the
following:

$$\tau (D^{o}) = 2.5 \pm 1.0 \quad \times 10^{-13} \text{ s}$$

$$\tau (D^{\pm}) = 9 \quad \pm 3 \qquad \text{"}$$

$$\tau (\Lambda_{c}) = 2.4 \pm 1.0 \qquad \text{"}$$

$$\tau (F) = 2.0 \pm 2.0 \qquad \text{"}$$

With the recent progresses made in this field and reported at
the workshop, one may think that the $D^{\pm}$ and D^{o} life-time will be
reasonably well established in a near future, and that the main
pending question will be the life-time of the B meson.

Of course, the life-times alone are not enough to answer all
the questions which can be asked on the decay properties of heavy
favours, and new more detailed experimental results will be
necessary to test our ideas.

In this sense, this workshop has been very useful to discuss
the new technical possibilities, to present a good and comprehen-
sive status of our knowledges and to single out the sectors in
which more work is needed. As the last speaker, I would like to
express my appreciation for the effort made by all the participants
to contribute so well to the success of this workshop, and, in the
name of all the participants, thank Prof. G. Bellini and the members
of his Committee for the nice atmosphere they have created in this
beautiful place.

PARTICIPANTS

ALTARELLI G.   Ist. di Fisica dell'Università
                                      P.le A.Moro 2, Roma, Italy

ANTREASYAN D.   Phys. Dept. Harvard Univ.
                                      42 Oxford Str., Cambridge, U.S.A.

APPEL J.   Fermilab
                                      Batavia, U.S.A.

BALTAY C.   Nevis Lab., Columbia Univ.
                                      Irvington, U.S.A.

BANERJEE S.   Inst. fur Kernphysic
                                      Kernforschungszentrum, Karlsruhe
                                      Fed. Rep. of Germany

BELLINI G.   Ist. di Fisica dell'Univeristà
                                      Via Celoria 16, Milano, Italy

BERTRAND COREMANS G.   Univ. Libre de Bruxelles
                                      Bd. du Triomphe, Bruxelles, Belgium

BETTINI A.   Ist. di Fisica dell'Università
                                      Via Marzolo 8, Padova, Italy

BIZZARRI R.   Ist. di Fisica dell'Università
                                      P.le A.Moro 2, Roma, Italy

BORREANI G.   Ist. di Fisica dell'Università
                                      C.so M. D'Azeglio 46, Torino, Italy

BRODSKI S.   S.L.A.C.
                                      Stanford, U.S.A.

BURGER P.   Enertec Schlumberger
                                      1 Parc des Tanneries, Lingolsheim
                                      France

CAPONE A.   Ist. di Fisica dell'Università
                                   P.le A.Moro 2, Roma, Italy

CELANI F.   Lab. Naz. dell'INFN
                                   Frascati, Italy

COLEMAN R.   Phys. Dept., Univ. of Rochester
                                   Rochester, U.S.A.

CONFORTO G.   Ist. di Fisica dell'Università
                                   P.le A. Moro 2, Roma, Italy

D'ALI' G.   Ist. di Fisica dell'Università
                                   Via Irnerio 46, Bologna, Italy

D'ANGELO P.   Ist. Naz. di Fisica Nucleare
                                   Via Celoria 16, Milano, Italy

D'ETTORRE PIAZZOLI B.   Lab. Naz. dell'INFN
                                   Frascati, Italy

DEUTSCHMANN M.   III Phys. Inst. R.W.T.H.
                                   Physikzentrum
                                   Aachen, Fed. Rep. of Germany

DIAMBRINI PALAZZI G.   EP Div. CERN
                                   Geneva, Switzerland

DI CAPORIACCO G.   Ist. di Fisica dell'Università
                                   L.go E. Fermi 2, Firenze, Italy

DICK L.   EP Div. CERN
                                   Geneva, Switzerland

DI CORATO M.   Ist. di Fisica dell'Università
                                   Via Celoria 16, Milano, Italy

FABBRI F.   Lab. Naz. dell'INFN
                                   Frascati, Italy

FOA' L.   Lab. di San Piero
                                   Via Livornese, San Piero a Grado
                                   Italy

FOCARDI E.   Ist. di Fisica dell'Università
                                   P.za Torricelli 2, Pisa, Italy

FOSTER B.   Rutherford and Appelton Lab.
                                   Chilton, U.K.

HECK W.   Albert Ludwigs Univ.
                                 H. Herder Strasse 3, Freiburg
                                 Fed. Rep. of Germany

HEIJNE E.   EP Div. CERN
                                 Geneva, Switzerland

HOLMGREN S.   Inst. of Phys. Univ. of Stockholm
                                 Vanadisvagen 9, Stockholm, Sweden

HRISOHO A.   Univ. de Paris Sud
                                 Orsay, France

KITAGAKI T.   Tohoku Univ., Phys. Lab.
                                 Aramaki Aoba, Sendai, Japan

KOBAYASHI T.   Univ. of Tokyo c/o D.E.S.Y. F22
                                 Hamburg, Fed. Rep. of Germany

KORBEL V.   D.E.S.Y.
                                 Hamburg, Fed. Rep. of Germany

LECOQ P.   EP Div. CERN
                                 Geneva, Switzerland

LIELLO F.   Ist. di Fisica dell'Università
                                 Via A. Valerio 2, Trieste, Italy

MANFREDI P.F.   Ist. di Elettronica Univ. di Pavia
                                 Via Strada Nuova 106/c, Pavia, Italy

MARGOLIS B.   Phys. Dept. McGill Univ.
                                 3600 University Str.
                                 Montreal, Canada

MARTIN A.   TH Div. CERN
                                 Geneva, Switzerland

MERONI E.   Ist. di Fisica dell'Università
                                 Via Celoria 16, Milano, Italy

MEYER H.   Univ. of Wuppertal, Gaussstr. 20
                                 Wuppertal, Fed. Rep. of Germany

MONACELLI P.   Ist. di Fisica dell'Università
                                 P.le A.Moro 2, Roma, Italy

MONETI G.   Syracuse Univ., 201 Syracuse Bld.
                                 Syracuse, U.S.A.

MONTANET L.   EP Div. CERN
                                Geneva, Switzerland

MUSSET P.   EP Div. CERN
                                Geneva, Switzerland

NATALI S.   Ist. di Fisica dell'Università
                                Via Amendola 173, Bari, Italy

NEWMAN H.   D.E.S.Y. Group F13, Notkestr. 85
                                Hamburg, Fed. Rep. of Germany

O'DONNEL P.   Dept. of Phys. Univ. of Toronto
                                Toronto, Canada

ODORICO R.   Ist. di Fisica dell'Università
                                Via Irnerio 46, Bologna, Italy

PETERSON V.   Dept. of Phys. Univ. of Hawaii
                                2505 Correa Rd., Honolulu, U.S.A.

POTTER D.   Phys. Dept. Rutgers Univ.
                                Busch Campus, Piscataway, U.S.A.

PROSPERI G.   Ist. di Fisica dell'Università
                                Via Celoria 16, Milano, Italy

RAMELLO L.   EP.Div. CERN
                                Geneva, Switzerland

RATTI S.   Ist. di Fisica dell'Università
                                Via A.Bassi 6, Pavia, Italy

REAY N.   Ohio State Univ. Phys. Dept.
                                174 W. 18th Ave. Columbus, U.S.A.

ROE B.   Dept. of Phys. Univ. of Michigan
                                Ann Arbour, U.S.A.

ROMANO G.   Ist. di Fisica dell'Università
                                P.le A.Moro 2, Roma, Italy

ROOSEN R.   Vrye Univ. Brussel
                                Pleinlaan 2, Brussels, Belgium

ROSSI L.   Ist. Naz. di Fisica Nucleare
                                V.le Benedetto XV 5, Genova, Italy

RUCHTI R.   Dept. of Phys. Univ. of Notre Dame
                                Notre Dame, U.S.A.

J. SACTON   Univ. de Bruxelles
                                       Boul. du Triomphe, Bruxelles, Belgium

SARTORELLI G.   Ist. di Fisica dell'Università
                                       Via Irnerio 46, Bologna, Italy

SIEDMAN A.   EP Div. CERN
                                       Geneva, Switzerland

STREIT K.P.   EP Div. CERN
                                       Geneva, Switzerland

STROOT J.P.   EP Div. CERN
                                       Geneva, Switzerland

SZAWLOWSKI M.   Inst. of Nucl. Res.
                                       Hoza 69, Warsaw, Poland

TREILLE D.   EP Div. CERN
                                       Geneva, Switzerland

VANDERHAEGHE G.   EP Div. CERN
                                       Geneva, Switzerland

WEBSDALE D.   Imperial College Phys. Dept.
                                       London, U.K.

ZAMBRA A.   Laben Div. Siel
                                       Via Bassini 15, Milano, Italy

ZICHICHI A.   EP Div. CERN
                                       Geneva, Switzerland

ZUMERLE G.   Ist. di Fisica dell'Università
                                       Via Marzolo 8, Padova, Italy